BIOCHEMICAL ENGINEERING III

ANNALS OF THE NEW YORK ACADEMY OF SCIENCES
Volume 413

BIOCHEMICAL ENGINEERING III

Edited by K. Venkatasubramanian, A. Constantinides, and W.R. Vieth,

The New York Academy of Sciences
New York, New York
1983

The cover illustration shows a suspension culture of *Daucus carota* (see page 385).

Library of Congress Cataloging in Publication Data

Main entry under title:

Biochemical engineering III.

(Annals of the New York Academy of Sciences; v. 413 (Dec. 27, 1983))
Papers from the Third Biochemical Engineering Conference held by the Engineering Foundation in Santa Barbara, Calif., Sept. 19–24, 1982.
Bibliography: p.
Includes index.
1. Biochemical engineering—Congresses.
I. Venkatasubramanian, K. II. Constantinides, A.
III. Vieth, W. R. (Wolf R.) IV. Biochemical Engineering Conference (3rd : 1982 : Santa Barbara, Calif.)
V. Engineering Foundation (U.S.) VI. Title: Biochemical engineering 3. VII. Title: Biochemical engineering three.
VIII. Series: Annals of the New York Academy of Science; v. 413. [DNLM: 1. Biochemistry—Congresses. 2. Chemical engineering—Congresses. W1 AN626YL v. 413/TP 248.3 B615]
Q11.N5 vol. 413 [TP248.3] 500s [660'.63] 83-26869
ISBN 0-89766-220-2
ISBN 0-89766-221-0 (pbk.)

SP
Printed in the United States of America
ISBN 0-89766-220-2 (cloth)
ISBN 0-89766-221-0 (paper)

ANNALS OF THE NEW YORK ACADEMY OF SCIENCES

Volume 413

December 27, 1983

BIOCHEMICAL ENGINEERING III[a]

Editors

K. VENKATASUBRAMANIAN, A. CONSTANTINIDES, and W. R. VIETH

CONTENTS

[a]This volume is the result of a conference entitled Biochemical Engineering III, held by The Engineering Foundation with the support of The New York Academy of Sciences and The National Science Foundation on September 19–24, 1982 in Santa Barbara, California.

Preface

This volume contains the papers presented at the Third Biochemical Engineering Conference held at Santa Barbara, California in September 1982. The theme of the conference was "Advances in Biochemical Engineering." The following topics, representing the forefront of Biochemical Engineering developments, were covered at this conference: Applications of Genetic Engineering; Bioenergetics; Process Modeling, Dynamics and Control; Separation and Purification Processes; Tissue Culture; Microbial Production of Fuels and Chemicals; and Design and Operation Strategies for Bioprocesses.

The Third Biochemical Engineering Conference was organized by The Engineering Foundation with the financial support of The National Science Foundation and The New York Academy of Sciences. The support of these organizations is gratefully acknowledged. We are also deeply appreciative of the financial contributions made by several companies including: Biocon Limited, BioTechnica International, Bio-Technical Resources, Inc., Boehringer Mannheim GmbH, Cetus Corporation, Chemapec, Inc., Corning Glass Works, Fermco Biochemics Inc., Genentech, Inc., Gist Brocades, N.V., Heinz Foundation, H.J. Heinz Company, Ionics, Inc., Kellogg Company, LSL Biolafitte, Inc., Merck & Company, Inc., Miles Laboratories/Biotechnology Group, Miller Brewing Company, Novo Industri, Pharmacia, Inc., Worthington Diagnostic Systems, Inc.

We are grateful to Dr. Sanford Cole, Director; Mr. Harold Comerer, Associate Director; and the Staff of the Engineering Foundation for their help in organizing the conference. It is a pleasure to acknowledge the impeccable editorial assistance of Ms. Joyce Hitchcock and Mr. Bill Boland of the Editorial Department of The New York Academy of Sciences. We warmly appreciate the singular devotion and the superb organizational skills of K. Venkatasubramanian's secretary, Ms. Sue Kowalski, in bringing this Conference from conception to fruition. We also thank Mrs. Gladys Dennison for secretarial assistance.

K. Venkatasubramanian
A. Constantinides
W. R. Vieth

Overproduction of Proteins in Recombinant Organisms

LYNN C. KLOTZ

BioTechnica International, Inc.
85 Bolton Street
Cambridge, Massachusetts 02140

INTRODUCTION

Often the aim of commercial applications of genetic engineering is to construct a microorganism that makes very large quantities of a particular protein. In simple terms, an organism that makes ten times as much of a particular protein as another organism will cut costs of fermentation ten times, everything else being equal.

In the following, the methods and materials of recombinant deoxyribonucleic acid (DNA) available to the genetic engineer for maximizing the expression of proteins (called overproduction here) will be presented. The successful construction of an overproducing organism can cause problems, however, for the fermentation engineer concerned with scale-up of these organisms for commercial production. In particular, the most serious problem is plasmid instability or shedding in recombinant organisms. The status of overproduction genetic engineering in the three common microorganisms *Escherichia coli, Bacillus subtilis,* and *Saccharomyces cerevisiae* will be discussed along with specific approaches for solving the plasmid instability problem.

THE METHODS OF OVERPRODUCTION

The course of production of protein in living organisms can be summarized as:

$$\text{DNA} \xrightarrow{\text{transcription}} \text{mRNA} \xrightarrow{\text{translation}} \text{protein} \qquad \textbf{(1)}$$

The DNA in the cells of prokaryotic microorganisms (for example, bacteria such as *E. coli* and *B. subtilis*) mainly exists in two forms, the chromosomal DNA and smaller pieces of DNA called plasmids. Usually, the chromosomal DNA contains all the information necessary for the microorganisms to survive under "normal" circumstances. The plasmid DNA often contains information necessary for the organism to survive under special circumstances. For example, the plasmid DNA is often the location of the genetic information for the microorganism to resist antibiotics.

It is usually the plasmid DNA that is the target of genetic engineering constructions. One reason for this is that plasmid DNA is much lower in molecular weight than chromosomal DNA. Typically, chromosomal DNAs of bacteria are in the 3×10^9 dalton range, whereas plasmids are in the 1×10^6 to 60×10^6 dalton range. This small plasmid size makes it easier for the genetic engineer to locate plasmid genes and other pieces of DNA necessary for the genetic engineering of overproduction. Furthermore, the small size of plasmid DNA makes it possible to transfer DNA into a host cell, so that the DNA can be propagated from generation to generation. Thus, the genetic engineering can be carried out *in vitro,* and the resulting engineered plasmids can be transferred into a living cell to survive and propagate indefinitely. It is the use of plasmids as the final recipient of recombinant DNA constructions that is anticipated to cause problems in fermentation scale-up.

The first step in the production of a protein is to make a messenger RNA (mRNA) copy of the region of the DNA (the gene) that contains the genetic information for the protein. This copying step is appropriately called transcription.

The mRNA copy then diffuses to the ribosomes, the protein construction site in the cell. At the ribosomes, the genetic information encoded in the mRNA nucleotide sequence is translated into the amino-acid sequence of the particular protein, and the protein is synthesized.

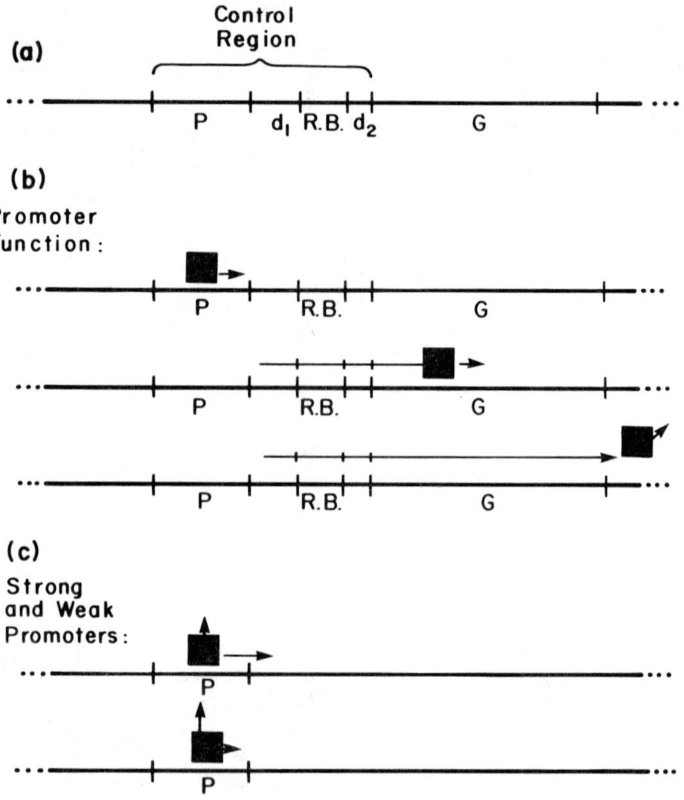

FIGURE 1. Schematic illustration of the control of genes in microorganisms. See text for details.

In order to overproduce a desired protein, the genetic engineer need only increase the rate of transcription or translation for that protein relative to other proteins in the cell. For example, a greater rate of transcription means more mRNA molecules are produced, so more of the ribosome protein synthesis sites will be occupied producing the desired protein instead of others in the cell.

The rates of transcription and translation of a gene are governed largely by the control DNA nucleotide sequences that reside on the DNA upstream from the gene. This situation for a typical bacterial or yeast gene is pictured in schematic fashion in FIGURE 1.

In FIGURE 1a, is pictured a piece of DNA (thick horizontal line) containing a gene (G), and to the left of the gene (upstream), a control region for that gene is shown. The DNA sequences in various parts of the control region affect both the rate of transcription and translation. (The various sections of the DNA are delineated by vertical hash marks.)

The section of the control region called the promoter (P) is an important factor in determining the rate of transcription. How a promoter functions is shown in FIGURE 1b. The DNA sequence in the promoter is "recognized" by an enzyme called RNA polymerase. Recognition involves the binding of the RNA polymerase to specific DNA sequences within the promoter. Once the polymerase is bound, transcription is initiated and the polymerase begins a downstream movement along the DNA toward the gene. The bound polymerase (solid square) is shown beginning its movement downstream (direction indicated by arrow) toward the gene in the top drawing in FIGURE 1b. (The molecular mechanism by which the RNA polymerase moves in one dimension along the DNA is not known. It is certainly an interesting physical biochemical phenomenon.)

In the middle drawing of FIGURE 1b, the mRNA copy is pictured as the thinner horizontal line. From the drawing, it can be seen that the mRNA copy is begun upstream from the sequence labeled R.B. in the drawing (R.B. stands for ribosome binding site, and its function will be discussed below). The copying of the DNA sequence continues through the end of the gene (the last vertical hash mark) until a terminator DNA sequence signal is encountered, at which point the RNA polymerase detaches itself from the DNA. This is shown in the bottom drawing of FIGURE 1b. The newly synthesized mRNA molecule is now free to diffuse to the ribosomes where its sequence is translated into a specific protein molecule.

The number of mRNA copies of a gene that are made in a given time depends in part on the DNA sequence of the promoter region. There are so-called strong and weak promoters. A strong promoter is one in which the RNA polymerase binds strongly to the DNA recognition sequence and efficiently initiates transcription of the mRNA and begins its downstream course to copy the gene. A weak promoter is one in which the binding is not strong or transcription initiation is not efficient, so that the RNA polymerase may diffuse away from the promotor before it has a chance to move downstream to copy the gene. The concept of strong and weak promoters is illustrated schematically in FIGURE 1c, where the lengths of the arrows indicate the probabilities of the promoter moving downstream or diffusing off the DNA.

The existence of both strong and weak promoters in nature is easy to understand. The protein products of different genes are required in different amounts in the cell. For example, a structural protein that is required in large quantities would be expected to have a strong promoter in front of its gene to insure an adequate rate of synthesis of that protein. In contrast, an enzyme that catalyzes a rare reaction in a cell is needed in only small amounts; consequently, a weak promoter might be found upstream from its gene. The state of the technology is such that today the genetic engineer has available several promoters of varying strengths, at least for *E. coli* (to be discussed below).

Another DNA sequence of importance for protein overproduction is the ribosome binding site (denoted R.B. in FIG. 1). This DNA sequence is copied into the mRNA (FIG. 1b, bottom). In order for the mRNA gene sequence to be translated into protein, the mRNA must bind to the ribosome at the ribosome binding site. The strength of the binding, controlled by the nucleotide sequence at the ribosome binding site, determines the efficiency of translation: with strong binding, the mRNA will bind to the ribosome long enough for translation to begin; with weak binding, the mRNA may diffuse off the ribosome before translation can begin. Thus, another DNA sequence the genetic engineer must have in hand for overproduction is a strong ribosome binding site.[1]

Also pictured in FIGURE 1a are two sequences labeled d_1 and d_2. These regions of DNA separating the promoter from the ribosome binding site and the ribosome binding site from the gene can influence the rate of transcription and translation too. For example, the number of DNA nucleotides and the sequence in the d_2 region can influence how strongly the mRNA binds to the ribosome. The influence of these nucleotides on ribosome binding can be used to the genetic engineer's advantage. By controlling the number of nucleotides in the d_2 region, the genetic engineer can control

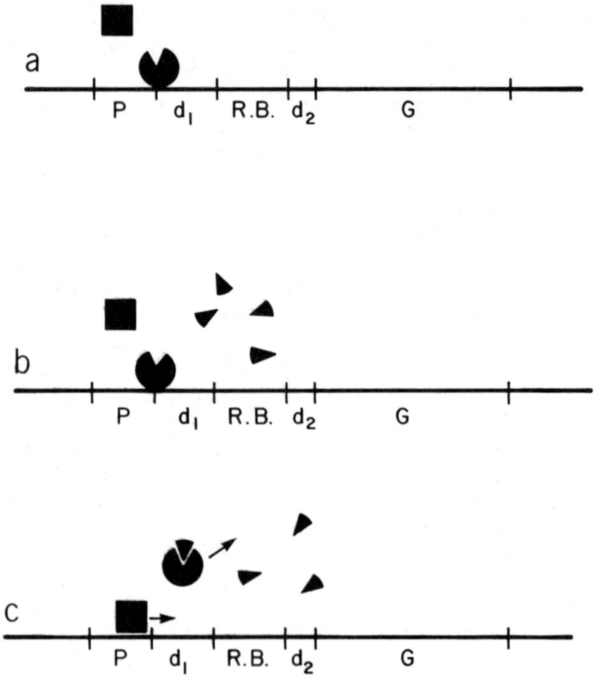

FIGURE 2. Schematic illustration of repressor function. See text for details.

the rate of translation and therefore the amount of protein made.[2–5] For purposes of maximum overproduction, the genetic engineer would vary the number of nucleotides in that region until the maximum rate of translation is achieved.

In the d_1 region and overlapping into the promoter region, the genetic engineer can use specific DNA sequences called operators to which protein molecules called repressors bind. With a repressor protein bound to the DNA in the d_1 region, the RNA polymerase is physically blocked from binding to the promoter and from moving downstream to copy the gene. Thus the genetic engineer can prevent the protein product from being made from a gene simply by engineering an operator sequence upstream from the gene. This situation is illustrated schematically in FIGURE 2a, where the RNA polymerase (solid square) is blocked from binding to the promoter by a repressor (solid sphere with wedge cut out) bound to the operator DNA region.

Operator-repressor combinations can be viewed as molecular switches; that is, they can be used to turn transcription on and off under the genetic engineer's control. For example, one type of repressor (the *E. coli lac* repressor)[6] will release its binding to the operator DNA region when certain sugars present in the cell bind to the repressor. This situation is illustrated in FIGURE 2, where the sugar inside the cell is pictured as wedges (FIG. 2b). When a sugar molecule binds (FIG. 2c), the repressor diffuses from the DNA, allowing the RNA polymerase to bind and move downstream to copy the gene.

Repressor binding to DNA can be controlled by other substances, temperature changes, and other parameters, depending on the repressor-operator combination used. Thus the genetic engineer has at his disposal means for turning genes on and off during commercial fermentations. How the use of such molecular switches can be used to reduce the problems caused by plasmid instability will be discussed below.

Finally, it should be noted that rates of transcription are often increased using another convenient method. The number of plasmid DNA molecules in a microorganism is not restricted to a single molecule; in fact, cells containing 5 to 50 copies of the same plasmid are typical. More plasmids containing the desired gene means that at any one time more mRNA from that plasmid will be transcribed, leading to more protein translated from the mRNA. The number of plasmid DNA molecules of a given type is governed in part by the DNA replication origin on that plasmid: an "active" origin of replication will lead to the plasmid DNA being replicated often during the cell division cycle, producing many plasmid molecules per cell, whereas an "inactive" origin of replication will cause that plasmid to be replicated less often. Genetic engineers have available DNA fragments that contain origins of DNA replication of varying activities that they can use in their constructions.

In summary, in order to overproduce a protein of commercial interest, the genetic engineer needs to possess the gene for the protein, strong promoters, strong ribosome binding sites, origins of DNA replication, and possibly operator-repressor combinations. These DNA components can be spliced together using the splicing tools of the trade (restriction, ligase, and nuclease enzymes). Keeping in mind that the distances between spliced components may be important too, the genetic engineer can splice all these components together on an appropriate plasmid DNA, and this plasmid can then be inserted into an appropriate host microorganism (for example, *E. coli, B. subtilis* or yeast) to be used for large-scale fermentation.

Using such plasmid constructions, it is routinely possible to make levels of desired proteins between 2% and 20% of the cell's total protein. Since an organism such as *E. coli* contains typically 2500 different proteins, a single protein is present at 0.04% of the total protein, on average. Thus levels of 2% to 20% represent 50-fold to 500-fold overproduction.

While such overproduction levels cannot always be achieved because of other factors, such as protein degradation, these levels are often achieved, as examples to be presented later will show.

STATUS OF OVERPRODUCTION IN MICROORGANISMS OF POTENTIAL COMMERCIAL IMPORTANCE

The state of genetic engineering technology is well advanced in *E. coli*. Because *B. subtilis* and yeast have many potential commercial advantages over *E. coli* for large-scale fermentations and for certain commercial applications, genetic engineering in these organisms is being developed at a rapid pace. The purpose of this section is to review the state of genetic engineering technology in these three organisms with respect to the practice of overproduction.

E. coli

In this microorganism, the methods of overproduction are highly developed. Several strong promoters, such as the *lac*,[2,3,7] λP_L,[8] *trp*,[9] and TAC[10] promoters are commonly used. Using the *lac* promoter, levels of 1% to 5% of the total cell's protein are typically achieved. The *trp*, λP_L, and TAC promoters are considered to be about 5 to 10 times stronger than the *lac* promoter. With these promoters, levels of expression of 5% to 20% can be achieved.

Ribosome binding sites can vary somewhat in *E. coli*. Most sites contain the minimal sequence . . .AGGA. . . . A consensus sequence gleaned from many ribosome binding sites is the so-called Shine-Dalgarno[1] sequence . . .TAAGGAGGTG. . . .

The best distance, d_2, for positioning the ribosome binding site for overproduction is known.[11] Finally, active origins of plasmid DNA replication have been identified, so that, overall, all the elements for maximum overproduction of proteins are widely available.

The most serious remaining problem hindering overproduction of proteins in *E. coli* is protein degradation. Many foreign proteins will be degraded by the endogenous protease enzymes in *E. coli,* so that the usable yield of protein can be low even under conditions of overproduction. Several approaches are now being tried to solve this problem. The brute force approach of increasing overproduction levels even further can create a situation where the protein is in such a high concentration in the cell that it precipitates and is inaccessible to proteases. Another approach to eliminating proteases is to engineer, or to select for by classical genetic procedures, *E. coli* strains that are deficient in protease activity.

While solution of the protein degradation problem in *E. coli* will make it a more attractive host for some genetic engineering applications, *B. subtilis* and yeast may well become the future hosts of choice. *B. subtilis* is known to contain fewer animal toxins than *E. coli,* and proteins potentially can be excreted from *B. subtilis,*[12,13] making purifications easier and potentially increasing yields. Also, there is more large-scale fermentation experience with this organism. Yeast, because it is a eukaryote, might be a preferred host for various eukaryotic proteins. Because there is a long history of commercial fermentations with yeast, it would be a natural host for genetically engineered variations of current processes.

B. subtilis

The state of genetic engineering methodology in this microorganism is not nearly so well developed as that in *E. coli*. However, rapid progress is being made, and it is possible that within two years the technology will arrive at the current state of sophistication in *E. coli*.

Several promoters (more than 12) have been isolated and characterized from *B. subtilis*.[14] Since *B. subtilis* forms spores when a culture reaches the stationary phase of growth where some of the proteins made by the cell differ from those made in the vegetative (log) phase, it is convenient to characterize the promoters for the various genes as either vegetative or sporulation promoters. (This characterization has practical implications, as will be seen below). Both strong vegetative and sporulation promoters have been isolated and are available for genetic engineering in *B. subtilis*.[12,14–16] The levels of overproduction using the stronger promoters are thought to be on the order of a few percent (J. Pero, personal communication); however, their production levels have not yet been as well characterized as those in *E. coli*.

Ribosome binding is generally stronger in *B. subtilis*. A range of free energies between −14 to −23 kcal/mole is needed for *B. subtilis* mRNA to bind to the ribosome, whereas a range of binding energies between −4 to −22 kcal/mole has been found in *E. coli*.[17] The *B. subtilis* ribosome binding site does contain the characteristic . . .AGGA. . . sequence, however. Furthermore, promoters from *B. subtilis* will work in *E. coli*, but *E. coli* promoters generally will not work in *B. subtilis*.[18] A consequence of these facts is that genetic engineering constructions made for *B. subtilis* will generally produce mRNAs that will bind to and be translated at *E. coli* ribosomes, but constructions made for *E. coli* will not bind to *B. subtilis* ribosomes, so that the many plasmids constructed over the last several years for *E. coli* hosts will not produce protein in *B. subtilis*.[19]

At present, little is known about how overproduction levels of protein are dependent on the d_1 and d_2 distances, although the same procedures for adjusting these distances in *E. coli* can be used in *B. subtilis* genetic engineering applications.

The sporulation promoters in *B. subtilis* can be used as molecular switches. Since they are activated only in stationary phase, any gene downstream from them will not be transcribed during the vegetative or log phase of cell growth. Only when the cells approach maximum density in stationary phase will the gene be transcribed. The mechanism of this molecular switch is very different from the operator-repressor switches discussed earlier, and it will not be discussed here. (See reference 16 for a description of how sporulation promoter switches function on a promoter level.)

The existence of sporulation promoters allows the genetic engineer to construct a *B. subtilis* plasmid so that the cells will expend energy making that protein only after they have grown to maximum density. This could be useful when the protein product is poisonous to the cell or when the cell's growth rate is significantly reduced because of production of the protein.

Yeast

The state of development of genetic engineering in yeast is slightly ahead of that in *B. subtilis*. Especially rapid further development is expected because of the large number of industrial and academic laboratories working on genetic engineering technology in this industrially important microorganism.

In yeast, the enzymes involved in glycolysis and sugar metabolism are present at high levels, so the promoters upstream from the genes for those enzymes have been targeted for use in overproduction applications. Since these sugar metabolism and glycolysis enzymes typically are present as 1% to 2% of the total cellular protein, genes for proteins genetically engineered downstream from these promoters can be expected to be overproduced in similar amounts (for proteins of molecular weights similar to the enzymes). Some examples of these promoters that have been isolated and used in genetic engineering constructions are: the phosphoglycerate kinase (PKG),[20] triose phosphate isomerase (TPI),[21] and galactokinase (GAL1) (R. Yocum, unpublished results) promoters. On a multicopy plasmid, the triose phosphate isomerase promoter can drive expression of its own gene to 20% of the cell protein.[21]

The GAL1 promoter is not as strong, but it has the advantage of being regulated over 1000-fold by a repressor. The repressor is made by another gene, called the GAL80 gene. Thus a genetically engineered yeast strain with a temperature-sensitive

GAL80 repressor protein would have a molecular switch for the transcription of the gene for the desired protein. That is, a slight change in temperature would change the conformation of the GAL80 repressor protein so that it would not bind to its operator DNA sequence upstream from the gene of interest, thereby allowing transcription to begin.

The glycolysis promoters are constitutive (turned on all the time), so their use could cause problems such as slow cell growth associated with overproduction of a protein not necessary for the growth of the host yeast cell.

Levels of several proteins have been measured for overproducing yeast strains. Levels vary from 0.2% (for the bacterial protein β-galactosidase) to 6% (for the animal protein leukocyte interferon.)[20–23]

Yeast does not have a specific ribosome binding site, so genetic engineering of this DNA region is more empirical.

THE PROBLEM OF PLASMID INSTABILITY AND POTENTIAL GENETIC ENGINEERING SOLUTIONS

Microorganisms harboring plasmids will sometimes lose the plasmids. The instability of plasmids due to loss (shedding) is thought to occur during cell division. As a parent cell divides into two daughter cells, the plasmid DNA molecules are often not divided equally between the daughter cells. In the extreme case, one daughter cell may receive all the copies of one type of plasmid, the other daughter cell will receive none. If the daughter cells that have shed the plasmid are viable, they too will grow and divide in the fermenter.

Even at very low rates of plasmid loss, this can be a serious problem for cells harboring overproduction recombinant DNA plasmids. The reason is that cells containing recombinant plasmids will often grow at much slower rates than those not containing them because the plasmid-containing cells are usually producing large quantities of a protein that has no function in the cell. On the other hand, cells that have lost the plasmid will not be wasting energy on producing an unuseful protein, and therefore can devote more energy to growth and cell division.

Theoretical analyses of cell growth of a population of cells continuously shedding plasmids have been worked out.[24,25] From the results of these analyses, the potential seriousness of plasmid shedding can be illustrated. As a simple example, suppose the fraction of daughter cells shedding plasmid at any cell division is 0.01, and suppose further that the growth rate of the resulting plasmidless cells is twice that for cells harboring the recombinant plasmid. After 15 generations of growth (for the slower-growing cells), the cells that have shed their plasmids will have overgrown their plasmid-containing counterparts by a ratio of 1000 to 1. Thus, the growth vessel would be full of cells that are not making the desired protein product. Since the numbers used in this calculation are not unrealistic, it is clear that the plasmid shedding problem must be addressed before fermentation scale-up to commercial production levels is attempted.

It is fortunate that genetic engineering can provide a solution to this genetic engineering-created problem. Potential solutions to the plasmid shedding problem fall roughly into three categories:

(1) Selection against cells that lose their plasmids. The most commonly used selection is antibiotic resistance. Genes for resistance to specific antibiotics can be engineered onto recombinant plasmids containing the gene for the protein to be

overproduced. If the cells are then fermented in a broth containing an antibiotic to which they are normally sensitive, any cells that shed plasmids will die because they lose antibiotic resistance. Selection by antibiotic resistance is commonly practiced in the laboratory to insure the survival and propagation of only plasmid-containing cells; however, this type of selection may be too expensive for large-scale industrial fermentations because of the high price of antibiotics.

For yeast, a gene that allows the cells to manufacture their own uracil is often incorporated onto recombinant plasmids. In this case, the host cell is a mutant yeast that lacks a gene in the uracil metabolic pathway.

(2) Controlled overproduction (molecular switches). Plasmid shedding would not be such a serious problem if the cells that shed their plasmids did not grow faster than those containing recombinant plasmids. When the gene that has been engineered to overproduce a commercially important protein is not being transcribed, the growth rate of the cells harboring the plasmid may be nearly identical to the cells that have shed their plasmids. Thus, the plasmidless cells will not overgrow the plasmid-containing cells. Repressor-operator combinations are already being used in overproducing *E. coli* fermentations and are being developed for yeast fermentation. For *B. subtilis,* sporulation promoters accomplish the same result since they will cause transcription of the gene they control only when cell growth has achieved its maximum density in the fermenter.

(3) Integration of overproducing genes into the chromosomal DNA. The chromosomal DNA is absolutely essential for the cell to function and survive since most of the structural and metabolic activity of the cell is encoded on this DNA; therefore, it cannot be shed. Procedures currently exist for stably integrating DNA regions of genetically engineered plasmids into the main chromosomal DNA of *E. coli,*[26] yeast,[27] and *B. subtilis.*[28,29] Such constructions have been carried out, and the resulting recombinant DNA strain is stable.

Currently, many overproducing strains achieve much of their overproduction by increased plasmid copy number. For DNA integrated into the chromosome, the copy number of the overproducing gene region will usually be one, or a few at most. Thus, the ability to engineer strong promoters and strong ribosome binding sites, and to be able to position these elements carefully (d_1 and d_2 distances and nucleotide sequence) is important where chromosomal integration is to be used.

SUMMARY

In a qualitative way, the materials and methods available to the recombinant DNA genetic engineer for overproducing proteins have been explained. The status of technology development for overproduction using *E. coli, B. subtilis,* and yeast as host microorganisms has been briefly assessed. Potential and actual genetic engineering solutions to the plasmid-shedding problem have been outlined.

Since plasmid shedding presents a serious problem to the fermentation engineer responsible for scale-up to commercial production levels and since the ways around this problem appear mostly to have their solutions in the realm of genetic engineering coupled with appropriate fermentation protocol, the genetic engineer should work closely with the fermentation engineer to make scale-up realizable. Neither the genetic engineer nor the fermentation engineer can afford to be ignorant of the tools available to each profession if fermentation scale-up of genetically engineered microorganisms is to be accomplished economically.

ACKNOWLEDGMENTS

I wish to thank Drs. David J. Glass, Gail D. Lauer, Janice G. Pero, Thomas M. Roberts, and R. Rogers Yocum for sharing with me their knowledge and expertise in *E. coli, B. subtilis,* and yeast molecular biology and genetic engineering.

REFERENCES

1. SHINE, J. & L. DALGARNO. 1975. Determinant of cistron specificity in bacterial ribosomes. Nature **254:** 34–38.
2. BACKMAN, K. & M. PTASHNE. 1978. Maximizing gene expression on a plasmid using recombination *in vitro.* Cell **13:** 65–71.
3. ROBERTS, T. M., R. KACICH & M. PTASHNE. 1979. A general method for maximizing the expression of a cloned gene. Proc. Natl. Acad. Sci. USA **76:** 760–764.
4. GOLD, L., D. PRIBNOW, T. SCHNEIDER, S. SHINEDLING, B. S. SINGER & G. STORMO. 1981. Translational initiation in procaryotes. Ann. Rev. Microbiol **35:** 365–403.
5. SHEPARD, H. M., E. YELVERTON & D. V. GOEDDEL. 1982. Increased synthesis in *E. coli* of fibroblast and leukocyte interferons through alterations in ribosome binding sites. DNA **1:** 125–131.
6. BARKLEY, M. A. & S. BOURGEOIS. 1980. Repressor recognition of operator and effectors. *In* The Operon. J. H. Miller & W. S. Reznikoff, Eds. Cold Spring Harbor Laboratory Press. Cold Spring Harbor, N.Y. pp. 177–220.
7. GOEDDEL, D., H. HEYNEKER, T. HAZUMI, R. ARENTZEN, K. ITAHURA, D. G. YANSURA, M. J. ROSS, G. MIOZZARI, R. CREA & P. H. SEEBERG. 1979. Direct expression in *Escherichia coli* of a DNA sequence coding for human growth hormone. Nature **281:** 544–548.
8. SHIMATAKE, H. & M. ROSENBERG. 1981. Purified λ regulatory protein cII positively activates promoters for lysogenic development. Nature **292:** 128–132.
9. KLEID, D. G., D. YANSURA, B. SMALL, D. DOWBENKO, D. M. MOORE, M. J. GRUBMAN, P. D. MCKERCHER, D. O. MORGAN, B. H. ROBERTSON & H. L. BACHRACH. 1981. Cloned viral protein vaccine for foot and mouth disease: Responses in cattle and swine. Science **214:** 1125–1129.
10. DEBOER, H. A., L. J. COMSTOCK, D. G. YANSURA & H. L. HEYNEKER. 1982. Construction of tandem *trp lac* promoter and a hybrid *trp lac* promoter for efficient and controlled expression of the human growth hormone gene in *Escherichia coli. In* Promoters, Structure & Function. R. L. Rodriguez & M. J. Chamberlin, Eds. Praeger Publishers. New York, N.Y.
11. SHEPARD, H. M., E. YELVERTON & D. V. GOEDDEL. 1982. Increased synthesis in *E. coli* of fibroblast and leukocyte interferons through alterations in ribosome binding sites. DNA **1:** 125–131.
12. CHANG, S., O. GRAY, D. HO, J. KROYER, S. Y. CHANG, J. MCLAUGHLIN & D. MARK. 1982. Expression of eukaryotic genes in *B. subtilis* using signals of penP. *In* Molecular Cloning and Gene Regulation Bacilli. A. T. Ganesan, S. Chang & J. A. Hoch, Eds. Academic Press. New York, N.Y. pp. 159–169.
13. PALVA, I., M. SARVAS, P. LEHTOVAARA, M. SIBAKOV & L. KAARIAINEN. 1982. Secretion of *Escherichia coli* β-lactamase from *Bacillus subtilis* by the aid of α-amylase signal sequence. Proc. Natl. Acad. Sci. USA **79:** 5582–5586.
14. MORAN, C. P., JR., N. LANG, S. F. J. LEGRICE, G. LEE, M. STEPHENS, A. L. SONENSHEIN, J. PERO & R. LOSICK. 1982. Nucleotide sequences that signal the initiation of transcription and translation in *Bacillus subtilis.* Mol. Gen. Genet. **186:** 339–346.
15. PALVA, I., R. F. PETTERSSON, N. KALKKINEN, P. LEHTOVAARA, M. SARVAS, H. SODERLUND, K. TAKKINEN & L. KAARIAINEN. 1981. Nucleotide sequence of the promoter and NH_2-terminal signal peptide region of the α-amylase gene from *Bacillus amyloliquefaciens.* Gene **15:** 43–51.
16. LOSICK, R. 1982. Sporulation genes and their regulation. *In* The Molecular Biology of the Bacilli. D. A. Dubnau, Ed. Academic Press. New York, N.Y. pp. 179–201.

17. McLaughlin, J. R., C. L. Murray & J. C. Rabinowitz. 1981. Unique features in the ribosome binding site sequence of the gram-positive *Staphylococcus aureus* β-lactamase gene. J. Biol. Chem. **256:** 11283–11291.
18. Lee, G., C. Talkington & J. Pero. 1980. Nucleotide sequences of a promoter recognized by *B. subtilis* RNA polymerase. Mol. Gen. Genet. **180:** 57–65.
19. Ehrlich, S. D. 1978. DNA cloning in *Bacillus subtilis*. Proc. Natl. Acad. Sci. USA **75:** 1433–1436.
20. Hitzeman, R. A., F. E. Magie, H. L. Levine, D. V. Goeddel, G. Ammerer & B. D. Hall. 1981. Expression of a human gene for interferon in yeast. Nature **293:** 717–722.
21. Alber, T. & G. Kawasaki. 1982. Nucleotide sequence of the triose phosphate isomerase gene of *Saccharomyces cerevisiae*. J. Mol. Appl. Genet. In press.
22. Hitzeman, R., D. Leung, F. Magie, C. Chen, J. Perry, W. Kohr, M. Levine, R. Wetzel & D. Goeddel. 1982. Expression, processing and secretion of heterologous gene products by yeast. Eleventh Int. Conf. on Yeast, Abstr.: 112.
23. Guarente, L. & M. Ptashne. 1980. Fusion of *Escherichia coli lacZ* to the cytochrome c gene of *Saccharomyces cerevisiae*. Proc. Natl. Acad. Sci. USA **78:** 2199–2203.
24. Imanaka, T. & S. Aiba. 1981. A perspective on the application of genetic engineering: Stability of recombinant plasmid. Ann. N.Y. Acad. Sci. **369:** 1.
25. Ollis, D. F. & H.-T. Chang. 1982. Batch fermentation kinetics with (unstable) recombinant cultures. Biotechnol. Bioeng. **24:** 2583–2586.
26. Murray, K. & N. E. Murray. 1975. Phage lambda receptor chromosomes for DNA fragments made with restriction endonuclease III of *Haemophilus influenzae* and restriction endonuclease I of *Escherichia coli*. J. Mol. Biol. **98:** 551–564.
27. Strathern, J., F. Jones & J. Broach. 1981. The Molecular Biology of the Yeast *Saccharomyces*. Cold Spring Harbor Laboratory Press. Cold Spring Harbor, N.Y.
28. Haldenwang, W. G., C. D. B. Banner, J. F. Ollington, R. Losick, J. A. Hoch, M. B. O'Connor & A. L. Sonenshein. 1980. Mapping a cloned gene under sporulation control by insertion of a drug resistance marker into the *Bacillus subtilis* chromosome. J. Bacteriol. **142:** 90–98.
29. Yoneda, Y., G. Scott & F. E. Young. 1979. Cloning of a foreign gene coding for α-amylase in *Bacillus subtilis*. Biochem. Biophys. Res. Commun. **91:** 1556–1564.

Applications of Genetic Engineering to the Pharmaceutical Industry

MICHAEL KONRAD

Cetus Corporation
600 Bancroft Way
Berkeley, California 94710

INTRODUCTION

One of the major tasks of genetic engineers is to take a gene from one organism and transfer it to another. The motive for gene transfer is usually that the host cell is difficult, expensive, or impossible to grow. It may also be that the level of expression (rate of synthesis per cell or per unit time) is unpractically low, or that a rather exotic induction procedure is required to cause the cell to make the protein. In the case of a protein with pharmaceutical use, the donor cell is most likely to be human. The choice of the cell species in which this gene is best transferred and expressed may require knowledge of the structure, and in some cases the function, of the particular protein.

INSULIN

An extreme example, in terms of biological difference between donor and recipient cell type, is the production of human insulin in the bactrium *E. coli*. The donor cell is one of the major cell classes present in the pancreas, the Islets of Langerhans. To my knowledge, no one has proposed culturing these cells as a source of insulin, although a varient cell type, from insulinoma tumors, continues to make insulin even though transformed, and is relatively easy to grow in tissue culture.

The first step in the tranfer of genetic information is to obtain a copy of the coding region of the gene that can be read by the bacterial cell. Although the genetic code is universal (almost) the majority of eukaryotic genes, including insulin,[1] contain nucleotide segments called introns, imbedded in the sequence that codes for the amino acid sequence of the protein. These segments are removed from the messenger RNA by an enzymatic processing system that is missing in bacteria. One must then isolate messenger RNA from the insulinoma and copy this enzymatically *in vitro* to produce a cDNA fragment.

The second problem is that the promoter that caused the insulin gene to be transcribed in the human cell doesn't work in bacterium. In a sense, this is no problem if we have a cDNA gene, since the promoter sequence is not represented in the RNA anyway. The solution is then to splice a bacterial promoter in front of the insulin gene. If the protein is going to be toxic to the bacterial cell, and it may be for direct or indirect reasons, it is important to use a promoter that can be turned off. In any case, it is best to use a promoter that can be controlled, because until you have the gene expressed in the cell you will not know if it's toxic or not. If it is toxic, it will prevent you from isolating the desired bacterial clone and you may never guess the reason.

The next problem is that the gene codes not for insulin, but for preproinsulin (see FIGURE 1). Let's take this problem in two steps, first assuming we had proinsulin. We can now take advantage of the fact that for several decades, in addition to the attention given to this hormone because of its widespread clinical importance, a significant number of protein chemists have taken the insulin system as a model. Proinsulin is available as almost a byproduct of the purification of insulin from bovine and porcine pancreas tissue, and biochemists have been able to develop procedures and conditions for the enzymatic cleavage of the protease-sensitive regions at both ends of the C peptide.[2,3] After this, separation of insulin from C peptide is straightforward.

The next problem is how to remove the "pre" or leader sequence. it is this polypeptide region that causes insulin to be secreted through the cell membrane, after which it is removed.[4] In the case of insulin, the secretion process is complicated by the

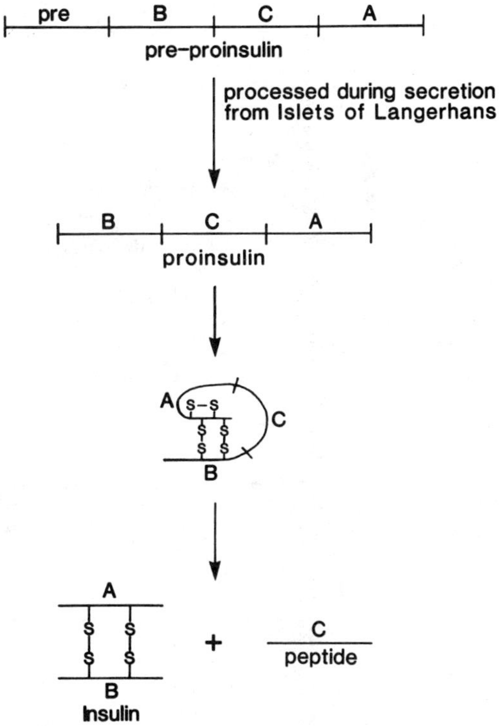

FIGURE 1. The conversion of pre-proinsulin to insulin in the pancreas. The precursor to insulin is the single polypeptide chain at the top of the figure. The NH_2-terminal "pre" segment is removed during secretion, yielding proinsulin, which contains three intramolecular disulfide bonds as illustrated in the third structure. An enzyme system with specificity similar to trypsin cuts the peptide at both ends of the C region to produce insulin (the third disulfide bond remains, but is not shown in this diagram).

fact that most is transported to localized storage areas where as a complex with zinc, it awaits subsequent release. The presence of a leader sequence is a quite general feature of most proteins that are of pharmaceutical interest. This is because such proteins will usually be active outside the cell or at the cell surface or be able to be taken up by cells (if not they wouldn't be effective after being administered extracellularly). Thus they must have been secreted by cells and have had a leader sequence.

One solution is to construct the codon for methionine just before the desired peptide. It is then likely, but not assured, that the normal bacterial processing system that removes N-formyl-methionine from bacterial proteins will remove it also.

A second solution is to add the proinsulin sequence to a bacterial protein with an amino acid that can be selectively cleaved. This was the approach used by the Genentech-Lilly group when in taking advantage of the absence of methionine in proinsulin, they constructed a chimera of the NH_2-terminal end of tryptophan synthetase, joined to proinsulin at a methionine residue. Treatment of this protein with cyanogen bromide, which cleaves the polypeptide chain after methionine residues, gives proinsulin.

A third approach is to utilize the secretion system of the bacterium. In the case of gram-negative organisms like *E. coli,* secretion is into the periplasmic space, which is defined as the compartment whose contents are released by osmotic shock. There is a convenient DNAase reaction site in the middle of the gene for beta lactamase, into which the cDNA insulin gene has been inserted by our colleagues at the University of Chicago, Prof. Donald Steiner and Dr. Shu Chan (see FIG. 2). However, this plasmid did not produce insulin presumably because the insulin gene was out of phase with the beta-lactamase gene. In such a situation, the ribosomes reading the chimeric mRNA will produce nonsense protein as they move into the insulin gene, and will be released when they hit one of the three termination condons.

Drs. Chander Bahl and David Mark at Cetus Corporation, utilized two convenient restriction sites in the beta-lactamase sequence upstream or 5′ to the insulin gene to remove a DNA fragment, and an exonuclease acting at each of the ends to enlarge the deletion even further. The two ends were then joined back to regenerate a circular plasmid and the DNA was used to transform bacteria. We now had a heterogenous collection of plasmids, some of which must be expressing at least a chimeric peptide containing the insulin sequence. Thousands of bacterial colonies were then screened by Dr. Tom White, using antibodies raised to insulin,[5] and a group producing insulin antigen were identified. Analysis of the size of the deletion in the positive clones allowed us to pick two that were producing a peptide close to the size of proinsulin. Small cultures were then grown in medium containing high specific activity ^{35}S, which would label only the 6 cysteine residues in insulin, and the small fraction of the bacterial protein that had been secreted into the periplasmic space was released by osmotic shock and shipped to our colleagues in Chicago. This material was purified by precipitation with antibody to insulin, and subjected to Edmond degradation, a process that strips amino acids sequentially from the NH_2-terminal end of the peptide. Two peaks of activity were seen at the 7th and 19th cycle corresponding to the two cysteine residues in the B chain of insulin.[6]

After determination of the DNA sequence of the plasmid, it was evident that what

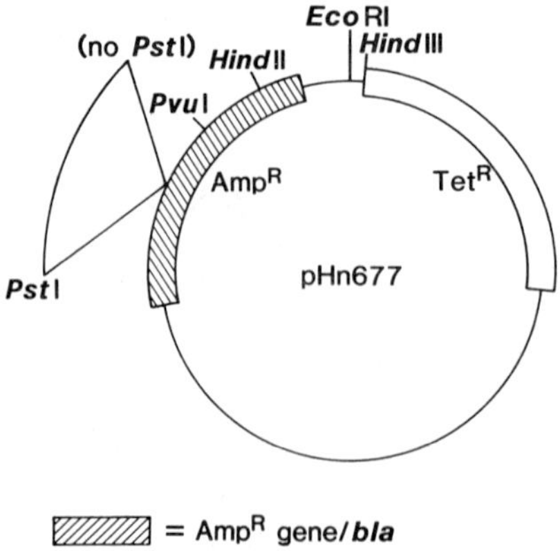

FIGURE 2. Insertion of the insulin cDNA copy into the pBR322 plasmid. The plasmid was cut at the single *Pst*I DNAase restriction site which is in the middle of the gene conferring resistance to ampicillin (the gene codes for the enzyme beta-lactamase). Poly dG segments were added to both 3′ ends of the cut plasmid, and poly dC segments were added to the corresponding ends of the insulin cDNA copy. Complementary base pair hydrogen bonding between the insulin and plasmid ends generated the plasmid pHn677. The 3′ *Pst*I site was regenerated as expected, while the 5′ one was not, presumably due to contaminating nuclease activity.

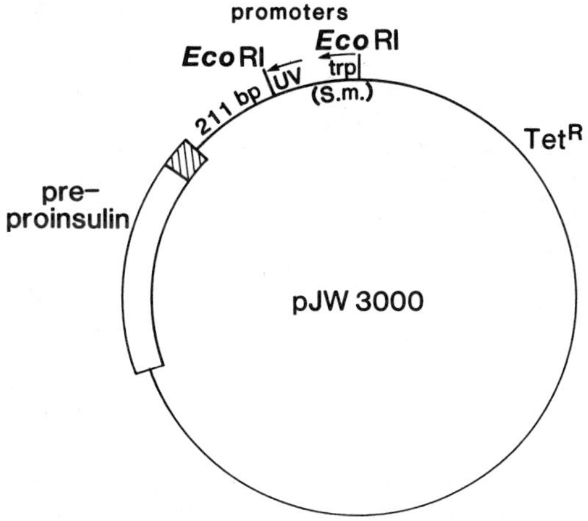

FIGURE 3. An intermediate in the construction of a plasmid expressing high levels of human insulin. In this plasmid, pJW3000, a large deletion has been made in the Amp gene 5′ to the insulin segment. The 36 base pairs coding for the first 12 amino acids of the Amp leader sequence have been fused to the last coding for the last 13 amino acids of the insulin gene. This chimeric peptide is cleaved correctly by the bacterial cell to produce proinsulin. A small DNA fragment containing the lac UV promoter and the tryptophan promoter (from *S. marcescens*) has been inserted at the *Eco*R1 site 211 base pairs upstream from the insulin gene.

we had done was to fuse the first 12 amino acids of the beta-lactamase leader sequence to the last 13 amino acids of the insulin leader sequence, providing a chimeric leader that was still correctly recognized by the bacterial secretion-protease system to produce proinsulin.[6] Results similar to these have been reported by Talmadge *et al.* using the gene for rat preproinsulin.[7]

The signal to start RNA synthesis, or transcription, and the signal to start protein synthesis, the ribosomal binding site (RBS), were still those belonging to beta lactamase, and the resulting level of expression was low. We had used a small fragment containing the corresponding sites for the tryptophan synthetase operon to express interferons in *E. coli,* and as I will show later, had obtained quite high levels of expression. Unfortunately, there were no convenient restriction sites close to the NH_2-terminal end of our insulin construction to insert this fragment. There are several tactics one can use in a case like this, and I'll briefly describe the one used by Dr. Anne Emerick to produce a clone with a higher level of expression.

The plasmid was opened at the *Eco*R1 site about 200 base pairs upstream of the insulin sequence (FIG. 3), and DNA was removed from both ends by random exonuclease action. The deletion in the distal direction was repaired by first cutting at the Bam H1 site, and then adding back the intact fragment. Finally, the plasmid was closed by blunt-end ligation. Most of the time the *Eco*R1 site will be lost, but there is about one chance in four that the nucleotide on the insulin end will be a C, and the specific site will be regenerated. Such a construction was isolated that contained the site as close as possible to the start of the coding sequence, only 14 nucleotides away. This construction was then opened, and the insulin-proximal end digested with the DNA polymerase produced by the bacterial virus T4 in the presence of dCTP. Under these conditions, the exonuclease activity of this enzyme allows it to chew back the 3′ chain until the first C is removed. When that happens, the polymerase activity incorporates dCTP in the chain to regenerate the damage. This cycle then goes on indefinitely, but under our conditions designed to remove only a few nucleotides, generated a series of plasmids with the RBS being various distances upstream of the ATG starting the insulin sequence. The best clone from this series when placed after the *trp* promoter and RBS produced 0.16 percent of its protein as insulin, a 16-fold increase over the original clone, pJW2172.

INTERFERON

Interferons are a group of proteins that protect cells against viral infections. They are typically produced by cells that have been infected and are then secreted by these cells. They then bind to cell membrane receptors on neighboring cells and trigger the synthesis of several special enzymes that inhibit viral expression and replication. Thus while the original cell may not escape the virus, it has altruistically protected its neighbors. Interferons also have growth-inhibitory activity, and there are many who believe that they may be involved in normal growth regulation and are one of the organism's defenses against cancer cells.[8] Indeed, Cetus Corporation and its partner in this venture, Shell Oil Company, plan to start clinical trials with cancer patients quite soon.

Interferons can be divided into three groups. The alphas number at least 12, and typically show 70% amino acid homology. Only one member of the beta and gamma class have been isolated, and they are both glycosylated, whereas the alphas appear not to be. Beta is clearly related to the alphas, with about 30% amino acid homology. The genes for beta and the alpha interferons do not contain introns. Gamma has no obvious sequence homology to the other interferons, and there are several introns in its genomic sequence.[9]

The function of the sugar residues on beta and gamma interferons (IFNs) are not understood. Some experiments with beta interferon suggest that the terminal carbohydrate residues may be sialic acid, which when enzymatically removed expose galactose. The galactose residue may be recognized by specific receptors in the liver, and the protein is removed from circulation.[10] Thus one function of the sugar may be to provide a mechanism for removing IFN from the body after a definite time. However, the residence time in the serum of the alpha IFNs, which are thought not to

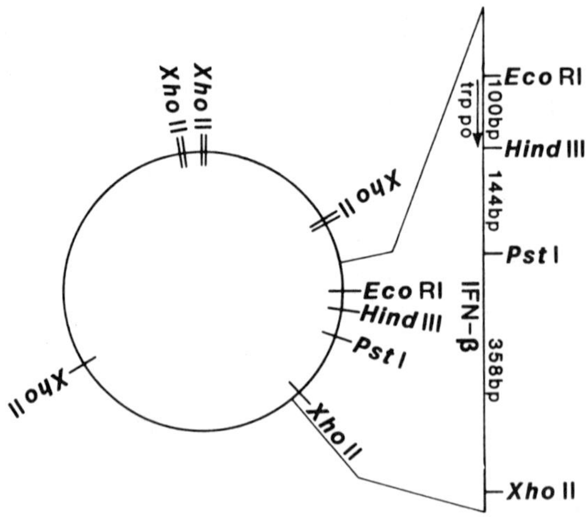

FIGURE 4. The gene coding for the human beta interferon gene inserted into the plasmid pBR322. The DNA fragment diagramed linearly at the top of the figure has been inserted in place of the *Eco*R1 to *Xho*II segment seen at the top of the circular pBR322 plasmid map. The inserted fragment contains the promoter and ribosomal binding site of the tryptophan synthetase operon on a 100 base pair long *Eco*R1 to *Hind*III fragment at the top left of the figure. This segment promotes transcription and translation of the gene coding for interferon extending to the *Xho*II site.

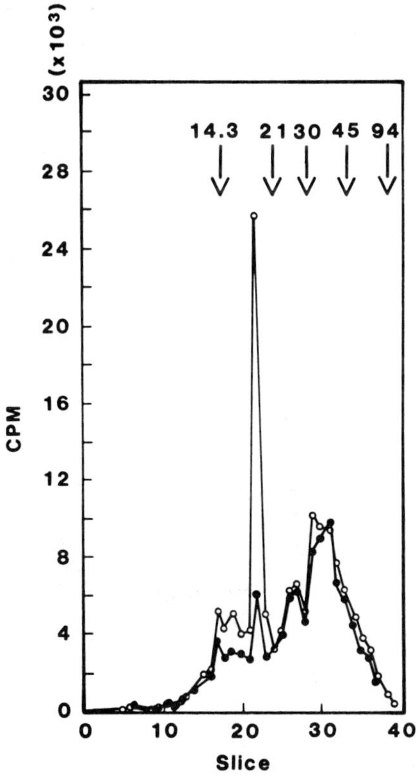

FIGURE 5. Induction of *E. coli* to produce beta interferon. Bacteria containing the plasmid described in FIGURE 4 were grown in the presence (repressed) and absence (induced) of tryptophan [^{14}C]leucine was added to the repressed culture and [^{3}H]leucine was added to the induced culture. Both cultures were chilled and the cells lysed after one minute of amino acid incorporation. The lysates were mixed and the proteins separated according to molecular weight by a sodium dodecylsulfate gel electrophoresis system. The distribution of both ^{3}H from the induced culture (open circles) and ^{14}C from the repressed culture (closed circles) of 2 mm slices of the gel are compared to the positions of molecular weight standard proteins indicated at the top of the figure (weights in kilodaltons). The peak at an apparent molecular weight of 19 kd is interferon.

be glycosylated, is rather modest, thus there must be other mechanisms for clearance. We have obtained preliminary pharmacokinetic data on beta produced in bacteria, and thus not glycosylated, and it certainly does not stay in the circulation indefinitely. Hopefully one of the contributions of genetic engineering will be to make experiments possible that will more completely elucidate the role of the carbohydrate residues. IFNs are certainly not unique in being glycosylated. Of the major proteins in the blood, only serum albumin is not glycosylated.

The construction of a plasmid expressing beta IFN was accomplished by a group at Cetus Corporation headed by Dr. David Mark. The structure of the plasmid is seen in FIGURE 4. The cDNA fragment containing the coding sequence was placed behind the tryptophan promoter mentioned previously in connection with the expression of insulin. The portion of the total cell protein synthesis that has been converted to

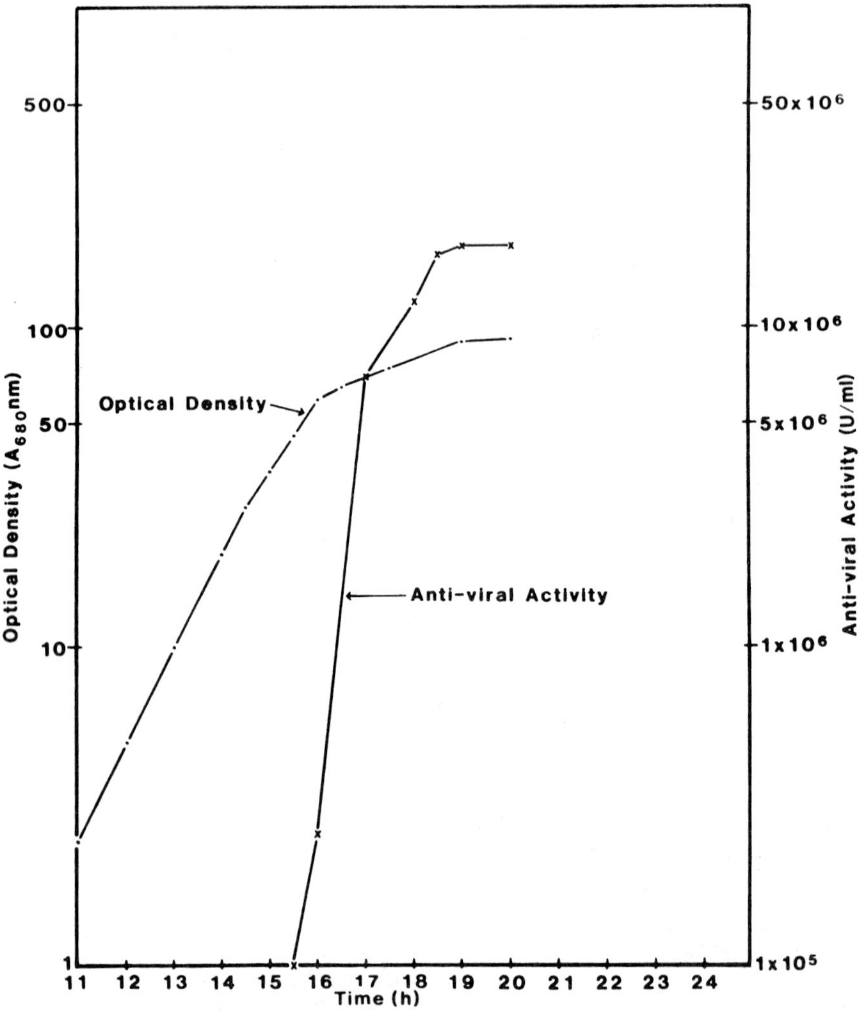

FIGURE 6. Production of human β_1 interferon by *E. coli* K-12 in a fermenter. The optical density of the culture is indicated on the left. At a value of about 60 and a run time of 16 hours, induction of interferon starts. The induction can be followed by the rapid appearance of antiviral activity in the cells as plotted on the right axis. Antiviral activity is assayed on human fibroblast cells challenged with VSV and is relative to an NIH standard interferon preparation.

making beta IFN is considerable, and is illustrated in FIGURE 5. A small culture of cells was suspended in tryptophan-free medium and the proteins made under these conditions were radioactively labeled for 1 minute with [^{3}H]leucine. A similar culture of the same cells, but growing in the presence of free tryptophan, was labeled with [^{14}C]leucine. The two cultures were chilled, mixed, sonicated, and then analyzed by SDS PAGE, which separates proteins by molcular weight. The distribution of both isotopes along the gel is seen in FIGURE 5. The very noticeable peak of ^{3}H at an apparent molecular weight of 18,000 daltons is beta IFN, and represents about 15% of the total instantaneous rate of protein synthesis. In FIGURE 6 we see the turbidity and total IFN concentration in a 10-liter fermentation run producing IFN to be used in the development of the purification process, and for biological characterization. In FIGURE 7 we see stained protein in an SDS PAGE analysis of cells from such a fermentation run. The band corresponding to beta can be clearly seen even in the crude extract, although it is not so conspicuous as seen in the isotopically labeled protein where only the proteins made after induction are visible.

A number of constructions producing beta interferon in *Bacillus subtilis* have been made by a group headed by Dr. Shing Chang. That the beta expressed in bacteria is not glycosylated can be seen by a technique developed to determine the molecular weight distribution of nanogram amounts of proteins present in mixtures. The proteins are first resolved by SDS PAGE, and transferred to activated paper where they become covalently bound. After extensive washing, the paper is incubated in a solution of antibody, in our case to beta IFN, and then with radioactive protein A, which selectively binds to antibody. The protein distribution is then determined by autoradiography. Bands corresponding to the beta produced in *E. coli* and *B. subtilis* that are about 20% lower in molecular weight than the band of beta induced in human fibroblast cells in tissue culture by poly I:C, which mimics viral infection.

In order to produce beta interferon that is glycosylated, the plasma seen in FIGURE 8 was constructed and characterized by Drs. Michael Innis and Frank McCormick at Cetus Corporation and a colleague at Stanford University, Prof. Gordon Ringold. The plasmid contains a segment of pBR322, which was used in the bacterial systems

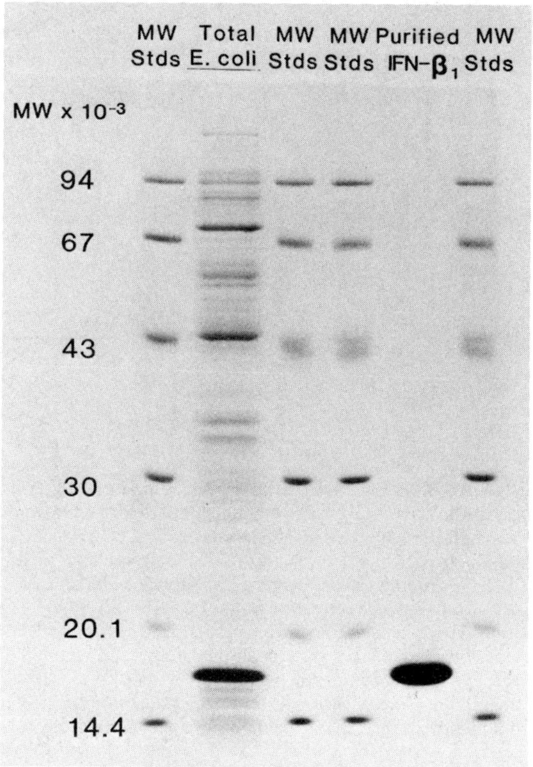

FIGURE 7. Purification of human β interferon produced in *E. coli*. The gel electrophoresis system described in FIGURE 5 was used to determine protein distribution in a whole cell lysate at the end of a fermentation, and a purified preparation of interferon. The proteins were visualized by staining, which is approximately proportional in intensity to the concentration of proteins in the bands. The positions of six proteins of known molecular weights are seen in lanes 1, 3, 4, and 6. The original bacterial lysate is seen in lane 2, while the purified material is seen in lane 5.

described above. In that segment is the origin of replication, which allows growth of the entire plasmid in *E. coli* in which it was constructed, and the ampicillin resistance gene, which allows easy selection of bacterial cell transformed by the plasmid. In another segment is the origin of replication from the animal virus SV40, and promoter from the mouse mammary tumor virus that allows transcription and expression of a gene coding for the enzyme dihydrofolate reductase. This gene allows selection in the animal cell host, if that host is dhfr−. Finally, it contains the gene for IFN-b_1 with its own promoter.[11] The host animal cell chosen is one of the oldest cell lines, derived from fibroblast Chinese hamster ovary cells (CHO) by Puck several decades ago. It grows rapidly, and through continuous passage has become transformed so that it no longer requires a solid support, but can be cultured in a spinner flask. Fortunately for us this cell line has lost the ability to produce its own IFN, and is not sensitive to human β-IFN. Finally, Drs. Chasen and Urlaub at Columbia University had isolated a dhfr− derivative of this line.[12]

After transformation and selection, cells were obtained in which the plasmid had been integrated into the chromosome producing a stable cell line. Some of these cells secreted IFN-β_1 into the tissue culture medium at the level of several hundred U/ml, but when subjected to induction by poly I:C produced 100,000 U/ml. This IFN was analyzed by the SDS electrophoresis technique discussed previously, and the material binding to beta interferon antibody had a molecular weight indistinguishable from the native material produced by fibroblasts, that is, it is glycosylated.

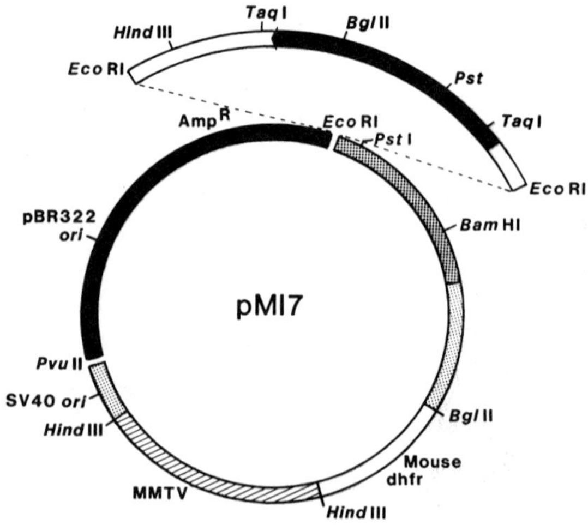

FIGURE 8. A plasmid that produces human beta interferon in hamster cells. Starting from the top of the circle and going counter-clockwise the plasmid pM17 contains: **Amp R:** The gene for ampicillin resistance allowing the selection of bacteria containing the plasmid. The fragments making up pM17 were assembled in a bacterial host. **pBR322 ori:** The origin of replication needed for DNA replication in bacteria. **SV40 ori:** A DNA origin of replication from the SV40 animal virus that enables replication in the hamster cell. **MMTV:** A promoter from the mouse mammary tumor virus, needed for transcription of the next gene. **Mouse dhfr:** The gene from the mouse that codes for the essential enzyme dihydrofolate reductase. The hamster cells into which this plasmid was introduced could not make this enzyme, thus cells containing the plasmid had a selective advantage. **Human beta interferon:** The structure of the genomic fragment that contains the gene and its promoter is diagramed on the upper semicircle.

Choice of the dhfr gene as a selection marker was motivated by the fact that selective pressure can be exerted on the cell to increase the amount of dhfr above that produced by one gene copy. This is accomplished by growing the cells in ever-increasing concentrations of a competitive inhibitor of the enzyme methotrexate. A concentration is picked where growth is inhibited but not stopped. Spontaneous derivatives that have increased enzyme levels due to the creation of multiple copies of the gene are able to replicate more rapidly and outgrow the other cells. Genes that are nearby often are also carried along piggyback by this process.[13] We are in the early stages of this process, but have obtained derivatives that have greater resistance to methotrexate and that also produce correspondingly higher levels of β-IFN.

While we are encouraged and excited by this work, it seems unlikely that it will be the ultimate, final system used for the production of IFN-β. For one thing, the induction protocol (or superinduction as it is usually called), the same one used to obtain IFN-β from human fibroblasts, is rather involved and not cheap. It consists of treating the cells with poly I:C in the presence of cycloheximide, which inhibits protein synthesis, but allows the build-up of IFN-β mRNA. The drug is removed but replaced with actinomycin D, which now allows transcription of the RNA producing IFN-β_1, but blocking induction of synthesis of new mRNA. This protocol has been rationalized as preventing the induction of a protein by IFN-β, which would then inhibit the synthesis of beta, that is, circumventing a natural negative feedback loop. This model has never been directly verified, and is doubted by some, but in fact it does produce higher levels of IFN than poly I:C alone. On the optimistic side however, we have spent no time trying to improve this induction scheme. If it is really necessary to block protein and RNA synthesis, there are probably easier ways to do it, reversible temperature-sensitive mutations, for example. If we know more about what was really happening, there might be easy ways to obtain the same result. However, it seems likely that the easiest and most general approach will be to replace the beta promoter with one that is stronger, and perhaps easily controllable. Such systems are under active development at Cetus Corporation.

However, even in its present form, this cell line produces levels that are an order of magnitude higher than that produced from regular fibroblasts. It is unlikely that the pattern of sugar residues will be exactly the same as that produced by human fibroblasts, although it may be quite close. It will enable us to proceed more rapidly in investigations of just what the sugar means to the biochemical properties of this kind of IFN.

SUMMARY

A frequent motive for genetic engineering is to transfer the gene for the desired protein from a cell that is difficult or impossible to grow into a cell which is easy and inexpensive to handle. In the pharmaceutical field this usually means isolation of a gene coding for a protein secreted by human cells, the protein often being glycosylated. While bacterial cells are an inexpensive host in which to express genes, they do not produce proteins which are glycosylated, nor necessarily properly processed. Thus while there are examples of successful bacterial systems, there is also an intense interest in expression in animal cells. The situation is often complicated by our lack of knowledge of the function, or even the necessity, of glycosylation. Examples of successes and problems in production of insulin and interferons in several hosts have been presented.

REFERENCES

1. Bell, G. I., R. L. Pictet, W. J. Rutter, B. Cordell, E. Tischer & H. M. Goodman. 1980. Nature (London) **284:** 26–32.
2. Steiner, D. F. & J. L. Clark. 1968. Proc. Natl. Acad. Sci. USA **60:** 622–629.
3. Kemmler, W., J. D. Peterson & D. F. Steiner. 1971. J. Biol. Chem. **246:** 6786–6791.
4. Milstein, C., C. G. Brownlee, T. M. Harrison & M. B. Mathews. 1972. Nature (London) New Biol. **239:** 117–120.
5. Broome, S. & W. Gilbert. 1978. Proc. Natl. Acad. Sci. USA **75:** 2746–2749.
6. Chan, S. J., J. Weiss, M. Konrad, T. White, C. Bahl, S-D. Yu, D. Mark & D. F. Steiner. 1981. Proc. Natl. Acad. Sci. USA **78:** 5401–5405.
7. Talmadge, K., J. Kaufman & W. Gilbert. 1980. Proc. Natl. Acad. Sci. USA **77:** 3988–3992.
8. Lengyel, P. 1982. Ann. Rev. Biochem. **51:** 251–282.
9. Gray, P. W. & D. V. Goeddel. 1982. Nature (London) **298:** 859–863.
10. Bocci, V., A. Pacini, G. P. Pessina, V. Barigli & M. Russi. 1977. J. Gen. Virol. **35:** 525–534.
11. Ohno, S. & T. Taniguchi. 1981. Proc. Natl. Acad. Sci. USA **78:** 5305–5309.
12. Urlaub, G. & L. A. Chasin. 1980. Proc. Natl. Acad. Sci. USA **77:** 4216–4220.
13. Ringold, G., B. Dieckmann & F. Lee. 1981. J. Mol. Appl. Genet. **1:** 165–175.

Using Genetically Engineered Bacteria for Vaccine Production

DENNIS G. KLEID[a]

Genentech, Inc.
460 Point San Bruno Boulevard
South San Francisco, California 94080

INTRODUCTION

All vertebrates possess an immune system that protects them from foreign invaders such as viruses or other disease-causing microorganisms. The system works by recognizing foreign invaders and mounting a response that neutralizes the infection. One response involves the production of antibodies that bind to the invader's surface, enabling the host to inactivate it. With certain viruses or microorganisms, the infection can be quite severe before the host mounts a neutralizing response. However, the immune response is significantly enhanced by either a previous infection with the same organism or by the use of vaccines made up of either attenuated or inactivated organisms. Vaccines stimulate the immune system principally to proteins on the invader's surface. Because of their protective effect, vaccines are a major medical tool used to help prevent or fight infections.

The large-scale production of safe and effective vaccines is a difficult and at times a tricky problem, dependent in part on the organism from which the vaccine is made. Edward Jenner found, for example, that inoculation of individuals with cowpox virus would protect them for life from smallpox and vaccines based on this principle have recently led to the worldwide eradication of the disease. Attenuated vaccines made from yellow fever virus, rubella virus, polio virus, measles virus and others have also been highly successful. For many diseases, such as herpes, syphilis, and malaria, no suitable vaccine product exists, and for others such as influenza, pertussis, and hepatitis, the vaccine products are either not very efficacious, have side reactions, or are expensive.

When it became possible to splice genes from one organism into another, it was recognized that vaccines might be made using this technique that would be safer and more effective, or for diseases where none are currently available. The genes that code for the antigenic proteins could be located, isolated, and by recombinant DNA techniques, introduced into relatively harmless bacteria for their expression as proteins. The protein antigens isolated from these new organisms could be used as vaccines, or possibly the whole organism could be genetically modified or attenuated.

Over the last five years, a number of approaches have been described envisioning the use of *E. coli* for the production of antigenic material: secretion of the antigenic

[a]This paper reviews research done at Genentech, Inc., South San Francisco, California 94080 by Dennis Kleid, Daniel Yansura, Donald Dowbenko, Barbara Small, Gregory Weddell, Maureen Hoatlin, Neal Clayton, Steven Shire, Lawrence Bock, Eric Patzer, Stuart Builder, and John Ogez and at the United States Department of Agriculture, Plum Island Animal Disease Center (PIADC), Greenport, New York 11944 by Douglas M. Moore, Marvin J. Grubman, Peter D. McKercher, Donald O. Morgan, Betty H. Robertson, Karl Axelson, Tom Fischer, Nicholas Shuot, JoAnn Henry, Edward Whittle, Howard L. Bachrach and Jerry J. Callis.

protein into the growth media,[1] production of the antigen inserted in the surface of the bacteria,[2] or production of the immunogenic protein inside the bacteria.[3] None of the studies have yet to lead to a vaccine, although small amounts of antigenic material have been observed.[4]

The problem may be the physical properties of the immunogenic surface proteins. Often these proteins are glycosylated or contain hydrophobic amino acid sequences that enable the protein to assemble into a virus coat or cellular membrane.[5] Apparently, a number of these proteins, when synthesized in *E. coli* by means of recombinant plasmids, are not expressed efficiently. They are either rapidly degraded by bacterial proteases or inhibit bacterial growth, possibly by interfering with some metabolic or cell division processes.

One successful solution has come from the use of fusion proteins. Fusion proteins are coded by structural genes obtained by ligating a part of one gene from the host bacteria with the gene of the desired antigenic protein in the same codon reading frame. By carefully choosing the sequences coded by the host and the gene for the desired antigen, a protein product can be expressed at high levels that is both stable in the cell and retains high immunogenic potency. What follows is an example of how this was accomplished for one disease-causing virus, foot-and-mouth disease virus.[6]

FOOT-AND-MOUTH DISEASE (FMD) AND THE FMD IMMUNOGENIC PROTEIN

Foot-and-mouth disease is a highly contagious disease that afflicts primarily cloven-hoofed animals. The disease has had a major economic impact on agricultural industries in many parts of the world owing to its rapid spread principally among cattle, swine, and sheep. Control of the disease, by extensive use of vaccines and limitations on cattle transport, as well as careful monitoring of outbreaks, has begun to limit the affected areas.[7] However, export of fresh meat and livestock from countries that have the disease to disease-free countries is severely restricted, limiting the market for animals and animal products.

The FMD virus (a picornavirus) is composed of a positive-stranded RNA genome of about 8,000 nucleotides containing a covalently linked protein (VP_G) at its 5′ end and a polyadenylate tract at its 3′ end. The genome RNA is contained within an icosahedral particle about 20–25 nm in diameter composed of four different virus structural proteins (called VP_1, VP_2, VP_3 and VP_4). One of these proteins, VP_1 (formerly designated VP_3,[8,9]), is located on the surface of the virus and when purified and injected into animals induces the production of virus-neutralizing antibody[9,10] and provides protection in swine[9] and cattle[11] from the viral infection. These activities were considered due to the presence in the isolated capsid protein of antigenic sites similar to those on the surface of the virus that elicit immune response. The antigenic sites are created by specific squences of amino acids in the VP_1 protein.

This knowledge was used to isolate the structural gene for VP_1 and to subsequently express the immunogenic VP_1 in *E. coli* by using the biosynthetic capacity of the bacteria. Procedures for producing and cloning copies of a gene from an RNA molecule have been described in detail.[6,12] The steps are outlined in FIGURE 1. The FMDV RNA genome from each strain required was isolated and annealed to a small thymidylate deoxyoligonucleotide primer, which hybridizes to the polyadenylate residues at the 3′ end of the viral RNA. In the presence of the enzyme reverse transcriptase, the RNA was copied into DNA (cDNA) beginning with the primer (see FIG. 1, step A), and subsequently into double-stranded cDNA (ds-cDNA) by

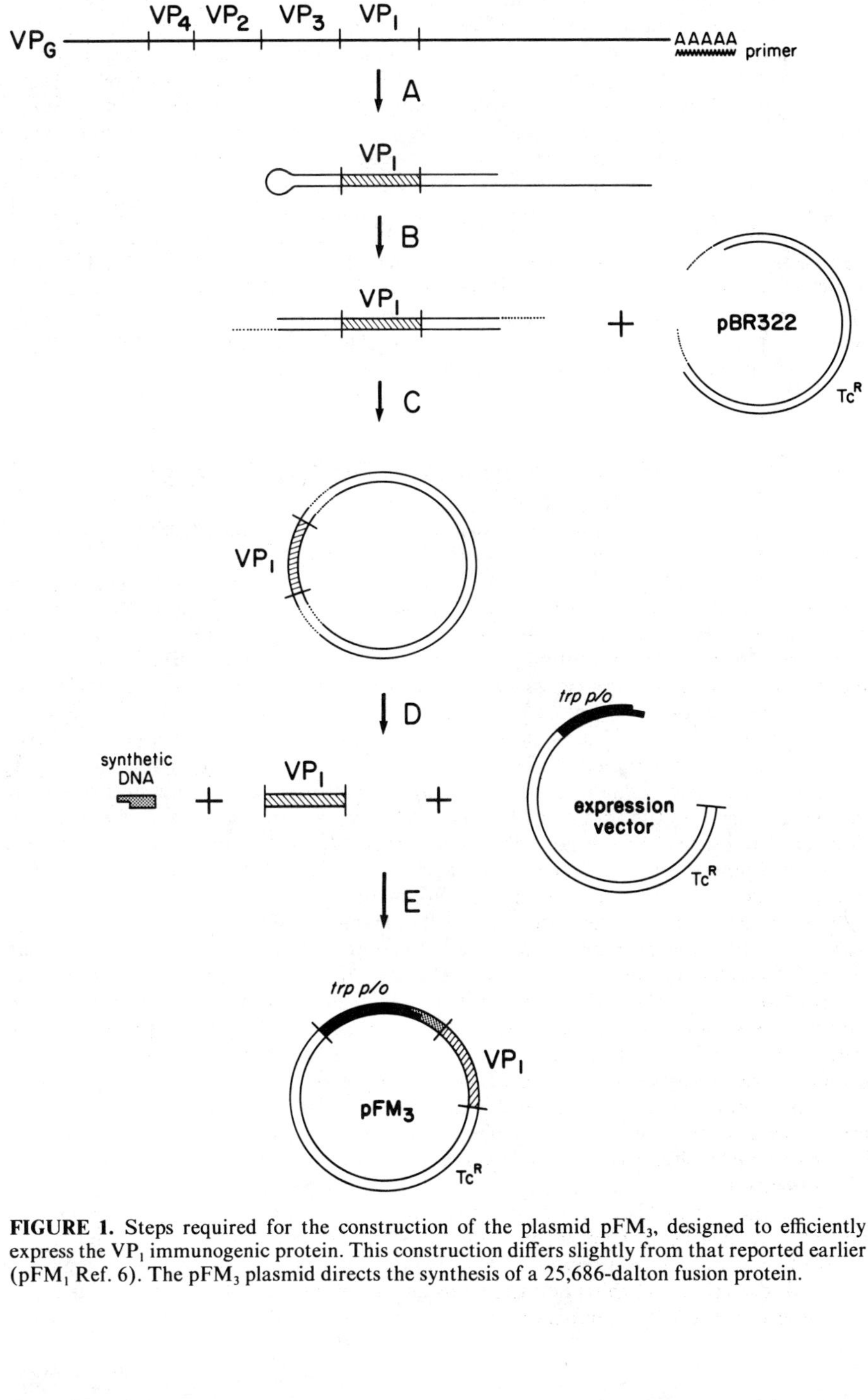

FIGURE 1. Steps required for the construction of the plasmid pFM_3, designed to efficiently express the VP_1 immunogenic protein. This construction differs slightly from that reported earlier (pFM_1 Ref. 6). The pFM_3 plasmid directs the synthesis of a 25,686-dalton fusion protein.

treatment with DNA polymerase. The ds-cDNA was treated with the enzyme S_1, which digests single-stranded regions, and the ds-cDNA was isolated by polyacrylamide gel electrophoresis (FIG. 1, step B). Segments of ds-cDNA greater than 2,000 base pairs were treated with an enzyme, deoxynucleotidyl transferase, in the presence of dCTP (2′-deoxycytidine 5′-triphosphate), to create polymers of cytidine on both ends of the DNA molecules. These segments were annealed to linearized plasmid pBR322 (FIG. 1, step C) previously treated to give polymers of guanosine on both ends. The plasmid and the ds-cDNA now annealed together were mixed with *E. coli* in the presence of calcium ions and briefly heated, causing some plasmids to enter the bacteria. The bacteria were plated on media containing tetracycline (Tc). This antibiotic eliminates bacteria not "transformed" to tetracycline resistance (Tc^r) with the plasmid, and upon replication allows growth of selected bacterial colonies each containing numerous copies of a single plasmid with its ds-cDNA insert. Plasmids with incorporated ds-cDNA inserts, were isolated from individual bacterial colonies and were analyzed by restriction enzyme mapping and nucleotide sequencing. Some of the inserts contained the gene coding for the VP_1 protein (FIG. 2).

EXPRESSION OF THE IMMUNOGENIC VP_1 IN *E. COLI*

Once the gene for the immunogenic protein was cloned, isolated, and its nucleotide sequence determined, the next step was to put the coding sequence back into *E. coli* so that it is efficiently copied into a mRNA and protein. This was done with another plasmid, called an expression vector, that contains five essential genetic elements: (1) an RNA polymerase binding site—the promoter; (2) a control element for the level of mRNA produced—the operator; (3) a sequence for directing protein synthesis—the ribosome binding sequence; (4) a codon for the beginning of translation of mRNA into protein—the initiation codon; and (5) a gene coding for part of an *E. coli* protein that will comprise a portion of a stable fusion protein. The expression vector used in this study contained these genetic elements from the *E. coli* tryptophan operon. This operon has a control element (i.e., the operator) that responds to the concentration of tryptophan in the growth medium. It also contains an efficient origin of replication (derived from pBR322) so that many copies of the plasmid reproduce in the host cell, and an antibiotic resistance gene (Tc^r) so that transformed *E. coli* can be selected simply by adding the antibiotic to the growth media.

The VP_1 gene was isolated from the plasmid containing the cloned gene of interest (FIG. 1, step D) by treatment with restriction endonucleases. Cleavage sites for the enzymes were located near the ends of the VP_1 gene. The 5′ end of the gene was reconstructed using chemically synthesized DNA fragments to link the structural gene and the expression vector so that exactly the correct relationship to the genetic control elements results. The coding sequence for a number of amino acids from the *trp*LE peptide (derived from *E. coli trp*ΔLE1413[13]) was arranged to produce a product containing amino acids from the *trp*LE peptide linked to the VP_1 protein (FIG. 2). The three DNA fragments were joined together with DNA ligase (FIG. 1, step D) to give the plasmid pFM_3 that was used for expression of the fusion protein VP_1 immunogen in *E. coli* (FIG. 2).

When plasmid pFM_3 was incorporated into *E. coli,* the rate of transcription was controlled by the level of tryptophan in the growth medium. Therefore cells containing the pFM_3 plasmid were grown in a tryptophan-containing medium until a high cell density was reached. At this point the tryptophan was depleted from the culture and the synthesis of the VP_1 immunogen became very efficient (FIG. 3). The fusion protein

```
1                             10                            20                            30
met lys ala ile phe val leu lys gly ser leu asp arg asp pro glu phe met thr thr ala thr gly glu ser ala asp pro val thr
ATG AAA GCA ATT TTC GTA CTG AAA GGT TCA CTG GAC AGA GAT CCA GAA TTC ATG ACT ACC GCT ACT GGT GAA TCT GCA GAC CCT GTC ACC
                                                            EcoRI                                   PstI
                              40                            50                            60
thr thr val glu asn tyr gly gly glu thr gln val gln arg arg his his thr asp val ser phe ile met asp arg phe val lys ile
ACC ACC GTG GAG AAC TAC GGT GGT GAG ACA CAA GTC CAG AGA CGT CAC CAC ACG GAC GTC AGT TTC ATC ATG GAC AGA TTT GTG AAG ATA

                              70                            80                            90
lys ser leu asn pro thr his val ile asp leu met gln thr his gln his gly leu val gly ala leu leu arg ala ala thr tyr tyr
AAA AGC TTG AAC CCC ACA CAC GTC ATT GAC CTC ATG CAG ACC CAC CAA CAC GGG CTG GTG GGT GCG TTG TTG CGT GCA GCC ACG TAC TAC

                              100                           110                           120
phe ser asp leu glu ile val val arg his asp gly asn leu thr trp val pro asn gly ala pro glu ala ala leu ser asn thr gly
TTC TCC GAC TTG GAG ATT GTT GTG CGG CAC GAT GGC AAT CTG ACC TGG GTG CCC AAC GGT GCC CCC GAG GCA GCC CTG TCA AAC ACC GGC

                              130                           140                           150
asn pro thr ala tyr asn lys ala pro phe thr arg leu ala leu pro tyr thr ala pro his arg val leu ala thr val tyr asn gly
AAC CCC ACT GCC TAC AAC AAG GCA CCG TTC ACG AGG CTT GCT CTC CCT TAC ACT GCG CCA CAC CGC GTG TTG GCA ACT GTG TAC AAC GGG

                              160                           170                           180
thr asn lys tyr ser ala ser gly ser gly val arg gly asp phe gly ser leu ala pro arg val ala arg gln leu pro ala ser phe
ACG AAC AAG TAC TCC GCG AGC GGT TCG GGA GTG CGA GGC GAT TTT GGG TCT CTC GCG CCG CGA GTC GCG AGA CAA CTT CCT GCT TCT TTC

                              190                           200                           210
asn tyr gly ala ile lys ala glu thr ile his glu leu leu val arg met lys arg ala glu leu tyr cys pro arg pro leu leu ala
AAC TAC GGT GCA ATT AAG GCC GAG ACC ATC CAC GAG CTT CTC GTG CGC ATG AAA CGG GCT GAG CTC TAC TGC CCC AGG CCA CTG CTG GCA

                              220                           230   233
ile glu val ser ser gln asp arg his lys gln lys ile ile ala pro gly lys gln asn ser his val OP
ATA GAG GTG TCT TCG CAA GAC AGG CAC AAG CAG AAG ATC ATT GCA CCC GGA AAA CAG AAT TCT CAT GTT TGA
                                                                             EcoRI

Translated Mol. Weight = 25686.40
```

FIGURE 2. The nucleotide sequence and the amino acid sequence of the VP_1 protein *E. coli trp* fusion protein as expressed from the vector pFM_3. The first 17 amino acids are contributed by the tryptophan operon followed by an incorporated *Eco*RI site. Between the *Eco*RI site and the *Pst*I site is a synthetic DNA fragment coding for the amino acid methionine and the first 7 amino acids of the VP_1 protein. This is followed by amino acid codons 8–211 of the FMDV A_{12} 119ab VP_1 protein linked at the COOH-terminal end to 4 amino acids contributed by the vector pBR322. The FMDV gene fragment (approximately 610 base pairs bounded by PstI and EcoRI sites) was obtained from the cloned VP1 gene.

VP1 →
FUSION PROTEIN

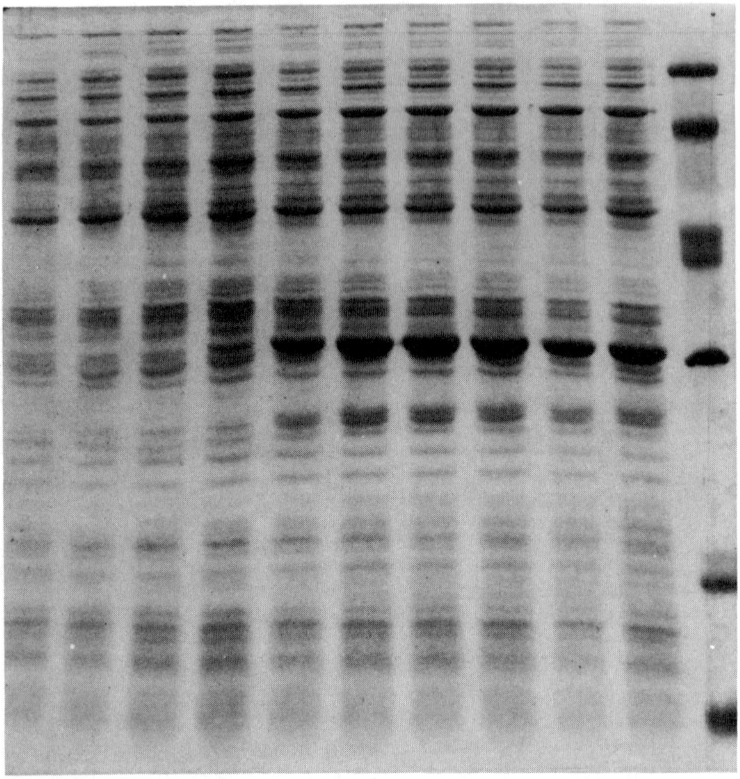

FIGURE 3. SDS-polyacrylamide gel of the Coomassie brilliant blue stained proteins derived from bacterial samples taken from a 10-liter fermentation during a five-hour period. The photo shows the induction of the FMDV VP_1 fusion protein when the culture medium becomes depleted in tryptophan.

expressed in these cells appeared as visible refractile bodies when viewed under a phase contrast microscope, accounted for greater than 20% of the total cell protein, and was recovered efficiently after lysing the cells.

RESPONSES IN CATTLE TO THE *E. COLI* EXPRESSED FMDV IMMUNOGENIC VP_1 PROTEIN

The VP_1 fusion protein from *E. coli*/pFM_3 was purified by ion-exchange chromatography,[9,14] combined with an incomplete Freund-type oil adjuvant and administered subcutaneously to cattle. (A cattle test was previously described in Ref. 6. In that study, 250 μg of VP_1 fusion protein from *E. coli*/pFM_1 isolated from polyacrylamide SDS gels was used as a test vaccine and found to protect cattle from FMDV challenge.) Here we present results on the injection of fusion protein (purified from *E. coli*/pFM_3) at different doses into 36 cattle, divided into four groups of 9 animals each. The following doses of protein were used: Group 1, 10 μg; Group 2, 50 μg; Group 3, 250 μg; and Group 4, 1250 μg. The level of neutralizing antibody was measured by the Skinner suckling mice assay.[15]

The results of the vaccination and challenge are presented in detail elsewhere (P.D. McKercher *et al.*, manuscript in preparation). Briefly, in the higher doses (250 μg and 1250 μg) levels of antibody were obtained and sustained for 7 months that were indicative of immunity. The lower doses (10 μg and 50 μg) produced significant, but lower levels of antibody. The lower dose groups were revaccinated at 3½ months and the higher dose groups at 7½ months. Following revaccination, antibody levels in all groups rose to very high levels. Challenge of immunity by contact with infected animals[17] was done at 7 months for groups one and two (lower dose) and at 10½ months for groups three and four (higher doses). The following protection levels were obtained: Group 1, 5 of 9; Group 2, 7 of 9; Group 3, 8 of 9; and Group 4, 9 of 9.

SUMMARY

We concluded from this and our earlier work[6] that biosynthetically produced FMDV VP_1-specific fusion proteins are effective vaccines. Whether this method of vaccine production can be extended to many other immunogenic proteins from other organisms is not known. Some problems that could be expected to occur with bacterially produced antigens are that the immunogenic site may not be properly exposed or the peptide sequence(s) within that site may not be able to form into the correct configuration. This could be caused by hydrophobic or hydrophilic interactions in the fusion protein that do not occur in the protein at the virus surface. Also, the immunogenic site may require disulfide bonding to bring two distant parts of a protein or two different peptide chains into close proximity to form an antigenic site, as demonstrated by the studies of Atassi *et al.*[16] for lysozyme-using synthetic peptides.

In summary, the use of genetically programmed bacteria is a promising avenue to vaccine manufacture. For FMD, biosynthetic protein vaccines have significant advantages over current whole-virus technology.

REFERENCES

1. VILLA-KOMAROFF, L., A. EFSTRATIADIS, S. BROOME *et al.* 1978. Proc. Natl. Acad. Sci. USA **75:** 3727–3731.
2. FRASER, T. H. & B. J. BRUCE. 1978. Proc. Natl. Acad. Sci. USA **75:** 5936–5940.
3. EMTAGE, J. S., W. C. A. TACON, G. H. CATLIN *et al.* 1980. Nature **283:** 171–174.
4. BURRELL C. J., P. MACKAY, P. J. GREENAWAY *et al.* 1979. Nature **1979:** 43–47.
5. SIMONS, K., H. GAROFF & A. HELENIUS. 1982. Sci. Am. **246:** 58–78.
6. KLEID, D. G., D. YANSURA, B. SMALL *et al.* 1981. Science **214:** 1125–1129.
7. European Commission for the Control of Foot-and-Mouth Disease. Report of the 24th Session (FAO Rome, April 1981).
8. STROHMAIER G., *et al.* Report of the Session of the Research Group of the Standing Technical Committee of the European Commission for the Control of Foot-and-Mouth Disease. June 1978, Brussels, Belgium.
9. BACHRACH, H. L., D. M. MOORE, P. D. MCKERCHER & J. POLATNICK. 1975. J. Immunol. **115:** 1636–1641.
10. LAPORTE, J., J. GROSCLAUDE, J. WANTYGHEM *et al.* 1973. C. R. Acad. Sci. **276:** 3399–3401.
11. BACHRACH, H. L., D. O. MORGAN, P. D. MCKERCHER, D. M. MOORE & B. H. ROBERTSON. 1982. Vet. Microbiol. **7:** 85–86.

12. MANIATIS, T., E. F. FRITSCH & J. SAMBROOK. 1982. Molecular Cloning, A Laboratory Manual. Cold Spring Harbor Laboratory.
13. MIOZZARI, G. F. & C. YANOFSKY 1978. J. Bacteriol. **133:** 1457–1466.
14. BERNARD, S., J. WANTYGHEM, J. GROSCLAUDE & J. LAPORTE. 1974. Biochem. Biophys. Res. Commun. **58:** 624–632.
15. SKINNER, M. M. 1953. Proc 15th Int. Vet Congr Stockholm **1:** 195–199.
16. ATASSI, M. Z. 1978. Immunochemistry **15:** 909–936.
17. GRAVES J. H., P. D. MCKERCHER, H. E. FARRIS, JR. & K. M. COWAN. 1968. Res. Vet Sci. **9:** 35–40.

Uses of Recombinant DNA for Analyses of *Streptomyces* Species

C. L. HERSHBERGER, J. L. LARSON, AND
S. E. FISHMAN

Eli Lilly and Company
307 East McCarty Street
Indianapolis, Indiana 46285

INTRODUCTION

The advent of recombinant DNA techniques has provided investigators with a new arsenal to attack problems of gene structure, genetic regulation, and genome organization. Already the intellectual rewards have been spectacular in providing solutions to problems that were intractable previously. However, the most important rewards are derived from applications to produce biological products that benefit mankind directly. The first-stage efforts have been mobilized for production of single proteins. September 20, 1982, is a historic date in stage one because Eli Lilly and Company marketed human insulin, which is the first clinically significant product from recombinant DNA that is available to the public. A plethora of additional protein products from recombinant DNA are in various stages of development or research. Undoubtedly, the results will have profound effects on the practice of medicine during the coming decades.

Perhaps the next important milestone of pharmaceutical applications will be the production of metabolites such as antibiotics. Essential use of life-saving antibiotics is a cornerstone of modern medicine. Successful development of recombinant DNA methods will provide tools that will help increase the fermentation yields of antibiotics and generate new antibiotic structures that are not available through traditional routes of discovery.[1] Full realization of these benefits necessitates developing recombinant DNA systems for *Streptomyces (S.)* species because two-thirds of all known antibiotics are produced by *S.* species.[2] Researchers, led by David Hopwood and Keith Chater at the John Innis Institute, have set the tempo for development of recombinant DNA systems in *S.* species: Hopwood working on plasmid vectors and Chater working on actinophage vectors. Their progress has been described recently and readers are referred to these excellent reviews.[2–5] Investigators with Stanley Cohen have reported progress in development of plasmid vectors,[6,7] Isogai *et al.*[8] have potential actinophage vectors and Richardson *et al.*[9] have published additional plasmid vectors. Multiple and parallel development of recombinant DNA systems provides the greatest likelihood for success because no one system is likely to work with all *S.* species or be applicable to all problems.

This report describes progress in our program to apply recombinant DNA technology to *S.* species. Steps are described in the construction of the plasmids pJL197 and pJL198, which should be particularly useful as recombinant DNA cloning vectors. Discovery of amplified DNA in *Streptomyces fradiae (S. fradiae)* and application of recombinant DNA techniques in the characterization of amplified DNA are described in the latter portion of the report. The features of this high-level DNA amplification are unique to *Streptomyces* among bacterial species.

EXAMPLE OF A MODEL VECTOR

Desired properties for a vector can be illustrated by examination of a useful vector. The plasmid, pBR322, has been used extensively for cloning DNA in *Escherichia coli K12 (E. coli).*[10] It is a small DNA plasmid of approximately 4.4 kilobases (Kb) that contains very little extraneous sequence that does not contribute to its utility as a cloning vector. Three functional regions define the important features of the plasmid. The replication region (Rep) contains all of the information that is needed for replication and maintenance of the plasmid in a suitable *E. coli* host. Two regions encode resistance to antibiotics. Ap^r designates resistance to ampicillin which is mediated by the enzyme, β-lactamase, and Tc^r designates resistance to tetracycline. Antibiotic resistance markers are important because they provide positive selection for transformed hosts that carry the plasmid. pBR325 is a useful cloning vector that was derived from pBR322.[11] The important difference between pBR322 and pBR325 for this discussion is the inclusion of a third marker for antibiotic resistance on pBR325. The marker, Cm^r, designates resistance to chloramphenicol that is mediated by chloramphenicol acetyl transferase.

DNA molecules are cut by restriction endonucleases at specific recognition sites. Two linear fragments of DNA can be joined together to form a circular molecule by the enzyme, DNA ligase, if the linear fragments have homologous termini. Useful cloning vectors should contain single recognition sites for restriction enzymes so that they can be converted to linear molecules without losing essential functions of the vector. Only a portion of the transformed hosts contain recombinant DNA molecules in a typical cloning experiment. The remainder of the transformed population will contain the vector that has been self-ligated without insertion of a new DNA fragment. One of the antibiotic resistance markers is needed to select transformed cells from those that do not contain plasmid; however, the second resistance marker can be used to identify cells that contain a recombinant plasmid by the phenomenon of insertional inactivation. The cloned fragment may interrupt expression of antibiotic resistance if insertion occurs in the gene that encodes the resistance. pBR322 contains a single recognition site for *Pst*I, which occurs in the gene for Ap^r. Recombinants of pBR322 will be recognized as Ap^s (ampicillin sensitive) and Tc^r transformants if DNA is cloned into the *Pst*I site. Similarly, recombinants of pBR322 will be recognized as Tc^s (tetracycline sensitive) and Ap^r transformants if DNA is cloned into the *Sal*I or *Bam*HI site. Recombinants of pBR325 will be recognized as Cm^s (chloramphenicol sensitive), Ap^r, and Tc^r if DNA is cloned into the *Eco*RI site.

DESIGN OF A VECTOR FOR *STREPTOMYCES* SPECIES

A desirable cloning vector for *S.* species should contain plasmid replication functions that are expressed in *S.* species, two markers for antibiotic resistance with a unique recognition site for a restriction enzyme in each resistance marker and functions for self-transmissability to facilitate rapid screening for complementation of a variety of mutations. The vector may be more useful if it can be shuttled between *S.* species and *E coli*[12] because plasmids are easily manipulated and isolated in *E. coli* while similar manipulations are more difficult in *S.* species. Therefore, the shuttle vector will contain plasmid Rep from both *E. coli* and *S.* species, at least one resistance marker that is expressed in *E. coli,* and at least two resistance markers that are expressed in *S.* species.

The plasmid, SCP2*, was chosen as the source of plasmid Rep functions for *S.* species in our research program. SCP2* is a sex factor that was isolated from

Streptomyces coelicolor (S. coelicolor).[13–22] The presence of SCP2* in a host is identified by the observation of pocks on an indicator strain. A pock is a zone of growth inhibition or development inhibition that surrounds a colony containing SCP2* when it grows on a confluent lawn of the same strain that does not contain SCP2*.

The use of shuttle vectors requires hosts that do not present barriers to introduction of foreign DNA. Many bacteria contain active restriction systems that degrade DNA if it is not modified according to the host's specificity. *Streptomyces griseofuscus* C581 (*S. griseofuscus*) was selected as the *S.* host because it does not express an active restriction system.[23] *E. coli* K12 C600$R_k^-M_k^-$ was selected as the *E. coli* host because it does not express an active restriction system.[24]

CHIMERIC PLASMIDS AND DEMONSTRATION OF THE SHUTTLE SYSTEM

The plasmids, SCP2* and pBR325, were combined at the *Eco*RI site in each plasmid and transformed into *E. coli.* Two plasmids, pJL120 and pJL121, were recovered. They differ from each other only in the orientation of the component plasmids that were used in the construction (FIG. 1). Both plasmids confer the phenotypes, Ap^r, Tc^r, and Cm^s in *E. coli* and form pocks in *S. griseofuscus.*

An assay was developed so that reciprocal shuttle between *E. coli* and *S. griseofuscus* could be detected with a minimum investment of time and effort. *S. griseofuscus* was grown for 20 hours in Trypticase Soy Broth from Difco Laboratories that was supplemented with 0.4% glycine. The culture was homogenized and inoculated at 5% (v/v) into the same medium and grown for 18 hours. The mycelial mass from the culture was employed to prepare protoplasts.[25] *E. coli,* containing the test plasmid, was grown with antibiotic selection and employed to isolate DNA "mini-preparations."[26] Twenty microliters (μl) of "mini-preparation" DNA was transformed into *S. griseofuscus* protoplasts that were then regenerated on modified R2 medium.[17] Individual colonies, isolated pocks or spore-lawns from the regeneration plates were inoculated into Trypticase Soy Broth that was supplemented with 0.4% glycine. The cultures were grown for 20 hours and employed for rapid isolation of total DNA. Fifty μl of the crude DNA was transformed into *E. coli* that had been rendered competent for transformation.[27] The transformed *E. coli* cells were isolated by growth of colonies on media containing an appropriate antibiotic. Recovery of back-transformed *E. coli* with appropriate antibiotic resistance markers was taken as a positive result for the shuttle assay. Representative colonies were chosen to verify the structure of the plasmid.

The plasmids, pJL120 and pJL121, shuttle reciprocally between *E. coli* and *S. griseofuscus.* The plasmids are isolated from either host by conventional procedures, and they are purified by equilibrium density gradient ultracentrifugation in propidium iodide–CsCl gradients. The frequency of plasmid transformation into *S. griseofuscus* is the same for plasmid DNA purified from either *E. coli* or *S. griseofuscus.* Also, the frequency of plasmid transformation into *E. coli* is the same for plasmid DNA purified from either *E. coli* or *S. griseofuscus* (data not shown).

IDENTIFICATION OF FUNCTIONS RETAINED BY MUTANTS WITH DELETIONS IN SCP2*

SCP2* contains approximately 30 Kb of DNA; therefore, it is beneficial to identify the portion of the plasmid that is required for Rep, self-transmissability, and pock

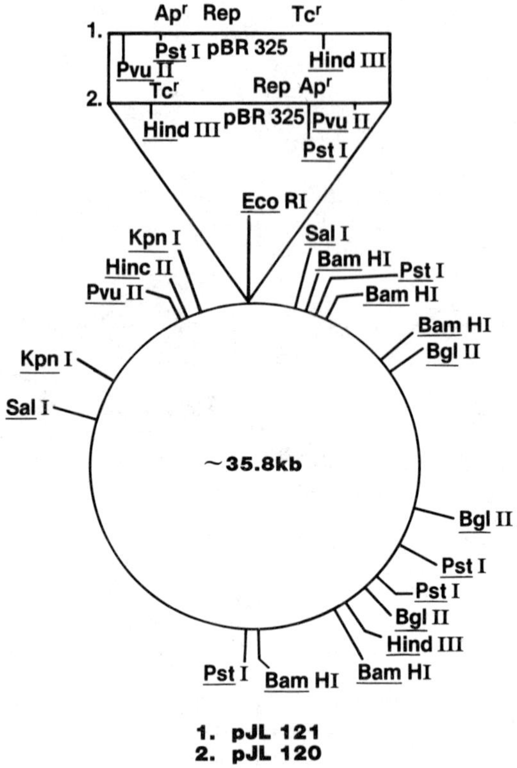

FIGURE 1. Restriction maps of pJL120 and pJL121. The circle shows the restriction map of SCP2*. The linear inserts show the restriction maps of pBR325 that were inserted at the *Eco*RI site of SCP2*. pJL120 and pJL121 contain identical DNA components; however, the orientations of pBR325 are inverted. Nomenclature for restriction endonuclease sites is described by Roberts.[43] Restriction maps in this figure and subsequent figures are not necessarily drawn to scale. They are intended to define the general organization and sequence of sites in the plasmids. Nomenclature for phenotypic markers on the plasmids is described by Novick.[44] Resistance phenotypes are indicated by two-letter abbreviations. The designations identify approximate locations of phenotypic markers in all of the figures. The designations are Ap^r for ampicillin resistance determined by β-lactamase in *E. coli*, Tc^r for tetracycline resistance in *E. coli*, Cm^r for chloramphenicol resistance determined by chloramphenicol acetyl transferase in *E. coli*, Nm^r for neomycin resistance determined by neomycin phosphotransferase in *S.* species, Ts^r for thiostrepton resistance determined by 23S rRNA methylase in *S.* species, and Rep for plasmid replication functions whether they are functional in *E. coli* or *S.* species.

formation so that the nonessential sequence of DNA can be omitted in the construction of useful cloning vectors. The chimeric plasmids, pJL120 and pJL121, were restricted separately with a variety of restriction enzymes and self-ligated to generate mutants that were deleted for various portions of SCP2*. Also, restriction digests of SCP2* were separately cloned into pBR322 or pBR325. (FIG. 2). The deleted plasmids were isolated in *E. coli* and transformed into *S. griseofuscus*. Plasmids that formed pocks in *S. griseofuscus* were considered positive for retention of SCP2* Rep. The plasmid, pJL125, contains the 5.4 Kb *Eco*RI–*Sal*I fragment of SCP2* and is deleted for the 1.89 Kb *Eco*RI–*Sal*I fragment of pBR325 (FIG. 3). It is the smallest plasmid that formed pocks when it was transformed into *S. griseofuscus*. pJL125 shuttles reciprocally between *S. griseofuscus* and *E. coli*. The 5.4 Kb fragment of SCP2* should be sufficient for construction of useful cloning vectors because it contains the sequence needed for Rep and pock formation in *S.* species.

A SITE THAT DETERMINES POCK MORPHOLOGY

The plasmids pJL120, pJL121, and all deletion mutants derived from them caused formation of small pocks, which are denoted as "mini-pocks." They are morphologi-

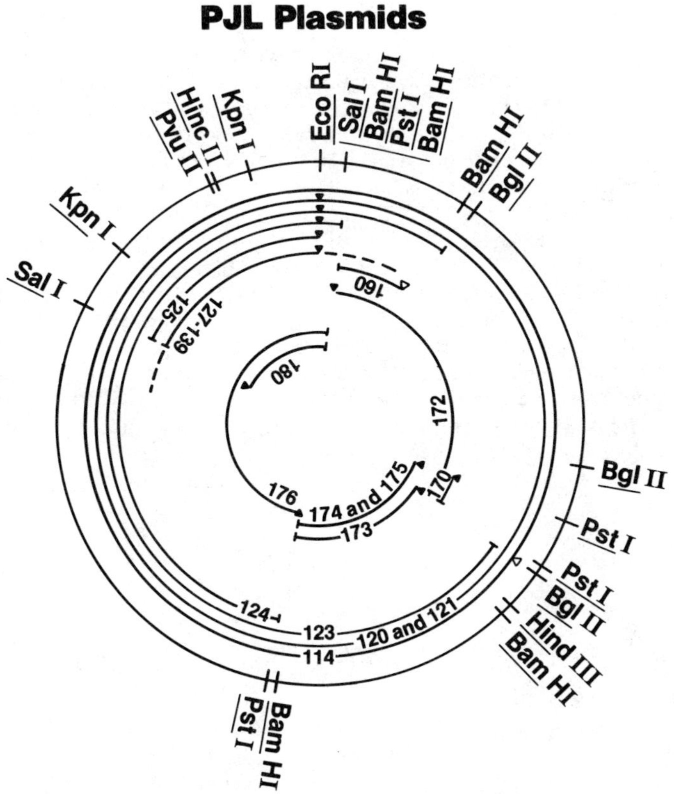

FIGURE 2. Restriction maps of chimeric plasmids with deletions in SCP2*. The outer circle is the restriction map of SCP2*. Δ identifies the position of pBR322 on the maps. ▲ identifies the position of pBR325 on the maps. The arcs, extrapolated to the outer circle, define the portions of SCP2* that are present in the deletion derivatives. The dashed arcs identify a family of plasmids deleted to different *Sal*I sites. The number on each arc or toward the center of the circle from the arc gives the pJL number assigned to each plasmid. Twenty-three chimeric plasmids are identified by this figure.

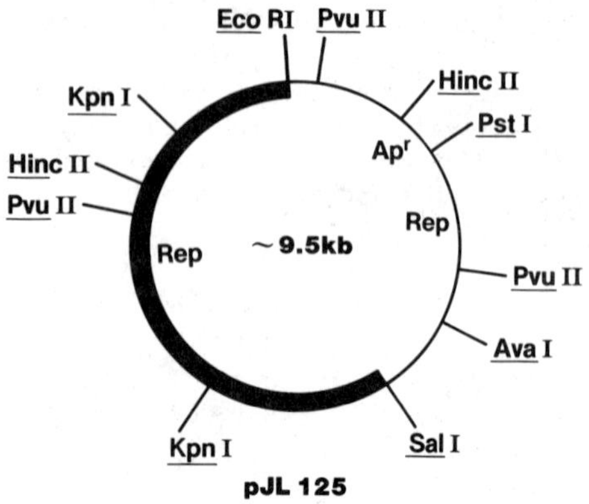

FIGURE 3. Restriction map of pJL125. Broad lines represent DNA from *S.* species and thin lines represent DNA from pBR322 or pBR325 in this figure and all subsequent restriction maps. The 5.4 Kb *Eco*RI–*Sal*I fragment of SCP2* is indicated by the solid line.

cally distinct from the normal pocks that are caused by SCP2* in *S. griseofuscus*. The plasmid, pJL176, was constructed by cloning the largest *Pst*I fragment of SCP2* into the *Pst*I site of pBR325 (FIG. 2). pJL176 causes formation of normal pocks that are indistinguishable from the pocks seen with SCP2*. These results suggest that the *Eco*RI recognition sequence defines a site that controls pock morphology. The plasmid, pJL180, was constructed to test this hypothesis (FIG. 4). pJL180 contains the 6.0 Kb *Sal*I fragment of SCP2* inserted into the *Sal*I site of pBR325. The 6.0 Kb *Sal*I fragment includes the 5.4 Kb *Sal*I–*Eco*RI fragment of pJL125 plus a 0.6 Kb *Eco*RI–*Sal*I fragment of SCP2* and provides continuity of the SCP2* sequence through the *Eco*RI site. Note that the SCP2* portion of pJL180 is an insertional isomer of the SCP2* portion of pJL130 (FIG. 2) that was generated by partial *Sal*I digestion of pJL121. pJL180 causes formation of normal pocks and pJL130 causes formation of "mini-pocks" when they are tested separately in *S. griseofuscus*. Therefore, the *Eco*RI recognition sequence does define a site that determines pock morphology. This property provides a convenient assay to identify recombinant plasmids that contain inserts at the *Eco*RI site.

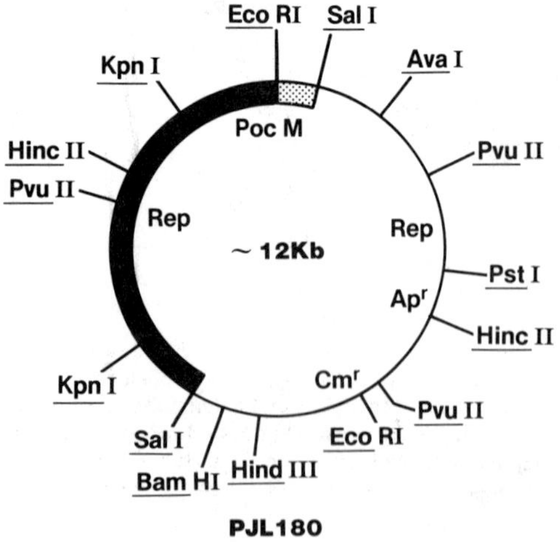

FIGURE 4. Restriction map of pJL180. The SCP2* portion of pJL180 is an insertional isomer of pJL130 (FIG. 2). The stippled line represents the 0.6 Kb *Eco*RI–*Sal*I fragment of SCP2*. PocM is at the *Eco*RI site and identifies that the pock morphology changes when DNA is inserted at the *Eco*RI site. The solid line represents the 5.4 Kb *Eco*RI–*Sal*I fragment of SCP2*. The thin line represents pBR325.

A NEOMYCIN RESISTANCE (Nm^r) MARKER

Morphological markers are useful for optimization of plasmid transfer. They can be used to identify recombinant plasmids. However, selectable resistance markers offer greater versatility in the use of cloning vectors. Thompson *et al.*[28,29] cloned the gene for neomycin phosphotransferase that specifies resistance to neomycin (Nm^r) in *S. fradiae*. The fragment containing the Nm^r marker was subcloned from PIJ2 into pBR322, thereby generating the plasmid pFJ165.[9] The 7.7 Kb *Eco*RI–*Hin*dIII fragment of FJ165 contains the Nm^r marker and all but the 30 bp *Eco*RI–*Hin*dIII fragment of pBR322. The 19 Kb *Eco*RI–*Hin*dIII fragment of pJL121 contains the 5.4 Kb Rep sequence and approximately 13.6 Kb of extraneous DNA from SCP2*. These two fragments were ligated together and transformed into *E. coli* to isolate the

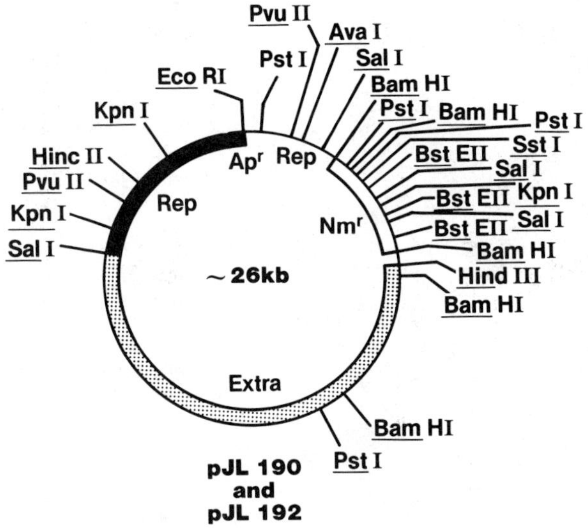

FIGURE 5. Restriction map of pJL190 and pJL192. The restriction maps of pJL190 and pJL192 are indistinguishable. The thin line represents DNA from pBR322. The solid line represents the 5.4 Kb *Eco*RI–*Sal*I fragment from SCP2*, the stippled line represents extraneous DNA from SCP2*. The open line represents the DNA fragment that carries the gene for Nm^r.

plasmid, pJL190 (FIG. 5) which imparts Ap^r in *E. coli*, Nm^r at 1.0 μg/ml in *S. griseofuscus*, "mini-pocks" in *S. griseofuscus* and shuttles reciprocally between *E. coli* and *S. griseofuscus*. The level of Nm^r is sufficient to select transformants with the plasmid because *S. griseofuscus* is sensitive to neomycin at approximately 0.2 μg/ml; however, a higher level of differential resistance is desirable for a useful cloning vector so that background colonies do not interfere with the isolation of recombinants. A mutant conferring a higher level of Nm^r was isolated by selecting a survivor of *S. griseofuscus* pJL190 on nutrient agar containing 10 μg/ml of neomycin. The mutant plasmid, pJL192, (FIG. 5), imparted the high level of resistance to virgin *S.* hosts when it was transferred by transformation or a sexual cross. The mutation did not cause a gross change in the plasmid because the restriction maps of pJL190 and pJL192 are indistinguishable.

Two plasmids (FIG. 6) were constructed to distinguish whether the mutation to elevated Nm^r occurred in the SCP2* Rep fragment or the fragment containing the gene for neomycin phosphotransferase. Both plasmids incorporate the 5.4 Kb *Eco*RI–*Sal*I fragment from pJL125. pJL195 contains the 6.7 Kb *Eco*RI–partial *Sal*I fragment from pFJ165 and pJL196 contains the 6.7 Kb *Eco*RI–partial *Sal*I fragment from pJL192. pJL195 confers low-level Nm^r on agar plates but does not allow growth in liquid culture containing 1.0 μg/ml of neomycin. pJL196 confers high-level Nm^r on agar plates and does allow growth in liquid culture containing 1.0 μg/ml of neomycin. Therefore, the mutation to increased Nm^r occurred in the fragment that contains pBR322 and the gene for neomycin phosphotransferase.

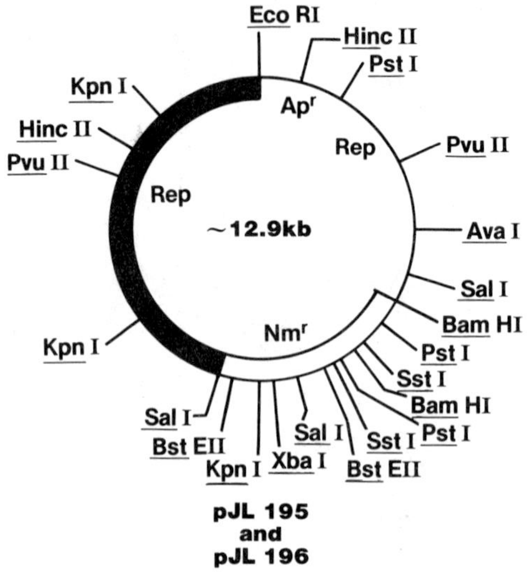

FIGURE 6. Restriction map of pJL195 and pJL196. The restriction maps of pJL195 and pJL196 are indistinguishable from each other. The solid line represents the 5.4 Kb *Eco*RI–*Sal*I fragment that was isolated from pJL125. The thin line represents DNA from pBR322 and the open line represents the DNA fragment that carries the gene for Nm^r. The 6.7 Kb *Eco*RI–partial *Sal*I fragment contains DNA from pBR322 and the fragment that carries the gene from Nm^r. pJL195 contains the corresponding 6.7 Kb fragment of pJL190 while pJL196 contains the corresponding 6.7 Kb fragment of pJL192.

A USEFUL VECTOR FOR *STREPTOMYCES*

Thompson *et al.*[28–30] cloned a gene for a 23S rRNA methylase that specifies resistance to thiostrepton (Ts^r) in *Streptomyces azureus (S. azureus)*. The fragment encoding Ts^r was subcloned in the construction of the plasmid pFJ105.[9] The 1.0 Kb *Bcl*I fragment from pFJ105 contains the gene for Ts^r. The enzymes, *Bam*HI and *Bcl*I, generate termini that can be ligated to each other. The hybrid junctions are resistant to cleavage by both enzymes when *Bam*HI and *Bcl*I termini are ligated to each other. pJL196 was partially digested with *Bam*HI to generate linear molecules that were cleaved at one of the two *Bam*HI sites. The linear molecules were ligated to the 1.0 Kb *Bcl*I fragment from pFJ105 transformed into *E. coli* and recombinants were selected for Ts^r after retransformation into *S. griseofuscus*. The plasmids, pJL197 and pJL198 (FIG. 7) impart Nm^r and Ts^r to *S. griseofuscus*. The vectors contain a single *Bam*HI site that provides insertional inactivation of Nm^r and a single *Cla*I site that provides insertional inactivation of Ts^r. Furthermore, the plasmids cause formation of "mini-pocks" when colonies are grown on a lawn of the host strain that does not contain the plasmid. pJL197 and pJL198 have inverted orientations of the *Bcl*I fragment that carries the gene for Ts^r, otherwise they are identical. pJL197 and several other

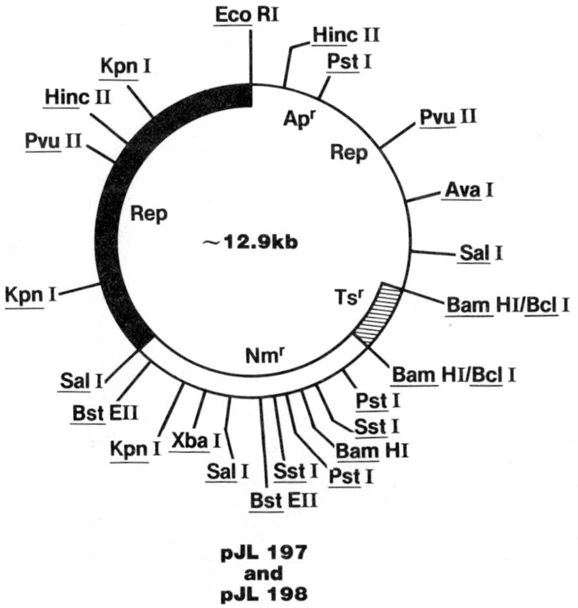

FIGURE 7. Restriction map of pJL197 and pJL198. These plasmids contain the *Bcl*I fragment, indicated by the striped line, that carries the gene for Ts^r. The *Bcl*I fragment has opposite orientations in pJL197 and pJL198, otherwise the plasmids are identical. The solid line represents the 5.4 Kb *Eco*RI–*Sal*I fragment of SCP2*, the open line represents the DNA fragment containing the gene for Nm^r, and the thin line represents DNA from pBR322.

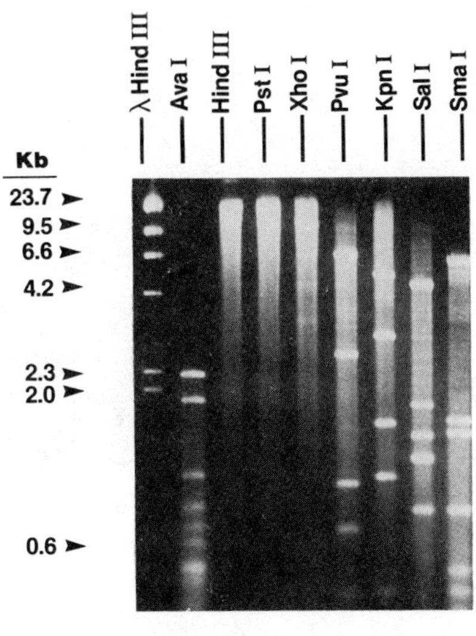

FIGURE 8. Restriction digests of DNA from *S. fradiae* JS85. Total genomic DNA was purified from *S. fradiae* JS85, separately digested with several restriction enzymes as indicated in the figure, and examined by electrophoresis in agarose gels. The lane, designated λ *Hin*dIII, contains DNA from bacteriophage λ that was digested with *Hin*dIII and employed as a set of reference markers. The sizes and positions of the reference bands are indicated at the left side of the figure.

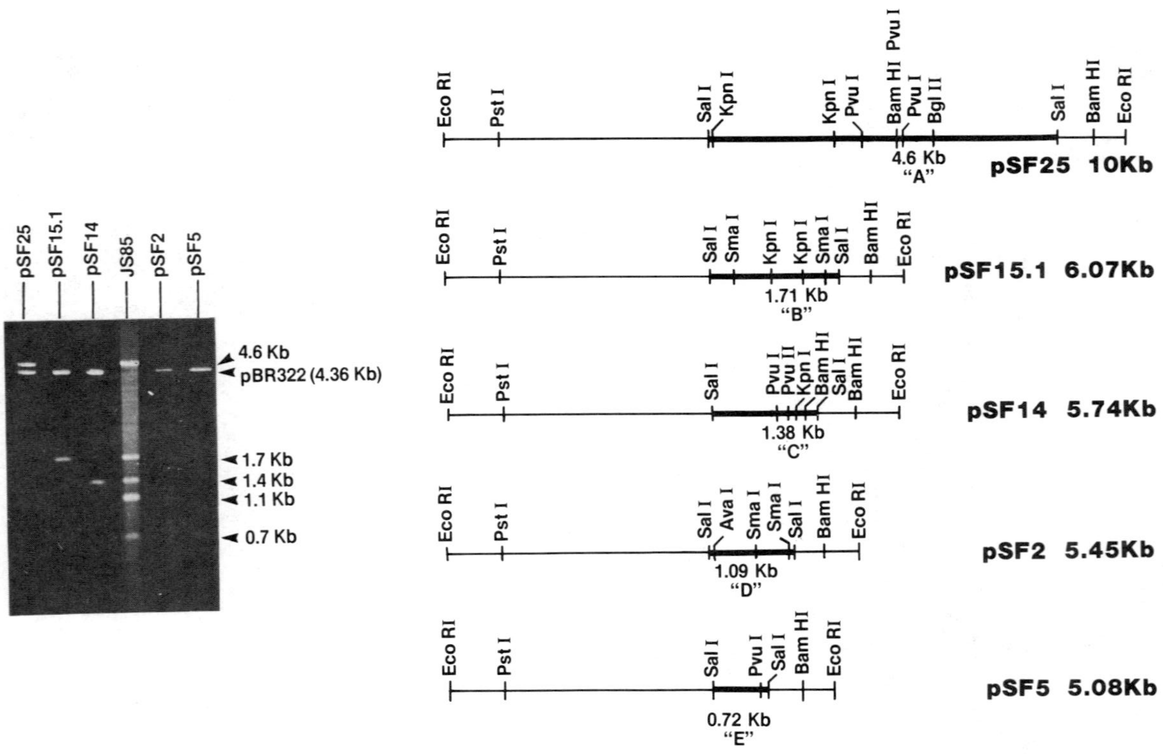

FIGURE 9. Cloned fragments from the amplified DNA of *S. fradiae* JS85. The left-hand panel shows agarose gel electrophoresis of *Sal*I digests of samples noted in the figure. The pSF number is the designation given to the recombinant plasmid. JS85 designates the lane that contains total genomic DNA from *S. fradiae* JS85. The positions and sizes of restriction fragments are indicated at the right side of the photograph of the gel. The position, designated pBR322, identifies the band produced by *Sal*I digestion of the vector, pBR322. The right-hand panel shows restriction maps of the recombinant plasmids. The thin line represents DNA from the vector, pBR322. The broad line represents DNA of the cloned fragment. The letters in quotations indicate designations of the *Sal*I fragments that are present in the recombinant plasmids. The size of each recombinant plasmid is indicated to the right of the plasmid designation.

plasmids in the lineage were tested to determine if they retained the property of self-transmissability that is characteristic of SCP2*. These tests were performed in *Streptomyces lividans (S. lividans)*. The plasmid was transformed into *S. lividans* 1326 and selected for Nm^r. *S. lividans* 1326 is sensitive to streptomycin, *S. lividans* CT2 is resistant to streptomycin and pJL197 does not provide cross resistance to streptomycin. *S. lividans* 1326/pJL197 was crossed with *S. lividans* CT2 and recombinants were selected for resistance to streptomycin, Nm^r and TS^r. The results indicate that pJL197 is self-transmissable. A streptomycin-resistant mutant of *S. griseofuscus was isolated and designated as S. griseofuscus* Str^r. Genetic crosses between *S. griseofuscus*/pJL197 and *S. griseofuscus* Str^r confirmed the results of crosses in *S. lividans*. The host range of pJL197 was examined by transformation into *S. lividans* 1326, *S. lividans* CT2 and *S. fradiae*. The plasmid transforms all hosts to antibiotic resistance. pJL197 and pJL198 should be useful as general cloning vectors for these *S.* species.

OBSERVATION OF AMPLIFIED DNA IN *STREPTOMYCES FRADIAE*

An unusual phenomenon in *S. fradiae* will be described rather than cloning experiments with pJL197 and pJL198. *S. fradiae* produces the macrolide antibiotic, tylosin[31]; therefore, investigations in this species may provide both practical and theoretical benefits. *S. fradiae* JS85 is a mutant that is blocked in the biosynthesis of tylosin.[32] Simultaneously it has lost resistance to tylosin. The total genomic DNA of *S. fradiae* JS85 was digested separately with several restrictions enzymes and analyzed by agarose gel electrophoresis. The results (FIG. 8) reveal a series of intense bands that are not present in the parental strain and suggest the presence of a DNA sequence in high copy number. *Sal*I fragments of DNA from *S. fradiae* JS85 were cloned into the *Sal*I site of pBR322 and isolated in *E. coli*. Plasmids from Ap^r and Tc^s recombinants were digested with *Sal*I and examined by agarose gel electrophoresis. Approximately 50% of the recombinant plasmids contain a *Sal*I fragment that coelectrophoreses with one of the five intense bands in the total genomic DNA from *S. fradiae* JS85. This observation supports the interpretation that the intense bands contain DNA fragments that are present in high copy number. Typical results are shown in FIGURE 9A for five of the recombinant plasmids. Restriction maps of the five plasmids are shown in FIGURE 9B. The plasmids were labeled with ^{32}[P] by nick-translation with DNA polymerase I[33] and hybridized to "Southern Blots" of the restricted DNA from *S. fradiae* JS85.[34] The results (data not shown) verified that the cloned *Sal*I fragments were homologous to the corresponding intense bands.

The DNA present in high copy number is designated as amplfied DNA. Restriction fragments from the amplified DNA have a combined size of approximately 10.5 Kb, indicating that the amplified DNA contains a 10.5 Kb reiterated sequence that is repeated many times. The number of copies of the reiterated sequence in the amplified DNA was determined by analyzing the kinetics of hybridization. ^{32}[P] was incorporated into the recombinant plasmids by nick-translation with DNA polymerase I. The radioactive probes were hybridized with total DNA from *S. fradiae* JS85 to measure the rates of reaction. (FIG. 10).[35] Kinetics of the reactions were analyzed as described previously[36] and employed to calculate copy number according to the procedure of Wetmur.[37] The calculations indicate that the amplified DNA in the genome of *S. fradiae* JS85 contains approximately 500 copies of the reiterated sequence.

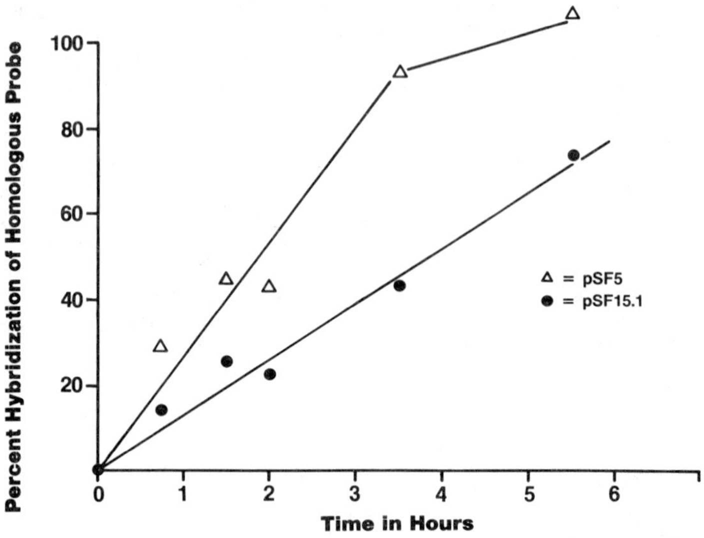

FIGURE 10. Kinetics of hybridization. Total genomic DNA of *S. fradiae* JS85 was randomly fragmented by sonication and hybridized to radioactive plasmids as indicated in the figure.

REITERATED SEQUENCE ORGANIZATION IN THE AMPLIFIED DNA

The amplified DNA could contain tandemly arranged repeats of the reiterated sequence, or it could contain copies of the reiterated sequence dispersed at many different sites in the genome. The presence of discrete restriction fragments suggests that the reiterated sequence is arranged as tandem repeats; however, two experimental approaches were employed to test the suggestion. First, the restriction map was identified for the amplified DNA. The restriction map of the reiterated sequence is shown in FIGURE 11. Analysis of the amplified DNA indicates that this restriction map is continuous in a head-to-tail fashion; therefore, the amplified DNA contains either circular molecules or tandem repeats of the reiterated sequence. Numerous attempts failed to identify or isolate a plasmid from *S. fradiae* JS85. These results suggest, but do not prove, that the amplified DNA contains tandem repeats of the reiterated sequence. Tandem repeats can be demonstrated by analyzing fragments generated by partial digestion of the amplified DNA with restriction enzymes. This approach is illustrated by analysis of fragments generated by digestion with *Bam*HI (FIG. 12A). Complete digestion of genomic DNA from *S. fradiae* JS85 generates two intense bands that are designated with A and B, respectively. Partial digestion with *Bam*HI generates a series of several moderately intense bands. The latter bands were verified to contain the reiterated sequence by preparation of "Southern Blots" that were hybridized to pSF25 (FIG. 12B). pSF25 is homologous to both A and B; therefore, it hybridizes to partial and complete digestion fragments generated with *Bam*HI if they contain fragments A or B from the reiterated sequence. Partial digestion products were

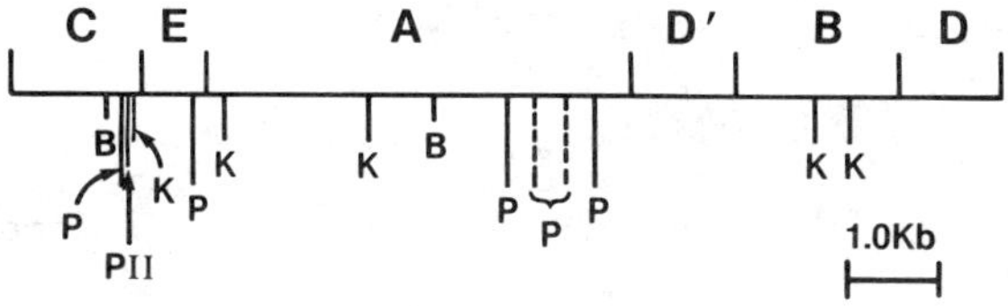

FIGURE 11. Restriction map of the reiterated sequence. Lines extending above the map indicate the sites for digestion with *Sal*I. The letters denote the designations of the *Sal*I fragments as seen in FIGURE 9. Other restriction sites are indicated by lines that extend below the map. Restriction enzymes are designated as B for *Bam*HI, P for *Pvu*I, PII for *Pvu*II, and K for *Kpn*I.

observed that contain ABA and BAB. Fragments of B with two copies of A and of A with two copies of B can result only if the reiterated sequence is arranged in tandem repeats. Also, fragments were observed that contain two or more complete copies of the reiterated sequence; therefore, confirming that the reiterated sequence is arranged in tandem repeats. The data are not sensitive enough to exclude the possibility that a few blocks of tandemly repeated reiterated sequence are located at different sites in the genome; however, they do demonstrate that most copies of the reiterated sequence in the amplified DNA occur in tandemly repeated units rather than single copies.

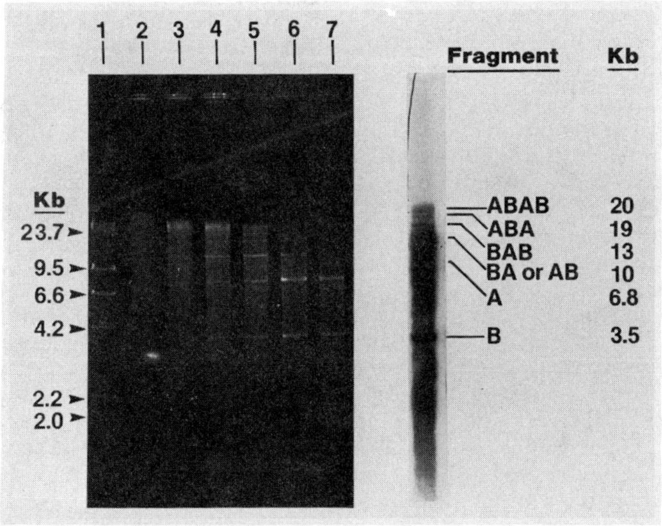

FIGURE 12. *Bam*HI digestion of DNA from *S. fradiae* JS85. The left-hand panel shows total genomic DNA from *S. fradiae* JS85 that was digested with *Bam*HI and electrophoresed in agarose gels. Lane 1 contains DNA from bacteriophage λ that was digested with *Hin*dIII and included as reference markers. The sizes and positions of the reference markers are indicated at the left side of the photograph of the gel. Lanes 2–6 contain samples that were digested for 0, 2.5, 5.0, 7.5 and 10.0 minutes, respectively. Lane 7 contains a sample that was digested for 1.0 hour and shows the result of completed digestion. The right-hand panel shows an autoradiograph of lane 5 after it was transferred to a sheet of nitrocellulose and hybridized with pSF25.

STABILITY OF AMPLIFIED DNA

The stability of amplified DNA was ascertained by examination of *Bam*HI restriction digests by agarose gel electrophoresis. Cultures were grown in Trypticase Soy Broth containing 0.4% glycine for one or several sequential transfers and total genomic DNA was isolated. Also, mycelia were fragmented and plated for isolation of single colonies. Single colonies were subcultured in Trypticase Soy Broth containing 0.4% glycine and total genomic DNA was isolated. The samples were digested with *Bam*HI and analyzed by agarose gel electrophoresis. All of the samples exhibit intense bands as seen in the *Bam*HI digest of FIGURE 8. These results indicate that the amplified DNA is stably maintained in the cells.

DISCUSSION

The first part of this report describes the series of experiments that led to the construction of the chimeric plasmids, pJL197 and pJL198. These plasmids are useful cloning vectors that have many advantages for applications to *S.* species. The plasmids are bifunctional shuttle vectors for *E. coli* and *S.* species because they contain two sets of plasmid replication functions. The Rep of pBR322 is a particular advantage in the shuttle vector because chloramphenicol amplification in *E. coli* allows accumulation of the plasmid until it represents approximately 40% of the total DNA;[38] therefore, it yields about 1000-fold more plasmid DNA than conventional low copy plasmids. The Rep of SCP2* allows replication and maintenance of the shuttle vectors in *S.* species. The vector DNA transforms *S. griseofuscus* at the same frequencies whether it is isolated from *E. coli* or *S. griseofuscus.*

Cells, transformed with pJL197 or pJL198, are selectable in both *E. coli* and *S.* species because the plasmids carry separate antibiotic resistance markers that are expressed in each genera. Insertional inactivation is available for both resistance markers in *S.* species because each gene for antibiotic resistance contains a unique recognition site for a restriction enzyme that does not cleave the plasmid at any other site. Digestion with *Bam*HI cleaves the gene for Nm^r and generates termini with a GATC overhang. Several restriction enzymes including *Mbo*I, *Sau*3A, *Bgl*II, *Xho*II, *Bcl*I and *Bam*HI generate the same overhang so that fragments generated with any of these enzymes can be cloned with insertional inactivation of Nm^r. Two restriction enzymes, *Mbo*I and *Sau*3A, are particularly useful because they recognize a four base pair sequence and cleave *Streptomyces* DNA extensively. Therefore, partial digestion can be used to generate the random distributions of DNA fragments that are preferred for contruction of genomic libraries. Digestion with *Cla*I cleaves the gene for Ts^r and generates termini with a CC overhang. Restriction enzymes, *Hpa*II and *Taq*I, recognize four base pair sequences and generate fragments with the same overhang as *Cla*I; therefore, the *Cla*I site is readily available for cloning DNA fragments with insertional inactivation of Ts^r. *Cla*I frequently gives partial digestion. However, this need not be a limitation because a large quantity of the vector can be isolated from *E. coli* and the linear product of *Cla*I digestion is easily purified away from the undigested plasmid. *Cla*I does not cleave some of the recognition sites because the DNA is methylated with the Dam specificity of *E. coli;* however, the plasmid can be prepared from *dam* mutants of *E. coli* that will allow cleavage of the DNA at all of the *Cla*I sites.

The plasmid, SCP2*, is present at low copy number in *S.* species. Incorporation of this replicon into a cloning vector offers a subtle, yet distinct advantage. Many of the interesting genes in *S.* species code for enzymes that catalyze the synthesis of antibiotics that are toxic to the producing organism when the antibiotics are present in high concentrations. High levels of expression of the enzymes are likely when the genes are cloned on high copy number vectors. The resultant toxicity could preclude recovery of recombinants with the genes for antibiotic synthesis. A low copy number vector should produce less toxicity because the enzymes may be expressed at lower levels. Thus, vectors such as pJL197 or pJL198 provide greater opportunity to clone the desired genes.

Self-transmissability in sexual crosses between *S.* strains is another advantage of pJL197. Recombinant plasmids may be transmitted in sexual crosses by replica-plating recombinant colonies onto confluent lawns of selected recipient strains. Thus, large numbers of recombinant plasmids can be screened rapidly for complementation of a variety of mutations. This property may be especially important because screening a genomic library can be the most difficult and time-consuming component of cloning interesting genes.

The second part of this report described an unusual phenomenon that has been discovered in *S. fradiae.* A reiterated sequence of 10.5 Kb is amplified to approximately 500 copies per genome. The amplified DNA contains tandem repeats of the reiterated sequence. Several investigators independently reported similar observations in other *S.* species while the experiments were in progress to characterize the amplified DNA in *S. fradiae.*[39–42] Thus, amplification of DNA may be a general phenomenon in *S.* species. The amplified DNA appears to be stable because it is not lost or rearranged at significant frequencies when cultures are grown in liquid or solid media. This stability is unique because relatively short tandem repetitions are unstable in other bacterial species.[45] The sudden single step appearance and stability of amplified DNA suggest that DNA amplification represents a genetically controlled event. Understanding the mechanisms and regulation of DNA amplification may provide new insights into the special genetic behavior of *S.* species and may provide important theoretical and practical methods for strain construction.

ACKNOWLEDGMENTS

Appreciation is expressed to R. H. Baltz for making mutants of *S. fradiae* available, F. D. Miller for assistance in measuring the kinetics of hybridization, J. Fayerman for making the plasmids pFJ165 and pFJ105 available, and L. A. Heneghan for typing the manuscript. The pJL plasmids are officially designated as pHJL and the pSF plasmids are officially designated as pSFH by the Plasmid Reference Center.

REFERENCES

1. Queener, S. Q. & R. H. Baltz. 1979. Annu. Rep. Ferm. Proc. **3:** 5–45.
2. Chater, K. F., D. A. Hopwood, T. Kieser & C. J. Thompson. 1982. Curr. Top. Microbiol. Immunol **96:** 69–95.
3. Bibb. M. J., J. M. Ward & D. A. Hopwood. 1980 Dev. Ind. Microbiol. **21:** 55–64.

4. HOPWOOD, D. A. & K. F. CHATER. 1982. Genet. Eng. Principles and Methods. **4:** 119–142.
5. HOPWOOD, D. A., C. J. THOMPSON, T. KIESER, J. M. WARD & H. M. WRIGHT. 1981. Microbiol. **1981:** 376–379.
6. SCHOTTEL, J. L., M. J. BIBB & S. C. COHEN. 1981. J. Bacteriol. **146:** 360–368.
7. BIBB, M. J., J. L. SCHOTTEL & S. N. COHEN. 1980. Nature **284:** 526–531.
8. ISOGAI, T., H. TAKAHASHI & H. SAITO. 1981. J. Gen. Appl. Microbiol. **27:** 373–379.
9. RICHARDSON, M. A., J. A. MABE, N. E. BEERMAN, W. M. NAKATSUKASA & J. T. FAYERMAN. 1982. Gene **20:** 451–457.
10. BOLIVAR, F., R. L. RODRIGUEZ, P. J. GREENE, M. C. BETLACH, H. L. HEYNEKER & H. W. BOYER. 1977. Gene **2:** 95–113.
11. BOLIVAR, F. 1978. Gene **4:** 121–136.
12. HERSHBERGER, C. L. 1982. Ann. Rep. Ferm. Proc. **5:** 101–126.
13. BIBB, M. J., R. F. FREEMEN & D. A. HOPWOOD. 1977. Mol. Gen. Genet. **154:** 155–166.
14. BIBB, M. J. & D. A. HOPWOOD. 1978. Microbiol. 1978. 139–141.
15. BIBB, M. J. & D. A. HOPWOOD. 1981. J. Gen. Microbiol. **126:** 427–442.
16. BIBB, M. J., J. L. SCHOTTEL & S. N. COHEN. 1980. Nature **284:** 526–531.
17. BIBB, M. J., J. M. WARD & D. A. HOPWOOD. 1978. Nature **274:** 398–400.
18. BIBB, M. J., J. M. WARD & D. A. HOPWOOD. 1980. Microbiol. **21:** 55–64.
19. SCHREMPF, H., H. BUJARD, D. A. HOPWOOD & W. GEOBEL. 1975. J. Bacteriol. **121:** 416–421.
20. SCHREMPF, H. & W. GOEBEL. 1977. J. Bacteriol. **131:** 251–258.
21. TROOST, T. R., V. N. DANILENKO & N. D. LOMOVSKAYA. 1979. J. Bacteriol. **140:** 359–368.
22. KIRBY, R. & S. WOTTON. 1979. FEMS Microbiol. Lett. **6:** 321–323.
23. COX, K. & R. H. BALTZ. Manuscript in preparation.
24. MESELSON, M. & R. YUAN. 1968. Nature **217:** 1110–1114.
25. BALTZ, R. H. & P. MATSUSHIMA. 1981. J. Gen Microbiol. **127:** 137–146.
26. KLEIN, R. D., E. SELSING & R. D. WELLS. 1980. Plasmid **3:** 88–91.
27. COHEN, S. N., A. C. Y. CHANG & L. HU. 1972. Proc. Natl. Acad. Sci. USA **69:** 2110–2114.
28. THOMPSON, C. J., R. H. SKINNER, J. L. THOMPSON, J. M. WARD, D. A. HOPWOOD & E. CUNDLIFFE. 1982. J. Bacteriol. **151:** 678–685.
29. THOMPSON, C. J., J. M. WARD & D. A. HOPWOOD. 1980. Nature **286:** 525–527.
30. THOMPSON, J. & E. CUNDLIFFE. 1981. J. Gen. Microbiol. **124:** 291–297.
31. SENO, E. T., R. L. PIEPER & F. M. HUBER, 1977. Antimicrob. Agents Chemother. **11:** 455–461.
32. BALTZ, R. H., E. T. SENO, J. STONESIFER, P. MATSUSHIMA & G. M. WILD. 1981. Microbiol. **1981:** 371–375.
33. RIGBY, P. W. J., M. DIECKMAN, C. RHODES & P. BERG. 1977. J. Mol. Biol. **113:** 237–251.
34. SOUTHERN, E. M. 1975. J. Mol. Biol. **98:** 503–517.
35. SLAVIK, N. S. & C. L. HERSHBERGER. 1976. J. Mol. Biol. **103:** 563–581.
36. SLAVIK, N. S. & C. L. HERSHBERGER. 1975. FEBS Lett. **52:** 171–174.
37. WETMUR, J. G. 1976. Annu. Rev. Biophys. Bioeng. **5:** 337–361.
38. BAZARAL, M. & D. R. HELINSKY. 1968. J. Mol. Biol. **36:** 185–194.
39. ONO, H., G. HINTERMANN, R. CRAMERI, G. WALLIS & R. HUTTER. 1982. Mol. Gen. Genet. **186:** 106–110.
40. ROBINSON, M., E. LEWIS & E. NAPIER. 1981. Mol. Gen. Genet. **182:** 336–340.
41. SCHREMPF, H. 1982. J. Chem. Tech. Biotechnol. **32:** 292–295.
42. SCHREMPF, H. 1982. J. Bacteriol. **151:** 701–707.
43. ROBERTS, R. J. 1982. Nucleic Acids Res. **10:** r117–r144.
44. ANDERSON, R. P. & J. R. ROTH. 1977. Ann. Rev. Microbiol. **31:** 473–505.

Gene Cloning in the Actinomycetes

DEAN P. TAYLOR, THOMAS ECKHARDT, AND
LOUIS R. FARE

Molecular Genetics
Smith, Kline and French Laboratories
P.O. Box 7929
Philadelphia, Pennsylvania 19101

INTRODUCTION

The actinomycetes are a group of microorganisms known primarily for their ability to produce antibiotics. These antibiotic producers are usually isolated from soil with *Streptomyces* and *Micromonospora* being the most abundant and the most intensively examined genera. Other genera that also produce antibiotics include *Actinoplanes, Nocardia, Simulosporangium,* and *Sreptosporangium* to name a few, but these genera are less prevalent in most soils and are less often studied. The ability to produce antibiotics represents the diversity of the metabolic capability these organisms possess. Antibiotics are designated secondary metabolites because they are not required for growth and are usually produced after rapid growth has ceased in what is known as the idiophase.

One of our goals is to be able to manipulate the genes involved in secondary metabolism in these organisms, since many of the biosynthetic pathways are sufficiently complex so as to discourage trying to produce these antibiotics in another host. The genetic manipulation of the biosynthetic pathways may permit the generation of new antibiotic leads, improvement of yield, stabilization of production, alteration of physiological regulation, and elimination of undesirable side reactions. In this way, the use of recombinant DNA should complement other approaches that have similar goals. One approach is mutasynthesis,[1] in which blocked mutants are fed unusual precursors resulting in the production of novel analogues. This approach is limited to those precursors that can be taken up by cells or require use of *in vitro* systems. Another approach is referred to as protoplast fusion,[2] in which protoplasts of different antibiotic producers are fused and the genetic material permitted to reassort, hopefully resulting in a new combination of enzymes involved in secondary metabolism capable of synthesizing a novel antibiotic.

Another goal of this work is to elucidate the basic elements involved in controlling gene expression in *Streptomyces*. This knowledge will facilitate realizing the goals already mentioned but may permit *Streptomyces* to be used as an alternative to the more popular host-vector systems for the expression of heterologous genes already developed for *E. coli, Bacillus,* and yeast. Use of *Streptomyces* as a producer organism capitalizes on the wealth of experience and safety gained from scaling up the production and isolation of *Streptomyces* products, especially antibiotics.

This discussion will review the progress toward meeting the basic requirements for gene cloning in the *Streptomyces* and the problems that can be encountered. Other reviews on this subject are recommended[3,4] and describe additional examples of gene cloning in *Streptomyces.*

TRANSFORMATION OF *STREPTOMYCES*

DNA manipulated *in vitro* by recombinant DNA techniques must be returned to a suitable host for replication and amplification as well as to assay for biological function. The transformation of *E. coli, Bacillus* and yeast with DNA is almost taken for granted, and so it is with certain *Streptomyces* species.[2,5] To transform *Streptomyces,* protoplasts must be obtained by treating cells with lysozyme. The lysozyme treatment works well if the cells are grown in subinhibitory concentrations of glycine with the concentration empirically determined for each species. The transformation with DNA is facilitated by the use of polyethylene glycol (PEG), which also causes protoplast fusion to occur. The transformed protoplasts must then be plated out on a medium that permits regeneration of the cell wall. Modification of the regeneration medium for each new species is required to optimize the transformation frequency. Recently the use of liposomes has been found to stimulate transformation and transfection by DNA without the DNA being encapsulated in the liposomes.[6]

VECTORS

To clone DNA fragments, a cloning vehicle is required. For some *Streptomyces* species, a number of plasmid and actinophage vectors are now available and the number and utility of these vectors is increasing at a rapid pace. The plasmid vectors are based on the plasmids naturally occurring in *Streptomyces,* the ones from the *S. coelicolor* and *S. lividans* being the most studied and therefore the most frequently used. SCP2 was the first plasmid isolated from *S. coelicolor* and characterized.[7] It was followed by the discovery of the SLP1 series[8] and more recently the pIJ101 series[9] of plasmids. These plasmids all are capable of mediating the phenomenon of lethal zygosis, which is associated with fertility.[10] This phenotypic character was used to follow transformation in early experiments. More recent derivatives have been constructed that carry antibiotic resistance determinants that are readily selected.

SCP2 Vectors

The SCP2* plasmid[11] was first used as a cloning vector in the cloning of methylenomycin resistance.[12] A number of other plasmid derivatives have been constructed many of which have already been described (C. Hershberger, this symposium). This work has resulted in the assignment of various genetic functions to different regions on the plasmid map. Of importance to remember is that SCP2 is a naturally occurring, autonomous fertility plasmid found in *S. coelicolor* and has a narrow host range limited to *S. coelicolor, S. lividans,* and *S. parvulus*. The copy of SCP2 has been estimated to be low, 1–2 copies per chromosome.[10]

SLP1 Vectors

The family of vectors derived from SLP1 plasmids includes plasmids with resistance markers for methylenomycin,[12] neomycin, thiostrepton, erythromycin, and viomycin individually and in pairwise combinations.[13,14] These plasmids are derived from the *S. coelicolor* chromosome but exist autonomously in *S. lividans.*[8] The copy number of SLP1 plasmids has been estimated at 5 or so copies per chromosome, but these plasmids are not stably maintained.[8]

pIJ101 Vectors

More recently, a series of vectors have been described that are derived from the plasmid pIJ101.[13] These plasmids were discovered by screening *Streptomyces* species for small, high copy-number plasmids. Derivatives with different combinations of antibiotic resistance markers have been constructed.[14] The organization for genetic functions and useful sites for cloning have been determined for this plasmid. The broad host range and high copy number make this a very good vector for cloning.

Phage Vectors

The temperate actinophage ϕC31 has been characterized and mutants derived from it have become the basis for some cloning vectors.[15,16] Although the size of the fragments that can be cloned with ϕC31 vectors is small, this phage has the potential for becoming as useful for cloning in *S. coelicolor* as lambda has proven to be in *E. coli*.

TABLE 1. Construction of *E. coli–Streptomyces* Hybrid Plasmids

Designation of Hybrid Plasmid	Site on pACYC184 Used in Fusion	Site on pIJ2 Used in Fusion
pSKC-1	*Bam*HI	*Bgl*II
pCNB-24	*Bam*HI	*Bcl*I
pCNS-2	*Sph*I	*Sph*I

Shuttle Vectors

Vectors capable of being selected and maintained in both *Streptomyces* as well as *E. coli* have been generated by fusing together vectors for cloning in each host. The *E. coli* vectors pACYC177 and 184 have been fused with SLP1.2 derivatives[17] and the DNA transformed into *S. lividans*. Interestingly, the fused plasmids conferred on the streptomycete host resistance to kanamycin and chloramphenicol, presumably by expression of the *E. coli* plasmid genes. Similar hybrid plasmids have been constructed by cloning pIJ101 derivatives on pBR322.[9] In *E. coli* the streptomycete gene for viomycin resistance was found to be expressed.[9] These bifunctional vectors should continue to prove useful for intergeneric gene transfer and the study of gene expression in heterologous hosts.

PROBLEMS ENCOUNTERED

In similar experiments, we have fused the *E. coli* vector pACYC184 to the streptomycete plasmid pIJ2, an SLP1.2 derivative bearing neomycin resistance, in three different arrangements (TABLE 1). The fused plasmids were constructed in *E. coli* and then returned to *Streptomyces*. In the absence of selection, the maintenance of these plasmids is remarkably unstable; after three successive rounds of sporulation, virtually none of the spores were resistant to antibiotics and had lost the hybrid plasmid. In shuttling these plasmids between *E. coli* and *S. lividans,* no sequence rearrangements have been observed.

This experiment provides an example of one problem that must be dealt with in the construction of improved plasmid cloning vectors, namely the problem of stable plasmid maintenance. Other problems that have been recognized include the problem of sequence stability once a gene has been cloned and the host range of the vectors. Although vectors are available for cloning in certain *Streptomyces* species and interspecies gene transfer is permitted using certain vectors that have been described as broad host range vectors, there remain *Streptomyces* species for which suitable cloning vectors will have to be developed.

In actinomycetes other than *Streptomyces,* the potential for gene cloning has just begun to be explored. Our experience with the genus *Streptosporangium* is presented as an example.[18] Protoplasts could be produced and regenerated and an endogenous plasmid (FIG. 1) was isolated and characterized. Hybrid plasmids were constructed, but all attempts to transform *Streptosporangium* have been unsuccessful. Although progress in developing vectors for some *Streptomyces* species has been rapid and is expected to continue, considerable effort will be required for the potential of gene cloning in other actinomycetes to be realized.

GENE CLONING IN *STREPTOMYCES*

The examination of particular cloned genes should provide insight into the regulation of gene expression in these organisms and permit the manipulation of the gene expression signals to improve the production of rate-limiting enzymes involved in

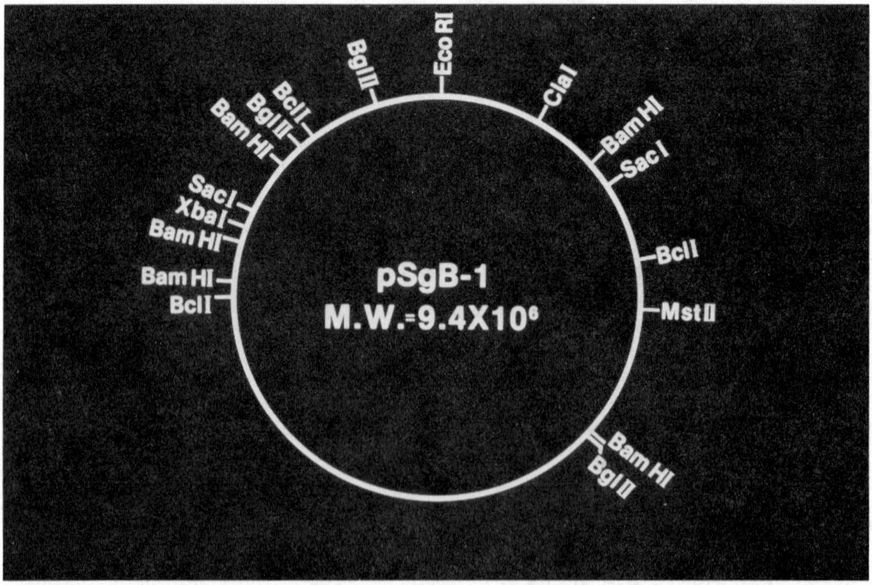

FIGURE 1. Restriction endonuclease map of the *Streptosporangium brasiliense* plasmid, pSgB-1.

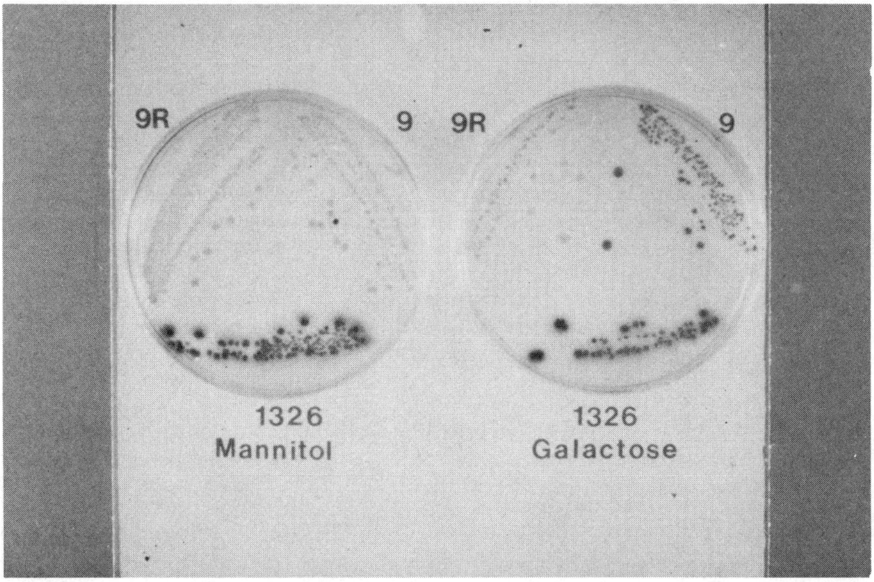

FIGURE 2. Growth of *S. lividans* strain 1326 and two mutants 1326-9 and 1326-9R which are unable to hydrolyze X-gal. The medium is a minimal basal salts medium containing X-gal and mannitol (A) or galactose (B) as carbon sources. On galactose-containing medium, the strain 1326-9 appears green while the 1326 strain appears blue after prolonged incubation at 28°C.

antibiotic biosynthesis. This insight will be essential in facilitating the expression of heterologous genes in *Streptomyces*.

Using the vectors already described, a number of genes have been cloned including genes for antibiotic resistance that have already been mentioned as well as genes for tyrosinase, para-amino benzioc acid synthetase, and amino acid biosynthetic genes.[14] Recently a gene with an activity similar to the β-galactosidase activity from *E. coli* has been cloned from *Streptomyces lividans*[19] and results derived from studies of this cloned gene will be presented. The various substrates available and the knowledge pertaining to gene expression and its regulation gained by studying β-galactosidase in *E. coli*[20] make the search for an analogous activity in *Streptomyces* a logical step.

S. lividans strain 1326 was found to hydrolyze the substrate 5-bromo-4-chloro-3-indolyl-β-D-galactoside (X-gal) to its chromogenic product in a medium containing various sugars (FIG. 2) with the activity most pronounced on medium containing galactose. Mutants of strain 1326, designated 1326-9 and 1326-9R, lacking this activity were isolated following nitrosoguanidine mutagenesis. The activity is seen as a halo of blue surrounding the colonies on solid medium (FIG. 3) or can be found as X-gal or *o*-nitrophenyl phosphate (ONPG) hydrolyzing activity in liquid medium (TABLE 2) indicating that the enzyme is excreted. DNA isolated from the wild-type parent was digested with the restriction endonucleases *Pst*I, *Kpn*I or *Sph*I and cloned onto the thiostrepton-resistant vector pIJ6. The recombinant plasmids carrying the gene for X-gal hydrolysis were isolated by transforming the mutants and picking the blue colonies (FIG. 4). The map of one such plasmid, pSKL-1, carrying an *Sph*I fragment on the pIJ6 vector, is shown in FIGURE 5. Comparison of the maps of the different

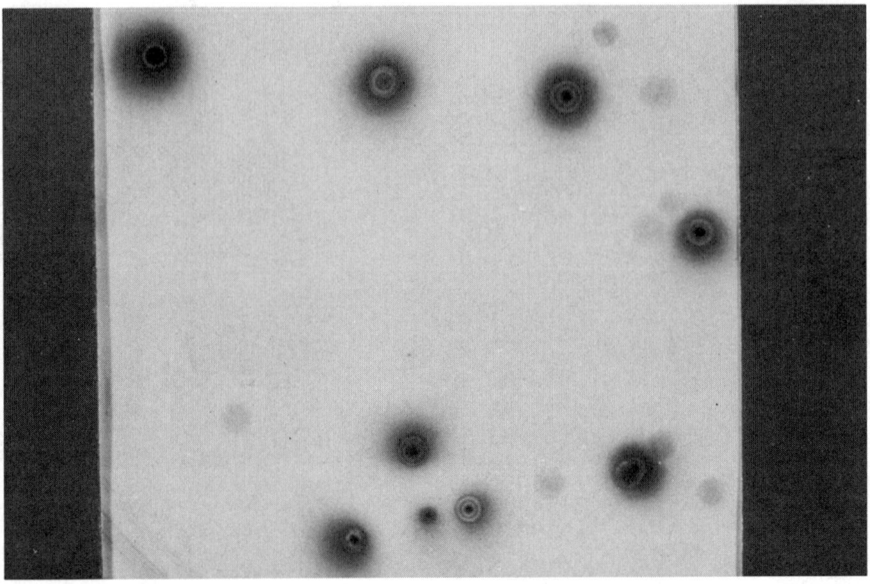

FIGURE 3. Colonies of the wild-type *S. lividans* strain 1326 surrounded by a halo of hydrolyzed X-gal or the mutant strain 1326-9 lacking the halo; growth was on medium containing X-gal and glucose as the carbon source.

clones permits the localization of the gene for X-gal hydrolysis to the 8 Kb region between the *Sph*I and *Pst*I sites on pSKL-1. When transformed into the wild-type host 1326, overexpression of the gene on pSKL-1 causes these transformants to appear blue even on medium containing glucose.

Subsequent investigation showed the plasmid to be unstable in the mutant host; that is, plasmid DNA isolated from a mutant host and transformed into that host gave rise to thiostrepton-resistant transformants that were unable to hydrolyze X-gal at a frequency of up to 40%, while transformation of the wild-type host resulted in a much lower frequency of transformants unable to hydrolyze X-gal. However, when plasmid DNA was isolated from the wild-type transformant able to hydrolyze X-gal and transformed back into the wild-type host, only blue thiostrepton-resistant transformants were observed. These results can be explained if homologous recombination were to occur resulting in the plasmid acquiring the mutant gene from mutant host chromosome. If this were the case, then isolation of plasmid DNA from a "white"

TABLE 2. Levels of the *Streptomyces* β-Galactosidase-like Enzyme[a]

Strain	Glucose	Lactose	Galactose
1326	12	76	184
1326-9	7	24	302
1326-9 (pSKL-1)	372	843	1242

[a]Enzyme levels were determined by the method of Miller[20] using cell-free extracts of cultures grown in minimal medium containing the carbon source indicated at a concentration of 1%; enzyme levels are given in nmoles of ONPG hydrolyzed/mg protein/min.

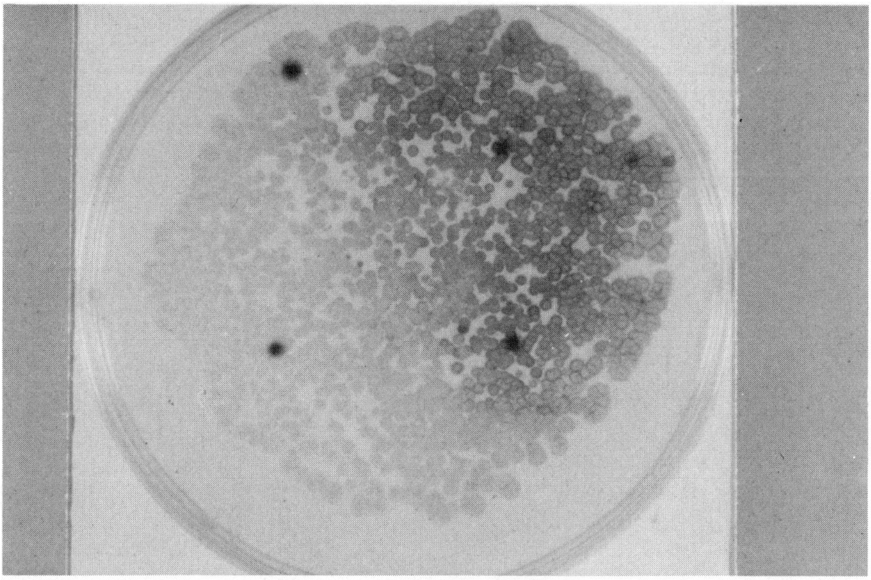

FIGURE 4. Cloning of the β-galactosidase-like gene. *Sph*I-cut chromosomal DNA from strain 1326 was ligated to *Sph*I-cut pIJ6 vector, transformed into strain 1326-9 and thiostrepton-resistant transformants selected. The desired recombinants were identified as blue (dark in this figure) colonies appearing on the medium containing X-gal.

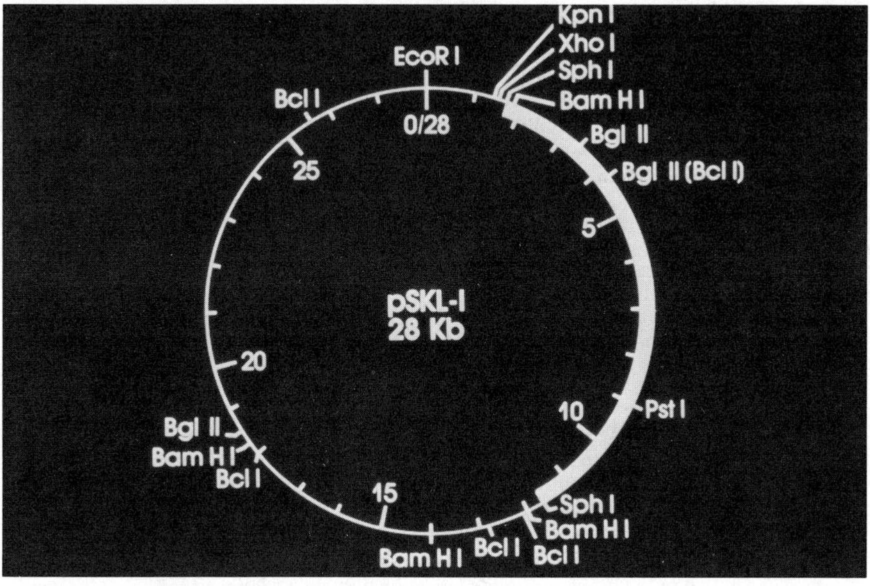

FIGURE 5. Restriction endonuclease map of the plasmid pSKL-1.

thiostrepton-resistant colony and transformation of the mutant host should produce only "white" thiostrepton-resistant transformants, but transformation of the wild-type host could result in occasional "white" thiostrepton-resistant transformants if recombination were to replace the wild-type chromosomal gene with the mutant gene from the plasmid. This result has been observed. The plasmid DNA isolated from these transformants are indistinguishable from the original recombinant plasmid as the mutation is probably a point mutation. This situation is representative of the rearrangements that can occur to produce sequence instability and indicates that

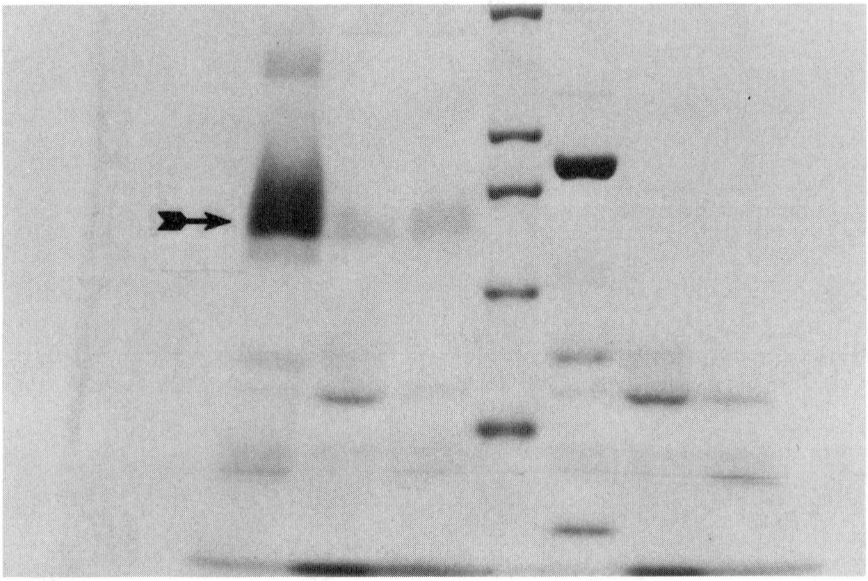

FIGURE 6. Electrophoresis of proteins in the culture supernatant. Tenfold concentrated supernatants were electrophoresed on an 8% polyacrylamide-0.2% SDS gel and the β-galactosidase activity visualized by X-gal hydrolysis to produce an aquablue (darker gray; indicated by arrow) stain of the gel and the remaining proteins were stained with Coomassie Blue. *Lane 1* (reading left to right) contains supernatant from strain 1326 bearing pSKL-1; *lane 2* is from a white transformant of strain 1326-9 with pSKL-1; *lane 3* is from strain 1326 bearing only the cloning vector; these supernatants were heated to 60°C to retain the β-galactosidase activity. *Lane 4* contains molecular weight standards of 200,000; 136,000; 92,000; 68,000; and 43,000 daltons. *Lanes 5–7* contain the same samples as lanes 1–3, but the samples were heated to 100°C (which inactivated the β-galactosidase activity).

homologous recombination may occur at a high frequency in *Streptomyces*. The problem may be turned to our advantage if this process of homogenotization will allow the insertion of *in vitro* generated deletions into the *Streptomyces* chromosome.

The X-gal hydrolyzing enzyme has been examined using SDS-polyacrylamide gel electrophoresis of tenfold concentrated culture supernatants (FIG. 6). The enzyme is active in the presence of 0.2% SDS and can be identified by its ability to hydrolyze X-gal *in situ*. The protein has an apparent molecular weight of 150,000. The nature of the protein excretion mechanism and the regulation of gene expression by carbon source are currently being examined. Examples of the levels of enzyme produced in the presence of different carbon sources and comparing the levels in the wild-type and mutant hosts and when the plasmid is carried on the plasmid compared to the chromosome are shown in TABLE 2.

SUMMARY

The production of antibiotics by soil-borne micro-organisms, the actinomycetes, has considerable economic importance. The manipulation of antibiotic producers has become a prime target for the application of recombinant DNA technology. Certain technical requirements have had to be met for gene cloning to be successful in the actinomycetes. These requirements, including the development of cloning vectors and transformation procedures, have been satisfied, in part, for some members of the *Streptomyces* genus. Some problems including sequence rearrangement and stability of plasmid maintenance are now being recognized. A number of genes have been cloned in *Streptomyces* and some preliminary results characterizing the gene for a *Streptomyces*-derived β-galactosidase-like activity were described.

REFERENCES

1. DAUM, S. J. & J. R. LEMKE. 1979. Mutational biosynthesis of new antibiotics. Ann. Rev. Microbiol. **33:** 241–265.
2. HOPWOOD, D. A. 1981. Genetic studies with bacterial protoplasts. Ann. Rev. Microbiol. **35:** 237–272.
3. CHATER, K. F., D. A. HOPWOOD, T. KIESER & C. J. THOMPSON. 1982. Gene cloning in *Streptomyces*. Curr. Top. Microbiol. Immunol. **96:** 69–95.
4. HOPWOOD, D. A. & K. F. CHATER. 1982. Cloning in *Streptomyces*: Systems and strategies. *In* Principles in Genetic Engineering. J. K. Setlow & A. Hollaender, Eds. Vol. 4. Plenum Press. New York.
5. BIBB, M. J., J. M. WARD & D. A. HOPWOOD. 1978. Transformation of plasmid DNA into *Streptomyces* at high frequency. Nature **274:** 398–400.
6. RODICO, M. R. & K. F. CHATER. 1982. Small DNA-free liposomes stimulate transfection of *Streptomyces* protoplasts. J. Bacteriol. **151**(3): 1078–1085.
7. SCHREMPF, H., H. BUJARD, D. A. HOPWOOD & W. GOEBEL. 1975. Isolation of covalently closed circular deoxyribonucleic acid from *Streptomyces coelicolor* A3(2). J. Bacteriol. **121**(2): 416–421.
8. BIBB, M. J., J. M. WARD, T. KIESER, S. N. COHEN & D. A. HOPWOOD. 1981. Excision of chromosomal DNA sequences from *Streptomyces coelicolor* forms a novel family of plasmids detectable in *Streptomyces lividans*. Mol. Gen. Genet. **184:** 230–240.
9. KIESER, T., D. A. HOPWOOD, H. M. WRIGHT & C. J. THOMPSON. 1982. pIJ101, a multicopy broad host-range *Streptomyces* plasmid: Functional analysis and development of DNA cloning vectors. Mol. Gen. Genet. **185:** 223–238.
10. BIBB, M. J., R. F. FREEMAN & D. A. HOPWOOD. 1977. Physical and genetical characterization of a second dex factor, SCP2, for *Streptomyces coelicolor* A3(2). Mol. Gen. Genet. **154:** 155–166.
11. BIBB, M. J. & D. A. HOPWOOD. 1981. Genetic studies of the fertility plasmid SCP2 and its SCP2* variants in *Streptomyces coelicolor* A3(2). J. Gen. Microbiol. **126:** 427–442.
12. BIBB, M. J., J. L. SCHOTTEL & S. N. COHEN. 1980. A DNA cloning system for interspecies gene transfer in antibiotic-producing *Streptomyces*. Nature **284:** 526–531.
13. THOMPSON, C. J., J. M. WARD & D. A. HOPWOOD. 1980. DNA cloning in *Streptomyces*: Resistance genes from antibiotic-producing species. Nature **286:** 525–527.
14. THOMPSON, C. J., J. M. WARD & D. A. HOPWOOD. 1982. Cloning of antibiotic resistance and nutritional genes in streptomycetes. J. Bacteriol. **151**(2): 668–677.

15. SUAREZ, J. E. & K. F. CHATER. 1980. DNA cloning in *Streptomyces:* A bifunctional replicon comprising pBR322 inserted into a *Streptomyces* phage. Nature **286:** 527–529.
16. CHATER, K. F., C. J. BRUTON, A. A. KING & J. E. SUAREZ. 1982. The expression of *Streptomyces* and *Escherichia coli* drug-resistance determinants cloned into the *Streptomyces* phage ϕC31. Gene **19:** 21–32.
17. SCHOTTEL, J. L., M. L. BIBB & S. N. COHEN. 1981. Cloning and expression in *Streptomyces lividans* of antibiotic resistance genes derived from *Escherichia coli.* J. Bacteriol. **146**(1): 360–368.
18. FARE, L. R., D. P. TAYLOR, M. J. TOTH & C. H. NASH III. Physical characterization of plasmids isolated from *Streptosporangium.* Plasmid. In press.
19. ECKHARDT, T. G. & L. R. FARE. 1982. β-galactosidase genes in *Streptomyces lividans.* Fourth Int. Symp. Genet. Ind. Microorganisms (Abstr.) p. 36.
20. MILLER, J. H. 1982. Experiments in molecular genetics. Cold Spring Harbor Laboratory. Cold Spring Harbor, New York.

Transformation of *Bacillus stearothermophilus* with Plasmid DNA and Its Application to Molecular Cloning

SHUICHI AIBA, TADAYUKI IMANAKA,
AND JUN-ICHI KOIZUMI

Department of Fermentation Technology
Faculty of Engineering
Osaka University
Yamada-oka, Suita-shi, Osaka 565, Japan

INTRODUCTION

The well-documented transformation procedure of *Bacillus subtilis* with plasmid DNA using either competent cells or protoplasts[1,2] has permitted the cloning of penicillinase genes of *B. licheniformis* in *B. subtilis*[3] and *B. licheniformis*,[4] and that of α-amylase gene of *B. amyloliquefaciens* in *B. subtilis*.[5] The cloning of a specific gene(s) in the genus *Bacillus* would not only be tantamount to breeding an industrially important strain(s) but also to revealing a mechanism(s) for secretion of extracellular enzymes.

The above-mentioned transformation in the genus *Bacillus* has been concerned with mesophilic bacteria. Hence, if the transformation system for a thermophile with plasmid could be established, the cloning of specific genes of thermostable enzymes would be made possible and the mode of gene expression at higher temperatures could be examined.

In view of the soaring attention being paid these days by many workers to thermostable enzymes from thermophilic bacteria, the establishment of a transformation of *B. stearothermophilus* with plasmid DNA would be more significant. The objective of this paper is first, to demonstrate the transformation procedure using various vector plasmids. Secondly, this paper deals with the transfer of penicillinase genes that have been cloned separately[4] in our laboratory in appropriate vector plasmids into *B. stearothermophilus* as an application of the transformation to molecular cloning. Lastly, some characteristic features in the culture of thermophiles will be discussed in comparison with mesophilic bacteria. Solving for the enthalpy balance equation that is coupled with simple equations for both growth and substrate utilization, temperature profiles in culture media for mesophilic and thermophilic bacteria will be simulated, respectively, with and without temperature control. Owing to the current paucity of growth data on thermophilic bacteria, data on specific growth rates are for *Thermus thermophilus* and *Bacillus caldotenax*, whereas those of a mesophile are for *Escherichia coli*.

TRANSFORMATION OF *BACILLUS STEAROTHERMOPHILUS*

The procedure of transforming *B. stearothermophilus* with plasmid DNA using protoplasts followed basically that procedure established earlier by Chang and Cohen for the transformation of *B. subtilis* protoplasts by plasmid DNA.[2] Since the transformation procedure of *B. stearothermophilus* is described elsewhere in detail,[6] and moreover, taking for granted that the principal purpose of this paper is not in reiterating here the sophisticated procedure of transformation but in arguing what is meant in biochemical engineering by transformation of a thermophile, the description on the procedure will be minimized.

B. stearothermophilus IFO 12550 (ATCC 12980) that is the type strain was subjected to spontaneous mutation (resistant to streptomycin) and also, cured spontaneously of the plasmid pBSO1. The mutant *B. stearothermophilus* CU21 thus obtained (see TABLE 1) was used throughout as host cells. Plasmids used are also shown in TABLE 1. Since the maximum and minimum temperatures for growth of *B. stearothermophilus* IFO 12550 (ATCC 12980) were 70 and 40°C, respectively,[10] and also referring to the fact that cells died quickly at room temperature but were stably maintained at 4°C, the host cells were exposed to 48°C and/or 4°C when required in the process of transformation.

Preparation of protoplasts of *B. stearothermophilus* that had been grown overnight in L broth at 55°C with lysozyme, mixing of plasmid DNA with the protoplast suspended in SMM-LG (sucrose-magnesium-maleate buffer mixed with L broth + glucose) medium in the presence of polyethylene glycol at 48°C, and finally, regeneration of protoplasts in regeneration top agar (RGTA) poured onto regeneration agar (RGA) plate that had been prewarmed at 48°C were the principal three consecutive steps in the transformation. Antibiotic, either kanamycin (Km) or tetracycline (Tc), was added into both RGA and RGTA to the extent of 25 μg/ml or 5 μg/ml, respectively for direct selection of transformants. Transformation frequency was scored after incubation of the protoplasts at 48°C for 5 days.

For compositions of L broth, SMM-LG medium, RGA, RGTA, and for those of various buffers needed in the transformation and the subsequent digestion of plasmid DNA with restriction endonucleases, if required, see reference 6. It should be mentioned in this connection that the use of an extremely low concentration of lysozyme, 1 μg/ml in contrast to 2 mg/ml for the protoplast formation of *B. subtilis* was to enhance the regeneration frequency of protoplasts. Even under the extremely low concentration of the enzyme, the conversion ratio of intact cells to protoplasts was more than 99.99 percent, if colonies of intact cells on LGA agar before and after the lysozyme treatment were counted. Regeneration frequencies of protoplasts were around 10 percent.

TABLE 1. Bacterial Strains and Plasmids Used

Strain or Plasmid	Mol Wt (10^6)	Characteristics of Phenotype	Origin or Reference
B. stearothermophilus			
IFO 12550 (ATCC 12980)		Wild-type, plasmid-carrier (pBSO1)	IFO[a]
S1		Sm^r, plasmid-carrier (pBSO1)	Spontaneous mutant of wild type
CU21		Sm^r, cured spontaneously of pBSO1	S1
Plasmids			
pUB110	3.0	Km^r	(7, 8)
pTB 19	17.2	Km^rTc^r	(9)
pTB 90	6.7	Km^rTc^r	(6)

[a]IFO, Institute for Fermentation, Osaka, Japan.

TABLE 2 Transformation of *B. stearothermophilus* CU21 with Plasmid DNA[6]

Plasmid	Copy Number per Chromosome in *B. stearothermophilus* CU21	Antibiotic (μg/ml)	Transformants per μg of DNA	Transformation[a] Frequency per Regenerant
pUB110	~50[b]	Km (25)	5.9×10^5	5.9×10^{-3}
pTB19	~ 1	Km (25)	1.0×10^5	1.0×10^{-3}
pTB19		Tc (5)	5.5×10^5	5.5×10^{-3}
pTB90	5 ~ 18[c]	Km (25)	1.3×10^7	1.3×10^{-1}
pTB90		Tc (5)	2.0×10^7	2.0×10^{-1}

[a]Number of regenerated protoplasts was 1.0×10^8 per ml.
[b]Taken from the data in *B. subtilus* 168 (For details, see Reference 2).
[c]Copy number ≃ 5 in LG broth containing Km (5 μg/ml).[6]
Copy number ≃ 18 in LG broth containing Tc (5 μg/ml).[6]

Transformation frequencies of *B. stearothermophilus* CU21 with regard to plasmids pUB110, pTB19, and pTB90 are shown in TABLE 2.[6] These plasmids were all prepared from *B. stearothermophilus* CU21. The third column from left in TABLE 2 refers to the antibiotic (concentration) used in the regeneration, while the last column pertains to transformation frequency per regenerant. Each plasmid used could also be transferred to *B. subtilis* MI113 using competent cells.[6] If the source of preparation of plasmid were *B. subtilis* MI113 instead of *B. stearothermophilus* CU21, transformation frequencies for plasmids pTB19 and pTB90 were reduced to an order of 10^{-2} and 10^{-3}, respectively; however, those of pUB110 remained almost unchanged despite the difference of the plasmid source.

Although not mentioned here, pTB90 was the deletion plasmid of pTB19; it has been constructed from *Eco*RI digests of pTB19 in *B. stearothermophilus* CU21. In fact, the difference of transformation frequencies depending on the source of plasmid, either *B. subtilis* MI113 or *B. stearothermophilus* CU21 is evidence that both pTB19 and pTB90 from *B. subtilis* MI113 suffered restriction in transformation of *B. stearothermophilus,* whereas pUB110 was not restricted regardless of the plasmid source. However intriguing the elaborate picture of transformation frequencies for these plasmids might be, the high frequency and good efficiency of transformation in *B. stearothermophilus* protoplasts with pTB90 (20 percent transformants per regenerant and 2×10^7 transformants per μg of plasmid DNA, respectively) that were comparable to those in *B. subtilis* with plasmid (pC194 or pUB110),[2] justify the usefulness of pTB90 as vector plasmid for molecular cloning of a specific gene(s) in *B. stearothermophilus.*

APPLICATION TO MOLECULAR CLONING[11]

A schematic diagram on cloning of penicillinase genes in *B. stearothermophilus* CU21 using pTB90 as vector plasmid is shown in FIGURE 1.[11] pTTE11 (Tc^r *penP*$^+$ *penI10*) and pTTE21 (Tc^r *penP*$^+$ *penI*$^+$) in the figure are recombinant plasmids constructed previously in the cloning of penicillinase genes from the constitutive and wild-type strains of *B. licheniformis* 9945A, respectively in *E. coli* C600-1[4] using pMB9 as vector plasmid, provided: superscript r stands for resistance to antibiotic; *penP* and *penI* are the structural gene and the repressor gene for penicillinase,

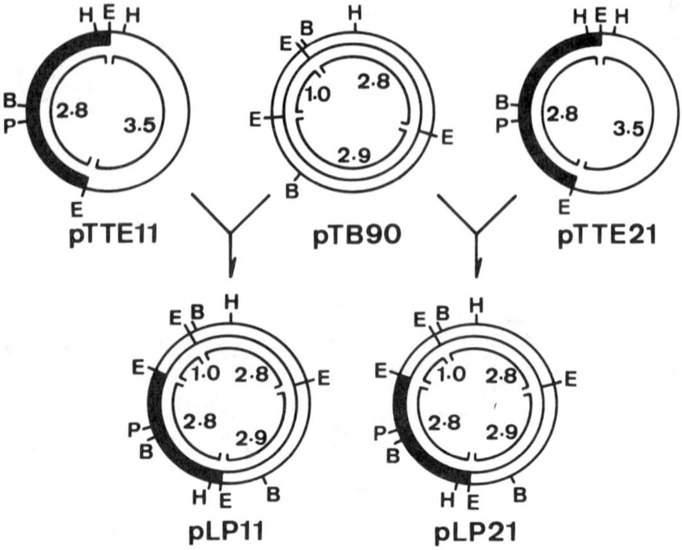

FIGURE 1. Schematic diagram of cloning (penicillinase genes) in *B. stearothermophilus* CU21.

respectively. B, E, H, and P in FIGURE 1 indicate cleavage sites of *Bgl*II, *Eco*RI, *Hin*dIII and *Pst*I, respectively. Thick arcs represent *Eco*RI fragment containing penicillinase genes from *B. licheniformis,* while arabic numbers inside circles indicate molecular size of *Eco*RI fragments in megadaltons.

Partially digested pTB90 DNA with *Eco*RI was mixed with an *Eco*RI digest of pTTE11 (*penP*$^+$ *penI*$^-$) or pTTE21 (*penP*$^+$ *penI*$^+$), followed by ligation. The ligation mixtures (pTTE11 + pTB90 and pTTE21 + pTB90) were used to transform *B. stearothermophilus* CU21 by the protoplast procedure. Penicillinase-positive (PCase$^+$) transformants could be detected quickly by transferring Km^r Tc^r transformants (replica plating) on L agar containing polyvinyl alcohol and an iodine solution as described by Sherratt & Collins.[12] Several Km^r Tc^r PCase$^+$ transformants were obtained with each ligation mixture. Plasmids were isolated from these transformants and examined by agarose gel electrophoresis. The smallest recombinant plasmids obtained from the ligation mixture of pTTE11 + pTB90 and pTTE21 + pTB90 were designated as pLP11 and pLP21, respectively. Their restriction maps are shown in FIGURE 1.

B. stearothermophilus strains carrying the cloned penicillinase genes were grown at constant temperatures (48, 55, and 60°C) in LG broth to late exponential phase. Km (5 μg/ml) was added to the culture medium to warrant the presence of the recombinant plasmid. Penicillinase was assayed in both the culture broth directly and the cell-free supernatant by the iodometric method[13] as described previously.[4] Penicillinase activities measured with and without the recombinant plasmids (as control) in the absence of inducer (cephalosporin C) are shown in TABLE 3.[11] Apparently, no enzyme activities were detected from either *B. stearothermophilus* or the strain carrying only pTB90. *B. stearothermophilus* cells carrying pLP21 (*penP*$^+$ *penI*$^+$) produced nearly the same amount of penicillinase (90 to 120 U/mg of cells) at 48 and

50°C, whereas considerably less activity was detected by cultivation at 60°C (17 U/mg of cells). On the other hand, a large amount of penicillinase (3,300 U/mg of cells) was observed for the strain harboring pLP11 (*penP*$^+$ *penI*$^-$).

It is evident from comparison of penicillinase activities between the strains carrying pLP21 and pLP11 in TABLE 3 that both genes, *penP* and *penI*, were expressed in *B. stearothermophilus*. The fact that the reduced penicillinase activity at higher temperatures for *B. stearothermophilus* (pLP21) was less marked than that of pLP11 carrier strain would suggest the thermal inactivation of the repressor, although ample room is left open for explaining the deterioration of penicillinase expressed at higher temperatures.

As was referred to earlier in TABLE 2, the copy number of pTB90 in *B. stearothermophilus* CU21 is amplified in the presence of Tc. The copy number of pLP11 should therefore increase if Tc were added to the medium. As shown in TABLE 3, penicillinase activities of the strains harboring pLP11 were enhanced with an increase in the amount of Tc added. The highest penicillinase activity (6,000 U/mg of cells) was observed when the bacterium was cultivated in the presence of 3 μg/ml of Tc, about twofold increase of penicillinase compared with the absence of Tc (3,300 U/mg of cells). Lastly, it would be worthwhile to mention that about 10 to 20 percent of the total penicillinase was secreted into the culture medium of *B. stearothermophilus* CU21 at 48°C in contrast to about 30 and 60 percent secretion of the enzyme for *B. subtilis* and *B. licheniformis*, respectively.[4]

SIGNIFICANCE OF TEMPERATURE CONTROL IN THE CULTIVATION OF MESOPHILIC AND THERMOPHILIC BACTERIA

Besides the various advantages expected of a thermophilic cultivation over the culture of a mesophile such as the growth-rate enhancement, the less susceptibility to

TABLE 3. Penicillinase Activity in Transformants of *B. stearothermophilus* CU21[11]

Plasmid	Antibiotic (μg/ml)	Growth Temp (°C)	Penicillinase activity[a] (U/mg of cells): Total	Supernatant
None	None	48	ND[b]	—[c]
pTB90	Km (5)	48	ND	—
pLP21	Km (5)	48	90[d]	7.4
pLP21	Km (5)	55	120	—
pLP21	Km (5)	60	17	—
pLP11	Km (5)	48	3,300	820
pLP11	Km (5)	55	610	—
pLP11	Km (5)	60	39	—
pLP11	Km (5) + Tc (1)	48	3,700	510
pLP11	Km (5) + Tc (2)	48	4,700	680
pLP11	Km (5) + Tc (3)	48	6,000	700
pLP11	Km (5) + Tc (4)	48	5,800	750

[a]Similar results were obtained by another series of experiments.
[b]ND, not detectable (0.01 U/mg of dry cell weight).
[c]—, not tested.
[d]One unit of penicillinase was defined as the quantity required to hydrolyze 1 μmol of penicillin G per hr at 30°C.

contamination, and so forth, it could be inferred intuitively that the amount of cooling water required for the run would be reduced in comparison with the mesophilic cultivation. Although several observations on thermophilic cultivation are needed before the mode of temperature control, economy of running cost around the reduction of cooling water, and so forth, could be thoroughly discussed, some preliminary assessment of cooling water required for both cultivations will be made below, citing from relevant references the data of specific growth rate (as a function of temperature) on *Thermus thermophilus* (an extreme thermophile),[14] *Bacillus caldotenax* (a moderate thermophile)[15] and *E. coli* (a mesophile).[16]

Supposing that a cylindrical fermenter vessel, whose depth of culture liquid is equal to the inside diameter, be aerated with a constant rate of dry air in complete mixing

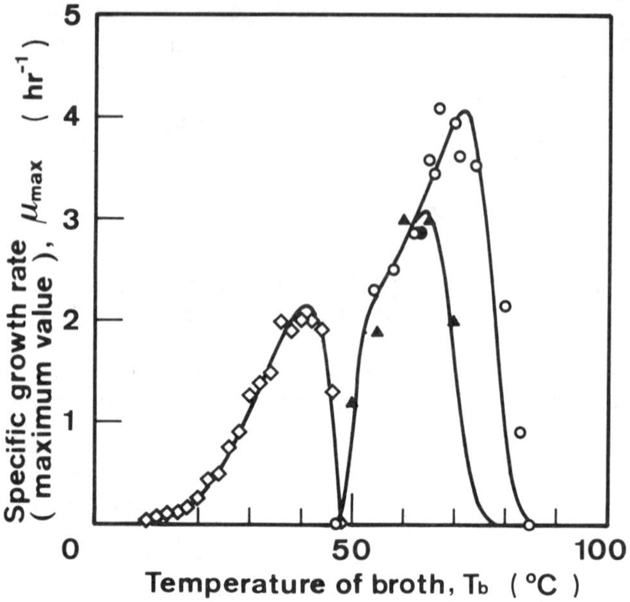

FIGURE 2. Maximum specific growth rate as function of temperature ◇, *E. coli;* ▲, *B. caldotenax;* ○, *T. thermophilus.*

with the liquid, only the growth of each bacterium in batch will be considered under the following assumptions:

(1) specific growth-rate data on *T. thermophilus,*[14] *B. caldotenax,*[15] and *E. coli*[16] are commensurate with maximum values, and these data are formulated as a weighted Arrhenius equation, respectively (see Appendix C and also refer to solid lines in Figure 2);

(2) no metabolites other than the cells, carbon dioxide and water are considered in each cultivation; in other words, complete combustion of carbon source (glucose) is assumed to facilitate calculation of the fermentation heat;

(3) no lag-phase in each growth;

(4) batch culture is terminated when the initial concentration of glucose is reduced to its one-thousandth;

(5) overall growth yield based on glucose and the saturation constant in growth kinetics remain unchanged, respectively, regardless of the difference in bacterial species used; and

(6) physical properties of culture broth such as liquid density and specific heat are taken as those of water.

Differential equations used to assess the temperature profile in batch culture (in the absence of temperature control) are:

$$\rho_b C_b \frac{dT_b}{dt} = (-\Delta H^*)\frac{dX}{dt} - \frac{UA(T_b - T_r)}{V} - a(N_h + \lambda_w N_k) \quad (1)$$

$$\frac{dX}{dt} = \mu X = \mu_{\max}\frac{XS}{K_s + S} \quad (2)$$

$$-\frac{dS}{dt} = \frac{1}{Y_{X/S}}\frac{dX}{dt} \quad (3)$$

provided:

$$aN_h = \frac{Gk_G a}{G + k_G a}(C_H T_b - C_g T_r) \quad (4)$$

$$aN_k = \frac{Gk_G a}{G + k_G a} H_s \quad (5)$$

$$H_s = \frac{M_v}{M_g}\frac{p_s}{P - p_s} \quad (6)$$

For nomenclature used, see APPENDIX A. For Equations 4 and 5, see APPENDIX B.

With respect to unit volume of culture broth, the first term on the right-hand side of Equation 1 represents the heat-evolution rate due to cell growth, while the second and the last terms are the rate of heat loss from the outside surface of the vessel without insulation and that due to aeration, respectively. The last term, indeed, is composed of both terms representing the sensible and latent heat of water vapor.

By using numerical values listed in APPENDIX C, (initial concentration of glucose, $S_o = 10$ kg · m^{-3} and inoculum size, $X_o = 0.01$ kg · m^{-3}), Equations 1 to 3 were solved for various geometrical figures of the vessel. ($\Delta t = 0.001$ hr; ACOS 6, SYSTEM 1000; Computation Center, Osaka Univ.) An example of $R = 1.5$ m, that is, $A/V = 2$ m^{-1} is given in FIGURE 3, in which cell concentrations, X kg · m^{-3} and temperatures of culture broth, T_b°C are plotted against time, t hr, with regard to *E. coli* (top diagram), *B. caldotenax* (middle) and *T. thermophilus* (bottom). The left arrow in each diagram shows the end of batch culture, if temperature of cultivation were controlled ideally from $t = 0$ throughout at $T_{\text{opt}} = 40.77$°C for *E. coli*, $T_{\text{opt}} = 64.12$°C for *B. caldotenax*, and $T_{\text{opt}} = 71.65$°C for *T. thermophilus*, respectively, where T_{opt} implies broth temperature, at which maximum values of specific growth rate for each bacterium are optimized (see FIG. 2 and also, see broken lines for both growth—cell concentration—and temperature in FIG. 3).

On the other hand, if temperature were not controlled, it is interesting to note from the top diagram (mesophilic culture) in FIGURE 3 that broth temperature turns out eventually to be a constant due to balance between heat evolution and heat loss in this particular example, whereas the growth is considerably delayed as demonstrated by a solid curve for X. The arrow on the right-hand side corresponds to the end of delayed batch in this instance. By the same token, right arrows in the lower diagrams (FIG. 3) show the termination of batch run for *B. caldotenax* and *T. thermophilus* without temperature control, respectively. It is noted that broth temperatures in the thermophilic cultures without the control are unbalanced throughout the batch, entailing, however, no marked time delays in winding up each batch as compared to the run of temperature control.

Designating the right-hand side of Equation 1 as Q kcal · m^{-3} · hr^{-1}, the ideal control of temperature in batch culture that is assumed to start at T_{opt} for each bacterium (see broken lines parallel to abscissae in FIG. 3) is to supply and/or remove

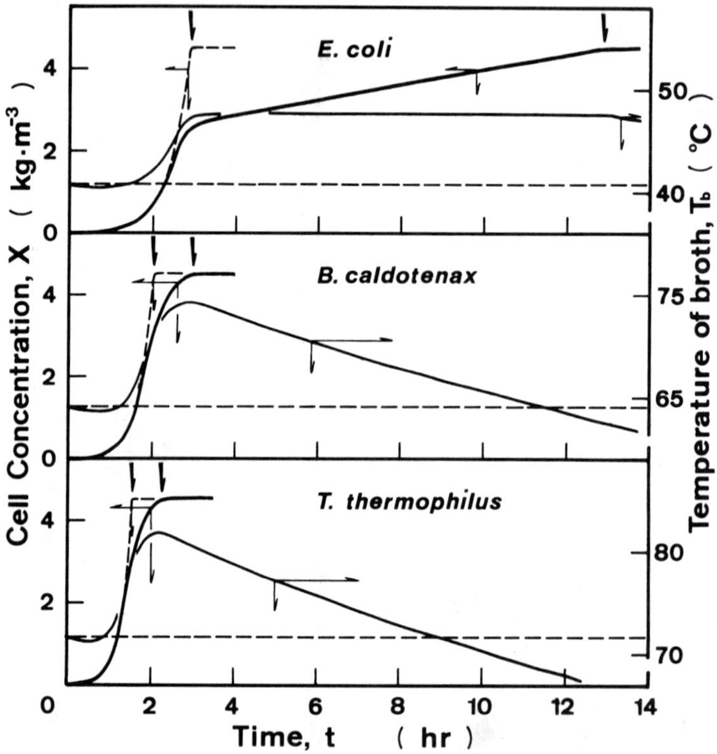

FIGURE 3. Bacterial growth with and without temperature control for *E. coli* (top), *B. caldotenax* (middle) and *T. thermophilus* (bottom). Vessel radius, $R = 1.5$ m

heat energy, p kcal · m^{-3} · hr^{-1} in such a fashion that the following equation is always satisfied throughout batch run.

$$p = -Q \tag{7}$$

If Q values that are assessable at each time during the batch culture (from Equations 1 to 3 with reference to APPENDIX C) and $\Delta t = 0.001$ hr (loc. cit.) were positive, p becomes negative in value, implying the necessity of heat removal by cooling, while heating is required whenever Q values are negative, that is, wherever the value of p in Equation 7 is positive.

As could be imagined from temperature profiles without the control in FIGURE 3, broth temperatures decrease temporarily below the level of T_{opt} due to the heat loss that predominates during the early period after the start of batch. Actually, it is in this brief period that heating is required to maintain the level of T_{opt} in each bacterial culture, and then the period wherein cooling is needed ensues. Although total energy required for heating per unit volume of broth decreases, apparently with the increase of fermenter size simply due to the scale effect that is defined by A/V m^{-1}, the cumulative energy for heating in the example of FIGURE 3 was only a fraction ($1.3 \sim 4.2\%$) of that energy required for cooling (data not shown here). Consequently, only the amount of cooling water needed to maintain the batch culture at T_{opt} throughout will be mentioned below.

The rate of cooling-water supply, W_c kg · hr^{-1} at each moment ($\Delta t = 0.001$ hr) can be estimated by the following equations:

$$V \cdot (-p) = A_c U_c \Delta T_{av} \tag{8}$$

$$= W_c \cdot C_c \cdot \Delta T_c \tag{9}$$

where

$$\Delta T_c = T_{c_2} - T_{c_1} \tag{10}$$

$$\Delta T_{av} = \frac{\Delta T_1 - \Delta T_2}{\ln (\Delta T_1 / \Delta T_2)}$$

$$= \frac{(T_{\text{opt}} - T_{c_1}) - (T_{\text{opt}} - T_{c_2})}{\ln (\Delta T_1 / \Delta T_2)}$$

$$= \frac{\Delta T_c}{\ln (\Delta T_1 / \Delta T_2)} \tag{11}$$

(see APPENDIX A).

Taking $A_c/V = 1$ m^{-1} and $U_c = 800$ kcal · m^{-2} · hr^{-1} · °C^{-1} in Equation 8, values of $(-p)$ assessed earlier at each moment of the batch give the values of ΔT_{av}. Since values of T_{opt} are given for each bacterium and taking T_{c_1} as 10°C at the inlet of cooling water, values of T_{c_2} (outlet temperature of cooling water) is searched in a region from T_{opt} to T_{c_1} (=10°C) by the method of bisection. Once T_{c_2} that satisfies Equation 11 is determined, it is feasible to estimate W_c from Equation 9, and the cooling water required in each batch is given by integrating appropriately W_c values with respect to time (compare FIG. 3). The result of computation on cooling water required per unit volume of broth, ton · m^{-3} regarding various sizes of the fermenter vessel is summarized in TABLE 4.

As far as this example of calculation is concerned, it is noted from TABLE 4 that the amount of cooling water, ton · m^{-3} for the mesophilic culture (*E. coli*) is about threefold of those needed for the thermophilic cultures (*B. caldotenax* and *T. thermophilus*), irrespective of geometry of the fermenter vessel. However, the requirement of cooling water for each culture tends to increase with the increase of vessel size; this trend in each bacterium reflects the decrease of heat loss (second term on the right-hand side of Equation 1) when the vessel size increases, culminating in the increase of heat to be removed by forced cooling.

An odd finding that the amount of cooling water for *T. thermophilus* is larger than that for *B. caldotenax* to an extent of about 15% might have resulted from the larger value of μ_{max} at T_{opt} (=71.65°C) for *T. thermophilus* than that for *B. caldotenax* at T_{opt} (=64.12°C) (see FIG. 2, first term on the right-hand side of Equation 1, and Equation 2). In other words, the larger value of dX/dt for a specific period of time with respect to *T. thermophilus*, suggesting the larger value of heat evolution rate for that period, might account for the larger amount of cooling water required for *T. thermophilus*.

In light of previous assumptions ((1) ~ (5)) for the calculation, it is urged that cumulative fermentation heat due to cell growth remain unchanged irrespective of the bacterial species. Hence, the total energy to be removed by cooling would not vary appreciably depending on the bacterial species and the vessel size. Actually in this example, the amount of total energy in the batch run to be removed by cooling ranges from $1.1 \times 10^4 \sim 1.3 \times 10^4$ kcal · m^{-3} regardless of the vessel size and bacterial species studied here. The only factor to contribute to the difference in the amount of cooling water is found in the difference of growth-rate pattern of each bacterium, in that of temperature at which the culture is maintained by an appropriate control, in the different magnitude of heat loss due to natural convection that depends on the vessel size, and so forth.

It might also be urged by intuition that the amount of cooling water in thermophilic cultivation would be reduced, without resorting to the calculation, compared to the culture of mesophiles. To cite an extreme case in connection with this example of

TABLE 4. Assessment of Cooling Water Required during Batch Cultures of Mesophilic and Thermophilic Bacteria[a]

Fermenter	R (m)	0.6	0.75	1	1.5	3
Vessel size	V (m^3)	1.4	2.7	6.3	21	170
Cooling water required, $\int W_c dt$	*E. coli*	0.93	0.99	1.07	1.15	1.25
(ton · m^{-3}) for each	*B. caldotenax*	0.27	0.28	0.30	0.31	0.33
batch culture	*T. thermophilus*	0.31	0.33	0.35	0.37	0.40

[a]Fermenter vessel radius, R = 0.6 ~ 3 m

mesophilic and thermophilic cultures, the values of ΔT_c for *E. coli, B. caldotenax*, and *T. thermophilus* would become 30.77°C, 54.17°C, and 61.65°C, if the coolant capacity ($C_c \cdot \Delta T_1$) were exhausted. Taking for granted that the amount of heat energy to be removed by cooling is nearly the same without respect to the bacterial species, it would be inferred that the amount of cooling water in *B. caldotenax* and *T. thermophilus* is of about 57% and 50%, respectively, of that needed for *E. coli* as unity.

It is now worthwhile emphasizing the fact, as shown in TABLE 4, that the amount of cooling water in the culture of thermophiles is around 30% of that in *E. coli*, that is, more than the above-mentioned figures (57 ~ 50%). The discrepancy between the intuitive estimate and the actual calculation might have originated from various factors relating to the growth rate pattern, the nonlinearity of the temperature difference, and so forth, as is pointed out earlier. Herein lies one of the points that the cultivation of a thermophile is rewarding compared to that of a mesophile.

CONCLUSION

(1)Using plasmids such as pUB110, pTB19, and pTB90, *B. stearothermophilus* CU21 could be transformed by the protoplast procedure. A high frequency of transformation of *B. stearothermophilus* protoplast with pTB90, that is, 20% transformants per regenerant and 2×10^7 transformants per μg of plasmid DNA was comparable to that in *B. subtilis* with plasmid pC194 or pUB110.

(2)Penicillinase genes were cloned in *B. stearothermophilus* CU21 using pTB90 as vector plasmid; both genes, *penP* (structural gene) and *penI* (repressor gene) were expressed, although reduced penicillinase activities at higher temperatures (55 ~ 60°C) required further studies on its mechanism of the reduction.

(3)It was confirmed by computation that the amount of cooling water required for a specific cultivation of a thermophile was reduced, in comparison with a mesophile, to an extent of more than that assessable by intuition.

REFERENCES

1. Anagnostopoulos, C. & J. Spizizen. 1961. J. Bacteriol. **81:** 741–746.
2. Chang, S. & S. N. Cohen. 1979. Molec. Gen. Genet. **168:** 111–115.
3. Gray, O. & S. Chang. 1981. J. Bacteriol. **145:** 422–428.
4. Imanaka, T., T. Tanaka, H. Tsunekawa & S. Aiba. 1981. J. Bacteriol. **147:** 776–786.
5. Palva, I., R. F. Pettersson, N. Kalkkinen, P. Lehtovaara, M. Sarvas, H. Söderlund, K. Takkinen & L. Kääriäinen. 1981. Gene **15:** 43–51.
6. Imanaka, T., M. Fujii, I. Aramori & S. Aiba. 1982. J. Bacteriol. **149:** 824–830.
7. Gryczan, T. J., S. Contente & D. Dubnau. 1978. J. Bacteriol. **134:** 318–329.
8. Keggins, K. M., P. S. Lovett & E. J. Duvall. 1978. Proc. Natl. Acad. Sci. USA **75:** 1423–1427.
9. Imanaka, T., M. Fujii & S. Aiba. 1981. J. Bacteriol. **146:** 1091–1097.
10. Gordon, R. E., W. C. Haynes & C. H.-N. Pang. 1973. *In* Agriculture Handbook No. 427. U. S. Department of Agriculture, Washington D. C. p. 214.
11. Fujii, M., T. Imanaka & S. Aiba. 1982. J. Gen. Microbiol. **128:** 2997–3000.
12. Sherratt, D. J. & J. F. Collins. 1973. J. Gen. Microbiol. **76:** 217–230.
13. Sargent, M. G. 1968. J. Bacteriol. **95:** 1493–1494.
14. Ohshima, T. 1978. Private communication.
15. Kuhn, H. J., S. Cometta & A. Fiechter. 1980. Eur. J. Appl. Microbiol. Biotechnol. **10:** 303–315.
16. Ingraham, J. L. 1958. J. Bacteriol. **76:** 75–80.
17. Bagnoli, E. 1973. *In* Chemical Engineers' Handbook, 5th ed. R. H. Perry & C. H. Chilton, Eds. Section 12-2. McGraw-Hill Kogakusha, Ltd. Tokyo.
18. Payne, W. J. 1970. Ann. Rev. Microbiol. **24:** 17–52.
19. Pratt, F. R. 1929. *In* International Critical Tables, E. W. Washburn, Ed. Vol. **5:** 166. McGraw-Hill, Inc. New York.
20. Liley, P. F. & W. R. Gambill. 1973. *In* Chemical Engineers' Handbook, 5th ed. R. H. Perry & C. H. Chilton, Eds. Section 3-45. McGraw-Hill Kogakusha, Ltd. Tokyo.
21. Sherwood, T. K., R. L. Pigford & C. R. Wilke. 1975. *In* Mass Transfer. McGraw-Hill, Inc., New York. p. 262.

APPENDIX A
NOMENCLATURE

A = outer surface area of cylindrical vessel ($=6\pi R^2$, provided: R = internal radius of cylinder; height = $2R$), m^2

A_c = area of cooling surface in fermenter vessel, m^2

a = interfacial area between gassed bubbles and broth per unit volume of broth, m^{-1}

C_b = specific heat of culture broth, kcal · kg of broth^{-1} · °C^{-1}

C_c = specific heat of cooling water, kcal · kg of water^{-1} · °C^{-1}
C_g = dry specific heat of air, kcal · kg of dry air^{-1} · °C^{-1}
C_H = wet specific heat of air, kcal · kg of dry air^{-1} · °C^{-1}
= $C_g + C_v \cdot H$
C_v = specific heat of water vapor, kcal · kg of vapor^{-1} · °C^{-1}
G = aeration rate, kg dry air · m^{-3} · hr^{-1} (=wt/vol)
H_s = absolute humidity (saturated), kg vapor · kg of dry air^{-1}
H = absolute humidity (unsaturated), kg vapor · kg of dry air^{-1}
ΔH = difference in absolute humidity between inlet air ($H_{in} = 0$) and outlet air ($H_{out} = H$), kg vapor · kg of dry air^{-1}
$-\Delta H_a$ = combustion heat of cells, kcal · kg of dry cells^{-1}
$-\Delta H_g$ = combustion heat of glucose, kcal · kg of glucose^{-1}
$-\Delta H^*$ = fermentation heat, kcal · kg of dry cells^{-1}
= $-\Delta H_g \cdot Y_{X/S}{}^{-1} - (-\Delta H_a)$
h = heat-transfer coefficient, kcal · m^{-2} · hr^{-1} · °C^{-1}
K_s = saturation constant, kg of glucose · m^{-3}
k_g = mass (water vapor)-transfer coefficient, kg vapor · m^{-2} · hr^{-1} · ΔH^{-1}
M_g = molecular weight of air, taken as 28.9
M_v = molecular weight of water (=18)
N_h = heat-transfer rate between broth and air bubbles per unit interfacial area, kcal · m^{-2} · hr^{-1}
N_k = mass (water vapor)-transfer rate between broth and air bubbles per unit interfacial area, kg vapor · m^{-2} · hr^{-1}
P = total pressure of air, taken as 760 mm Hg
p = heat energy supply or removal rate, kcal · m^{-3} · hr^{-1}
p_s = saturated (water) vapor pressure, mm Hg
Q = total terms on the right-hand side of Equation 1, kcal · m^{-3} · hr^{-1}
S = substrate (glucose) concentration, kg · m^{-3}
S_o = initial concentration of substrate (glucose), kg · m^{-3}
T_b = temperature of culture broth, °C
T_g = gas (air) temperature at exit of fermenter vessel, °C
T_{opt} = temperature of culture, at which maximal value of specific growth rate is optimized, °C
T_r = room temperature, °C
ΔT_{av} = logarithmic mean of temperature difference, °C
ΔT_1 = temperature difference between T_{opt} and inlet temperature, T_{c_1} of cooling water, °C
ΔT_2 = temperature difference between T_{opt} and outlet temperature, T_{c_2} of cooling water, °C
t = time, hr
U = overall heat-transfer coefficient with respect to outer surface of cylindrical vessel due to natural convection, kcal · m^{-2} · hr^{-1} · °C^{-1}
U_c = overall heat-transfer coefficient regarding cooling surface inside fermenter vessel, kcal · m^{-2} · hr^{-1} · °C^{-1}
V = volume of culture broth in fermenter, m^3
W = air supply rate, kg dry air · hr^{-1}
W_c = supply rate of cooling water, kg water · hr^{-1}
X = cell concentration, kg of dry cells · m^{-3}
X_o = inoculum size, kg of dry cells · m^{-3}
$Y_{x/s}$ = overall growth yield, kg of dry cells · kg glucose^{-1}
λ_w = latent heat of water vapor, kcal · kg of vapor^{-1}
μ = specific growth rate, hr^{-1}

μ_{max} = maximum value of specific growth rate, hr^{-1}
ρ_b = density of culture broth, $kg \cdot m^{-3}$

APPENDIX B

With regard to heat and mass transfer at interfacial area in the vessel,

$$N_h = h(T_b - T_g) \quad \textbf{(4-1)}$$

$$N_k = k_G(H_s - H) \quad \textbf{(5-1)}$$

Assuming that the vessel is aerated in complete mixing with dry air, whose temperature at the inlet is equal to room temperature, heat- and mass-balance equations between input and output with respect to the vessel are:

$$a\,N_h = G(C_H T_g - C_g T_r) \quad \textbf{(4-2)}$$

$$a\,N_k = G(H - 0) \quad \textbf{(5-2)}$$

Canceling out T_g from Equations 4-1 and 4-2 and rearranging,

$$a\,N_h = \frac{G \cdot h \cdot a}{G \cdot C_H + h \cdot a}(C_H T_b - C_g T_r) \quad \textbf{(4-3)}$$

Substituting the following approximation,[21]

$h/k_G \doteqdot C_H$ into Equation 4-3

$$a\,N_h = \frac{G \cdot k_G \cdot a}{G + k_G \cdot a}(C_H T_b - C_g T_r) \quad \textbf{(4)}$$

Likewise, canceling out H from Equations 5-1 and 5-2, and rearranging,

$$a\,N_k = \frac{G \cdot k_G \cdot a}{G + k_G \cdot a} \cdot H_s \quad \textbf{(5)}$$

APPENDIX C
LIST OF CONSTANTS

(1)

A_c/V = 1 m^{-1}
$C_b = C_c$ = 1 $kcal \cdot kg\ of\ liquid^{-1} \cdot {}^\circ C^{-1}$
C_g = 0.24 $kcal \cdot kg\ dry\ air^{-1} \cdot {}^\circ C^{-1}$ [17]
C_v = 0.45 $kcal \cdot kg\ water\ vapor^{-1} \cdot {}^\circ C^{-1}$ [17]
G = 77.41 $kg\ dry\ air \cdot m^{-3} \cdot hr^{-1}$(1 vvm)
$-\Delta H_a$ = 5.3×10^3 $kcal \cdot kg\ of\ dry\ cells^{-1}$ [18]
$-\Delta H_g$ = 3.74×10^3 $kcal \cdot kg\ glucose^{-1}$ [19]
K_s = 0.1 $kg\ glucose \cdot m^{-3}$
$k_G a$ = 2 $kg\ water\ vapor \cdot m^{-3} \cdot hr^{-1} \cdot \Delta H^{-1}$
M_g = 28.9 $kg \cdot kmol\ of\ air^{-1}$
M_v = 18 $kg \cdot kmol\ of\ water\ vapor^{-1}$

$P = 760$ mm Hg
$S_o = 10$ kg glucose · m^{-3}
$T_{c_1} = 10$ °C
$T_{opt} = 71.65$ °C for *T. thermophilus*[14]
$T_{opt} = 64.12$ °C for *B. caldotenax*[15]
$T_{opt} = 40.77$ °C for *E. coli*[16]
$T_r = 25$ °C
$U = 10$ kcal · m^{-2} · hr^{-1} · $°C^{-1}$
$U_c = 800$ kcal · m^{-2} · hr^{-1} · $°C^{-1}$
$X_o = 0.01$ kg cell · m^{-3}
$Y_{X/S} = 0.45$ kg dry cell · kg $glucose^{-1}$
$\rho_b = 1 \times 10^3$ kg broth · m^{-3}

(2)

μ_{max} for *E. coli*:[16]

$$\mu_{max} = W(T_b) \exp\{a_1 + a_2/(T_b + 273.15)\}$$

$$W(T_b) = \begin{cases} 1 \,;\, T_b < (a_3 - 1)/a_4 \\ a_3 - a_4 T_b \,;\, (a_3 - 1)/a_4 \leq T_b \leq a_3/a_4 \\ 0 \,;\, a_3/a_4 < T_b \end{cases}$$

μ_{max} for *B. caldotenax*[15] and *T. thermophilus*:[14]

$$\mu_{max} = W(T_b) \exp\{b_1 + b_2/(T_b + 273.15)\}$$
$$W(T_b) = 1/4 \operatorname{erfc}\{b_3(T_b - b_4)\}[1 + \operatorname{erf}\{b_5(T_b - b_6)\}]$$

provided:

$a_1 = 46.87$
$a_2 = 14.13 \times 10^3$
$a_3 = 2.33$
$a_4 = 48.85 \times 10^{-3}$
$b_1 = 13.96$
$b_2 = 4.31 \times 10^3$
$b_3 = 0.22$
$b_4 = 69.2$ (for *B. caldotenax*)
$= 76.8$ (for *T. thermophilus*)
$b_5 = 0.5$
$b_6 = 50$

(3)

$p_s = \exp(p_1 T_b^2 + p_2 T_b + p_3)$[20]

where

$p_1 = -156.4 \times 10^{-6}$
$p_2 = 65.76 \times 10^{-3}$
$p_3 = 1.622$
$\lambda_w = \lambda_1 + \lambda_2 T_g$[20]

where

$\lambda_1 = 597.6$
$\lambda_2 = -0.581$

Kinetics of Product Formation and Plasmid Segregation in Recombinant Microbial Populations

J. E. BAILEY, M. HJORTSO, S. B. LEE, AND F. SRIENC

Department of Chemical Engineering
California Institute of Technology
Pasadena, California 91125

INTRODUCTION

The advent of recombinant DNA technology provides the means of manufacturing a wide spectrum of valuable proteins in fermentation processes using bacteria, yeast, or mammalian cell cultures. At the cellular level, this objective is frequently achieved by expression of cloned genes contained in multiple copies of small, autonomously replicating DNA molecules called plasmids. The introduction of recombinant plasmids into cells leads to interactions between host cell and plasmid functions that determine the rates of plasmid replication and plasmid gene expression. This paper summarizes recent research in the authors' laboratory aimed at experimental characterization and mathematical description of the primary features of recombinant cells that determine overall reactor productivity.

In this paper, it will be assumed that the process objective is manufacture of a single protein obtained by expression of a cloned gene. In such a case, it is convenient to describe the overall rate of production formation as shown in FIGURE 1. As indicated there, the overall rate of product synthesis may be written as the summation of product synthesis rates in individual cells containing different numbers of plasmids. In this representation, it is assumed that the rate of product synthesis in an individual cell is a function only of the number of plasmids in the cell and the condition of the cell's environment. The former quantity is designated p in FIGURE 1, while the environmental state is indicated by s. Cell growth properties determine the cell number concentration n. The frequency function $f(p, s)$ is determined by the population balance equation, which requires for calculation of f information on cell division regulation, the kinetics of plasmid replication in an individual cell, and the rules that govern plasmid segregation to daughter cells upon cell division [$f(p, s)dp$ is the fraction of cells with plasmid content between p and $p + dp$]. The single-cell rate of gene expression, r_e, is taken here to depend upon plasmid content and the environment, since often environmental manipulation is employed to increase or decrease the level of expression of a particular cloned gene. Not included in FIGURE 1 but important in determining the *net* rate of product formation, is the rate of product deactivation and decomposition. The formalism in FIGURE 1 may be readily extended to incorporate these phenomena.

As FIGURE 1 indicates, the interrelated processes that determine the various factors on the right-hand side of the overall rate of product synthesis equation both depend on genetic and environmental parameters. In order to optimize catalyst (the cell containing recombinant plasmids) and process design, it is important to understand in quantitative terms the relationships among host cell genotype, plasmid construction, cell culture environment, and process productivity. The new experimental and kinetic

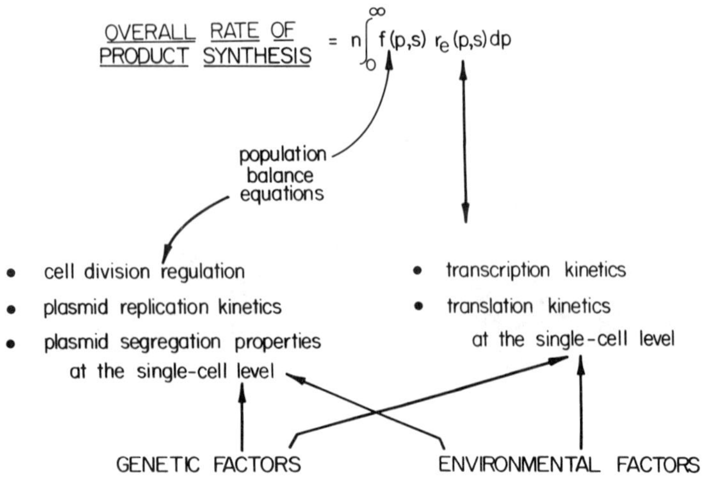

FIGURE 1. Schematic diagram of the influences of genetic and environmental factors on overall product synthesis through single-cell macromolecular synthesis kinetics and controls.

modeling approaches summarized below should contribute to achievement of this goal.

RAPID EXPERIMENTAL CHARACTERIZATION OF PLASMID SEGREGATION

A major consideration in production of a specific protein by cells containing recombinant plasmids is loss of productivity due to loss of plasmids. Plasmid-free cells may be present in the culture either due to impurities in the inoculum that do not contain plasmids or due to poor segregation of plasmids upon cell division, the latter event generating daughter cells entirely free of plasmids. Thus the fraction of the population containing plasmids F_p and the fraction of the population free of plasmids F_0 are of major interest and may be written in terms of f as follows:

$$F_p = \int_1^\infty f(p)dp \tag{1a}$$

$$F_0 = 1 - F_p \tag{1b}$$

Understanding of the genetic and environmental influences upon loss of plasmids will be greatly facilitated by the availability of efficient experimental methods for characterizing plasmid segregation.

Currently conventional methodology involves including a gene for a selection marker in the plasmid so that cells with the plasmid can grow on some suitable agar medium while cells without plasmids do not grow. This method, while quite straightforward, requires that cells with plasmid be able to grow for a sufficient number of generations to form a visible colony. Also the time required to conduct the assay, a period of hours or days, is longer than desirable for research applications and certainly undesirably long from the viewpoint of process monitoring and control in a production situation.

A new experimental methodology based upon flow cytometry has been developed in this laboratory for study of plasmid segregation.[1] The marker employed in this method is the enzyme β-galactosidase from *Escherichia coli* or an enzymatically active fragment of this protein that is provided in the host cell by expression of the *E. coli lacZ* gene or an appropriate fragment of this gene. This gene is widely available in many current plasmids and on restriction fragments convenient for cloning into other plasmids.

FIGURE 2 provides a schematic map of the plasmid pRB73 (gift of M. Rose) that has been employed in the development of the flow cytometry method. This plasmid is capable of expressing bacterial β-galactosidase activity at a high intracellular level in *Saccharomyces cerevisiae*.[2] In particular, when this plasmid is introduced into *S. cerevisiae* strain DBY689, a uracil and leucine auxotroph, transformed cells (strain YT76) become prototrophic for leucine, require uracil for growth, and contain an active fragment of *E. coli* β-galactosidase. This plasmid is, however, quite unstable. Even under selection conditions in leucine-free medium, cultures of YT76 contain a significant fraction of plasmid-free cells.

The flow cytometric detection of β-galactosidase activity in individual yeast cells is provided by the combined use of fluorogenic substrate, Naphthol AS-BI β-(D)galactopyranoside (NASBIG, Sigma), and the trapping reagent Fast Red TR (Sigma). The nonfluorescent substrate, NASBIG, is converted to Naphthol AS-BI (NASBI), a fluorescent product, in the presence of β-galactosidase. Access of the substrate to the intracellular enzyme is provided by permeabilization of the cells with an isopropanol solution. The trapping reagent reacts with the soluble and diffusible NASBI product to give an intracellular precipitate that is also fluorescent. The trapping reagent step is critical to prevent crosstalk between cells containing the enzyme and cells without the enzyme by diffusion of the fluorescent product from one cell type to the other in the liquid incubation mixture. The cells treated by this procedure are then analyzed in a flow cytometer in which single cells can be examined for fluorescence of the trapped NASBI product, and hence, presence of β-galactosidase, in less than 1 msec. The overall protocol is summarized in FIGURE 3, which also includes a summary of a procedure without the trapping agent that may be used for average enzyme content assays using a spectrofluorometer.

FIGURE 4 illustrates some of the results obtained by this measurement method.

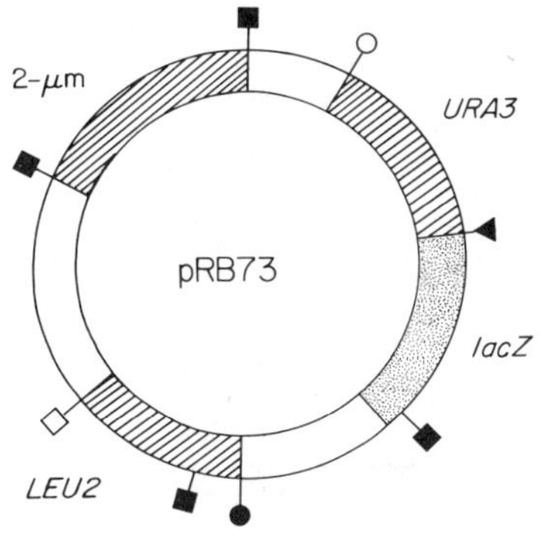

FIGURE 2. Structure of the plasmid pRB73 (gift of M. Rose). The fragment of the *E. coli lacZ* gene (stippled) fused to the yeast *URA3* gene and promoter gives β-galactosidase activity in yeast. Autonomous replication function in yeast is provided by 3.8 Kb from the B-form of the 2-μm plasmid of yeast. Hatched regions are yeast sequences; white areas are fragments of pBR322 including the origin of replication and the amp^R gene (restriction enzyme sites: ■ , *Eco*RI; ● , *Sal*I; ▼ , *Bam*HI; ○ , (*Bam*Hi/*Bgl*II); □ , (*Sal*I/*Xho*I); see Rose *et al.*[2] for discussion of similar plasmids).

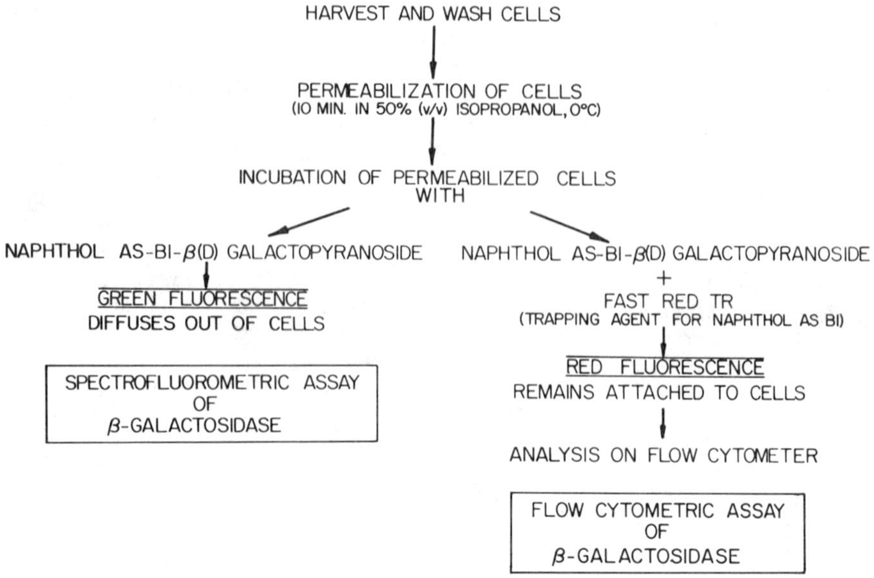

FIGURE 3. Summary of protocols for spectrofluorometric (population-average) and flow cytometric (rapid single-cell) assays of β-galactosidase activity in *S. cerevisiae.*[1]

These data show clearly that increased single-cell fluorescence (proportional to channel number) appears as a result of incubation of enzyme-containing cells in substrate-trapping agent solution. The increase in fluorescence intensity is related to substrate concentration as expected according to Michaelis-Menten kinetics. Measurements for cells without enzyme after incubation are identical to those shown in FIGURE 4 for zero substrate. Consequently, proportions of cells with and without enzyme in mixed populations may be readily determined by computational analysis to resolve the components of the combined overlapping distributions obtained in such cases.[1]

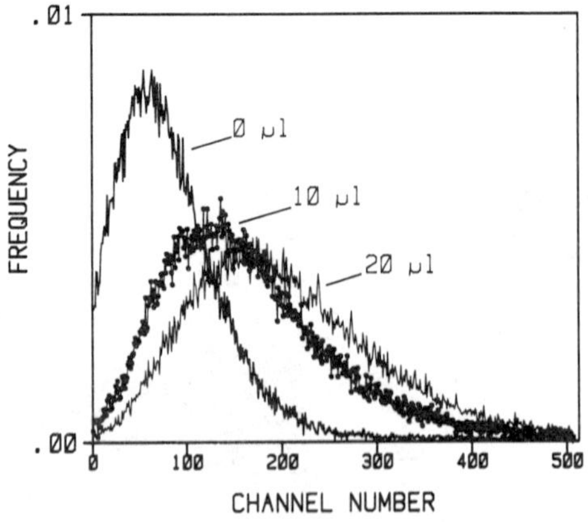

FIGURE 4. Flow cytometric demonstration of β-galactosidase activity in single cells of a population of *S. cerevisiae* strain YT76 containing the plasmid pRB73. The histograms represent measurements of 10^5 permeabilized cells after exposure for 5 min to the incubation mixture (a total volume of 3 ml containing 0, 10, and 20 μl of NASBIG substrate solution (10 mg/ml)). These data demonstrate dependence of the enzymatic reaction on substrate concentration. Channel number is proportional (linear scale) to single-cell fluorescence.

The permeabilization and incubation steps involved in the above assay require a total of thirty minutes, and this time could be reduced if needed for process-monitoring application. It should be possible by this approach to obtain much more information than a simple determination of the fraction of the population that contains enzyme. Distributions of signal intensities are readily available in flow cytometry measurements as FIGURE 4 shows. Consequently, there is excellent potential for extracting information from this type of measurement on the distribution of enzyme content in the culture and for relating this distribution to the distribution of plasmid molecules per cell. Such information would aid tremendously in further development of the approach to recombinant system productivity outlined in FIGURE 1. Research is currently in progress in this direction as well as on other single-cell assays for different genetic markers.

MATHEMATICAL DESCRIPTION OF PLASMID PROPAGATION IN CELL POPULATIONS

For design, control, and optimization of microbial reactors containing recombinant organisms and for interpretation of research data, it is desirable to formulate a mathematical description of the coordinated processes of cell growth, plasmid synthesis, cell division, and plasmid segregation at division. Previous efforts in this direction have used nonsegregated descriptions to describe the growth of plasmid-containing and plasmid-free cells viewed essentially as a competitive mixed culture.[3-5] These methods have the advantages of closed-form solutions with a small number of parameters needed to describe plasmid segregation, possible growth-modifying influences of plasmids, substrate utilization, and product formation in recombinant cultures.

Because the problem of plasmid segregation is only one aspect of a more general problem of the distribution of plasmid content within the culture, it may be advantageous to approach the mathematical description of plasmid content in a cell population with the objective of determining distributions of cell types in the culture. This requires a segregated view of the system. The population balance equations required for such a treatment are more difficult to formulate and to solve than overall nonsegregated model equations, but there are compensating benefits.

First, the framework provided by population balance equations encourages explicit recognition of the individual cell kinetic and regulatory properties on which overall population properties depend. These factors for the problem of segregated modeling of cellular plasmid content are summarized in FIGURE 1. Second, substantial information is often available in the biological sciences literature concerning these single-cell properties. Thus, population balance models provide a means to link basic information on cell-cycle regulation and plasmid replication and segregation obtained in the sciences with culture properties of central interest to the manufacturer of a plasmid-gene product. Mathematical models with more direct mechanistic bases are likely to be more reliable when extrapolating to new genetic or environmental conditions. Third, the population balance approach yields not only the relative numbers of cells in the population with and without plasmids but also provides the complete distribution of plasmid content in the culture. Since such results are obtained by transformations of an age distribution, other transformations of this same population property allow calculation of distributions of total cell mass and other cell states of possible interest. Finally, as discussed in more detail in a previous paper,[6] the population balance approach in conjunction with flow cytometric and other measurements of the distributions of cell properties of interest allow an inverse application of the population balance equations to deduce single-cell kinetic and regulatory properties based upon measured distributions. Thus, this strategy should eventually contribute to greater knowledge of plasmid replication and segregation.

Recently, the age, mass, and total DNA cellular content distributions for budding yeast populations in steady-state and transient growth conditions have been described.[7–10] These analyses have now been extended to consider the problem of plasmid propagation in budding yeast systems.[11] This work is of immediate practical interest because of increasing utilization of *S. cerevisiae* hosts and recombinant plasmids in basic and applied research, and because the multicopy recombinant plasmids in *S. cerevisiae* are often quite unstable. The effect of genetic and environmental factors on stability of yeast plasmids is not presently well understood.

As an illustration of how mathematics and cell biology join powerfully in this arena, one elementary result of the most recent analysis will be summarized here. Suppose that a budding yeast population, with a single-cell division cycle described by the budding cycle model,[12,13] grows in a chemostat in a selective medium so that cells born without plasmids do not divide. Then it can be shown that the maximum possible dilution rate, that is, the dilution rate at washout, is given by

$$D_{\max} = \frac{1}{P}(\ln 2 - \theta_D - \theta_M) \tag{2}$$

Here P denotes the time interval between mother cell birth and mother cell division. θ_D and θ_M are the fractions of daughter and mother cells, respectively, that have no plasmids at birth. If it is also assumed that the dividing cell contains N plasmids that segregate randomly at birth, the fractions θ_D and θ_M may be calculated from

$$\theta_D = f(N \mid N) \tag{3a}$$

$$\theta_M = f(0 \mid N) \tag{3b}$$

where the binomial distribution

$$f(x \mid N) = \frac{N!}{x!(N - x!)} \alpha^x (1 - \alpha)^{N-x} \tag{4}$$

gives the conditional probability that a newborn mother cell, arising from division of a cell with N plasmids, will contain x plasmids. In this formulation, the single parameter α, which denotes the probability that a plasmid remains in the mother cell at birth, is employed to characterize plasmid segregation properties.

Referring to Equation 2, it is clear that if no cells without plasmid are born at cell division ($\theta_D = \theta_M = 0$), the dilution rate at washout $D_{\max}$ is:

$$D_{\max_0} = \ln 2/P. \tag{5}$$

Thus, comparing Equation 2 and 5 shows that the dilution washout rate is reduced by plasmid instability in a selective medium. This is a rather obvious result, since, in a selective medium, birth of cells without plasmids is a nonproductive event analogous to cell death. The magnitude of this effect is displayed in FIGURE 5 in which the dilution rate at washout given by Equation 2 has been normalized by the dilution rate at washout for a completely stable plasmid $D_{\max_0}$.

This elementary calculation gives several interesting results. First, it shows that, even if there is no bias for preserving plasmids in the mother cell (probability parameter $\alpha = 0.5$), the maximum culture growth rate is significantly less than unity if the plasmid copy number is low. This clearly illustrates that efforts to control plasmid content of a production culture strictly by environmental means, namely by applying selection pressure, may not be completely effective. A combination of genetic manipulation and environmental adjustment is required to obtain the greatest productivity in this reactor configuration. Analogous arguments would also apply, of course, to batch production.

Another interesting conclusion arises from this analysis. Suppose that plasmid-free cells grown in a nonselective medium have exactly the same single-cell growth rate as cells containing plasmids grown in a selective medium. The above result shows that if the plasmid is unstable, that is, if either θ_D or θ_M is nonzero, the *apparent* overall growth rate of the plasmid-containing culture will be less than the overall growth rate of the plasmid-free cells in the nonselective medium, even though the true single-cell growth rates for both cell types are identical. This is significant because, in the analysis of simultaneous growth of plasmid-containing and plasmid-free cells in nonselective (or

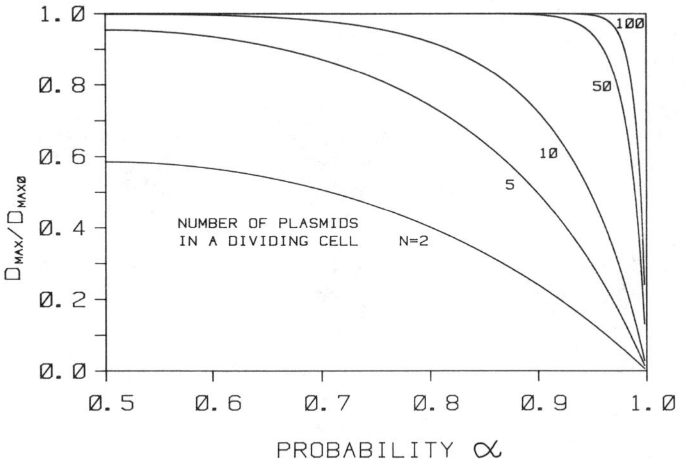

FIGURE 5. Dilution rate at washout (D_{MAX}) for yeast with random plasmid segregation grown in selective medium. (D_{MAX_0} = washout dilution rate for regular segregation; N = number of plasmids in a dividing cell; α = probability plasmid remains in mother cell at division.)

partially selective) medium, the relative growth rates of the two cell types is a very important parameter.

Practically all previous experiments designed to study the relative growth rates of plasmid-containing and plasmid-free cells have used selection pressure in the plasmid-containing growth measurement. Thus, it is possible that, because of plasmid instability, the reported growth rates in these experiments may underestimate significantly the true single-cell growth rates of the plasmid-containing cells, which are the growth rates required for the mixed-culture competition calculation in nonselective medium. Further research seems in order to explore the extent to which overall culture growth rates in selective medium truly reflect the single-cell growth properties of plasmid-containing cells.

Even in this relatively simple analysis, it is interesting to note the ability to connect single-cell properties such as the mother cell growth interval P and the number of plasmids in a dividing cell N with important overall process characteristics such as the dilution rate at washout in a continuous culture.

A DETAILED SINGLE-CELL MODEL FOR PLASMID REPLICATION

When calculating the distribution of plasmid content in a growing cell culture, it is necessary to know the kinetics of plasmid synthesis in individual cells. In order to provide this information, efforts have been initiated to formulate mathematical descriptions of the molecular control systems for plasmid replication. Such models, if successful, provide a robust framework for examining and representing quantitatively the effects both of changes in the nucleotide sequences of the regulatory elements in the plasmid and also of the cell's growth environment. In addition to providing useful information for segregated models of cell populations, the single-cell model may also be employed to describe an average cell in the system and serve on its own, from this viewpoint, as a structured nonsegregated model of the entire culture.[14]

The initial system selected for study is the λdv plasmid, a deletion derivative of phage λ, which replicates autonomously in *E. coli*. This particular system was chosen for this initial modeling effort because a wealth of information is available about regulation of replication of this plasmid, including quantitative data on interactions between the regulating molecular species and DNA sequences.[15–17] The model so obtained[18] and its properties are summarized next.

A schematic illustration of the important elements in regulation of replication of the λdv plasmid is provided in FIGURE 6. Important genes and regulatory sites in the neighborhood of the plasmid origin of replication are shown. At the outset, it should be emphasized that, like many other macromolecular synthesis processes in the *E. coli* cell, the kinetics of plasmid synthesis depend primarily upon the frequency of initiation of plasmid synthesis, because the replication process once initiated proceeds at a rate that is relatively independent of intracellular condition or growth environment. Thus it is appropriate to focus attention on the processes that determine when replication of a plasmid begins.

The dominant regulatory element in this origin of replication is the autorepressor system. A repressor protein R is synthesized by expression of the gene *cro* initiating at the promotor P_R. The R protein regulates transcription of the P_R promoter by binding to three operator sequences designated O_R1, O_R2, and O_R3. Thus, the repressor protein regulates its own synthesis. Approximately 80% of the transcription events initiated at P_R terminate at t_R1 in the wild-type plasmid. The remaining 20% of such transcriptions continue through the genes labeled O and P and through the origin of replication region

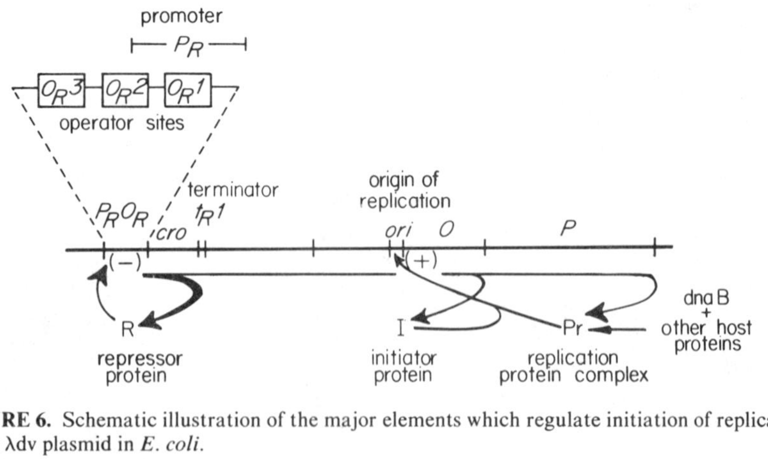

FIGURE 6. Schematic illustration of the major elements which regulate initiation of replication of the λdv plasmid in *E. coli*.

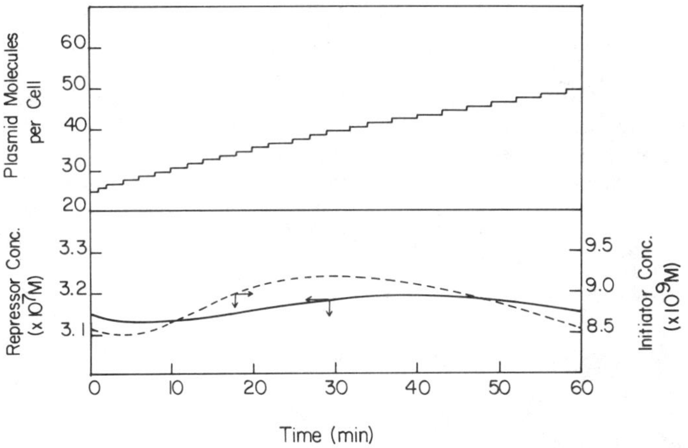

FIGURE 7. Simulated trajectories of number of λdv plasmid molecules (top), repressor (*R*) concentration (bottom solid) and initiator (*I*) concentration (bottom dashed) in an *E. coli* cell with a doubling time of one hour (adapted from Lee and Bailey[18]).

ori contained in the *O* gene. Transcription of the *ori* region activates this region in this model, allowing interaction of the activated origin sequence with an initiator protein I (the *O*-gene product), and a complex of host cell proteins Pr and the protein encoded in the plasmid gene *P*. Replication of the plasmid is presumed to initiate when the initiation complex of activated origin and activation proteins I and Pr reaches a critical value, at which point the amount of activated origin is reset to zero. The λdv genome size is so small relative to the length of the bacterial chromosome that, assuming similar orders of magnitude for plasmid and chromosomal strand synthesis, a new plasmid copy is synthesized essentially instantaneously (relative to the cell-cycle time) once replication initiates.

It is very important in developing the mathematical relations describing this system and its function in a growing *E. coli* cell to recognize that substantial experimental evidence exists for plasmid replication by random selection, which implies that only one plasmid at a time is participating in the origin activation process and in subsequent replication.[17] Also, it is important to write the equations in terms of intrinsic cell composition variables so that dilution by cell growth is properly represented.[19] With these points in mind, it is not difficult to develop the equations assuming that binding of repressor to the three O_R regions is at equilibrium, the binding of RNA polymerase to the P_R promoter is at equilibrium, and that gene expression proceeds by sequential steps of transcription and translation that are described by ordinary differential equations. Interactions of activating proteins with each other and with the *ori* region are all presumed to be described by equilibrium relationships.

The resulting model contains 12 parameters, nine of which may be estimated based upon information provided in fifteen different papers in the biological sciences literature. The calculated results are relatively insensitive to two of the unknown parameters. The most important unknown parameter, the critical concentration of activated origin-initiation protein complex at which initiation begins, has been chosen to provide agreement between the average number of plasmid molecules per cell calculated from the model and the experimentally determined value.

FIGURE 7 shows the calculated time trajectories of the number of plasmid molecules per cell and the concentrations of repressor (R) and initiator (I) proteins.[18]

The plasmid number increases in a stepwise fashion because of the combined effects of the random selection replication mechanism and the very rapid synthesis process that follows initiation of replication. While few measurements of this type have been made, data for other plasmids suggest a similar stepwise trajectory.[20] The average repressor concentration in the cell also matches the experimentally measured value for the wild-type plasmid. It is interesting to note that neither the repressor nor the initiator concentrations fluctuate substantially in these calculations. This suggests that the activator accumulation[21] and inhibitor dilution[22] models proposed for regulation of initiation of chromosomal DNA replication in *E. coli* may not be valid for plasmid replication.

Another interesting application of this model of λdv plasmid replication is study of the effects of irregular plasmid segregation. Simulations have been conducted in which

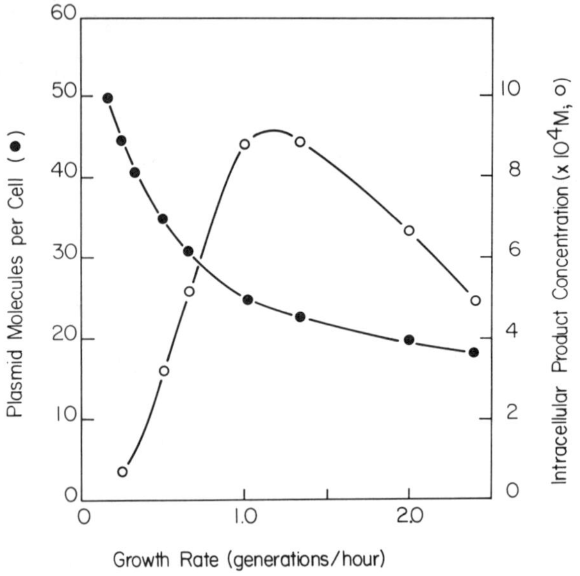

FIGURE 8. Influence of *E. coli* growth rate on plasmid and protein product content of the cells (calculations based on λdv plasmid replication mathematical model[24]).

perturbations from regular segregation are introduced, allocating to daughter cells at division perturbed repressor and plasmid concentration values. It has been found in these calculations that the λdv replication control system compensates for such perturbations by accelerating or decelerating the frequency of plasmid replication initiation in order to restore the normal wild-type cycle in three to five generations. This feature is important when calculating propagation of plasmids in cell populations. Work on a detailed mathematical description of regulation of another plasmid origin of replication is currently in progress to explore the generality of this result.

The λdv replication model has been extended to simulate the effects of mutations in the regulatory sequences.[23] Measured influences of altered nucleotide sequences in the O_R and t_Rl regions on average plasmid and repressor levels and on transcription efficiency initiated at P_R are represented well and consistently by the model, after appropriate modification of repressor binding and transcription termination efficiency parameters. Also, the model simulates properly the influence of plasmid oligomerization on plasmid copy number.

According to the above model, regulation of replication of this plasmid is closely coupled with the transcription rate. The latter quantity is known to be growth-rate dependent in *E. coli,* as is the cell size at division. Taking these factors into consideration, the model has been extended to calculate the effects of the growth rate on cellular plasmid content.[24] Also, a model of expression of a plasmid gene involving sequential transcription and translation steps, with degradation of message and protein presumed to follow first-order kinetics, has been added to the plasmid replication model to allow calculation of the effect of growth rate on the cellular content of a cloned gene.

FIGURE 8 illustrates some of the results of this research. The calculated decline in number of plasmid molecules per cell as growth rate increases is consistent with experimental data.[25] It is very interesting to note that the calculated protein product content of *E. coli* exhibits a maximum with respect to growth rate. Experimental data on this relationship are very limited but appear to be consistent with the general trend illustrated in FIGURE 8. This result exemplifies the opportunity of calculating important trends in fermentation processes employing recombinant microorganisms based on first principles of molecular biology and extended by mathematical modeling to the reactor level.

A MOLECULAR-LEVEL MODEL FOR REGULATION OF CLONED-GENE EXPRESSION

In the calculations just described, it was assumed that expression of the cloned gene is constitutive. However, in many applications, it may be desirable to construct the expression plasmid with a controlled promoter for the cloned production gene so that gene expression can be adjusted to a desired level by manipulation of the cell environment. Such capability is critical, for example, if the product protein is toxic to the host cell.

Several different regulated promoters are available, and, among these, the *lac* promoter-operator is the best characterized and one of the most widely used. As illustrated schematically in FIGURE 9, the regulation of initiation of transcription at the *lac* promoter-operator is quite complicated. The efficiency of transcription originating at this promoter depends upon the interaction of two different molecular species, RNA polymerase (RNP) and *lac* repressor (R), and one bimolecular complex, cyclic-AMP receptor protein (CRP) and cyclic AMP (cAMP), with three different binding regions in the promoter-operator. (For consistency with notation in related papers, the same symbol R has been used here for *lac* repressor as was used in the previous discussion of the *cro* repressor involved in λdv plasmid replication regulation; these are *not* the same molecular species.)

As shown in FIGURE 9, the intracellular concentrations of each of the regulatory species are affected by their interactions not only with the binding sequences in the *lac* promoter-operator but also with other molecular entities in the cell, namely the σ-factor for activation of the RNA polymerase apoenzyme, the inducer that binds to *lac* repressor, and the concentrations of cAMP and CRP species in the cell that determine the amount of cAMP-CRP complex level. Also, especially important in considering cloned-gene expression under regulation of this promoter, one must recognize that the concentration of promoter-operators and thus of the three types of binding sites depends upon the number of plasmids in the host cell.

Space does not permit complete description of the mathematical treatment of this complex regulatory system. Applying methods of statistical thermodynamics to describe the intermolecular interactions just summarized and carefully searching the molecular biology literature for the equilibrium constants and intracellular levels of the different participating molecular species, an algebraic equation can be derived for the overall efficiency of transcription initiating at the *lac* promoter-operator.[26] Several previous models of this control system, present as a single copy in the bacterial chromosome, have been presented.[27-31] The new model mentioned here is more complete, involves more extensive use of experimentally determined parameters, and has been extended to describe operation of the promoter-operator in multicopy plasmids.

This new model, which accounts for all the interactions described above and which can represent the effects of genetic alterations in the host cell or in the promoter-operator sequences with respect to any part of the regulatory system just described, contains 26 constants. Because of the extensive previous research on the system, however, values for 25 of these model parameters are available, at least in an order of magnitude sense, from 30 different papers in the molecular biology literature.

Most experimental studies of the function of this system have measured expression of the *lacZ* gene under control of the *lac* promoter-operator. Assuming that transcription is the rate-limiting step in gene expression,[32] the transcription efficiency of the *lac* promoter-operator is proportional to the rate of synthesis of the *lacZ* gene product, the enzyme β-galactosidase. Thus, the single remaining model parameter has been chosen to provide agreement between the model results and measured β-galactosidase levels for the induced and uninduced wild-type promoter-operator functioning in the chromosome of *E. coli*.

The model was formulated including all of the complexities mentioned above in order to allow greatest opportunity for comparison with experimental data. Many of

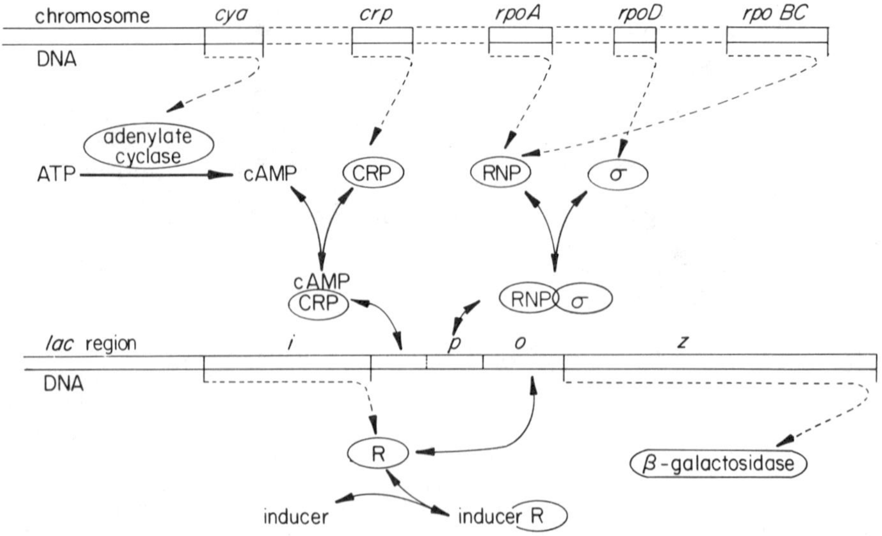

FIGURE 9. Schematic illustration of the molecular controls of transcription in the *lac* promoter-operator. Proteins are circled, gene expression is indicated by dashed lines, and equilibria are shown with heavy lines. Abbreviations are defined in the text.

TABLE 1. Calculated and Experimentally Measured Values of *lacZ* Chromosomal Gene Expression under Control of the *lac* Promoter-Operator for Wild-Type *E. coli* and for Three Mutants (i^-, i^q, i^{sq}) Producing Different Intracellular Levels of *lac* Repressor R($[R]_0 = 2 \times 10^{-8}$ M[a])

Genotype	Repressor Concentration	Inducer	β-galactosidase Activity[b] Calculated	Experimental	Reference
i^-	0	−	113	100–140	33
i^-	0	+	113	100–130	33
i^+	$[R]_0$	−	0.1	<0.1	33
i^+	$[R]_0$	+	100	100	33
i^q	$10[R]_0$	−	0.011	0.01–0.014	34
i^q	$10[R]_0$	+	62	65	35
i^{sq}	$50[R]_0$	−	0.002	0.003–0.004	34
i^{sq}	$50[R]_0$	+	23	25	35

[a]Adapted from Lee and Bailey.[26]
[b]Normalized to 100% for induced wild-type (i^+).

the data available on operation of this promoter-operator have been obtained using mutants that have different levels of repressor, different binding-site affinities for regulatory species, and other alterations relative to the normal wild-type system. Model-experiment comparisons for one class of *E. coli* mutants will be highlighted here.

The model has been used to calculate the effect of changed repressor level on induced and uninduced transcription rates.[26] Corresponding data are available for *E. coli* mutant strains i^-, i^q, and i^{sq}. Since the intracellular repressor concentrations have been measured for these mutants, these measured values may be used in the model directly, and none of the remaining model parameters are changed. As shown in TABLE 1, the model successfully represents the effects of changes in repressor content of the cells under inducing and noninducing cultivation environments.

Once the strategy of describing the function of a complex system of molecules and regulatory binding regions at the molecular level has been adopted, the complexity of the resulting model and the number of parameters required is directly related to the scope of genetic variation that one wishes to describe. That is, if the model must account for mutations that change the intracellular level of, say, σ-factor, one must consider the σ-factor-RNA polymerase apoenzyme equilibrium and include that equilibrium constant in the model. On the other hand, if the planned application of the model spans a narrow range of genetic variation, say the alteration in intracellular repressor protein concentration by mutation, the model is greatly simplified and far fewer parameters are required. Considering the latter example, for instance, the algebraic expression for the transcription efficiency η, which describes both the effect of inducer and of total repressor concentration in the host cell, is

$$\eta = \frac{1}{Q} \frac{1}{1 + K[R]_0} \tag{6}$$

where Q is a scaling constant and K is the following function of inducer concentration:

$$K = \frac{K_A + K_1 [I]}{1 + K_2 + K_3 [I]} \tag{7}$$

The parameter values in Equation 7 are given by $K_A = 2 \times 10^{12}\ M^{-1}$, $K_1 = 2 \times 10^{16}\ M^{-2}$, $K_2 = 40$, and $K_3 = 6 \times 10^9 M^{-1}$.

The *lac* promoter-operator model has been applied to calculate transcription rates of cloned genes under control of this promoter in multicopy plasmids in *E. coli*.[36] The first cases considered corresponded to F plasmids containing different promoter-operator and *lacZ* sequences that had been studied experimentally. The major question at this juncture of the modeling effort was the validity of extending the model from description of transcription of a single copy gene in the chromosomal DNA to multiple gene copies in plasmids. The model produced results uniformly consistent qualitatively and quantitatively with the available experimental results. With this foundation established, the model has been applied to calculate cases, less thoroughly characterized experimentally, in which the *lac* promoter-operator is used to regulate gene expression in multicopy systems.

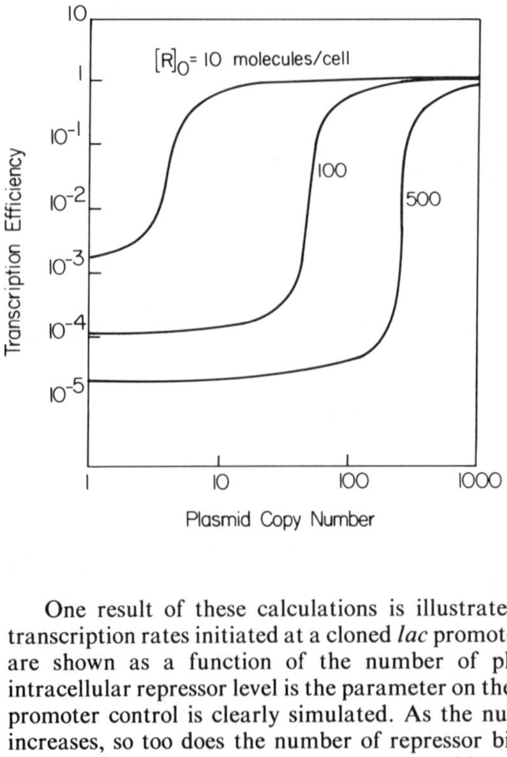

FIGURE 10. Calculated dependence of cloned *lac* promoter-operator transcription efficiency on plasmid copy number and host cell repressor (R) concentration in the absence of inducer. Transcription efficiency is here normalized by the fully induced state (=1.0 on this scale). The increasing repressor levels shown correspond to i^+, i^q, and i^{sq} strains, respectively (adapted from Lee and Bailey[36]).

One result of these calculations is illustrated in FIGURE 10. Here the relative transcription rates initiated at a cloned *lac* promoter-operator in the absence of inducer are shown as a function of the number of plasmid molecules in the cell. The intracellular repressor level is the parameter on the different curves shown. Here loss of promoter control is clearly simulated. As the number of plasmid molecules per cell increases, so too does the number of repressor binding sites available, such that the total available amount of repressor is titrated by the available binding sites. Eventually, when the repressor binding-site number approaches and exceeds the total number of repressor molecules in the cell, control of plasmid gene expression is lost. By changing to a different host that produces a higher basal level of repressor, this loss of transcription control is postponed until higher copy numbers are reached. The result is entirely consistent qualitatively with early experience with expression of cloned genes,[37-39] and further illustrates the scope of molecular design features that can be addressed quantitatively and specifically using mathematical models of the molecular control system. Future developments in such models should lead to systematic strategies for molecular design to maximize reactor productivity.

DISCUSSION

The paradigm

molecular level control systems
↓
single-cell kinetics
↓
microbial population dynamics
↓
reactor productivity

described in this paper is a powerful approach, especially for protein products synthesized using recombinant plasmid host-cell systems. Alternatively, as mentioned above, the single-cell model can be used itself as a model for overall kinetics of the culture. The latter approach may be required in situations where the single-cell description is relatively complicated because of the inclusion of a variety of genetic and environmental influences.

The presentation above has summarized research in several areas that contribute to different components of the overall paradigm. An important feature of this approach is the systematic connection of reactor behavior to the molecular systems that, in conjunction with reactor operating and design strategies, determine process productivity. It is hoped that this structure will provide a continuum for synergistic interaction between molecular biologists, microbiologists, and biochemical engineers to accelerate the optimization of genetically engineered production systems and to maximize their flexibility. While significant advances have been made in several areas, much more research remains to be done to complete this general and powerful strategy for optimizing productivity in recombinant systems.

SUMMARY

Experimental and mathematical analysis of productivity of cultures containing recombinant plasmids is based in this work on the following paradigm: molecular controls → single cell kinetics → cell population dynamics → reactor productivity. Mathematical models have been developed for replication control of the λdv plasmid and for efficiency of the *lac* promoter-operator based on the molecular control mechanisms of these systems in *Escherichia coli*. A special and important attribute of these models is their ability to describe quantitatively a wide range of genetic effects as well as environmental influences on molecular control function. Equations describing plasmid maintenance and distribution in growing cell cultures have been determined based on the population-balance equations applicable to a segregated culture model. The washout dilution rate for continuous cultivation of plasmid-containing *Saccharomyces cerevisiae* in selective medium is given in terms of single-cell division-cycle parameters and plasmid copy number for one single-cell model of plasmid replication and segregation. A new experimental method based on flow cytometry for rapid characterization of heterogeneity of single-cell accumulation of a plasmid gene product is also described. Generalization of these methods and of the overall strategy should provide a useful framework for synthesis of biological and engineering principles and methods to optimize organisms and processes based on recombinant DNA technology.

ACKNOWLEDGMENTS

The authors gratefully acknowledge the support of the National Science Foundation, the Energy Conversion and Utilization Technology (ECUT) Program of the United States Department of Energy, the Korea Science and Engineering Foundation (KOSEF), and the Federal Economic Chamber of Austria. Judith L. Campbell and Frank A. Dolbeare provided important assistance to this research.

REFERENCES

1. SRIENC, F., J. L. CAMPBELL & J. E. BAILEY. 1983. Detection of bacterial β-galactosidase activity in individual *Saccharomyces cerevisiae* cells by flow cytometry. Biotechnol. Lett. **5**:43–48.
2. ROSE, M., M. J. CASADABAN & D. BOTSTEIN. 1981. Yeast genes fused to β-galactosidase in *Escherichia coli* can be expressed normally in yeast. Proc. Natl. Acad. Sci. USA **78**(4): 2460–2464.
3. IMANAKA, T. & S. AIBA. 1981. A perspective on the application of genetic engineering: Stability of recombinant plasmid. Ann. N.Y. Acad. Sci. 369: 1–14.
4. OLLIS, D. F. 1982. Industrial fermentations with (unstable) recombinant cultures. Philos. Trans. R. Soc. London B **297:** 617–629.
5. OLLIS, D. F. & H. T. CHANG. 1982. Batch fermentation kinetics with (unstable) recombinant cultures. Biotechnol. Bioeng. **24:** 2583–2586.
6. BAILEY, J. E. 1983. Single-cell metabolic model determination by analysis of microbial populations. *In* Foundations of Biochemical Engineering: Kinetics and Thermodynamics in Biological Systems. E. T. Papoutsakis, H. W. Blanch & G. N. Stephanopoulos, Eds. ACS Symposium Series. **207**:135–157.
7. HJORTSO, M. A. & J. E. BAILEY. 1982. Steady-state growth of budding yeast populations in well-mixed, continuous-flow microbial reactors. Math. Biosci. **60:** 235–263.
8. HJORTSO, M. A. & J. E. BAILEY. 1983. Transient responses of budding yeast populations. Math. Biosci. **63**:121–148.
9. ALBERGHINA, L., E. MARTEGANI & L. MARIANI. 1982. Analysis of protein distribution in populations of budding yeast based on a structured model of cell growth. IFAC Workshop on Modeling and Control of Biotech. Proc. August, 1982. Helsinki.
10. ALBERGHINA, L., L. MARIANI, E. MARTEGANI & M. VANONI. 1983. Analysis of protein distribution in budding yeast. Biotechnol. Bioeng. In press.
11. HJORTSO, M. A. & J. E. BAILEY. 1983. Plasmid stability in budding yeast populations: Steady-state growth with selection pressure. Biotechnol. Bioeng. In preparation.
12. HARTWELL, L. H. & M. W. UNGER. 1977. Unequal division in *Saccharomyces cerevisiae* and its implications for the control of cell division. J. Cell Biol. **75:** 422–435.
13. LORD, P. G. & A. E. WHEALS. 1980. Asymmetrical division of *Saccharomyces cerevisiae*. J. Bacteriol. **142:** 808–818.
14. SHULER, M. L., S. LEUNG & C. C. DICK. 1979. A mathematical model for the growth of a single bacterial cell. Ann. N.Y. Acad. Sci. **326:** 35–55.
15. BERG, D. E. 1974. Genes from phage λ essential for λdv plasmids. Virology **62:** 224–233.
16. MATSUBARA, K. 1976. Genetic structure of a replicon of plasmid λdv. J. Mol. Biol. **102:** 427–429.
17. MATSUBARA, K. 1981. Replication control system in lambda dv. Plasmid **5:** 32–52.
18. LEE, S. B. & J. E. BAILEY. 1983. A mathematical model for λdv plasmid replication: Analysis of wild-type plasmid. Plasmid. Submitted.

19. FREDRICKSON, A. G. 1976. Formulation of structured growth models. Biotechnol. Bioeng. **18:** 1481–1486.
20. GUSTAFSSON, P., K. NORDSTRÖM & J. W. PERRAM. 1978. Selection and timing of replication of plasmids R1 *drd-19* and F'*lac* in *Escherichia coli*. Plasmid **1:** 187–203.
21. JACOB, F., S. BRENNER & F. CUZIN. 1963. On the regulation of DNA replication in bacteria. Cold Spring Harbor Symp. Quant. Biol. **28:** 329–348.
22. PRITCHARD, R. H., P. T. BARTH & J. COLLINS. 1969. Control of DNA synthesis in bacteria. Symp. Soc. Gen. Microbiol. **19:** 263–297.
23. LEE, S. B. & J. E. BAILEY. 1983. A mathematical model for λdv plasmid replication. Analysis of copy number mutants. Plasmid. Submitted.
24. LEE, S. B. & J. E. BAILEY. 1983. Analysis of growth rate effects on productivity of recombinant *Escherichia coli* populations using molecular mechanism models. Biotechnol. Bioeng. In preparation.
25. ENGBERG, B. & K. NORDSTRÖM. 1975. Replication of R-Factor R1 in *Escherichia coli* K-12 at different growth rates. J. Bacteriol. **25:** 179–186.
26. LEE, S. B. & J. E. BAILEY. 1983. Effects of genetic variations on expression of the *lac* operon. Biotechnol. Bioeng. In preparation.
27. VAN DEDEM, G. & M. MOO-YOUNG. 1973. Cell growth and extracellular enzyme synthesis in fermentation. Biotechnol. Bioeng. **15:** 419–439.
28. TODA, K. 1976. Dual control of invertase biosynthesis in chemostat culture. Biotechnol. Bioeng. **18:** 1117–1124.
29. IMANAKA T. & S. AIBA. 1977. A kinetic model of catabolite repression in the dual control mechanism in microorganisms. Biotechnol. Bioeng. **19:** 757–764.
30. GONDO, S., K. VENKATASUBRAMANIAN, W. R. VIETH & A. CONSTANTINIDES. 1978. Modeling the role of cyclic AMP in catabolite repression of inducible enzyme biosynthesis in microbial cells. Biotechnol. Bioeng. **20:** 1797–1815.
31. ROELS, J. A. 1978. Regulatory mechanisms and the modeling of fermentation processes. *In* "Biotechnology" Proc. 1st Eur. Congr. Biotechnol. Dechema. pp. 221–249.
32. MAALØE, O. 1979. Regulation of the protein-synthesizing machinery—ribosomes, tRNA, factors, and so on. *In* Biological Regulation and Development. Vol. 1. Gene Expression. R. F. Goldberger, Ed. Plenum Press. New York. pp. 487–542.
33. JACOB, F. & J. MONOD. 1961. Genetic regulatory mechanisms in the synthesis of proteins. J. Mol. Biol. **3:** 318–356.
34. JOBE, A., J. R. SADLER & S. BOURGEOIS. 1974. *Lac* repressor-operator interaction I.X. The binding of *lac* repressor to operators containing O^c mutations. J. Mol. Biol. **85:** 231–248.
35. GILBERT, W. & R. MÜLLER-HILL. 1970. The lactose repressor. *In* The Lactose Operon. J. R. Beckwith & D. Zipster, Eds. Cold Spring Harbor Laboratory. pp. 93–110.
36. LEE, S. B. & J. E. BAILEY. A model for cloned gene product formation under control of the *lac* operon in recombinant *Escherichia coli*. Biotechnol. Bioeng. In preparation.
37. BACKMAN, K., M. PTASHNE & W. GILBERT. 1976. Construction of plasmids carrying the *c*I gene of bacteriophage λ. Proc. Natl. Acad. Sci. USA **73** (11): 4174–4178.
38. O'FARRELL, P. H., B. POLISKY & D. H. GELFAND. 1978. Regulated expression by readthrough translation from a plasmid-encoded β-galactosidase. J. Bacteriol. **134:** 645–654.
39. SADLER, J. R., M. TECKLENBURG & J. L. BETZ. 1980. Plasmids containing many tandem copies of a synthetic lactose operator. Gene **8:** 279–300.

Production of Human Interferon in *E. coli* Under *lac* and *tryplac* Promoter Control

YUTI CHERNAJOVSKY, YVES MORY, BARUCH VAKS,
SHELDON I. FEINSTEIN, DAVID SEGEV,
AND MICHEL REVEL

Department of Virology
Weizmann Institute of Sciences
Rehovoth, Israel

INTRODUCTION

Many bacterial promoters are characterized by two homology sequences: the Pribnow TATAAT box at position -10 before the mRNA start, and a TTGACA sequence at position -35.[1,2] These sequences lie in the DNA segments with which the *E. coli* RNA polymerase molecule binds, and probably represent areas of contact between the DNA chain and the enzyme.[1] Not all bacterial promoters have these exact consensus sequences. For example, the *lac uv5* mutant promoter of *E. coli* has the TATAAT Pribnow box sequence but an altered -35 consensus sequence (TTTACA), while the *trp* promoter lacks the consensus Pribnow box sequence. By fusion of the distal part of the *trp* promoter to the proximal part of the *lac uv5* promoter at position -20, it is possible to reconsitute a hybrid promoter that contains both consensus sequences, and to investigate the efficiency of this hybrid promoter. Another feature of the hybrid tryplac (TL) promoter is that it lacks the binding site for the catabolite activator protein CAP,[3] but retains the binding site for the *lac* repressor (operator). The TL promoter should, therefore, remain inducible by *lac* (but not *trp*) inducers and be insensitive to catabolite repression. In this work, we have compared the efficiency at which two human interferon (IFN) genes are expressed under *lac uv5* and *tryplac* promoter control. De Boer *et al.*[4] have carried out a similar comparison for the human growth hormone gene.

Human genomic fragments containing either the coding sequence of mature IFN-β1 (the major fibroblast interferon),[5,6] or of mature IFN-αc (one of the numerous leucocyte interferon subspecies),[7,8] were introduced after the ribosomal binding site of the *lac uv5* promoter in a pBR322 vector. From these *lac*-IFN plasmids, we could construct the hybrid *tryplac* promoter-containing plasmids (TL-IFN) without changing the structure of the ribosomal binding site. We could, therefore, compare the efficiency of the *tryplac* promoter to that of the *lac uv5* promoter by measuring the amount of human IFN activity produced in bacterial cultures under various growth conditions. We found that the hybrid TL promoter gives a 10-fold higher expression than the *lac uv5* promoter, and, moreover, that the TL promoter is less sensitive to catabolite repression. Thus, in rich medium such as L-broth, the TL-IFN plasmid gives over 25-fold more IFN than the *lac uv5*-IFN vector.

RESULTS AND DISCUSSION

Construction of lac- and tryplac-IFN Recombinant DNAs

Lac uv5-IFN β1 Recombinant Plasmids

A tandem repeat of two *lac uv5* promoter regions of 95 bp each, separated by 95 bp of unrelated DNA, was excised from plasmid pLJ3[9] and introduced in the *Eco* R1 site of pBR322, oriented toward the *Bam* H1 site of the plasmid. The fact that *E. coli* RNA polymerase binds to the *lac* promoter,[3] was used to protect the *Eco* R1 site proximal to the *Bam* H1 site (FIGURE 1), while the distal *Eco* R1 site was filled in and thus converted to *Xmn* 1 (*Lac-Bam* vector, FIG. 1). A fragment of the 1.84 Kb human genomic DNA clone IFC 631 (FIG. 1) containing the IFN-β1 gene,[6] was excised with *Hinc* 2 and *Bgl* 2 to isolate the complete preinterferon coding sequence.[5] The mature IFN-β1 protein has an aminoterminal methionine,[5] whose ATG codon can be used to initiate translation. Among the 4 *Alu* 1 sites in the IFN-β 1 coding sequence, the closest to *Hinc* 2 is 8 nucleotides before this ATG. A partial *Alu* 1 digestion was used to cut only at this site and the fragment obtained inserted in the *Lac-Bam* vector (FIG. 1). The *Alu* 1 site of IFN-β1 was ligated to the filled-in *Eco* R1 site restoring the *Eco* R1 sequence, and the *Bgl* 2 site was ligated to the *Bam* H1 site of the vector, leaving a *Sau3a* 1 site. After verifying the sequence, the *Eco* R1 site was cut and *Bal* 31 nuclease digestion was used to shorten the distance between the Shine-Dalgarno sequence and the ATG codon (see FIG. 2). *E. coli* transformed by these plasmids were screened for IFN-β1 production and clone B-11 was thereby selected (FIG. 1).

A vector containing 3 *lac uv5* promoters in tandem (plasmid 11-C) was derived from B-11 as shown in FIGURE 3. This process also reinserted the *Hind* 3–*Bam* H1 segment of pBR322 and restored the tetracycline resistance gene. When the 11-C plasmid was used to transform *E. coli* JM101 (a *lac* i^q strain), growth in medium containing 5 μg/ml tetracycline was dependent on the induction of *lac* transcription by IPTG (not shown). In plasmid 11-C, the *lac uv5* promoter, therefore, controls not only the expression of the IFN-β1 gene (see below) but also that of the tetracycline resistance gene that follows it.

Tryplac-IFN β1 Recombinant Plasmid

To construct the hybrid *tryplac* (TL) promoter, the promoter of the *trp* operon[10] was cut out of the *trp* plasmid pEP121-221[11] shown in FIGURE 4. A *Taq* 1–*Taq* 1 fragment of 180 bp from this promoter region was cloned in the *Cla* 1 site of pBR322 and the clockwise orientation was selected (plasmid 121-1, FIG. 4). This procedure restores a *Cla* 1 site on the 3′ end of the *trp* promoter fragment. From the *lac*-IFN β1 recombinant 11-C plasmid, the IFN-β1 coding sequence linked to 55 bp of the *lac uv5* promoter, was excised by *Hpa* 2–*Sau3a* 1 cleavage (FIG. 4), and this fragment was inserted between the *Cla* 1 and *Bam* H1 sites of 121-1 to give plasmid TL-11. The relevant sequences of the *lac uv5*-IFN β1 (11-C) and *tryplac*-IFN β1 (TL11) constructions are shown in FIGURE 2. It should be noted that the distance between the -35 and -10 sequences is 18 bp in lac, but 16 bp in the TL promoter.

Lac and tryplac-IFN αc Recombinant Plasmids

A similar procedure was used for the human IFN-αc gene, which codes for one of the leucocyte IFN subspecies.[7,8] This gene was isolated from lambda Charon 4A

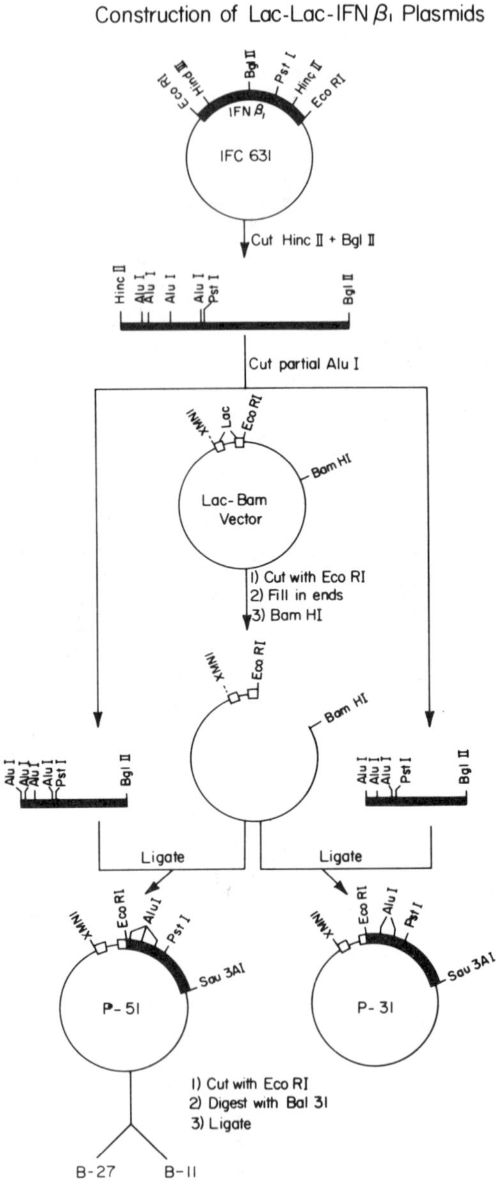

FIGURE 1. *Lac uv5*-IFN β1 recombinant plasmids. Insertion of the human IFN-β1 gene for direct expression in the *Lac-Bam* vector. Details are described in the text. Only clone B-11 was used in this work and the sequence of the *lac*-IFN junction is shown in FIGURE 2. Black areas: IFN-β1 gene sequences. Open boxes: *lac* promoter regions.

genomic clones 18-3 and 5-1 originating from our human adult gene library,[6] and which contains three identifiable IFN-α gene sequences (FIG. 5). One of these, IFN-αc* was found to contain the exact sequence of the cDNA-clone IFN-αc of Goeddel *et al.*[7] From the genomic clone, a 2-Kb *Eco* R1 fragment containing the IFN-αc gene was excised and recloned in pBR322 (clone 18-33). Mature IFN-αc has an aminoterminal cysteine TGT codon,[7] which is adjacent to the first *Sau3a* 1 of the gene sequence (FIG. 5). After subcloning a 600-pb *Alu* 1–*Fnud* 2 fragment, through *Hind* 3 linkers, we isolated by partial *Sau3a* 1 digestion the proper *Sau3a* 1–*Hind* 3 fragment, which was then cloned between the *Bgl* 2 and *Hind* 3 site of a pSV vector (FIG. 5) restoring a *Bgl* 2 site at the vector-IFN junction. This site was next ligated to a chemically synthesized DNA duplex adaptor containing the methionine and cysteine codons (ATGTGT). Through this adaptor the 5′ end of the IFN-αc sequence was ligated to the *Eco* R1 site of the *Lac-Bam* vector of FIGURE 1, while the *Hind* 3 end was ligated to the *Hind* 3 site of this vector (*lac uv5*-IFNαc, FIG. 5). To construct the *tryplac* derivative, this DNA was cut with *Hpa* 2, and the IFN-αc-containing fragment introduced in the *Cla* 1 site of the *trp* 121-1 vector (FIGS. 4 and 5). The sequence of the TL-IFNαc construction obtained is given in FIGURE 2.

Expression of the IFN Genes under lac and tryplac Promoter Control

Level of Expression

We compared the amount of human interferon activity that could be recovered from bacterial cultures harboring the *lac uv5*-IFN β1 plasmid (11-C with 3 *lac* promoters in tandem as shown in FIG. 4) or with the hybrid *tryplac*-IFN β1 plasmid (TL-11 of FIG. 4). The *E. coli* strains MM294 or minicell-producing P679-54[12] were transformed by these plasmids using ampicillin resistance and selection of blue colonies on X-gal (5 bromo-4 chloro-3 indolyl-B-D-galactoside) plates. The bacteria were grown in M9 medium with 0.5% glucose, 0.5–2% casamino acids and 100 μg/ml ampicillin, in shaker flasks or in a New Brunswick MultiGen 1-liter fermenter. When

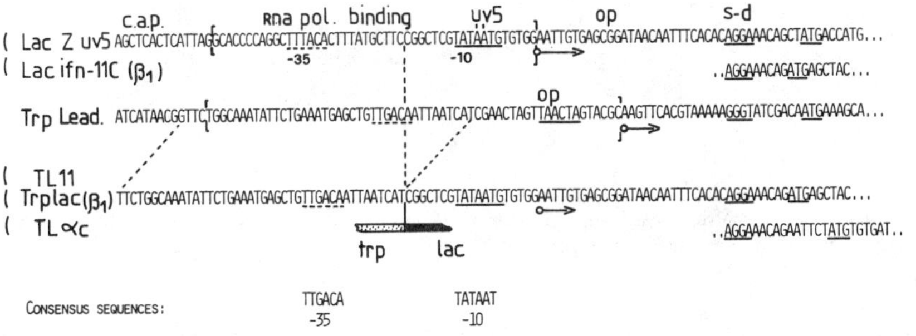

FIGURE 2. Nucleotide sequence of the promoter and ribosomal binding sites of the *lac* and *tryplac*-IFN recombinants. Line 1: *Lac uv5* promoter sequence.[3] CAP is the binding site of the catabolite-activated repressor protein; RNA polymerase binding site is shown between the brackets. The -35 and -10 sequences are underlined; *uv5* designates the 2 nucleotides that differ from wild type in *lac uv5*. The circled arrow shows the *lac* mRNA start site, op is the *lac* repressor binding site, and s-d is the Shine and Dalgarno signal (underlined). The ATG codon of the β-galactosidase Z cistron is underlined. Line 2: junction of the *lac* ribosomal binding site to the ATG of the mature IFN-B1 sequence. This sequence is the same for clones B-11 and 11-C. Line 3: sequence of the promoter and ribosomal binding site of the *E. coli* tryptophan operon.[10] Line 4: Hybrid *tryplac* (TL) promoter and ribosomal binding site in the case of IFN-β1 (TL-11). Line 5: same as line 4 but for the IFN-αc gene.

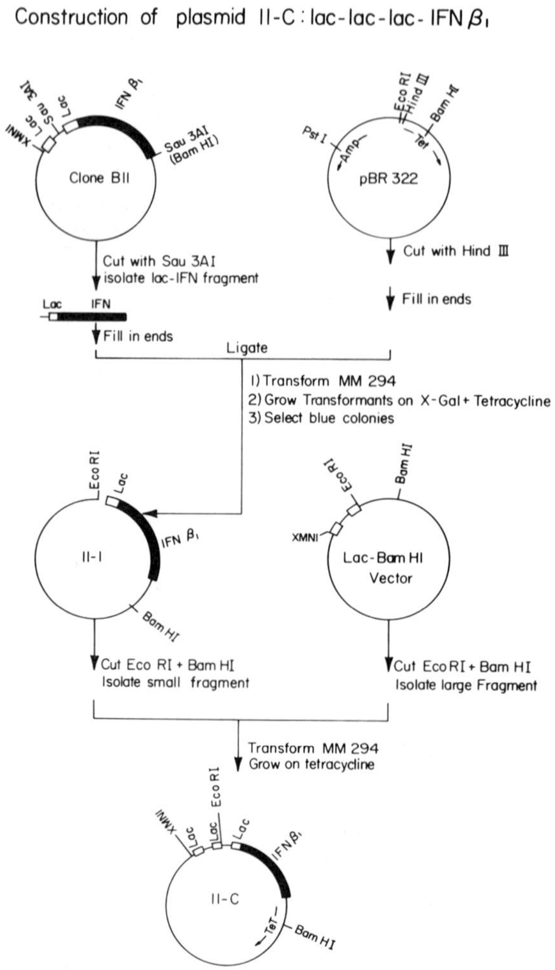

FIGURE 3. Construction of plasmid 11-C with 3 *lac uv5* promoters in tandem and a restored tetracycline resistance gene. For details see text.

the cells reached the early stationary phase, 1 ml of culture was centrifuged, the cells resuspended in 0.2 ml of 25% sucrose, 25 m*M* Tris-HCl, pH 7.5, 25 m*M* EDTA, 50 μg/ml phenylmethylsulfone fluoride (PMSF), 2 mg/ml lysozyme, overnight at 0°C. Then, 0.3 ml propylene glycol and 0.2 ml 5 *M* NaCl were added and dilutions of the extract were assayed for inhibition of the cytopathic effect of Vesicular Stomatitis Virus (VSV) on human diploid fibroblasts FS11, in comparison to a human IFN-β1 standard as described.[6] The bacterial IFN had all the properties expected from IFN-β1 (not shown). TABLE 1 indicates that the TL promoter gives 10 times more IFN activity than the *lac uv5* promoters. Higher levels of IFN could be obtained if two copies of the IFN gene with the TL promoter were present per plasmid, or if a *trp* transcription termination signal[13] was introduced in the *Sal* 1 site of the plasmid. When the *lac uv5* and the TL promoter were compared for the expression of the human IFN-αc gene, the TL-promoter was 20 times better than the *lac* promoter (TABLE 1).

In *E. coli lac* i^+ strains, the induction by IPTG was not essential, and the bulk of IFN accumulation took place during the early stationary phase (FIG. 6), with a smaller peak of IFN production during the exponential phase of growth. The escape from *lac* repression is probably a function of the plasmid copy number, that is, the number of copies of *lac* operator DNA. To measure *lac* derepression, we assayed the level of β-galactosidase. In a typical fermenter culture of *E. coli* MM294/TL11, the β-galactosidase activity (without IPTG induction) was about 25 units/OD of culture during the exponential phase and reached 45 units/OD during stationary phase. In the same experiment, the IFN activity was $1–2 \times 10^6$ units/OD/liter in exponential phase (at OD = 2) and reached 6×10^6 units/OD/liter in early stationary phase (at OD = 10). The increase in IFN levels per bacterium in stationary phase appears, therefore, larger than the increase in *lac* derepression.

When the *lac* repressor overproducing strain *E. coli* JM101 (Δ *lac-pro,* F'*lac* i^q, ZΔ M15, tra D) was used as host, the synthesis of human IFN with the TL-11 plasmid was IPTG inducible, showing that the TL promoter is still repressible provided enough *lac* repressor is present. In a typical experiment, when shaker flask cultures of JM101/TL11 reached the early stationary phase (OD = 2), the IFN activity produced (in units/liter) were 14×10^6 without IPTG and 56×10^6 with IPTG, while the MM294/TL11 strain gave approximately the same value with or without IPTG. The best time for adding the IPTG inducer to the JM101/TL11 cultures appears to be the

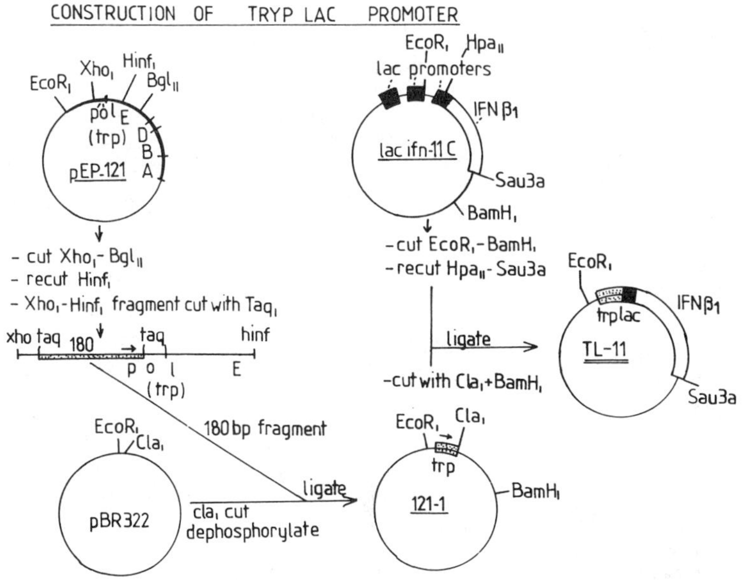

FIGURE 4. *Tryplac*-IFN β1 recombinant plasmid. Details of the construction of TL-11 are given in the text. Dotted box: *trp* segment of the promoter. Black areas: *lac* promoter fragment. Open box: coding sequence of mature IFN-β1.

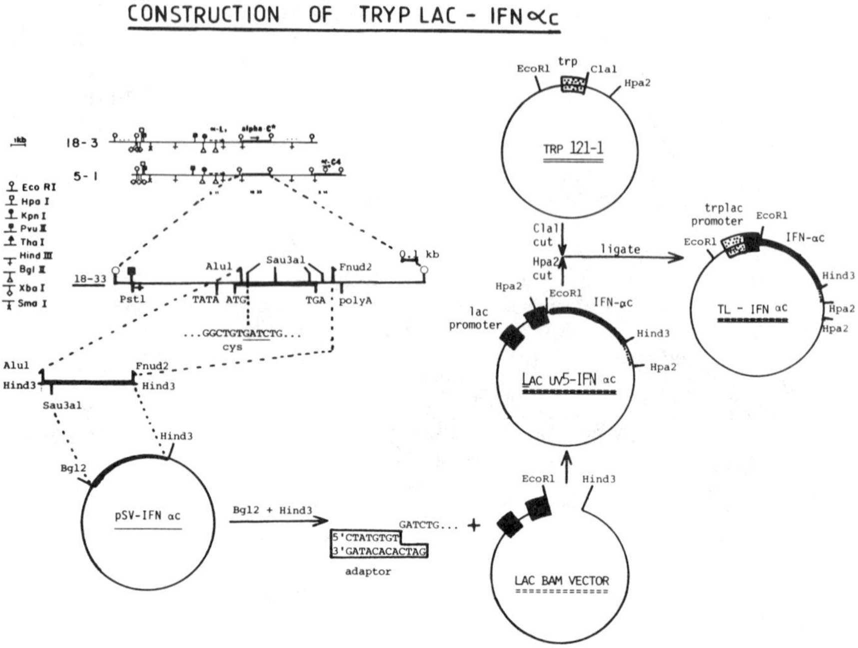

FIGURE 5. *Tryplac*-IFNαc recombinant plasmid. The IFN-αc was derived from the inserts of two overlapping human genomic clones in phage lambda: 18-3 and 5-1. These inserts represent roughly 16 Kb of genomic DNA and contain in addition to IFN-αc, another very similar gene IFN-αc4 and a more distant pseudogene IFN-αL2 (Feinstein *et al.*, in preparation). The map of the 2 Kb *Eco* R1 fragment 18-33 containing the IFN-αc coding sequence (heavy line) is shown enlarged. The *Sau3a* 1 site adjacent to the aminoterminal cysteine is indicated. The adaptor was chemically synthesized. Symbols are as in FIGURE 4. Details of the construction are given in the text.

TABLE 1. Expression of Human Interferon α and β Genes by *tryplac* Hybrid Promotor

Strain Characteristics	IFN Activity (Units/liter/O.D. bacterial culture)	
	IFN-β1	IFN-αc
3 *lac uv5* promoters	lac-11c: 0.5×10^6	Lac-αc: 0.4×10^6
1 *tryplac* hybrid promoter (TL)	TL-11: 5×10^6	TL-αc-647: 11×10^6
2 TL-IFN genes per plasmid	TL-218: 10^7	
TL promoter + trp terminator	TL-44: 2.4×10^7	

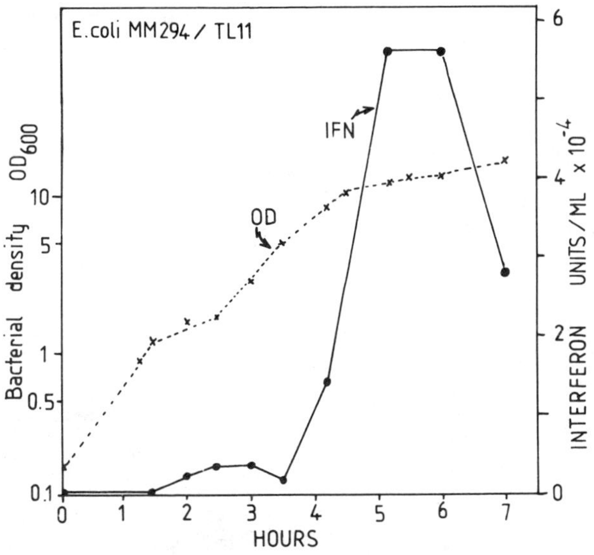

FIGURE 6. IFN-β1 production by *tryplac* plasmid TL-11. The 500-ml fermenter run was carried out without induction in M9 medium (× 2), 0.5% glucose, 2% casaminoacids, 100 μg/ml ampicillin. The IFN activity was assayed as described in the text. Dotted line: optical density at 600 mμ of the bacterial culture. Full line: IFN activity.

early stationary phase when the bulk of IFN production also occurs in the MM294 strain. With the TL-IFN plasmid we found no effect of *trp* operon inducers (data not shown).

Catabolite Repression

One of the features of the *lac* promoter is that it is sensitive to catabolite repression. Even the mutant *lac uv5* promoter is less efficiently expressed if the bacteria are grown in medium containing glucose. TABLE 2 shows that, compared to cultures in medium with glycerol (0.5 or 2%), the *lac uv5*-IFN β1 plasmid produced only 60% as much interferon in glucose-containing medium and 25% in L-Broth. In contrast, the hybrid *tryplac*–IFN β1 (TL-11) plasmid was expressed 1.6–1.8 times better in glucose or L-Broth than in glycerol medium. The difference between the TL and *lac uv5* promoters depended, therefore, on the medium used for culture: TL was 4 times better

TABLE 2. Effect of Bacterial Growth Medium on Interferon Production under the Control of *lac* or of the *tryplac* Hybrid Promoters[a]

			IFN activity, Units/liter/O.D. × 10^{-6}		
Strain	Promoter	Gene	Minimal Glycerol	Minimal Glucose	L-Broth
lac-11c	*lac uv5*	IFN-β_1	0.3	0.19	0.08
TL-11	*tryplac*	IFN-β_1	1.2	2.0	2.2

[a] For growth conditions, see text.

than *lac uv5* in glycerol medium, but over 10 times better in glucose and 27 times better in L-Broth. This feature of the TL promoter allows the bacterial host to grow in rich medium and to improve growth rate and interferon yields.

Why is the TL promoter less inhibited by catabolite repression than the *lac uv5* promoter? The *lac uv5* mutation, which changes the -10 sequence from TATGTT to the consensus TATAAT, probably alleviates catabolite repression by strengthening the promoter.[3] The TL hybrid promoter retains the *lac uv5* mutation and has the added advantage that it lacks the DNA fragment that binds the catabolite activator protein (CAP), eliminating more completely the requirement for this cAMP-binding CAP protein in transcription.

In conclusion, the hybrid TL promoter, or tacI of De Boer *et al.*,[4] appears at least 10 times more active than the *lac uv5* promoter for directing the expression of the human IFN genes. This is likely to be due to the fact that the hybrid TL promoter contains the two consensus -35 and -10 sequences, but the influence of other modifications must still be investigated. In addition, the removal of the CAP binding site of the *lac* promoter in the TL promoter, seems to eliminate the repression by glucose catabolites. This allows the use of rich growth media for the culture of the human interferon-producing bacteria.

ACKNOWLEDGMENTS

The excellent technical assistance of Ms. Perla Federman, Hanna Berissi, Raya Zwang and the collaboration of Dr. Luisa Chen are gratefully acknowledged. We thank Dr. I. Stabinsky for his help. Work supported in part by InterYeda Ltd, Israel.

[**Note added in proof:** After modification of the ribosomal binding site and growth of the host bacteria under optimal conditions, the present TL promoter has given 3×10^9 U/liter of IFN-αc and 10^9 U/liter of IFN-β1.]

REFERENCES

1. SIEBENLIST, U., R. B. SIMPSON & W. GILBERT. 1980. Cell **20:** 269–281.
2. ROSENBERG, M. & D. COURT. 1979. Ann. Rev. Genet. **13:** 319–353.
3. REZNIKOFF, W. S. & J. N. ABELSON. 1980. *In* The Operon. J. H. Miller & W. S. Reznikoff, Eds. Cold Spring Harbor Laboratories. Cold Spring Harbor, N.Y. pp. 221–243.
4. DE BOER, H. A., L. J. COMSTOCK & M. VASSER. 1983. Proc. Natl. Acad. Sci. USA. **80:** 21–25.
5. OHNO, S. & T. TANIGUCHI. 1981. Proc. Natl. Acad. Sci. USA **78:** 5305–5309.
6. MORY, Y., Y. CHERNAJOVSKY, S. I. FEINSTEIN, L. CHEN, U. NIR, J. WEISSENBACH, Y. MALPIECE, P. TIOLLAIS, D. MARKS, M. LADNER, C. COLBY & M. REVEL. 1981. Eur. J. Biochem. **120:** 197–202.
7. GOEDDEL, D. V., D. W. LEUNG, T. J. DULL, M. GROSS, R. M. LAWN, R. MCCANDLISS, P. H. SEEBURG, A. ULLRICH, E. YELVERTON & P. W. GRAY. 1981. Nature Lond. **290:** 20–26.
8. WEISSMANN, C. *In* Interferon 3. I. Gresser, Ed. Academic Press. New York. pp. 101–134.
9. JOHNSRUD, L. 1978. Proc. Natl. Acad. Sci. USA **75:** 5314–5318.
10. PLATT, T. 1980. *In* The Operon. J. H. Miller & W. S. Reznikoff, Eds. Cold Spring Harbor Laboratories. Cold Spring Harbor, N.Y. pp. 263–302.
11. ENGER-VALK, B. E., H. L. HEYNEKER, R. A. OOSTERBAAN & P. H. POUWELS. 1980. Gene **9:** 69–85.
12. ADLER, H. I., W. D. FISHER, A. COHEN & A. A. HARDIGREE. 1976. Proc. Natl. Acad. Sci. USA **57:** 321–326.
13. WU, A. M., A. B. CHAPMAN, T. PLATT, L. P. GUARENTE & J. BECKWITH. 1980. Cell **19:** 829–836.

Construction of Various Host-Vector Systems Carried Forward to Plants

K. SAKAGUCHI
Mitsubishi-Kasei Institute of Life Sciences
Machida-shi
Tokyo 194, Japan

INTRODUCTION

The construction of various host-vector systems other than *Escherichia coli* was studied in our laboratory because we believe the *E. coli* host vector system is not the best for the production of various substances including peptide hormones, antibiotics, food proteins, and others. I write here about the *Pseudomonas* system, the yeast system, and the studies for the construction of a plant host-vector system.

RESULTS

Pseudomonas *Host-Vector System and RSF 1010 (= R1162, Very Similar to R300B)*[1,2]

In *Pseudomonas* cells, the *E. coli* gene is expressed very efficiently, but not vice versa. The *Pseudomonas* gene can not be expressed in *E. coli* cells without the aid of adjacent *E. coli* promoter activity. Col-El-type plasmid like pBR322 cannot propagate in *Pseudomonas* cells, whereas RSF1010 propagated abundantly in *Pseudomonas* cells. Its copy number reached 170, occupying 22% of the total DNA amount. The plasmid was kept more stably in *P. putida* cells than in *E. coli* cells. This nontransferable plasmid has a wide host range, and is applicable to other species of *Pseudomonas, Rhizobium* and other gram-negative bacteria, as studied later by Dr. M. Bagdasarian & Dr. K. M. Timmis in Max-Planck Institute, Berlin. We believe the *Pseudomonas* host-vector system has some advantages over the *E. coli* system in the area of gene expression, and the plasmid RSF1010 is one of the best vectors among various gram-negative bacteria.

Linear Plasmids from Yeast[3]

Looking for new plasmids among 72 yeast strains, two linear plasmids were discovered in *Kluyveromyces lactis,* which is discussed in more detail in another chapter of this book by Dr. N. Gunge. The longer plasmid, pGKl 2 of 8, 4 md, was responsible for the replication. The shorter one, pGKl 1 of 5.4 md, was responsible for the production of killer factor against *Saccharomyces* and *Kluyveromyces* from its center area. The sequencing studies of the terminal structure and the structure of killer-factor-producing genes are now under investigation.

Chlorella *Protoplast Formation and Its Chloroplast DNA*[4]

It was well known that *Chlorella* had a tough cell wall and was not accessible to various lytic enzymes. In looking among various *Chlorella* species, it was found that the *Chlorella* species could be divided into two groups. One of them, lacking sporopollenin in the cell wall structure, could be lysed with a mixture of cellulase, pectinase, and macerozyme. The *Chlorella ellipsoidea* C-87 strain was efficiently converted into protoplast form. Its chloroplast, nucleus, and mitochondria were obtained intact, and also its DNA was obtained intact, since it had a different specific gravity. Its chloroplast DNA was obtained in a circular form of 170 Kb with 21 cleaving sites by *Eco*R1 restriction enzyme.

Propagation of Chlorella *Chloroplast and* Vinca Rosea-*Replicating Gene in Yeast Cells*

Chlorella chloroplast DNA or *Vinca rosea* chromosomal DNA were cleaved with *Eco*R1, ligated with YIp5 plasmid, which contained pBR322 and *Saccharomyces* Ura3 gene but lacked a yeast replication unit. The ligated DNA was transformed into *Saccharomyces cerevisiae* Ura3 strain. Transformants were obtained with the longer generation time of 13 hr and 17 hr on *Chlorella* ARS gene and *Vinca* ARS gene, respectively. Wild yeast strain replicated within 3 hr. The plasmid DNA thus obtained was transferred from yeast cells to *E. coli* cells for proliferation and their constitutions were analyzed. With the Southern blotting technique, the presence of *Chlorella* chloroplast DNA in the plasmid was assured.

N_2-fixing Chlorella

A N_2-fixing *Chlorella* strain was eventually discovered from a hot spring in Japan. After monocell culturing, it was found to be a unicellular eukaryotic green alga that possessed only chlorophyll a and b but no phycocyanin. It grew microaerophirically without a nitrogen source. It preferred an oxygen tension of 8%. The nitrogenase activity was demonstrated, which increased if illuminated. When the light was off, the nitrogenase production stopped. N^{15} gas incorporation into the cell was assayed. This is the first eukaryotic green plant that has been reported to possess nitrogen-fixing ability.

SUMMARY

The *Pseudomonas* host-vector system was constructed by introducing RSF 1010 into the cells of *P. putida.* Two linear plasmids were obtained from a cheese yeast, *Kluyveromyces lactis.* From a *Chlorella* species, spheroplast was deduced, and its chloroplast DNA was obtained. A fragment of the chloroplast DNA was replicated in *Saccharomyces cerevisiae* cells using *Chlorella* ARS gene. The tenuazonic, acid-degrading gene was constructed in *Bacillus subtilis* and introduced into *E. coli.* A nitrogen-fixing *Chlorella* was isolated and its nitrogenase activity was demonstrated.

REFERENCES

1. NAGAHARI, K. & K. SAKAGUCHI. 1978. J. Bacteriol. **133:** 1527–1529.
2. SAKAGUCHI, K. 1982. Curr. Top. Microbiol. Immunol. **96:** 31–44.
3. NIWA, O., K. SAKAGUCHI & N. GUNGE. 1981. J. Bacteriol. **148:** 988–990.
4. YAMADA, T. & K. SAKAGUCHI. 1981. Agric. Biol. Chem. **45:** 1905–1909.
5. YAMADA, T. & K. SAKAGUCHI. 1980. Arch. Microbiol. **124:** 161–167.

Available Electron and Energetic Yields in Fermentation Processes[a]

L. E. ERICKSON AND M. D. ONER

Department of Chemical Engineering
Kansas State University
Manhattan, Kansas 66506

INTRODUCTION

In aerobic microbial processes involving growth and product formation, the biomass energetic yield, η, and the product energetic yield, ξ_p, have been determined by assuming that these values are equal to the corresponding fraction of available electrons incorporated into biomass and products, respectively.[1–8] For aerobic processes, this assumption has been shown to be appropriate and reasonable.[1,9,10]

In this work, consideration is given to anaerobic processes as well as aerobic growth processes. Roels and coworkers[10–13] have considered the energetics of anaerobic growth processes and pointed out the need to consider the differences of the energy content per equivalent of available electrons of the substrate and the product. Roels[10,11] presents some results in which the thermodynamic efficiency associated with anaerobic growth has been calculated.

In earlier work[5,9,14,15] the yield parameters, η_{max} and ξ_p^{max}, and the maintenance coefficient, m_e, have been introduced and defined. As attention is directed to anaerobic processes, it should be clearly understood that the available electron balance is exact and that the energy balance differs significantly from the available electron balance. It is also important to carefully consider the basic definitions of quantities and parameters as new concepts are introduced.

THEORY

Consider a stoichiometric equation associated with growth and product formation of the form

$$CH_mO_l + aNH_3 + bO_2 = y_cCH_pO_nN_q + zCH_rO_sN_t + cH_2O + dCO_2 \quad (1)$$

where CH_mO_l, $CH_pO_nN_q$, and $CH_rO_sN_t$ give the elemental compositions of the carbon, hydrogen, oxygen, and nitrogen in the organic substrate, biomass, and extracellular product, respectively. Using the valences C = 4, H = 1, O = −2, and N = −3, the available electron balance is[1–4]

$$\frac{4b}{\gamma_s} + y_c\frac{\gamma_b}{\gamma_s} + z\frac{\gamma_p}{\gamma_s} = 1.0 \quad (2)$$

or

$$\epsilon + \eta + \xi_p = 1.0 \quad (3)$$

[a]This work was supported in part by National Science Foundation Grants CPE 69-18202 and CPE 81-20039.

where γ is the reductance degree[1–4] and the subscripts s, b, and p refer to substrate, biomass, and product, respectively. The carbon balance is

$$y_c + z + d = 1.0 \tag{4}$$

Equation 1 describes many aerobic processes and some anaerobic processes (b = 0 for the anaerobic processes).

Free Energy Yields

A free energy balance may also be written for Equation 1. If the free energy level is taken as zero for NH_3 in aqueous solution, H_2O liquid, and CO_2 gas, only the free energy of the substrate, biomass, and product need to be considered in Equation 1. That is,

$$\Delta G_s = y_c \Delta G_b + z \Delta G_p + \Delta G_D \tag{5}$$

where ΔG_s, ΔG_b, and ΔG_p are the free energies of the substrate, biomass, and product per g mole of carbon, respectively, and ΔG_D is the free energy dissipated in the process. Let us define a free energy yield for growth as

$$\eta_{th} = y_c \frac{\Delta G_b}{\Delta G_s} \tag{6}$$

and a free energy yield for product formation as

$$\xi_{th} = z \frac{\Delta G_p}{\Delta G_s} \tag{7}$$

Equation 5 may be written in the form

$$\epsilon_{th} + \eta_{th} + \xi_{th} = 1.0 \tag{8}$$

where

$$\epsilon_{th} = \frac{\Delta G_D}{\Delta G_s} \tag{9}$$

Free energy is used in the microbial process for growth, maintenance, and product formation. Conceptually one may define

$$\eta_{\mathrm{th}}^{\max} = \frac{\text{free energy incorporated into biomass}}{\text{free energy allocated for biomass production}} \tag{10}$$

and

$$\xi_{\mathrm{th}}^{\max} = \frac{\text{free energy incorporated into products}}{\text{free energy allocated for product formation}} \tag{11}$$

The allocation of free energy may now be written in the form

$$\begin{matrix}\text{Free} \\ \text{energy} \\ \text{input}\end{matrix} = \begin{matrix}\text{Free energy} \\ \text{allocated for} \\ \text{biomass production}\end{matrix} + \begin{matrix}\text{Free energy} \\ \text{for maintenance}\end{matrix} + \begin{matrix}\text{Free energy} \\ \text{allocated for} \\ \text{product formation}\end{matrix}$$

or dividing by the free energy input

$$1 = \frac{\eta_{th}}{\eta_{th}^{max}} + \epsilon_m + \frac{\xi_{th}}{\xi_{th}^{max}} \tag{12}$$

where the fraction of free energy consumed for maintenance is

$$\epsilon_m = \frac{m_{th}\eta_{th}}{\mu} \tag{13}$$

where m_{th} is the maintenance coefficient in free energy units; that is, kJ free energy consumed/kJ free energy in biomass per hour. The specific growth rate, μ, divided by the biomass free energy yield, η_{th}, is the specific rate of free energy consumption of organic substrate. Equation 12 may be written in the form

$$\frac{1}{\eta_{th}} = \frac{1}{\eta_{th}^{max}} + \frac{m_{th}}{\mu} + \frac{\xi_{th}}{\eta_{th}\xi_{th}^{max}} \tag{14}$$

If the free energy per equivalent of available electrons is g_s, g_b, and g_p for the substrate, biomass, and products, respectively, then Equations 6 and 7 become

$$\eta_{th} = y_c \frac{\gamma_b g_b}{\gamma_s g_s} = \frac{g_b}{g_s}\eta \tag{15}$$

and

$$\xi_{th} = z\frac{\gamma_p g_p}{\gamma_s g_s} = \frac{g_p}{g_s}\xi_p \tag{16}$$

Equation 14 may be written in the form[5,14,15]

$$\frac{1}{\eta} = \frac{1}{\eta_{max}} + \frac{m_e}{\mu} + \frac{\xi_p}{\eta\xi_p^{max}} \tag{17}$$

where

$$\eta_{th}^{max} = \frac{g_b}{g_s}\eta_{max} \tag{18}$$

$$\xi_{th}^{max} = \frac{g_p}{g_s}\xi_p^{max} \tag{19}$$

and

$$m_{th} = \frac{g_s m_e}{g_b} \tag{20}$$

where η_{max}, ξ_p^{max}, and m_e are yield and maintenance parameters associated with the available electron yields η and ξ_p.[5,14,15] From Equations 18, 19, and 20, it is clear that the new free energy yield and maintenance parameters are related to the corresponding available electron parameters. The methods of parameter estimation that have been presented previously,[7,8] to estimate the values of η_{max}, ξ_p^{max}, and m_e may be employed to obtain point and interval parameter estimates of these parameters from substrate, biomass, product, specific growth rate, oxygen, and carbon dioxide measurements. From these results and Equations 18, 19, and 20, point and interval estimates of η_{th}^{max}, ξ_{th}^{max}, and m_{th} may be obtained.

The numerical values of the free energy per equivalent of available electrons for the substrate and product may usually be found from tabulated literature data.[10,16] For substrates such as glucose and ethanol that do not contain nitrogen, the free energy of combustion per equivalent of available electrons should be used to find g_s and g_p. For biomass, the value of Roels[10] of $-67.1\ kJ/C$ mole for the free energy of formation of biomass may be used to obtain a value of

$$g_b = 113\ kJ/\text{equivalent}$$

for the free energy of the biomass relative to aqueous NH_3, CO_2 gas, and liquid H_2O. This value is in good agreement with the value of 114.7 kJ/equivalent reported by Minkevich.[17,18]

Equation 14 should be valid for both anaerobic and aerobic processes that follow Equation 1. To illustrate the application of these equations, let us examine the ethyl alcohol fermentation with glucose as the organic substrate. Under conditions of excess oxygen, aerobic cell growth and maintenance are important. Under oxygen-limited conditions, ethanol is produced and both substrate level phosphorylation and oxidative phosphorylation are important. Under strictly anaerobic conditions, ethanol is the electron acceptor and ATP production from substrate level phosphorylation provides all the energy for rearrangement of substrate-available electrons into biomass.

Under strictly anaerobic conditions, two moles of ATP are formed from ADP when one mole of glucose is converted to ethyl alcohol. The available electrons in the glucose are transferred to ethyl alcohol and stoichiometrically, $\xi_p^{max} = 1.0$; that is, the available electrons allocated for ethyl alcohol production are all incorporated into ethanol. The available electron balance for Equation 1 becomes

$$\eta + \xi_p = 1.0 \tag{21}$$

Equation 21 implies that available electrons that are not transferred to alcohol will be incorporated into biomass; that is, Equation 1 does not consider any loss of available electrons because of other products or biochemicals that might be present in small quantities. Equation 21 implies that $\eta_{max} = 1.0$; that is, all the available electrons allocated for biomass production are incorporated into biomass. These results also have implications for the free energy balance; that is,

$$\epsilon_{th} = 1 - \frac{g_b}{g_s}\eta - \frac{g_p}{g_s}\xi_p \tag{22}$$

or because of Equation 21

$$\epsilon_{th} = 1 - \frac{g_b}{g_s} + \xi_p\left(\frac{g_b - g_p}{g_s}\right) \tag{23}$$

In the process of anaerobic growth with alcohol production, the maximum possible thermodynamic efficiency would occur when all of the available electrons are transferred to ethanol and biomass. Under these conditions from Equations 18 and 19 and $\eta_{max} = 1.0$ and $\xi_p^{max} = 1.0$,

$$\eta_{th}^{max} = \frac{g_b}{g_s} \tag{24}$$

and

$$\xi_{th}^{max} = \frac{g_p}{g_s} \tag{25}$$

Consideration of the generation of ATP in product formation for use in growth would increase ξ_{th}^{max} and decrease η_{th}^{max}. The fraction of free energy dissipated must be at least as large as that given in Equation 23. In actual practice, some of the available electrons will not be transferred to ethanol and biomass; and the free energy dissipated will be larger than that given in Equation 23.

Anaerobic Conditions

The ethyl alcohol produced under anaerobic conditions may be viewed as a product that must be generated to produce the ATP that is required for growth and maintenance. If the available electrons that are transferred from glucose to ethanol are viewed as being expended to provide energy, another set of yield and maintenance parameters may be defined. A prime will be used with these parameters to distinguish them from those used earlier.

Under strictly anaerobic conditions, the fraction of the available electrons incorporated into biomass depends on the mass of cells per mole of ATP, Y_{ATP}^{max}, according to the relationship[15]

$$\eta'_{max} = \frac{\dfrac{\sigma_b \gamma_b}{12} Y_{ATP}^{max}}{\dfrac{\sigma_b \gamma_b}{12} Y_{ATP}^{max} + 12} \tag{26}$$

where 12 equivalents of available electrons are transferred to ethanol to produce 1 g mole of ATP from ADP. For $\sigma_b = 0.462$, $\gamma_b = 4.291$, and $Y_{ATP}^{max} = 28.8$ g cells per mole ATP, $\eta'_{max} = 0.284$. For $Y_{ATP}^{max} = 10.5$ g cells/mole ATP, $\eta'_{max} = 0.126$. That is, neglecting maintenance, the maximum fraction of available electrons of glucose that can be incorporated into biomass under these anaerobic conditions is 0.284 based on the value of 28.8 g cells/mole ATP of Stouthamer.[19] Using the average value of 10.5 g cells/mole ATP of Bauchop and Elsden,[20] $\eta'_{max} = 0.126$. To obtain a large fraction of ethanol and a small fraction of cells, a small value of Y_{ATP}^{max} or a large maintenance coefficient appears to be desirable. Based on Equation 1, for strictly anaerobic processes, the maintenance process may be viewed as the utilization of ATP, which is generated during ethanol formation and used for maintenance purposes. Growth yield and maintenance parameters for anaerobic growth may be written in the form

$$\frac{\mu}{\eta} = \frac{\mu}{\eta'_{max}} + m'_e \tag{27}$$

$$\frac{\mu(\eta + \xi_p)}{\eta} = \frac{\mu}{\eta'_{max}} + m'_e \tag{28}$$

$$\frac{\mu(y_c + d + z)}{\eta} = \frac{\mu}{\eta'_{max}} + m'_e \tag{29}$$

$$\frac{\mu}{Y_{ATP}} = \frac{\mu}{Y_{ATP}^{max}} + m_{ATP} \tag{30}$$

and

$$\frac{\mu}{\eta'_{th}} = \frac{\mu}{\eta_{th}^{max}} + m'_{th} \tag{31}$$

where η'_{max} and Y_{ATP}^{max} are related as shown in Equation 26. An equation analogous to Equation 26 may be used to relate η and Y_{ATP}. The maintenance coefficients are related according to the relationship

$$m'_e = \frac{144\, m_{ATP}}{\sigma_b \gamma_b} \tag{32}$$

where m_{ATP} is the moles of ATP expended per g of cells per hour for maintenance and m'_e is the maintenance coefficient in equivalents of available electrons expended per equivalent of available electrons of biomass per hour. Equation 32 is limited to fermentations in which 12 equivalents of available electrons are transferred per mole of ATP generated. Equations 27, 28, and 29 are similar and follow Pirt's model[21] for growth and maintenance. Equation 27 utilizes organic substrate, biomass, and specific growth-rate measurements. Equation 28 utilizes biomass, ethanol, and specific growth-rate measurements, while Equation 29 is designed for biomass, carbon dioxide, ethanol, and specific growth-rate measurements. Equations 28 and 29 may be viewed as alternate versions of Equation 27 where the available electron and carbon balance, respectively, have been inserted to enable other measurements to be used in data analysis. Equations 27, 28, and 29 may also be viewed as the result of derivations analogous to those presented elsewhere[4,5] in which glucose consumption, ethanol production, and CO_2 production, respectively, provide the starting form of the Pirt model.

In Equation 31, the free energy biomass yield, η'_{th}, must take into account the free energy remaining in the ethanol product; that is,

$$\eta'_{th} = \frac{g_b \eta}{g_s - g_p \xi_p} \tag{33}$$

The available electron balance may be used to eliminate ξ_p in Equation 33. The true free energy biomass yield, $\eta_{th}'^{max}$, is related to the corresponding available electron yield according to the relationship

$$\eta_{th}'^{max} = \frac{g_b \eta'_{max}}{g_s - g_p(1 - \eta'_{max})} \tag{34}$$

The free energy and available electron maintenance coefficients are related; that is,

$$m'_{th} = \frac{(g_s - g_p) m'_e}{g_b} \tag{35}$$

where m'_{th} is the free energy expended for maintenance per unit of biomass free energy per hour. Comparison of m'_{th} and m_{th} as well as $\eta_{th}'^{max}$ and η_{th}^{max} shows that these parameters are defined the same way in both cases. Thus, the numerical values should be similar. Note, however, that η_{th} is not the same as η'_{th}.

RESULTS AND DISCUSSION

The application of theory to anaerobic process is illustrated using data from the published literature.[13,22] The anaerobic data of Dekkers *et al.*,[13] for the growth of *Saccharomyces cerevisiae* CBS 426 on glucose in a chemostat is examined for consistency in TABLE 1. Except for the 12th data point, the data are reasonably consistent. The results of parameter estimation with the 12th data point excluded are presented in TABLE 2. The methods of data analysis have been described previously.[7,8,23,24] Equations 27, 28, and 29 are used with least squares parameter estimation to estimate the true growth yield, η'_{max}, and the maintenance coefficient, m'_e. Linear regression wih Equation 27 in the form presented is referred to as Form II. Division of every term of Equation 27 by the specific growth rate, μ, gives Form I. Methods to use all of the measurements simultaneously to obtain maximum likelihood estimates

TABLE 1. Data Consistency Analysis for the Growth of *Saccharomyces cerevisiae* CBS 426 on Glucose under Anaerobic Conditions[a]

No. of Data Point	μ	η	ξ_p	$\eta + \xi_p$	y_c	z	d	$y_c + z + d$
1	0.012	0.0634	0.7697	0.8332	0.0591	0.5269	0.2527	0.8387
2	0.014	0.0931	1.0194	1.1125	0.0867	0.6918	0.3881	1.1667
3	0.014	0.1197	0.8907	1.0105	0.1116	0.5953	0.4721	1.1791
4	0.017	0.1033	0.9245	1.0279	0.0963	0.6262	0.3179	1.0405
5	0.027	0.0916	0.7220	0.8137	0.0854	0.4813	0.3177	0.8844
6	0.030	0.0957	0.8398	0.9355	0.0892	0.5716	0.2885	0.9493
7	0.034	0.0999	0.8087	0.9087	0.0932	0.5495	0.2437	0.8864
8	0.051	0.0639	0.8297	0.8937	0.0596	0.5665	0.2059	0.8320
9	0.053	0.0949	0.8183	0.9132	0.0885	0.5607	0.2804	0.9296
10	0.056	0.0757	0.9148	0.9905	0.0706	0.6229	0.2647	0.9582
11	0.058	0.0932	0.8914	0.9846	0.0868	0.6080	0.2480	0.9428
12	0.058	0.0812	1.4507	1.4981	0.0847	0.9381	0.2971	1.4981
13	0.062	0.0909	0.9862	1.0674	0.0756	0.6708	0.2562	1.0027
14	0.067	0.0627	0.8799	0.9426	0.0584	0.5953	0.3249	0.9787
15	0.064	0.0836	0.8297	0.9133	0.0779	0.5547	0.2537	0.8864

[a]Data of Dekkers *et al.*, Reference 13.

(MLE) have been presented elsewhere.[7,23] Method III, with all the covariates included, gives the maximum likelihood estimate.[23] Method III, with only one covariate included, was also applied in this work. When one eigenvalue is much larger than the other, the covariate associated with the largest eigenvalue should be retained. A measure of the goodness of fit is[23]

$$J = \frac{\sigma^2}{N - P - 1},$$

where σ^2 is a mean square error, N is the number of data points, and P is the number of parameters to be estimated. For additional detail on the statistical methods, please refer to the earlier works.[7,8,23,24] The average values $\sigma_b = 0.462$ and $\gamma_b = 4.291$ are used in the analysis of the data.

For a known value of Y^{max}_{ATP}, the value of η'_{max} can be calculated using Equation 26.

TABLE 2. Point and 95% Confidence Interval Estimates of the Maintenance Parameter, m'_e, and the True Biomass Available Electron Yield, η'_{max}, for the Data of TABLE 1[a]

Equation and Method	Form[b]	Maintenance Parameter		True Biomass Yield		Convariates Included	J
		Point Estimate	95% Confidence Interval	Point Estimate	95% Confidence Interval		
Eq. 27, L.R.[c]	I	−0.0192	−0.0791, 0.0408	0.0796	0.0661, 0.100		
	II	−0.0431	−0.160, 0.0738	0.0751	0.0635, 0.0917		
Eq. 28, L.R.	I	−0.0181	−0.0737, 0.0341	0.0840	0.0706, 0.140		
	II	−0.0407	−0.154, 0.0731	0.0791	0.0667, 0.0973		
Eq. 29, L.R.	I	−0.0032	−0.0517, 0.0454	0.0876	0.0740, 0.107		
	II	−0.0261	−0.144, 0.0922	0.0823	0.0685, 0.103		
Eq. 27, 28, 29	I	0.0043	−0.0610, 0.0696	0.0896	0.0723, 0.1181	z_2, z_3	0.524
		−0.0064	−0.0602, 0.0474	0.0879	0.0721, 0.1125	z_2	0.448
Method III		−0.0104	−0.0712, 0.0502	0.0838	0.0701, 0.1043	z_3	0.495
	II	−0.0464	−0.179, 0.0866	0.0766	0.0608, 0.103	z_2, z_3	0.00123
		−0.0375	−0.159, 0.0841	0.0783	0.0633, 0.102	z_2	0.00103
		−0.0447	−0.168, 0.0788	0.0774	0.0643, 0.0972	z_3	0.00101

[a]The 12th data point is excluded.

[b]For Form I, the eigenvalues are 16.8 and 2.92 for z_2 and z_3; for Form II the eigenvalues are 0.0107 and 0.00658 for z_2 and z_3, respectively.

[c]L.R. refers to linear regression.

Dekkers *et al.* calculated a value of $Y_{ATP} = 8.6$ g dry wt/mole ATP for their anaerobic data and they reported a value of $Y_{ATP} = 15.1$ g dry wt/mole ATP for their aerobic data. Examination of the results in TABLE 2, together with Equation 26, shows that the values of η'_{max} in TABLE 2 correspond to smaller values of Y_{ATP}^{max} than 8.6 g dry wt/mole ATP if 12 equivalents available electrons transferred to ethanol produce 1 g mole of ATP. The value of $Y_{ATP}^{max} = 8.6$ corresponds to $\eta'_{max} = 0.11$. This value is at the upper end of the 95% confidence interval

$$0.0608 \leq \eta'_{max} \leq 0.118.$$

The results in TABLE 2 show that the maintenance coefficient, m'_e, is small. Only the positive portion of the 95% confidence interval is of interest since m'_e must be positive; that is, we have 95% confidence that

$$0 \leq m'_e \leq 0.0866.$$

Esener *et al.*[25] showed that m_{ATP} varies between 0.25 to 0.70 *m* moles ATP/g(hr),

TABLE 3. Data Consistency Analysis for the Growth of *Zymomonas mobilis* on Sucrose[a,b]

No. Data Point[c]	μ	η	ξ_p	$\eta + \xi_p$	y_c	z	d	$y_c + z + d$
1	0.0367	0.0113	0.9063	0.9176	0.0105	0.6223	0.2838	0.9167
2	0.0836	0.0138	0.7574	0.7712	0.0165	0.5286	0.2286	0.7737
3	0.1313	0.0216	0.8789	0.9005	0.0201	0.6189	0.2599	0.8989
4	0.1859	0.0243	0.9986	1.0230	0.0226	0.7028	0.2958	1.0212
5	0.2508	0.0285	0.9945	1.0230	0.0226	0.7100	0.2847	1.0211
6	0.0680	0.0168	0.8339	0.8509	0.0157	0.5597	0.2742	0.8496
7	0.1000	0.0201	0.8129	0.8329	0.0184	0.5412	0.2605	0.8202
8	0.1504	0.0320	0.9739	1.0060	0.0225	0.6787	0.2952	0.9964
9	0.2000	0.0355	0.9278	0.9633	0.0243	0.6498	0.2759	0.9499
10	0.2720	0.0286	0.9705	0.9991	0.0267	0.6880	0.2824	0.9972
11	0.3240	0.0339	0.9506	0.9845	0.0316	0.6889	0.2616	0.9822

[a]Data of Lee *et al.*, Reference 22.
[b]Yeast extract and levan (as side product) are included in calculations.
[c]First five points are from FIGURE 6. Last six points are from FIGURE 5.[22]

which corresponds to 0.018 to 0.051 hr^{-1} for m'_e. These values are within the 95% confidence interval presented above.

The continuous culture data of Lee *et al.*[22] for growth of *Zymomonas mobilis* on sucrose is examined for consistency in TABLE 3. The consistency is reasonably good. The data points in which the feed contained 100 g/l sucrose were combined with those in which the feed contained 150 g/l and analyzed together. *Zymomonas mobilis* has an absolute requirement for pentothenic acid.[26,27] It produces levan when it grows on sucrose.[20] In this analysis, yeast extract as substrate and levan as side product are included in analyzing the data. Levan is considered as product analogous to ethanol; that is, ATP is assumed to be produced in the process of converting glucose to levan. The yeast extract is included as a substrate and the values of $\sigma = 0.40$ and $\gamma = 4.0$ are used in estimating the carbon and available electron content of the yeast extract.

The parameter estimation results are presented in TABLE 4. The values of the

TABLE 4. Point and 95% Confidence Interval Estimates of the Maintenance Coefficient, m'_e, and the True Biomass Available Electron Yield, ν'_{max}, for the Data of TABLE 3

Equation and Method	Form	Maintenance Parameter		True Biomass Yield		Convariates Included	J
		Point Estimate	95% Confidence Interval	Point Estimate	95% Confidence Interval		
Eq. 27, L.R.[a]	I[b]	2.54	1.74, 3.33	0.0415	0.0303, 0.0659		
	II[c]	2.81	1.42, 4.20	0.0458	0.0341, 0.0697		
Eq. 28, L.R.	I	2.07	1.54, 2.60	0.0413	0.0331, 0.0549		
	II	1.96	0.68, 3.24	0.0400	0.0314, 0.0522		
Eq. 29, L.R.	I	2.07	1.52, 2.62	0.0416	0.0331, 0.0560		
	II	1.95	0.64, 3.25	0.0401	0.0313, 0.0559		
Eq. 27, 28, 29	I	1.97	1.41, 2.52	0.0369	0.0305, 0.0468	z_2, z_3	2.97
		1.96	1.25, 2.67	0.0415	0.0323, 0.0566	z_2	4.37
Method III		2.30	1.61, 2.99	0.0402	0.0303, 0.0582	z_3	5.28
	II	2.52	0.94, 4.10	0.0392	0.0308, 0.0539	z_x, z_3	0.0776
		2.05	−0.015, 4.12	0.0406	0.0294, 0.0656	z_2	0.127
		2.93	0.99, 4.86	0.0451	0.0333, 0.0699	z_3	0.111

[a] L.R. refers to linear regression.
[b] For Form I, the eigenvalues are 209 and 5.74 for z_2 and z_3, respectively.
[c] For Form II, the eigenvalues are 1.09 and 0.182 for z_2 and z_3, respectively.

maintenance parameter, m'_e, are much larger for this organism while the values of the true growth yield are much smaller when they are compared to the results in TABLE 2. *Zymomonas mobilis* uses the Entner-Doudoroff pathway and produces 1 mole of ATP/mole of glucose converted to ethanol.[26] This organism has received considerable attention with respect to ethanol production as it is more ethanol tolerant and it can produce ethanol faster than conventional yeasts.[28–31] The high maintenance requirements of this organism have been considered by Cromie *et al.*[32] and Esener *et al.*[25] Esener *et al.*[25] pointed out that the high maintenance requirement requires rapid ethanol production by the organism to meet this need. Thus, larger values of m'_e should be found for this organism. Esener *et al.*[25] has estimated that $m_{ATP} = 10.3$ m moles ATP/g(hr) for this organism, which corresponds to 1.50 hr^{-1} for m'_e, assuming 24 equivalents of available electrons generate 1 ATP. (The coefficient in Equation 32 is 288). This value is within the 95% confidence interval

$$0.937 \leq m'_e \leq 4.10.$$

The 95% confidence interval for the true growth yield is

$$0.0305 \leq \eta'_{max} \leq 0.0539.$$

The interval may be compared to $Y^{max}_{ATP} = 8.3$, which is equivalent to $\eta'_{max} = 0.054$ found by Forrest.[33] Since 24 equivalents of available electrons yield 1 mole of ATP, Equation 26 becomes

$$\eta'_{max} = \frac{\dfrac{\sigma_b \gamma_b}{12} Y^{max}_{ATP}}{\dfrac{\sigma_b \gamma_b}{12} Y^{max}_{ATP} + 24}$$

for *Zymomonas mobilis*.

Values of Y_{ATP} ranging from 4.3 to 9.3 have been estimated by Swing and DeLey;[26] however, with the large maintenance requirement for this organism, the corresponding value of Y^{max}_{ATP} may be considerably higher.

The true growth yield in available electron units, η'_{max}, can be converted to a better thermodynamic measure of growth efficiency by using Equation 34 to obtain the true free energy biomass yield. In TABLE 5, values of the true free energy biomass yield, the true free energy product yield, and the corresponding maintenance parameter in free energy units are presented for several sets of data. The close agreement of the true free energy biomass yield from aerobic and anaerobic measurements for the data of Dekkers *et al.*[13] should be noted. For *Zymomonas mobilis,* the efficiency of growth is lowest and the maintenance coefficient is largest. Both of these contribute to the high rate of alcohol production by this organism.

CONCLUSIONS

The methods of parameter estimation in which all of the data are considered simultaneously to obtain maximum likelihood estimates of parameters have been extended to anaerobic growth processes. Parameter estimates have been obtained for two sets of literature data. For anaerobic growth where the product is the only electron acceptor, growth and maintenance are coupled with product formation. When product formation is desired, a high maintenance requirement is a desirable property.

TABLE 5. Comparison of Free Energy Maintenance Parameters, m_{th} and m'_{th}, True Growth Yields, η_{th}^{max} and $\eta_{th}^{\prime max}$, and, True Product Yields ξ_{th}, between Aerobic and Anaerobic Microbial Growth Processes[a]

Form of Equations	(m_{th} or m'_{th})	(η_{th}^{max} or $\eta_{th}^{\prime max}$)	(ξ_{th}^{max})		Organism	Substrate	Product	Source of Data
I II	0.00439 0.00465	0.619 0.600	0.885 1.07	Aerobic	*Rhodopseudomonas sphaeroides*	Glucose	Polysaccharides	Nishizawa *et al.*, Ref. 34
I II	0.00697 0.00404	0.597 0.562	0.783 0.860	Aerobic	*Rhizobium trifolii*	Glucose	Polysaccharides	de Hollander *et al.*, Ref. 35
I II	0.00360 −0.00940	0.504 0.473	0.998 1.03	Aerobic	*Saccharomyces cerevisiae* CBS 426	Glucose	Ethanol	Dekkers et al., Ref. 13
I II	0.00380 0.00370	0.546 0.545	1.05 1.05	Aerobic	*Saccharomyces cerevisiae*	Glucose	Ethanol	von Meyenburg, Ref. 36, 37
I II	−0.00055 −0.00402	0.510 0.475		Anaerobic	*Saccharomyces cerevisiae* CBS A 26	Glucose	Ethanol	Dekkers *et al.*, Ref. 13
I II	0.170 0.218	0.301 0.313		Anaerobic	*Zymomonos mobilis*	Sucrose	Ethanol	Lee *et al.*, Ref. 22

[a]Prime denotes anaerobic processes.

SUMMARY

Microbial growth and product yields have been investigated for several cultures. Available electron and free energy balances may be used to examine the efficiency of microbial energetics. Energetic yields based on free energy analysis were obtained and compared with yields from available electron analysis. Estimates of true growth yields under anaerobic conditions have been obtained using statistical methods and the results were compared with theoretical maximum yields.

ACKNOWLEDGMENTS

We thank B. O. Solomon and Dr. S. S. Yang for their help.

REFERENCES

1. MINKEVICH, I. G. & V. K. EROSHIN. 1973. Folia Microbiol. **18:** 376.
2. ERICKSON, L. E., I. G. MINKEVICH & V. K. EROSHIN. 1978. Biotechnol. Bioeng. **20:** 1595.
3. ERICKSON, L. E., S. E. SELGA & U. E. VIESTURS. 1978. Biotechnol. Bioeng. **20:** 1623.
4. ERICKSON, L. E., I. G. MINKEVICH & V. K. EROSHIN. 1979. Biotechnol. Bioeng. **21:** 575.
5. ERICKSON, L. E. 1979. Biotechnol. Bioeng. **21:** 725.
6. FERRER, A. & L. E. ERICKSON. 1979. Biotechnol. Bioeng. **21:** 2203.
7. SOLOMON, B. O., L. E. ERICKSON, J. L. HESS & S. S. YANG. 1982. Biotechnol. Bioeng. **24:** 633.
8. ONER, M. D., L. E. ERICKSON & S. S. YANG. 1983. Biotechnol. Bioeng. **25:** 631.
9. ERICKSON, L. E. 1980. Biotechnol. Bioeng. **22:** 451.
10. ROELS, J. A. 1980. Biotechnol. Bioeng. **22:** 2457.
11. ROELS, J. A. 1982. ACS-IEC 1982 Winter Symposium "Foundations of Biochemical Engineering" ACS Symposium Series. In press.
12. ROELS, J. A. 1981. Ann. N.Y. Acad. Sci. **369:** 113.
13. DEKKERS, J. G. J., H. E. DEKOK & J. A. ROELS. 1981. Biotechnol. Bioeng. **23:** 1023.
14. ERICKSON, L. E. & J. C. HESS. 1981. Ann. N.Y. Acad. Sci. **369:** 81.
15. ERICKSON, L. E. 1980. J. Ferment. Technol. **58:** 53.
16. WEAST, R. C. 1976. Handbook of Chemistry and Physics, 56th Edition. CRC Press.
17. MINKEVICH, I. G. 1980. *In* Limitation and Inhibition of Microbial Processes. V. K. Eroshin, Ed. Academy of Sciences Scientific Center for Biological Research. Pushchino, Moscow Region, USSR (In Russian). p. 55.
18. MINKEVICH, I. G. 1982. J. Theor. Biol. **95:** 569.
19. STOUTHAMER, A. H. 1979. Int. Rev. Biochem. **21:** 1.
20. BAUCHOP, T. & S. R. ELSDEN. 1960. J. Gen. Microbiol. **23:** 457.
21. PIRT, S. J. 1965. Proc. R. Soc. B **163:** 224.
22. LEE, K. G., M. L. SKOTNICKI, D. E. TRIBE & P. L. ROGERS. 1981. Biotechnol. Lett., **3:** 201–212.
23. SOLOMON, B. O., M. D. ONER, L. E. ERICKSON & S. S. YANG. 1982. Estimation of parameters where dependent observations are related by equality constraints. Presented at AIChE Los Angeles Meeting.
24. YANG, S. S., B. O. SOLOMON, M. D. ONER & L. E. ERICKSON. 1982. Technometrics. Submitted for publication.
25. ESENER, A. A., J. A. ROELS & N. W. F. KOSSER. 1981. Biotechnol. Lett. **3:** 1520.
26. SWINGS, J. & J. DECEY. 1977. Biotechnol. Rev. **41:** 1–46.
27. BELAICH, J. P., J. C. SENEZ & M. MURGIER. 1968. J. Biotechnol. **35:** 1750–1757.
28. ROGERS, P. C., K. J. LEE & D. E. TRIBE. 1978. Biotechnol. Lett. **1:** 165–170.
29. GROTE, W., K. J. LEE & P. C. ROGERS. 1980. Biotechnol. Lett. Vol. **2:**,No. 11: 481–486.
30. ROGERS, P. L., K. J. LEE & D. E. TRIBE. 1980. Process Biochem. **16** (No. 5): 7–11.

31. SKOTNICKI, M. L., K. J. LEE, P. E. TRIBE & P. L. ROGERS. 1981. Appl. Environ. Microbiol. **41:** 889–893.
32. CROMIE, S. & H. W. DOELLE. 1980. Biotechnol. Lett **2:** 375–361.
33. FORREST, W. W. 1967. J. Biotechnol. **94:** 1453–1463.
34. NISHIZAWA, Y., S. NAGAI & S. AIBA. 1974. J. Ferment. Technol. **52:** 526.
35. DE HOLLANDER, J. A., C. W. BETTENHAUSSEN & A. G. STOUTHOMER. 1979. Antonie van Leeuwenhoek **45:** 401.
36. VON MEYENBURG, H. K. 1960. Arch. Microbiol. **23:** 457.
37. VON MEYENBURG, H. K. 1963. Katabolit-Repression and der Sprossurg-szyklus von Saccharomyces cerevisiae. Dissertation. Eidgenonssischen Techischen Hochschule. Zurich. Viertelijohrschr. Naturforsch. Ges. Zurich.

APPENDIX

NOMENCLATURE

a	Moles of ammonia per quantity of organic substrate containing one g atom carbon, g mole/g atom carbon.
b	Moles of oxygen per quantity of organic substrate containing one g atom carbon, g mole/g atom carbon.
c	Moles of water per quantity of organic substrate containing one g atom carbon, g mole/g atom carbon.
d	Moles of carbon dioxide per quantity of organic substrate containing one g atom carbon, g mole/g atom carbon.
ΔG_b	Free energy of biomass per g mole of carbon in biomass, KJ/mole C.
ΔG_D	Free energy dissipated per g mole of carbon in substrate, KJ/mole C.
ΔG_p	Free energy in product per g mole of carbon in product, KJ/ mole C.
ΔG_s	Free energy in substrate per g mole of carbon in substrate, KJ/mole C.
g_b	Amount of free energy in biomass per g equivalent of available electron in biomass, KJ/g equivalent of a.e.
g_p	Amount of free energy in product per g equivalent of available electron in product, KJ/g equivalent of a.e.
g_s	Amount of free energy in substrate per g equivalent of available electron in substrate, KJ/g equivalent of a.e.
MLE	Maximum likelihood estimate.
m_e	Rate of organic substrate consumption for maintenance, g equivalent of available electron per g equivalent of available electrons in biomass (hr).
m_{ATP}	Moles of ATP expended per g of cells per hour for maintenance, moles ATP/g-hr.
m'_e	Same definition as m_e. Prime denotes anaerobic process.
m_{th}	Amount of free energy consumed per unit of free energy in biomass per hour, KJ free energy consumed/KJ free energy in biomass. (hr).
m'_{th}	Same definition as m_{th}. Prime denotes anaerobic process.
N	Number of data points.
P	Number of parameters to be estimated.
Y_{ATP}	Biomass yield based on ATP, g dry biomass/g mole ATP.
Y^{max}_{ATP}	Maximum biomass yield based on ATP, g dry biomass/g mole ATP.
y_c	Biomass carbon yield (fraction of organic substrate carbon in biomass), dimensionless.

z Fraction of organic substrate carbon in products, dimensionless.

z_i Covariates included in Method III.

γ_b Reductance degree of biomass, equivalents of available electrons per g atom carbon.

γ_p Reductance degree of products, equivalents of available electrons per g atom carbon.

γ_s Reductance degree of substrate, equivalents of available electrons per g atom carbon.

ϵ Fraction of energy in organic substrate that is evolved as heat, dimensionless.

ϵ_m Fraction of energy in organic substrate that is evolved as heat because of cell maintenance, dimensionless.

ϵ_{th} Fraction of free energy in substrate that is evolved as heat, dimensionless.

η Fraction of available electrons in organic substrate that is converted to biomass or biomass energetic yield, dimensionless.

η_{max} "True" biomass available electron yield, dimensionless.

η'_{max} "True" biomass available electron yield, dimensionless. Prime denotes anaerobic process.

η_{th} Fraction or free energy in organic substrate that is converted to biomass energetic yield, dimensionless.

η'_{th} Biomass free energy yield, dimensionless. Prime denotes anaerobic process.

η_{th}^{max} "True" biomass free energy yield, dimensionless.

$\eta_{th}^{\prime max}$ "True" biomass free energy yield, dimensionless. Prime denotes anaerobic process.

μ Specific growth rate hr^{-1}.

ξ_p Fraction of available electrons in organic substrate that is converted to products, dimensionless.

ξ_p^{max} "True" product energetic yield, dimensionless.

ξ_{th} Fraction of free energy in organic substrate that is converted to products, dimensionless.

ξ_{th}^{max} "True" product free energy yield, dimensionless.

Diffusional/Kinetic Analysis of the Neurotransmission Process at the Nerve-Muscle Junction

WOLF R. VIETH AND GOPAL CHOTANI

Department of Chemical and Biochemical Engineering
Rutgers–The State University of New Jersey
Piscataway, New Jersey 08854

INTRODUCTION

A uniquely detailed picture of the subtle and intricate pattern of phenomenological relationships governing the response to acetylcholine at the nerve-muscle junction has gradually emerged.[1–3,6,8] At this relatively early stage, however, quantitative description of the phenomenon amounts to an unsolved, or at best, partially solved problem that demands further effort. Progress could conceivably lead to a better understanding of neurotransmitter–receptor interactions, generally.

Previous efforts aimed at quantitative descriptions of the neutrotransmission process have attempted to isolate and model a single event that starts in the cleft and concludes in the postsynaptic membrane following release of Ach.[1–3] The common postulation is diffusive radial spreading away from a single vesicle, followed by binding of transmitter in a disk-shaped cleft. However, possible interaction effects of other vesicles in the local colony have not been analyzed and the transmitter is considered to appear instantaneously in the cleft. The latter assumption is a fundamental limitation that affects all subsequent estimates of diffusion times, and so forth. Under these assumptions, several authors have acknowledged that these mathematical characterizations of synaptic transmission ought to be regarded as tentative.[2,3] To explore another alternative, we have constructed *process* models of continuous and intermittent types, in contrast to *event* type models.

The synaptic cleft is recognized as a fluid-filled fibrous matrix containing immobilized Ach-ase, terminating in a layer of Ach receptors. The matrix itself is composed of collagen and mucopolysaccharide fibers in association with enzyme. Previously, a number of workers in our laboratory have investigated reconstituted collagen as a carrier for a large number of enzymes and have measured and analyzed penetrant transport in such enzyme membrane structures.[4] We draw upon this experience in approaching the present work.

MODEL

A thorough comparison of the analysis of this paper, with those of recent models, is shown in TABLE 1. Simplifying our previous model,[5] the synaptic cleft is represented as an infinitely thin slab or membranelike diffusion medium, with Ach-ase distributed uniformly within and/or at the postsynaptic surface. Likewise, receptors are located anisotropically at the postsynaptic membrane.

TABLE 1. Model Comparisons

Process Models—This Work	Event Models[1,2,3]
1. Structural picture of the synaptic cleft: Anisotropic membrane.	1. Thin, wide disc.
2. Transmitter release from a vesicle colony in an active zone forms unit pulse(s) at the presynaptic membrane, characteristic time of the pulse = θ_c, or characteristic number of intervals = n, each of time = $\Delta\theta_c$.	2. Transmitter appears instantaneously in the cleft. An explicit mechanism to describe the temporal release pattern from a local colony comprising many vesicles is not provided.
3. Infinite (planar) thin-slab diffusion medium. Transmitter diffuses axially across cleft. Because of the extreme thinness of the cleft plus postsynaptic membrane, axial gradients are very steep; the influence of lateral gradients is considered minimal, relative to the influence of axial gradients (see APPENDIX A for further discussion of the choice of process coordinates).	3. Transmitter spreads by radial diffusion away from an isolated vesicle, in the plane of the cleft. Axial diffusion is not emphasized because of the assumption that the cleft plus postsynaptic membrane dimension must be too thin to control the process timescale on the basis of a diffusion process.
4. Transmitter is removed by enzymatically catalyzed hydrolysis and diffusion away from the receptor locale.	4. Part of the transmitter is removed by enzyme reaction and another part by diffusion away from disc boundaries.
5. Transmitter–receptor binding and dissociation reactions can be at local equilibrium. Local binding/dissociation rates can be considered rapid compared to local diffusion jump frequency and/or jump length, in the presence of severe penetrant immobilization.[9,14]	5. Transmitter–receptor interactions occur under nonequilibrium conditions.
6. Analytical solutions are derived for bound neurotransmitter concentrations (both active and resting populations) at the receptor locale. By taking very short time intervals, $\Delta\theta_c$, the model can be broken down to apply to miniature end plate potentials and currents.	6. Numerical simulations illustrate trends in space integral of active transmitter population,[2] spreading away from a single vesicle, in the absence of axial transport. The overall process of neurotransmission from the vesicle colony is not modeled. An explicit technique for the build-up of a process from the behavior of single, isolated noninteracting vesicles is not shown.
7. epc = $g_{Na}(V - V_{Na}) + g_K(V - V_K)$ where V is the endplate potential; V_{Na}, V_K are the Nernst potentials and epc is the endplate current. g_{Na} and g_K are the sodium and potassium ion conductivities, respectively.	7. Miniature end plate current (mepc) = single channel conductance × electrochemical driving force.[1]

Then, the transient diffusion equation for Ach takes the form

$$\frac{\partial Y_M}{\partial \theta} = \frac{\partial^2 Y_M}{\partial X^2} - \mu Y_M$$
$$\theta = \frac{D_{eff} t}{\ell^2} \qquad X = \frac{x}{\ell} \qquad Y_M = \frac{S}{S_0} \tag{1}$$

The membrane thickness is ℓ and D_{eff} is the diffusion coefficient. S_o is the upstream concentration of Ach and μ is a reaction modulus, defined as $k\ell^2/D_{eff}$, where k is the first-order enzyme reaction velocity constant.

The boundary conditions for a unit pulse that commences at $\theta = 0$ and is cut off at $\theta_c = 1.0$ are:

$$Y_M(X, 0) = 0$$
$$Y_M(0, \theta) = 1, \qquad 0 \leq \theta \leq \theta_c \tag{2}$$
$$= 0, \qquad \theta \geq \theta_c \tag{3}$$
$$\frac{\partial Y_M}{\partial X} = -\mu Y_M, \qquad X = 1, \theta > 0 \tag{4}$$

For the enzymatically anisotropic structure of the cleft, Equation 1 becomes a simple diffusion equation,

$$\frac{\partial Y_M}{\partial \theta} = \frac{\partial^2 Y_M}{\partial X^2} \tag{5}$$

with hydrolysis described by Equation 4.

Equation 1 can be solved mathematically along with Equations 2 to 4 by a combination of separation of variables and convolution techniques, to obtain analytical solutions as follows:

At $X = 1.0$,

$$Y_M = (\cosh \sqrt{\mu} + \sqrt{\mu} \sinh \sqrt{\mu})^{-1} + 2 \sum_{n=1}^{\infty} \frac{\sin \lambda_n \cdot \frac{\lambda_n^2}{\delta_n^2}}{\cos \lambda_n \cdot \sin \lambda_n - \lambda_n} e^{-\delta_n^2 \theta}, \theta \leq \theta_c \tag{6}$$

$$Y_M = 2 \sum_{n=1}^{\infty} \left(\frac{\sin \lambda_n \cdot \frac{\lambda_n^2}{\delta_n^2}}{\lambda_n - \cos \lambda_n \cdot \sin \lambda_n} \right) (e^{-\delta_n^2(\theta - \theta_c)} - e^{-\delta_n^2 \theta}), \theta \geq \theta_c \tag{7}$$

where

$$\lambda_n \cdot \cos \lambda_n = -\mu \cdot \sin \lambda_n \text{ (eigenvalues)} \tag{8}$$

and

$$\delta_n^2 = \lambda_n^2 + \mu$$

The enzyme reaction is modeled as a first-order process. This is in agreement with the information that the Ach concentration in the cleft is at about the level of the Michaelis constant, K_M, for Ach ase.[6,7]

The Ach receptor is modeled as an allosteric protein with two sets of sites: one that corresponds to a conformation in a rest condition with a fractional penetrant saturation (Y_R) and another (Y_A) in which Ach is bound in an active (i.e., open channel) conformation.[4,8] For an allosteric protein having two oligomers,

$$R \rightleftharpoons A$$

$$2R_0 + S \rightleftharpoons R_1 \qquad\qquad 2A_0 + S \rightleftharpoons A_1$$

$$R_1 + S \rightleftharpoons 2R_2 \qquad\qquad A_1 + S \rightleftharpoons 2A_2$$

$$Y_A = \frac{S}{K_A + S}\overline{A} \tag{9}$$

where

$$\overline{A} = \frac{L\left(1 + \frac{S}{K_A}\right)^2}{L\left(1 + \frac{S}{K_A}\right)^2 + \left(1 + \frac{S}{K_R}\right)^2} \tag{10}$$

and

$$A_0 = LR_0 \tag{11}$$

The procedure is next to obtain S at $X = 1$, from Equation 6 and 7, thence substituting its value into Equation 9 to obtain Y_A at $X = 1$, the receptor locale.

In arriving at models for ion conductivity, we obtained guidance from an excellent review article by A. M. Monnier,[10] through an analogy of voltage-dependent gating effects in the axon membrane with transmitter-actuated gating effects in neurotransmission. Accordingly, we formulate ion conductivities as follows:

$$g_{Na} = \overline{g}_{Na} \cdot p_A \tag{12}$$

$$g_K = \overline{g}_K \cdot p_A \tag{13}$$

where

$$p_A = Y_A \tag{14}$$

and

$$\text{Cation Conductance ratios} = \frac{g_{Na}}{\bar{g}_{Na}} = \frac{g_K}{\bar{g}_K} = Y_A \tag{15}$$

where $\bar{g}_{Na}$, $\bar{g}_K$ are the saturation level conductivities for Na^+ and K^+, respectively. (All conductivity values are expressed relative to rest levels as baselines.) Equation 14

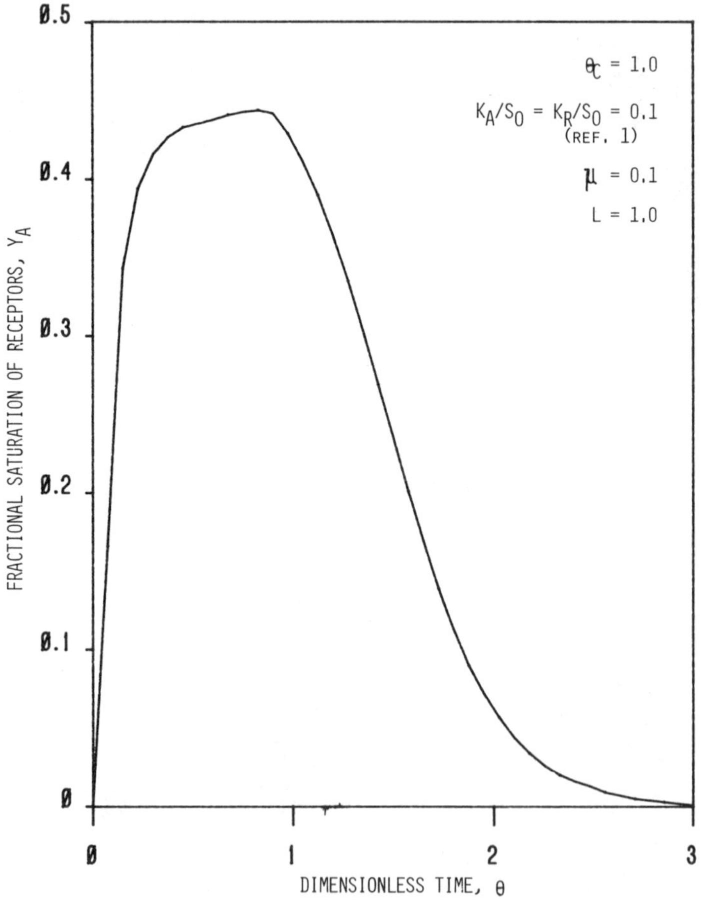

FIGURE 1. Response to a pulse: fraction of channels opened by transmitter.

expresses the probability that a cation channel is open that is just equal to the probability that an Ach is bound locally by the receptor in the active conformation.

The response of the model for a unit pulse is shown in FIGURE 1. Interestingly, regarding one measure of process efficiency, only about 40% of the channels are open at the conduction peak for the set of parameters shown.

APPLICATIONS TO ENDPLATE CURRENT

Under voltage clamp conditions, mepc = (constant)($Y_A(\theta)$). Therefore, the time variation of Y_A (open channel fraction) superposed with normalized miniature endplate current (i.e., mepc/$(\text{mepc})_{max}$) should be the same. Comparing theoretical results with the data supplied by Rosenberry,[3] we arrive at a value of $\theta_c = 0.04$, for a rise time of 120 μ sec. The time axis shift factor, (D_{eff}/ℓ^2), is then $(0.040/0.120) = 0.33$ msec^{-1}. The limiting value as $\mu \rightarrow 0$ for the decay constant is $(1/\tau) = (D_{eff}/\ell^2) \cdot \delta_1^2 = 0.33$ msec$^{-1} \times (\pi/2)^2 = 0.81$ msec^{-1}, in good agreement with limiting values from the literature for mepc[3] and epc.[11] For the mepc case, even though the pulse length is short, the maximum fraction of open channels during the response is nearly 0.20, in agreement with the estimate of Adams.[2]

To explore the other limit (i.e., high enzymatic activity, or $\mu \rightarrow \infty$) one may examine the first eigenvalues of the solution (see APPENDIX A, TABLE A1). The ratio, $[\delta_1^2\ (\mu \rightarrow \infty)/\delta_1^2\ (\mu \rightarrow 0)] = [(3.10)^2/(1.60)^2] = 3.75$, corresponds to nearly a fourfold increase in decay constant. Contributions of the other eigenvalues would make this ratio still larger.

Mathematically, for a train of mini-pulses, Duhamel's theorem can be applied to the transient solution for a unit step input, that is, Equation 6. The ultimate solution takes the form,

$$\text{Response} = \sum_{j=1}^{n} B_j \cdot \Delta f_{n+1-j} \tag{16}$$

where

$$\Delta f_j = Y_M(1, j\Delta\theta_c) - Y_M(1, \overline{j-1}\Delta\theta_c)$$

$$Y_M = \text{obtained from unit step response}$$

$$B_j = \text{amplitude of } j\text{th pulse} \begin{bmatrix} b_j = 1.0 \text{ firing} \\ = 0.0 \text{ rest} \end{bmatrix} \tag{17}$$

$$n = \text{total number of intervals in a train}$$

$$\Delta\theta_c = \text{time span of each interval} \left(= \frac{\theta_c}{n} \right)$$

For example, if $n = 10$, $\theta_c = 1.0$, $\Delta\theta_c = 0.1$, then the pulse train contains 5 pulse firings and 5 rests. For the various profiles shown in FIGURE 2, each pulse is of unit amplitude. For reasons of simplicity, firing and rest-state time intervals are assumed to be equal. When the time interval between pulses approaches zero or pulses overlap, the response will be the same as if a single unit pulse of θ_c were the input. With some modifications in Equations 16 and 17, response curves can be obtained with unequal time intervals of firing and rest states of vesicles. In this way, the continuous process model can be broken down into the very small intervals appropriate to multiple mepcs.

Generally, speaking, as soon as a pulse of S (i.e., Ach) sets in at the presynapse, it spreads diffusively through the cleft into the receptor sites at the postsynapse. As a result, active conformations come forth, ion channels open and the density of vacant receptor sites diminishes. This gives rise to the increase of conductances g_{Na}, g_K. Similarly, reverse phenomena take place as soon as the pulse is terminated upstream. The overall model is shown in the following sketch, TABLE 2. The time evolution of the sodium and potassium conductances provide the key components in the relation of endplate current and endplate potential, through $Y_A(\theta)$, which we have employed to model normalized mepc under voltage-clamp conditions (FIG. 3). Agreement of the theory with the data is exceptionally good in this case.

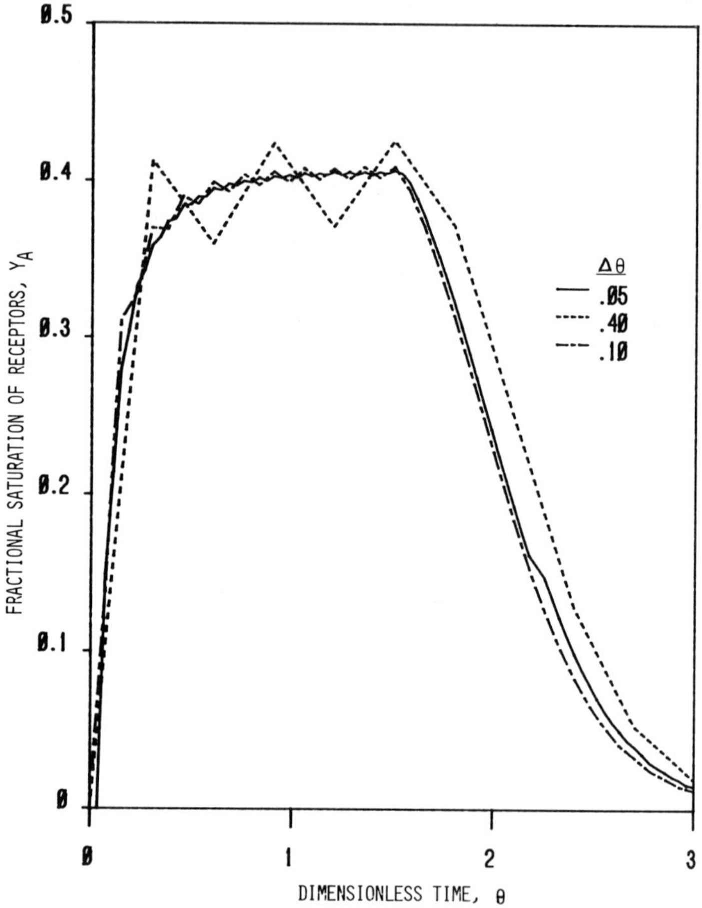

FIGURE 2. Response to a train of minipulses: fraction of channels opened by neurotransmitter.

ACH DIFFUSIVITY

The agreement of the theory with experimental data just noted provides evidence that the model properly weighs the counter-posing influences on the diffusion process of receptor interaction and enzyme reaction. In general, the former causes a retardation effect because a fraction of the penetrant population is immobilized and unavailable for diffusion at any instant during the process.[5] This effect is well known

TABLE 2. Outline of Process

Input Ach Released from Vesicles		Transport Diffusion of Ach Across Cleft, with Enzymatic Hydrolysis		Activation Allosteric Binding of Ach with Receptor Sites		Output Ionic End Plate Current
$Y_M = 0, \theta = 0$, all X				$R \rightleftharpoons A$		$g_{Na} = \bar{g}_{Na} Y_A$
Pulse $Y_M = 1, 0 \le \theta \le \theta_c, X = 0$ $= 0, \theta \ge \theta_c, X = 0$		$\dfrac{\partial Y_M}{\partial \theta} = \dfrac{\partial^2 Y_M}{\partial X^2} - \mu Y_M$		$Y_A = \dfrac{S}{K_A + S} \bar{A}$		
Pulse Train	→		→		→	$g_k = \bar{g}_k Y_A$
$Y_M = 1, \theta_{i-1} \le \theta \le \theta_i$ $(i = 1, 3, 5, \ldots)$ $= 0, \theta_{i-1} \le \theta \le \theta_i$ $(i = 2, 4, 6, \ldots)$		$\dfrac{\partial Y_M}{\partial X} = -\mu Y_M, X = 1, \theta > 0$		$\bar{A} = \dfrac{L\left(1 + \dfrac{S}{K_A}\right)^2}{L\left(1 + \dfrac{S}{K_A}\right)^2 + \left(1 + \dfrac{S}{K_R}\right)^2}$	→	epc $= g_{Na}(V - V_{Na}) + g_K(V - V_K)$

and widely documented by a number of researchers.[9,14] The latter effect causes an accelerating sharpening of local gradients[5] that partially offsets the retardation effect. This is formally identical to the case of diffusion with chemical reaction that has been widely investigated.

The combined influence of these effects can be estimated for a laminate membrane

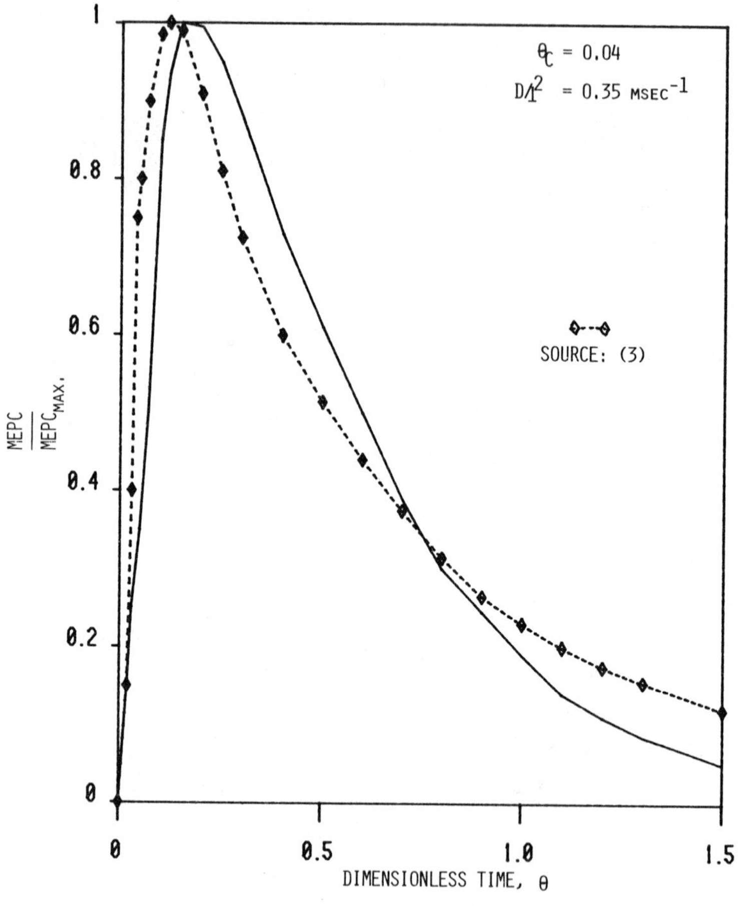

FIGURE 3. Miniature endplate current.

structure with one layer very thin, as shown below:

$$K_{12} = \frac{\ell_1 K_1 + \ell_2 K_2}{\ell_{12}}; \qquad \frac{D_1}{D_{12}} = 1 + \alpha \tag{18a, b}$$

where $\alpha = (\ell_2 K_{12})/\ell_1 K_1)$, K_{12} is the Ach distribution coefficient for the cleft postsynaptic membrane combination and K_1, K_2 are the values for the cleft and postsynaptic membrane, respectively. Now,

$$K_2 = \frac{(1-\epsilon) + \epsilon K_v}{(1-\epsilon) + \epsilon} = 1 + \epsilon(K_v - 1); \qquad \text{when } \epsilon = 1.0,\ K_2 = K_v \qquad \textbf{(19a, b)}$$

where K_v is the distribution coefficient for the receptor layer that can be estimated from measurements on pelleted vesicles purified from torpedo fish.[12] (Here ϵ is the vesicle volume fraction in a membrane laminate model system, comprising a collagen layer and a receptor-rich liquid membrane layer.[13])

Using available estimates[12–14] for the parameter ratios ($K_2/K_1 = 150$, $\ell_2/\ell_1 = 1/9$), we arrive at $(D_{12}/D_1) = (1/17.7)$. Taking D_1 ($\mu = 0$) as 1×10^{-6} cm^2/sec,[1,12,13] $D_{12} = 5.6 \times 10^{-8}$ cm^2/sec. Thus, the effective diffusivity for the laminate structure is estimated as 5.6×10^{-8} cm^2/sec. A major feature in this reduction of the diffusivity is the immobilization of a very large fraction of the penetrant population at receptor sites of both types, as reflected in the high value of K_2/K_1. Note that this estimate is consistent with an earlier one provided by Katz and Miledi[14] in their consideration of the effects of penetrant immobilization on the effective diffusivity.

The local equilibrium assumption means that at the downstream boundary, $X = 1.0$, the diffusion jump frequency and/or jump length are considered small compared to rates of receptor binding and dissociation. To test this assumption, one calculates a value for $2k_{-2}$, a measure of the influence of the slowest dissociation rate, and considers it in relation to $(D_{12})/\ell^2$; the value of the shift factor under the severe penetrant immobilization that prevails in the receptor-rich phase when the last fraction of the Ach front is exiting the scene. $(D_{12})/\ell^2$ was estimated from the mepc simulation. The value of $2k_{-2}$ was obtained from Reference 1. (The value of $1/\ell^2$ in $[(D_{12})/\ell^2]$ should be considered the transport mean value averaged over the cleft.)

The ratio, $\mu_R = [2k_{-2}\ell^2/(D_{12})]$, has the form of a reaction/transport modulus for the composite structure. We obtain a value of

$$\mu_R = (1.0 \text{ msec.}^{-1}) \left(\frac{1}{0.35} \text{ msec.} \right) = 2.85$$

which says that the slowest chemical reaction is still sufficiently rapid to permit diffusion control down to the last bit of Ach exiting the scene. This supports our assumption of local equilibrium for the binding/dissociation reaction set when $\mu \rightarrow 0$.

This assumption has been likewise verified for small molecule transport in microheterogeneous polymeric structures containing immobilizing sites.[9] The assumption can be relaxed to permit analysis of cases where the receptor binding/dissociation kinetics are not rapid. Analytical solutions are sometimes possible (e.g., see synopsis of work of Tshudy and Von Frankenburg[9] and APPENDIX B of this paper. However, for the mepc case just examined, it is unnecessary to resort to the assumption of reaction rate control to obtain the time axis shifts that are otherwise obtainable simply by allowing the pulse cut-off time, θ_c, to approach zero, as in the present work.

ENDPLATE POTENTIAL: EXCITATION AND CONDUCTION OF IMPULSES IN NERVES

The endplate system is considered to transmit signals by nerve pulses called endplate potentials (epp).

The notable features of such nerve pulses are: each rises to a level of approximately 20 mV in amplitude above the rest state and is of the order of msecs in duration, and pulses appear to be similar in shape. Therefore, the amount and type of information is decided by their sequence and the traversed pathway.[15]

The nerve membrane is the key component of the excitation mechanism. After the nerve impulse arrives at the terminal, it makes the inside portion more positive relative to the outside part by the stimulus. Having been stimulated over a time interval, τ_0, a chemical transmitter, for example, acetylcholine, is released from the vesicles (with a certain lag time). The larger the stimulus, the larger would be the release time, τ_c, and the endplate potential (epp) and the reverse is true for the lag time. For a stimulus below the threshold value, the lag time would be extremely large.

A continuous process of release of transmitter is an approximation of the quantal release pattern, and it is the number of released Ach quanta of unit size that dictate the size of the epp.[16] The released transmitter diffuses across the synaptic cleft, reacts with receptor molecules in the postsynaptic membrane, and alters the properties of the membrane. According to a simple picture of the ion channels, complex protein molecules (receptors) embedded in the membrane allow permeation of cations like Na^+, K^+, and Ca^{++} between the inside and outside of the membrane after spatial relocation of a charged region.[15] Under the effect, sodium ions flow across the postsynaptic membrane from an outside pool at high concentration and cause a further depolarization, thereby bringing about peaking of the endplate potential. At this point, the potassium flux rises so that potassium ions leave the synapse at a higher rate than that of the entering sodium ions. This brings the membrane back to its null potential. Under short time stimulus, the net membrane current density is approximately zero since capacity current balances out the potassium and sodium currents (assuming there is no leakage current involved). Therefore, following Hodgkin and Huxley,[17] the equation for the membrane potential is

$$-C_M \frac{dV}{dt} = g_K(V - V_K) + g_{Na}(V - V_{Na}) \tag{20}$$

This equation has been solved under the boundary condition:

$$t = 0,\ V = \text{rest potential}$$

The large potential change response in the form of endplate potential propagates away from the postsynaptic membrane without decrement in its amplitude. The well-known Cable equation for potential propagation along a continuous nerve fiber[17] can be applied in conjunction with Equation 20 under appropriate boundary conditions, as follows

$$-K \frac{d^2V}{dx^2} + \frac{dV}{dt} + \frac{1}{C_M}[g_K(V - V_K) + g_{Na}(V - V_{Na})] = 0 \tag{21}$$

Boundary conditions are:

(i) $t = 0,\ q/c_m = \Delta \mathcal{V}_0,\ x > 0$
(ii) $t > 0,\ q/c_m = \Delta \mathcal{V}_0,\ x = 0$
(iii) $t > 0,\ \Delta V = 0,\ x \rightarrow \infty$

Equations 20 and 21 can be put into more convenient forms, as follows:

$$\frac{-KC_M}{g_K + g_{Na}} \cdot \frac{d^2V}{dx^2} + \frac{C_M}{g_K + g_{Na}} \cdot \frac{dV}{dt} + \left(V - \frac{g_K V_K + g_{Na} V_{Na}}{g_K + g_{Na}}\right) = 0 \tag{22}$$

Let $\hat{V} = V - (g_K V_K + g_{Na} V_{Na})/(g_K + g_{Na})$

$$\lambda^2 = \frac{KC_M}{g_K + g_{Na}}; \qquad \tau_m = \frac{C_M}{g_K + g_{Na}}$$

Then,

$$-\lambda^2 \frac{d^2\hat{V}}{dx^2} + \tau_m \frac{d\hat{V}}{dt} + \hat{V} = 0 \tag{23}$$

At the membrane, that is, $x = 0$:

$$\tau_m \frac{d\hat{V}}{dt} + \hat{V} = 0 \tag{24}$$

Over a small interval of time,

$$\hat{V} = \hat{V}_0 e^{-t/\tau_m} \tag{25}$$

where $\hat{V}_0 = (g_K V_K + g_{Na} V_{Na})/(g_K + g_{Na})$ and τ_m is a function of time because of its dependence on $g_K(t)$, $g_{Na}(t)$, through $Y_{A1}(t)$ and $Y_{A2}(t)$.

Equation 24 can be solved by a high-accuracy, stepwise, numerical integration technique. In this paper, in order to generate membrane potential versus time curves, we used an analytical solution similar to Equation 25 by modifying the boundary condition and averaging parameters, that is, g_K and g_{Na}, over each time step.

$$\hat{V}(t + \Delta t) = \hat{V}(t) \cdot \exp(-\Delta t/\tau_m) \tag{26}$$

APPLICATIONS OF NEUROTRANSMISSION MODEL TO ENDPLATE POTENTIAL

Because of the accelerating effect of the enzyme reaction already alluded to, D_{12} ($\mu \rightarrow \infty$) can display an apparent increase of possibly fivefold; this result corresponds to an *apparent* increase of D_1 ($\mu = 0$) by a factor of five, bringing it close to the estimate of Rosenberry.[3] When this occurs, μ_R drops to a value of (2.85/5), or 0.57, and now one should investigate whether the slowest chemical reaction can exert partial rate control. For this purpose, it is necessary to analyze the coupling of the diffusion results with the receptor reaction kinetics, as shown in Appendix B, to obtain $A_1(t)$ and $A_2(t)$. The results are given below in Equations 27 and 28

$$\frac{A_1(t)}{A_0} = \frac{2k_{A1} S_0 Y_M}{k'_{A1}} \tag{27}$$

$$\frac{A_2(t)}{A_0} = \frac{(k_{A2})(S_0 Y_M)(2k_{A1} S_0 Y_M)}{(k'_{A1})(2k'_{A2})} \tag{28}$$

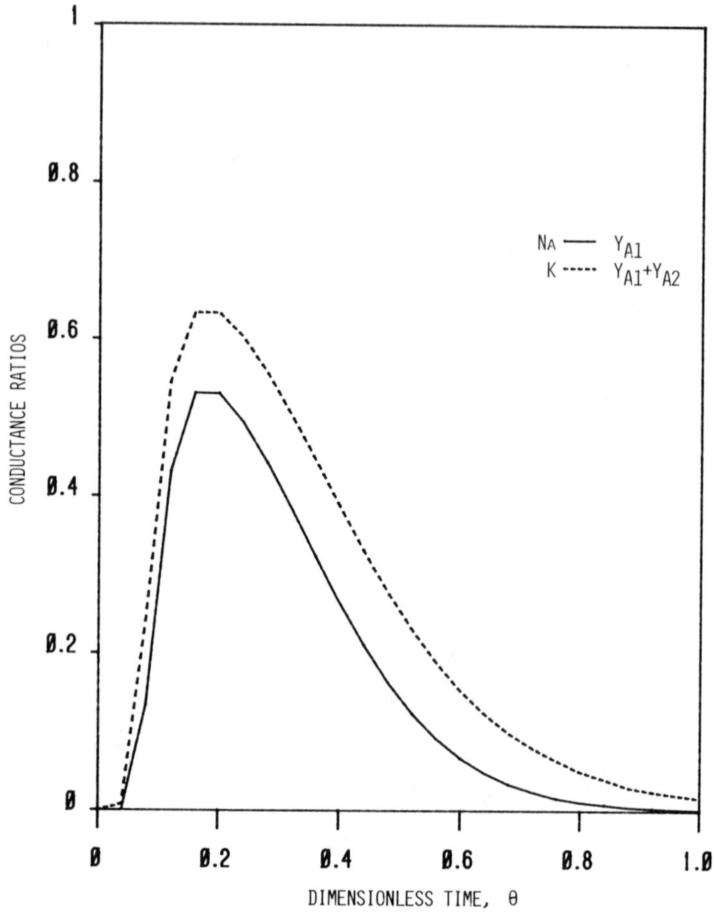

FIGURE 4. Ion conductance ratios.

In this instance, the results again correspond to the local equilibrium condition, that is, the time dependence of $A_1(t)$ and $A_2(t)$ is governed by the rate of arrival of neurotransmitter ($Y_M(t)$).

Now, writing equations for the fractional saturation of singly and doubly bound neurotransmitters, we obtain

$$Y_{A1}(t) = \frac{A_1(t)/2A_0}{\frac{1}{L}\left(1 + \frac{S}{K_R}\right)^2 + \frac{A_1(t)}{A_0} + \frac{A_2(t)}{A_0} + 1} \tag{29}$$

$$Y_{A2}(t) = \frac{A_2(t)/A_0}{\frac{1}{L}\left(1 + \frac{S}{K_R}\right)^2 + \frac{A_1(t)}{A_0} + \frac{A_2(t)}{A_0} + 1} \tag{30}$$

This leads to a realization that had previously eluded our notice, that is, the ion conductivities can best be individually formulated on the basis of the singly and doubly bound receptors:

$$g_{Na} = \bar{g}_{Na}(Y_{A2}) \tag{31}$$

$$g_K = \bar{g}_K(Y_{A1} + Y_{A2}) \tag{32}$$

That is, a time separation of the conductivity profiles, as is needed for epp simulation, can be effected on the basis of the probabilities of occurrence of singly and doubly bound species (Y_{A1}, Y_{A2}) (see FIG. 4). Note that limiting decay constants for Y_{A1} and Y_{A2} are 0.81 $msec^{-1}$ and 1.62 $msec^{-1}$, respectively. Therefore, epc process decay constants could be found within this range in many instances.[1]

Using the data provided in TABLE 3, Equations 31 and 32 are solved simultaneously with Equation 26, leading to the epp simulation results presented in FIGURE 5. The simulation was carried out in conformity with the known facts. That is, "the membrane potential rises by 15–20 mV from its rest value and returns with a half-decay time of a few msec."[8] (Therefore, τ at the peak should be set near 3.0 msec.) The enzyme reactivity is set at a relatively high value, that is, $\mu = 1.0$, as recommended by Rosenberry.[3] The same value of D/ℓ^2 that was shown earlier to fit the mepc data also fit the epp data. The time separation of the conductivity peaks (FIG. 4) is 0.04 units of θ. This is a very sensitive parameter, as is shown in FIGURE 5.

The agreement of the theory with the membrane potential trends, as shown in FIGURE 5, is very good. We are progressing in our efforts to simulate the propagating potential; results will appear shortly in a forthcoming paper from our laboratory. Preliminary results are shown in FIGURE 5. It seems appropriate now to recapitulate briefly.

Over the course of this work, models have been constructed, rigorously examined for logical inconsistencies, and successfully applied to existing data. Therefore, after critically considering all aspects of this approach and following tests of the theory with accepted standards of mepc and epp data, we conclude that our unified model is effective and useful in describing cation conduction through the synaptic junction. It is recommended that the model be thoroughly tested, explored and, as necessary, further refined!

TABLE 3. EPP Constants

Constant	Symbol	Value Chosen
Peak amplitude above rest (mV)	$V_{max} - V_r$	20[8]
Saturation conductivities:		
Sodium (mmho/cm^2)	$\bar{g}_{Na}$	2.0 (this work)
Potassium (mmho/cm^2)	$\bar{g}_K$	6.0 (this work)
Membrane capacitance (μF/cm^2)	C_M	1.0[17]
Mean decay constant, start of falling phase (msec)	$\frac{C_M}{(g_{Na} + g_K)}$	ca. 3.0[8]
Sodium equilibrium potential (mv)	$V_r - V_{Na}$	−139[19]
Potassium equilibrium potential (mv)	$V_r - V_K$	+10[19]
Time axis shift factor (msec)$^{-1}$	D/ℓ^2	0.35 (this work)
Dimensionless rise time	θ_R	0.35[18] (this work)
Dimensionless transmitter release time	θ_c	0.075 (this work)
Dimensionless enzyme reactivity	μ	1.0[3] (this work)
Propagation constant (cm^2 msec^{-1})	K	0.01[18] (this work)

SUMMARY

Transmission at the neuromuscular junction is known to comprise the following steps: vesicular release of Ach pulses; transport across the synaptic cleft; partial hydrolysis by Ach-ase; binding to Ach receptors; ion channel opening and closure, and,

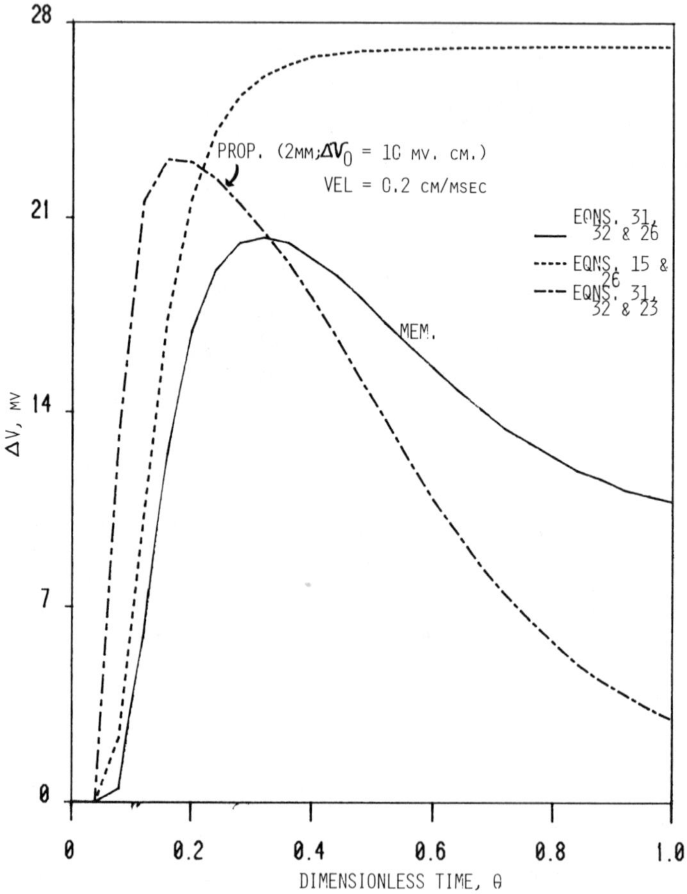

FIGURE 5. Endplate potential.

thereby, alteration of sodium and potassium conductances. Combining these elements, a unified model has been constructed that is effective in describing the relation between neurotransmitter arrival at receptor sites and channel openings, leading to simple relationships for sodium and potassium conduction at the neuromuscular junction. Application of the model to mepc and epp data has been successful.

ACKNOWLEDGMENTS

We thank Professors S. Gondo and H. Pedersen for some helpful discussions.

REFERENCES

1. WATHEY, J. C., M. N. NASS & H. A. LESTER. 1979. Biophys. J. **27:** 145–164.
2. ADAMS, P. R. 1980. Information Processing in the Nervous System, H. M. Pinsky & W. D. Willis, Jr., Eds. Raven Press. New York. pp. 108.
3. ROSENBERRY, T. L. 1979. Quantitative simulation of endplate currents at neuromuscular junctions based on the reaction of acetylcholine with acetylcholine receptor and acetylcholinesterase. Biophys. J. **26:** 263–289.
4. SODDU, A. & W. R. VIETH. 1980. Duality of substrate-binding site interactions in a structural protein. J. Mol. Catalysis. **7:** 491–510.
5. VIETH, W. R. & T. CIFTCI. 1981. Transport models of the neurotransmitter receptor interaction. Ann. N.Y. Acad. Sci. **369:** 99–112.
6. HEIDMAN, T. & J. P. CHANGEUX. 1978. Structural and functional properties of the acetylcholine receptor protein in its purified and membrane-bound states. Ann. Rev. Biochem. **47:** 317.
7. BARMAN, T. E. 1969. Enzyme Handbook, II. Springer-Verlag. New York, pp. 510.
8. CHANGEUX, J. P. 1981. The Acetylcholine Receptor: An Allosteric Protein, *In* Harvey Lectures.
9. VIETH, W. R., J. M. HOWELL & S. H. HSIEH. 1976. Dual Sorption Theory. J. Membr. Sci. **1:** 177–220.
10. MONNIER, A. M. 1977. Excitable and ion-selective artificial membranes; I. axon and bilayer membranes. J. Membr. Sci. **2:** 49.
11. KORDAS, M. 1977. On the role of junctional cholinesterase in determining the time course of the end-plate potential. J. Physiol. (London) **270:** 133.
12. HIROSE, S., W. R. VIETH & M. D. TAKAO. 1982. Sorption and transport of acetylcholine in reconstituted biomembrane structures. J. Mol. Catalysis. **18:** 11–22.
13. HIROSE, S. & W. R. VIETH. 1982. Combined transport and penetrant immobilization in a membrane laminate model of the neuromuscular junction. Appl. Biochem. Biotechnology. Humana Press. In press.
14. KATZ, B. & R. MILEDI. 1973. The binding of acetylcholine to receptors and its removal from the synaptic cleft. J. Physiol. (London) **231:** 549–574.
15. EHRENSTEIN, G. 1976. Ion channels in nerve membranes. Phys. Today. pp. 33–38.
16. KATZ, B. 1966. Nerve, Muscle, and Synapse. McGraw-Hill. pp. 140.
17. HODGKIN, A. L. & A. F. HUXLEY. 1952. A quantitative description of membrane current and its application to conduction and excitation in nerves. J. Physiol. (London) **117:** 500.
18. FATT, P. & B. KATZ. 1951. An analysis of the end-plate potential recorded with an intracellular electrode. J. Physiol. (London) **155:** 320–370.
19. TAKEUCHI, N. 1963. J. Physiol. (London) **167:** 141–155.

APPENDIX A

PROCESS COORDINATES

We have examined solutions to the problem of diffusion in a thin, wide disc for which the L/r ratio is $1/10$. Irrespective of the choice of point source or plane source, the unsteady-state solutions for pulse amplitude (Y_A) will depend upon $e^{-\delta_n^2 Dt/L^2}$

(our existing solution), $e^{-\alpha_n^2 Dt/r^2}$ (radial spreading analysis) and the product $e^{-\delta_n^2 Dt/L^2} \cdot e^{-\alpha_n^2 Dt/r^2}$. The following is a table of the eigenvalues, δ_n and α_n.

TABLE A1. EIGENVALUES

δ_n Very High Enzyme Activity i.e., $\mu \rightarrow \infty$	δ_n $\mu \rightarrow 0$	α_n	n
3.1	1.6	2.4	1
6.2	4.8	5.5	2
9.3	8.0	8.65	3
12.4	11.2	11.8	4
15.5	14.4	14.9	5

Now, when $\theta_r = (Dt/r^2)$ or $\theta_L = (Dt/L^2)$ is greater than 0.01, only one term $(\delta_1; \alpha_1)$ is needed in calculating amplitudes, Y_A, or decay constants for the falling phase. It is clear that, if α_n and δ_n are about the same, $(Dt/r^2) = (Dt/100L^2)$ is going to have a negligible influence compared to (Dt/L^2). Hence, our choice of the axial transport system as a focus for detailed analysis is clearly justified. Thus, solutions in the radial coordinate system describe cases where events are caused to move so slowly by diffusion that it becomes necessary to invoke a reaction limitation approach.

APPENDIX B

Partial Reaction Rate Control

$$2R_0 + S \rightleftharpoons R_1 \qquad 2A_0 + S \rightleftharpoons A_1$$

$$R_1 + S \rightleftharpoons 2R_2 \qquad A_1 + S \rightleftharpoons 2A_2$$

$$\frac{dA_1}{dt} = 2k_{A1} \cdot S \cdot A_0 - k'_{A1}A_1 - k_{A2} \cdot S \cdot A_1 + 2k'_{A2}A_2 \tag{B1}$$

$$\frac{dA_2}{dt} = k_{A2} \cdot A_1 \cdot S - 2k'_{A2}A_2 \tag{B2}$$

$Y_M = (S/S_0)$; $S = S_0 Y_M$; [$Y_M(t)$ is obtained from diffusion results; i.e., Equations 6 and 7].

Boundary Conditions: $A_1(0) = A_2(0) = 0$

Let $y_1 = A_1/A_0$; $y_2 = A_2/A_0$.

$$\frac{dy_1}{dt} = (2k_{A1} \cdot S_0 \cdot Y_M) - (k'_{A1})y_1 - (k_{A2} \cdot S_0 \cdot Y_M)y_1 + (2k'_{A2})y_2 \tag{B3}$$

$$\frac{dy_2}{dt} = (k_{A2} \cdot S_0 \cdot Y_M)y_1 - (2k'_{A2})y_2 \tag{B4}$$

Substitute

$$D = 2k_{A1} \cdot S_0 \cdot Y_M, \quad A = k'_{A1}, \quad B = k_{A2} \cdot S_0 \cdot Y_M \text{ and } C = 2k'_{A2} \tag{B5}$$

$$\frac{dy_1}{dt} = D - Ay_1 - By_1 + Cy_2; \quad \frac{dy_2}{dt} = By_1 - Cy_2, \; y_1 = \frac{1}{B}\left[\frac{dy_2}{dt} + Cy_2\right] \tag{B6}$$

$$\frac{d^2y_2}{dt^2} = B\frac{dy_1}{dt} - C\frac{dy_2}{dt} = B\left[D - \frac{A+B}{B}\left(\frac{dy_2}{dt} + Cy_2\right) + Cy_2\right] - C\frac{dy_2}{dt}$$

$$= B \cdot D - (A + B + C)\frac{dy_2}{dt} - (AC)y_2 \tag{B7}$$

or

$$\frac{d^2y_2}{dt^2} + E\frac{dy_2}{dt} + Fy_2 = G \tag{B8}$$

where $E = A + B + C$, $F = A \cdot C$, $G = B \cdot D$. Over a small interval of time, the solutions are:

$$y_2 = \frac{G}{F} + C_1e^{m_1t} + C_2e^{m_2t} \tag{B9}$$

where

$$m_{1,2} = \frac{-E \pm \sqrt{E^2 - 4F}}{2} \tag{B10}$$

and

$$E^2 - 4F > 0 \tag{B11}$$

$$t = 0, \quad y_2 = 0, \quad y_1 = 0, \quad \frac{dy_2}{dt} = 0 \tag{B12}$$

$$\frac{G}{F} = -(C_1 + C_2) \tag{B13}$$

$$m_1C_1 + m_2C_2 = 0; \quad C_2 = \frac{m_1}{m_2}C_1 \tag{B14}$$

$$C_2 = -\frac{m_1}{m_2}\frac{G}{F}\frac{m_2}{m_1 - m_2} \tag{B15}$$

$$y_1 = \frac{1}{B}[C_1m_1e^{m_1t} + c_2m_2e^{m_2t} + Cy_2] = m_2e^{m_2t}\left(-\frac{m_1}{m_2} \cdot \frac{D}{F} \cdot \frac{m_2}{m_1 - m_2}\right)$$
$$+ m_1e^{m_1t}\left(\frac{D}{F}\frac{m_2}{m_1 - m_2}\right) + \frac{D}{F} + c_1e^{m_1t} + c_2e^{m_2t} \tag{B16}$$

$$y_2(\theta) = \frac{G}{F} + C_1e^{\hat{m}_1\theta} + c_2e^{\hat{m}_2\theta} \tag{B17}$$

$$y_1(\theta) = \frac{1}{B}\,[C_1 m_1 e^{\hat{m}_1\theta} + C_2 m_2 e^{\hat{m}_2\theta} + Cy_2] \qquad \textbf{(B18)}$$

Now, because of the magnitudes involved (see TABLE B1), the terms $e^{\hat{m}_1\theta}$, $e^{\hat{m}_2\theta}$ drop out virtually immediately ($\theta \simeq 0$) and the solutions may be simplified to the equilibrium forms as follows:

$$y_2(\theta) \simeq \frac{G}{F} = \frac{BD}{F} \qquad \textbf{(B19)}$$

$$y_1(\theta) \simeq \frac{C}{B}\,y_2(\theta)) \simeq \frac{C}{B} \cdot \frac{BD}{F} = \frac{CD}{F} \qquad \textbf{(B20)}$$

$$\frac{y_1(\theta)}{y_2(\theta)} = \frac{C}{B} = \frac{2k'_{A2}}{k_{A2}S_0Y_M(\theta)} \qquad \textbf{(B21)}$$

TABLE B1. MAGNITUDES

$A = k'_{A1} = 7.5 \times 10^{3a}$
$B = k_{A2}S_0Y_M = 3 \times 10^4 Y_M{}^a$
$C = 2k'_{A2} = 10^{3a}$
$D = 2k_{A1}S_0Y_M = 3 \times 10^8 \times 10^{-3} Y_M = 3 \times 10^5 Y_M{}^a$
$E = A + B + C = 8.5 \times 10^3 + 3 \times 10^4 Y_M$
$F = AC = 7.5 \times 10^6$
$G = BD = 3 \times 10^4 \times 3 \times 10^5 \times Y_M^2 = 9 \times 10^9 Y_M^2$

[a]See reference 1.

Biochemical Energy Conversion by Immobilized Whole Cells

SHUICHI SUZUKI, ISAO KARUBE, HIDEAKI MATSUOKA, AND SATOSHI UEYAMA

Research Laboratory of Resources Utilization
Tokyo Institute of Technology
Nagatsuta-cho, Midori-ku
Yokohama, 227, Japan

HIROAKI KAWAKUBO, SATOSHI ISODA, AND TOSHIAKI MURAHASHI

Central Research Laboratories
Mitsubishi Electric Co.
Tsukaguchihoncho, Amagasaki
Hyogo, 661, Japan

INTRODUCTION

The biochemical production of energy is attractive as a new energy source and various approaches are now under investigation. Various bacteria and algae produce hydrogen under anaerobic conditions.[1] Hydrogen is known as one of the cleanest fuel resources, and exhibits excellent reactivity in electroactive materials.[2] Fuel cell systems are now attracting attention as efficient energy conversion systems. Hydrogen produced biochemically can be supplied to a fuel cell system. One of the first biochemical fuel cell systems designed to use hydrogen-producing bacteria was developed by Rohrback *et al.*,[3] who reported on the production of hydrogen from the fermentation of glucose by *Clostridium butyricum.* However, because the hydrogen evolution system, especially hydrogenase, in bacteria is unstable, it is difficult to use whole cells for continuous hydrogen production.[4]

Hydrogen-producing bacteria, *Clostridium butyricum,* were immobilized in polyacrylamide gel and the immobilized whole cells continuously produced hydrogen from glucose under aerobic conditions.[5,6] Therefore, immobilized hydrogen-producing bacteria have been applied to a fuel cell. However, the current obtained from the fuel cell was low.

Recently hydrogen-producing bacteria, *Clostridium butyricum,* were immobilized in an agar-acetylcellulose filter and the immobilized whole cells were employed for continuous production of hydrogen from an alcohol factory's wastewaters with a continuously stirred reactor system.[7] The immobilized whole cells continuously produced hydrogen (20 ml $\cdot$ min^{-1} $\cdot$ kg wet gel^{-1}) over a one-month period. Hydrogen produced was supplied to the hydrogen-oxygen (air) fuel cells. The fuel cell system was left on for 10 days and a current of 0.8 A and the cell voltage of 2.2 V (5 cells) were obtained over a 10-day period. However, a scale-up of the bacterial fuel cell system is still required for the development of a new energy conversion system.

HYDROGEN PRODUCTION BY VARIOUS BACTERIA

TABLE 1 shows the hydrogen-evolving activity of various bacteria. Hydrogen produced was determined by gas chromatography. *C. butyricum* IFO 3847 evolved the largest amount of hydrogen and was used thereafter.

Cultivation and Immobilization of Microorganisms

C. butyricum was cultivated in a medium containing 1.0% glucose, 0.5% peptone, 0.2% yeast extract, 0.2% meat extract, and 1.25% K_2HPO_4. After precultivating in a 3-l medium for 13 hr at 37°C, a large-scale cultivation was performed in a 180-l medium for 5 hr at 36–38°C. The microorganisms were collected by continuous centrifugation at 6500 × *g* and 150 ml · min^{-1}. After cultivation, about 800 g wet cells were obtained. The cells were suspended in a phosphate buffer solution (0.1 M, pH 7.5) and stored at 4°C under nitrogen gas.

C. butyricum (3 kg wet cells) was absorbed in a cellulose paper (350 sheets, 36 × 11 cm^2, thickness 1.2 mm). Then the surface of the sheets was coated with 1.5% agar solution at 50°C.

Analytical Methods for Various Organic Compounds

Glucose and sucrose were determined by enzyme electrodes prepared by the authors. The enzyme electrode for glucose consisted of immobilized glucose oxidase and an oxygen electrode. The enzyme electrode for sucrose consisted of immobilized invertase, mutarotase, glucose oxidase, and an oxygen electrode. Organic acids in a

TABLE 1. Screening of Hydrogen-producing Bacteria.[a]

Strains	Hydrogen Evolved [μmol]
Cladosporium butyricum IFO 3315	29
Clostridium butyricum IFO 3847	33
Cl. b. IFO 3852	31
Cl. b. IFO 3858	31
Cl. b. ss. *convexa* IAM 19021	2
Cl. b. ss. *saccharobutyricum* IAM 19005	28
Cl. caproicum IAM 19228	7
Cl. isopropylicum IAM 19102	0
Cl. valerianicum IAM 19236	3
Cl. pectinolyticum IAM 19232	1
Escherichia coli (wild type)	23
E. coli HB 101	20
E. coli RR 1	28
E. coli C 600	21
E. coli CGSC 4291	29
E. coli NS 428	20
E. coli NS 433	25

[a]Experimental conditions:
substrate: glucose 0.1%, 10 ml (55 μmol)
microorganisms: 0.05 g wet cells
temperature: 37°C, pH: 7.0, incubation: 20 hr

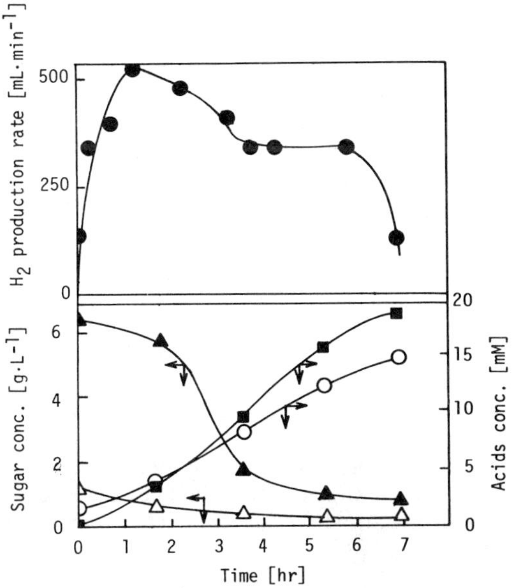

FIGURE 1. Hydrogen production from molasses in a batch system. Immobilized cells: 2 kg wet cells, pH: 6.3–6.7, Temperature: 36.5–37.7°C.

(○) acetate
(△) glucose
(■) butyrate
(▲) sucrose

reaction mixture were determined by high-pressure liquid chromatography (Shimadzu Seisakusho, Model LC-3A) (column: SCR-101 H). Gas components were analyzed by gas chromatography (Shimadzu Seisakusho, Model GC-3BT) (column: molecular sieve 5A, 60–80 mesh).

Hydrogen Evolution by Immobilized Whole Cells

Hydrogen production by immobilized whole cells was performed in a batch system using a 200-l reactor. About 3 kg of immobilized whole cells (wet) and 150 l of molasses (sugar content, 0.87%) were employed in the experiments. Immobilized whole cells continuously evolved hydrogen over 7 hours, as shown in FIGURE 1. Initially, the amount of hydrogen produced increased with increasing incubation time. Since molasses was used as a raw material, activation of the hydrogen-production system might occur in the medium. Main organic components in the reactor were analyzed after 7 hours of operation and summarized in TABLE 2. About 75% of the glucose and sucrose were consumed and organic acids such as acetate, butyrate, and lactate were accumulated in the reactor.

About 1.04 moles of hydrogen were produced from 1 mole glucose by the immobilized whole cells. The conversion of carbohydrates to hydrogen is achieved by a multienzyme system. The hydrogen-producing system in bacteria is believed to involve glucose conversion to 2 moles of pyruvate, with 2 moles of ATP and 2 moles of NADH formed by the Embden-Meyerhof pathway.[8,9] Then pyruvate may be oxidized through a pyruvate/ferredoxin oxidoreductase or through a pyruvate/formate lyase. The products of a pyruvate/ferredoxin oxidoreductase reaction are acetyl CoA, CO_2 and reduced ferredoxin. Pyruvate/formate lyase decomposes pyruvate to acetyl CoA and formate. Then formate is oxidized to CO_2 and ferredoxin is reduced. Furthermore, NADH/ferredoxin oxidoreductase also oxidizes NADH and reduces ferredoxin.[10]

TABLE 2. Change in Main Organic Compounds During Operation in a Batch System.

Components	Concentrations Before	After	Difference
Acetate[a]	1.8	15.0	13.2
Butyrate[a]	0	18.4	18.4
Lactate[a]	0	9.2	9.2
Glucose[b]	1.2	0.3	0.9
Sucrose[b]	6.3	0.7	5.6

[a]mM
[b]$g \cdot l^{-1}$

Reduced ferredoxin is reoxidized and hydrogen is formed by hydrogenase. As a result, 4 moles of hydrogen are produced from 1 mole of glucose under ideal conditions.[11]

As described above, of total glucose consumed, about 26% was converted to hydrogen in this immobilized whole-cell system under optimum conditions. Therefore, improvement of hydrogen productivity by genetic engineering techniques and screening of new hydrogen-producing bacteria are required for efficient energy conversion and are under investigation in our laboratory.

Storage Stability of the Immobilized Whole Cells

Immobilized whole cells were stored in a phosphate buffer solution (0.1 M, pH 7.5) at 4°C under air. The hydrogen productivity of the immobilized whole cells was examined by gas chromatography. The hydrogen productivity gradually decreased, and more than 50% of the initial activity was lost during a 50-day storage as shown in FIGURE 2. However, the immobilized whole cells could be activated by incubation in a culture medium containing peptone and other nutrients.

Phosphoric Acid Fuel Cell System

FIGURE 3 shows a schematic diagram of a phosphoric acid fuel cell. The fuel cell consisted of two carbon plate separators, a cathode, and an anode. The carbon plate

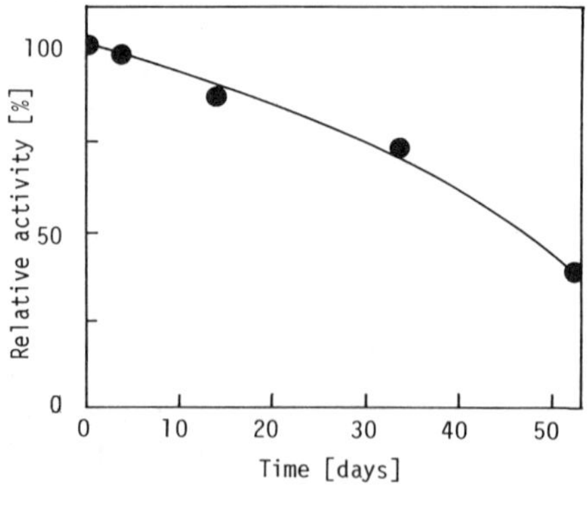

FIGURE 2. Storage stability of immobilized whole cells. Storage condition: 4°C in phosphate buffer (0.1 M, pH 7.5) under air.
Reaction conditions: 37°C, pH 6.3–6.7, 1-hr reaction.
Initial activity: 1.17 mmol hydrogen $\cdot$ $hr^{-1} \cdot$ g wet cell^{-1}

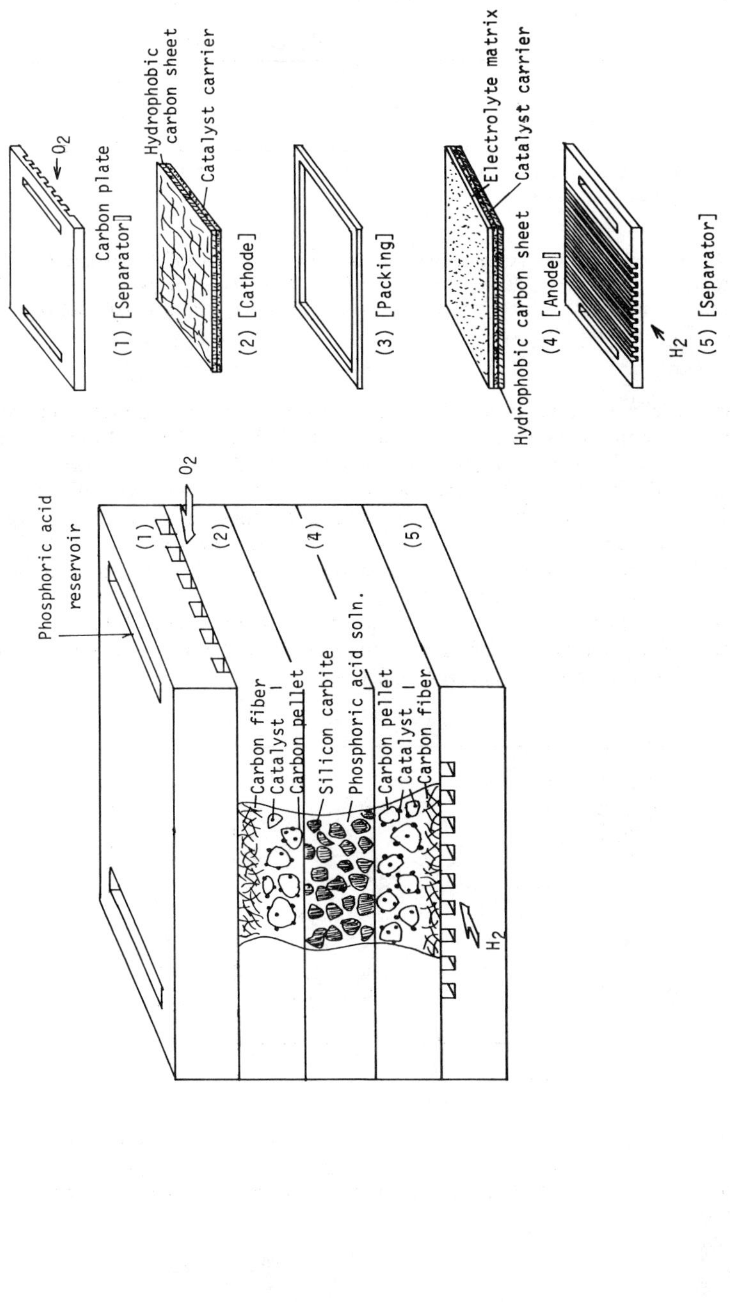

FIGURE 3. Schematic diagram of the phosphoric acid fuel cell.

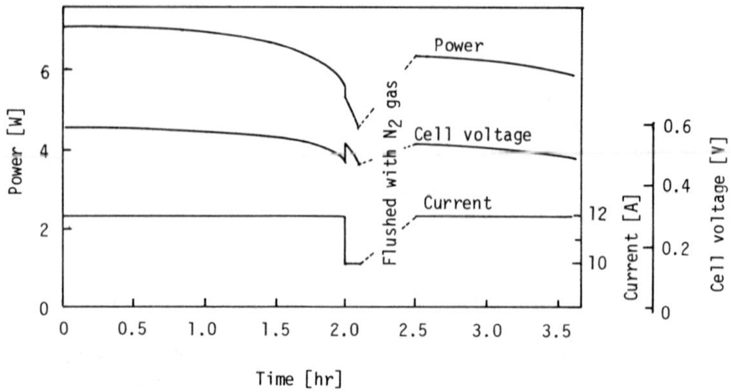

FIGURE 4. Time course of the bacterial fuel cell system in batch system. Operational conditions: (I) Bioreactor—Immobilized cells: ca. 3 kg wet cells, pH: 6.3–6.6, Temperature: 36.5–37.5°C. (II) Fuel cell—1 fuel cell was employed. Temperature: 200°C, H_2 flow rate: 400–800 ml · min^{-1}, O_2 flow rate: 1–1.5 l · min^{-1}.

separators have gas passages for hydrogen and oxygen. The anode and cathode consisted of hydrophobic carbon fibers, platinum black catalysts and carbon pellets. The matrix containing silicon carbite and phosphoric acid (100%) separated the anode and the cathode. Two fuel cells were connected in parallel. The fuel cell system was equipped with the electronic control system for a constant current. By adjusting an electronic load manually, a steady-state current could be maintained during the operation. The current and cell voltage were measured by an ammeter and an electrometer and displayed on a recorder.

The effect of temperature on cell performance was examined. The current density and the cell voltage increased with rising temperature of the fuel cell. In the following experiments, the fuel cell was operated at 200°C.

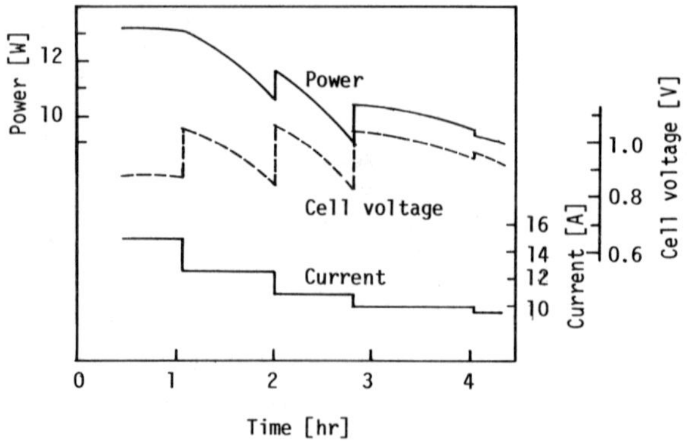

FIGURE 5. Improved performance of the bacterial fuel cell system. (I) Bioreactor—Pressure was controlled to the positive side. (II) Fuel cell—2 fuel cells were employed. H_2 flow rate: 400–800 ml · min^{-1} (through KOH solution and tap water traps). Other conditions were the same as FIGURE 4.

Bacterial Fuel Cell System

Biogas containing hydrogen and carbon dioxide was directly transferred to the anode at 400–800 ml · min^{-1}. Oxygen was also transferred to the cathode at 1.5 l · min^{-1}. FIGURE 4 shows the time course of the electric current, cell voltage, and electric power. After 2 hours of operation, the cell voltage suddenly decreased. When the fuel cells were flushed with nitrogen gas for about 30 min, the output voltage returned to the initial level.

Therefore, biogas might contain some poisonous components that inhibit the catalytic activity of platinum. In order to remove the poisonous components, biogas was passed through 2 traps containing concentrated KOH solution (30%) and tap water.

The performance of the fuel cell was considerably decreased by the oxygen contaminated when the gas pressure in the reactor was reduced. Therefore, control of the gas pressure was very important. With the control system described below, the gas pressure in the reactor was controlled to the positive side by supplying nitrogen gas. The improved performance of the bacterial fuel cell system is shown in FIGURE 5. The stable current from 14 to 10 A was obtained for 4 hr. Power of 9 to 13 W was observed during operation.

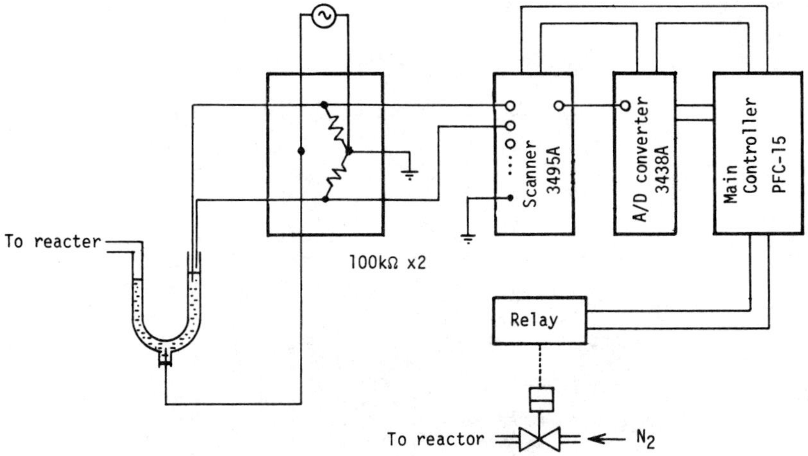

FIGURE 6. Gas pressure control system of the reactor.

Control System for a Bacterial Fuel Cell System

The bacterial fuel cell system was controlled with a microcomputer system composed of a main controller (Panafacom, PFC-15), scanner (Hewlett Packard, 3495A), AD converter (Hewlett Packard, 3438A), and a relay system.

pH in the reactor was monitored with a combined-type pH electrode and displayed on CRT. Concentrated NaOH was injected into the reactor automatically to adjust pH at 6.2–6.8.

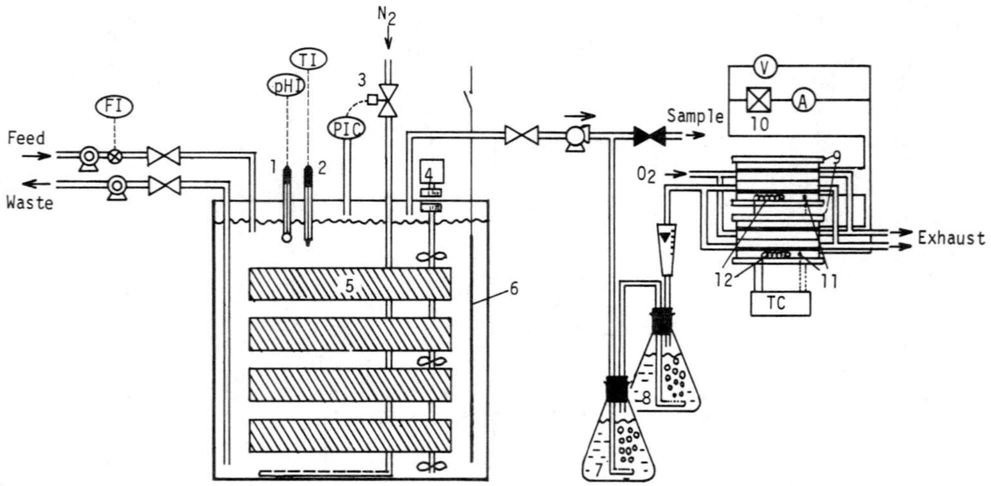

FIGURE 7. Schematic diagram of the bacterial fuel cell system for continuous operation.

(1) pH electrode, (2) thermister, (3) solenoid gas valve, (4) motors, (5) immobilized bacteria, (6) heater, (7) KOH solution, (8) water, (9) fuel cells, (10) electronic load, (11) thermocouples, (12) heaters.

pHI: pH indicator, TI: temperature indicator, PIC: pressure indicator controller, TC: fuel cell temperature controller, FI: feed rate indicator.

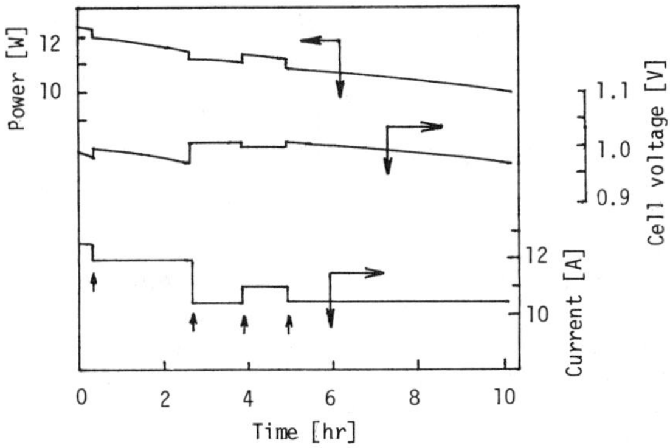

FIGURE 8. Continuous operation of the bacterial fuel cell system. Operational conditions: (I) Bioreactor—pH: 6.2–6.8, Temperature: 36.4–37.8°C, Immobilized cells: ca. 3 kg wet cells, Feed rate: 900 ml · min^{-1}. (II) Fuel cell—Temperature: 200°C, H_2 flow rate: 400–800 ml · min^{-1} (through KOH solution and tap water traps), O_2 flow rate: 1–1.5 l · min^{-1}.

Temperature in the reactor was monitored with a thermistor and displayed on CRT. Temperature was controlled at 40°C with a heater (5 kW).

Gas pressure in the reactor was monitored with a U-tube manometer with level switches as shown in FIGURE 6. The gas pressure was controlled at 0–2 cm of water by this system.

Continuous Operation of a Bacterial Fuel Cell

FIGURE 7 shows a schematic diagram of the bacterial fuel cell system for the continuous operation. Diluted molasses was supplied continuously at 900 ml · min^{-1} and the medium was removed from the reactor at the same flow rate. About 10–12 W was continuously obtained for 10 hr as shown in FIGURE 8. The electric current obtained was 10–12 A.

During the operation, the hydrogen productivity decreased gradually. This might be caused by the accumulation of organic acids and other trace compounds that might inhibit the hydrogen-producing system. The hydrogen productivity was recovered by the incubation of the immobilized whole cells in the cultivation medium.

TABLE 3. Hydrogen Energy Efficiency of the Reactor

Items	Equations	Results
(a) Consumed sugar Sucrose 5.6 g · l^{-1} Glucose 0.9 g · l^{-1} Fructose (0.9 g · l^{-1})	(5.6 + 0.9 + 0.9) × 150	1.11 kg
(b) Amount of sugar equivalent to glucose	(a)/180	6.17 mol
(c) Heat content of consumed glucose	(b) × 647	3990 kcal
(d) H_2 generated during 7-hour operation		157 l
(e) Heat content of H_2 generated	[(d)/24.6] × 68	434 kcal
(f) Energy efficiency	(e)/(c) × 100	10.9%
(g) Theoretical amount of H_2 generated from glucose	(b) × 4 × 24.6	608 l
(h) H_2 production efficiency	(d)/(g) × 100	26%

Energy Balance of the Reactor and the Bacterial Fuel Cell System

The reactor contained 150 l of molasses. The sugar content was about 0.87% (sucrose 0.63%, glucose 0.12%, fructose 0.12%). From the results of FIGURE 1, the energy efficiency was estimated and summarized in TABLE 3. The energy efficiency obtained was 10.9% as indicated by item (f). As described above, theoretically, 1 mole glucose generates 4 moles of hydrogen. Therefore, the actual productivity was 26% as shown by item (h).

On the other hand, electrical energy balance was calculated in TABLE 4. The bacterial fuel cell system required electric power for agitation, pumping (feeding of biomass and gas), heaters for fuel cells and control system. The total power consumed for the operation was about 1 kW. Hydrogen was converted to electric current with the fuel cells. The electric power obtained was about 11 W. Therefore, 1% of the energy consumed was recovered by the bacterial fuel cell system. However, no attempt has been made to optimize the biochemical fuel cell system. Therefore, the energy balance would be improved after optimization of the system.

TABLE 4. Electrical Energy Balance of the Bacterial Fuel Cell System.

	Input Power [W]		Output Power [W]	
	Nominal	Actual (total)	H_2 (for combustion)	Electricity
Motors for agitation	80			
Pumps for feeding	220	1070	120	11
Pumps for gas transfer	43	(100%)	(11%)	(1.0%)
Heaters for fuel cells	800		(600 ml · min^{-1})	
Control system	300			

SUMMARY

Hydrogen-producing bacteria, *Clostridium butyricum,* were immobilized in papers with agar (1.5%) gel. The immobilized whole cells (3 kg wet cells) were employed for continuous production of hydrogen from molasses. The immobilized whole cells continuously produced hydrogen (400–800 ml · min^{-1}) over a two-month period. Hydrogen produced was supplied to two phosphoric acid fuel cells connected in parallel. About 10–12 W and a current of 10–12 A were obtained for 10 hr. The energy balance of the bacterial fuel cell system was also discussed.

REFERENCES

1. THAUER, R. K., K. JUNGERMANN & K. DECKER. 1977. Energy conservation in chemotropic anaerobic bacteria. Bacteriol. Rev. **41:** 100–180.
2. DELDUCA, M. G., J. M. FRISCOE & R. W. ZURILLA. 1963. Direct and indirect bioelectrochemical energy conversion systems. Dev. Ind. Microbiol. **4:** 81–91.
3. ROHRBACK, G. H., W. R. SCOTT & J. H. CANFIELD. 1963. Biochemical fuel cells. Proc. 16th Annu. Power Sources Conf. pp. 18–21.
4. GRAY, C. T. & H. GEST. 1965. Biological formation of molecular hydrogen. Science **148:** 186–191.
5. KARUBE, I., T. MATSUNAGA, S. TSURU & S. SUZUKI. 1976. Continuous hydrogen production by immobilized whole cells of *Clostridium butyricum.* Biochim. Biophys. Acta **444:** 338–343.

6. SUZUKI, S., I. KARUBE, T. MATSUNAGA, S. KURIYAMA, N. SUZUKI, T. SHIROGAMI & T. TAKAMURA. 1980. Biochemical energy conversion using immobilized whole cells of *Clostridium butyricum*. Biochimie **62:** 353–358.
7. KARUBE, I., S. SUZUKI, T. MATSUNAGA & S. KURIYAMA. 1981. Biochemical energy conversion by immobilized whole cells. Ann. N.Y. Acad. Sci. **369:** 91–98.
8. THAUER, R. K., F. H. KIRCHNIAWY & K. A. JUNGERMANN. 1972. Properties and function of the pyruvate-formate-lyase reaction in *Clostridiae*. Eur. J. Biochem. **27:** 282–290.
9. RAEBURN, S. & J. C. RABINOWITZ. 1971. Pyruvate:ferredoxin oxidoreductase—1. The pyruvate-CO_2 exchange reaction. Arch. Biochem. Biophys. **146:** 9–20.
10. JUNGERMANN, K. A., R. K. THAUER, G. LEIMENSTOLL & K. DECKER. 1973. Function of reduced nucleotide-ferredoxin oxidoreductases in saccharolytic *Clostridiae*. Biochim. Biophys. Acta **305:** 268–280.
11. ADAMS, M. W. W., L. E. MORTENSON & J. S. CHEN. 1981. Hydrogenase. Biochim. Biophys. Acta **594:** 105–176.

A Simple Batch Fermentation Model: Theme and Variations

DAVID F. OLLIS

Chemical Engineering Department
University of California
Davis, California 95616

INTRODUCTION

The kinetics of microbial growth, substrate consumption, and product formation are routinely formulated in terms of equations that lead to coupling between the associated rates, as, for example, in the general case Equations 1a, b, and c.

$$\frac{dX}{dt} = f(X, P, S) \quad \textbf{(1a)}$$

$$\frac{dP}{dt} = g(X, P, S) \quad \textbf{(1b)}$$

$$\frac{dS}{dt} = h(X, P, S) \quad \textbf{(1c)}$$

In consequence, analytical solutions of these equations are not often possible, and resort is made (often usefully) to numerical solution techniques.

Batch microbial growth is characterized by two regions, that of exponential growth where $dX/dt = \mu_{max} \cdot X$ (μ_{max} = constant) and of stationary growth ($dX/dt = 0$). It is not uncommon for the transition region between these two kinetic regions of substrate-independent growth rates to occupy 10–20% or less of the total fermentation time. In consequence, for many batch fermentations, a simpler approach than Equation 1a is available by taking an autonomous growth form, that is, $dX/dt = f'(X)$. In the following examples, we illustrate the resulting simplifications of the kinetic system Equations 1a, 1b, 1c, which allow for both analytical solutions of these equations and convenient serial evaluation of the kinetic parameters.

EXAMPLE 1: EXOPOLYSACCHARIDE PRODUCTION

Microbial exopolysaccharides may be produced in growth and nongrowth phases, depending upon the microbial strain of concern. In consequence, a general kinetic system for microbial exopolysaccharide production is given by Equations 2a, b, and c.

Biomass X(t):

$$\frac{dX}{dt} = \mu_{max}X(1 - X/X_{max}) \quad \textbf{(2a)}$$

Product P(t):

$$\frac{dP}{dt} = nX + m\frac{dX}{dt} \qquad \text{(2b)}$$

(nongrowth associated) (growth associated)

Substrate S(t):

$$\frac{dS}{dt} = -\frac{1}{Y_x}\frac{dX}{dt} - \frac{1}{Y_p}\frac{dP}{dt} - k_eX \qquad \text{(2c)}$$

(biomass) (biopolymer) (maintenance)

This particular form, shown previously to describe production of the biopolysaccharides xanthan gum,[1] pullulan,[2] polyalginic acid,[2] and a biopolymer of a *Pseudomonas sp.*,[2] allows the following analytical solutions and parameter evaluations.

Biomass

$$X(t) = \frac{X_o e^{\mu_{max}t}}{1 - (X_o/X_{max})(1.0 - e^{\mu_{max}t})} \qquad \text{(3a)}$$

Rearrangement yields

$$\ell n\left[\frac{X(t)/X_{max}}{1 - X(t)/X_{max}}\right] = \mu_{max}t - \ell n\left[\frac{X_{max}}{X_o} = 1.0\right] \qquad \text{(3b)}$$

Parameter evaluation proceeds in the following way: from batch data, X_{max} is obtained by inspection; a plot of $\ell n[X/(X_{max} - X)]$ versus time will be linear if Equation 3a is valid, giving a slope of μ_{max} and an intercept at $t = 0$ of $\ell n[(X_{max}/X_o) - 1.0]$, from which the initial *viable* inoculum size, which went on to grow to X_{max}, is determined.

Product

The Luedeking-Piret[3] form (Equation 2b) parameters are evaluated as follows: in stationary conditions, $(dX/dt) = 0$, $X = X_{max}$ and Equation 2b provides

$$n = \frac{(dP/dt)_{\text{stationary phase}}}{X_{max}} \qquad \text{(4a)}$$

Integration and rearrangement of Equation 2b gives Equation 4b,

$$P(t) - P_o - n\left(\frac{X_{max}}{\mu_{max}}\right)\ell n\,[1 - (X_o/X_{max})(1.0 - e^{\mu_{max}t})] = m(X(t) - X_o) \qquad \text{(4b)}$$

from which a plot of the left-hand side versus $(X(t) - X_o)$ gives a line of slope m.

Substrate

Combination of Equations 2b and 2c yields Equation 5a

$$\frac{dS}{dt} = -\alpha\frac{dX}{dt} - \beta X \tag{5a}$$

where

$$\alpha \equiv \left(\frac{1}{Y_x} + \frac{m}{Y_p}\right) \tag{5b}$$

and

$$\beta \equiv \left(\frac{n}{Y_p} + k_e\right) \tag{5c}$$

The parameters α and β are evaluated in the same manner as for (m, n) of Equation 2b.

The comparison of the model of Equations 2a, b, and c with experimental data of Moraine and Rogovin[4,5] for a *Xanthamonas campestris* strain is given in FIGURE 1, along with the corresponding parameter values (legend). A subsequent paper[2] indicates similar quality agreement between model and data for fermentations producing other microbial extracellular biopolymers from *Pseudomonas* species,[6,7] *Azotobacter vinelandii,* and *Aureobasidium pullulans.*[8] Model parameters from these four fermentation examples are given in TABLE 1. These results show that exobiopolymer production may be growth-associated (pullulan and polyalginate), nongrowth associated (*Pseudomonas* sp. biopolymer) or mixed (xanthan gum). These results indicate that the two-term description for dP/dt assumed in Equation 2b is both necessary and sufficient to provide adequate kinetic descriptions of exopolysaccharide fermentations.

EXAMPLE 2: ETHANOL MULTIPRODUCT FERMENTATION

The Luedeking-Piret form for product rate dP/dt allows straightforward extension of Equations 2a, b, and c to a multiproduct fermentation.[9] Here, *each* product is modeled by an equation of the form of Equation 2b:

$$\frac{dP_i}{dt} = n_i X + m_i\frac{dX}{dt} \tag{6a}$$

The corresponding form for a generalization of Equation 2c is again a Leudeking-Piret form:

$$\frac{dS}{dt} = -\frac{1}{Y_x}\frac{dX}{dt} - \sum\left(\frac{1}{Y_{P_L}}\frac{dP_i}{dt}\right) - k_e X \tag{6b}$$

$$= -\alpha\frac{dX}{dt} - \beta X \tag{6c}$$

where

$$\alpha \equiv \left[\frac{1}{Y_x} + \sum_{i=1}^{N}\left(\frac{m_i}{Y_{P_i}}\right)\right] \tag{6d}$$

and

$$\beta \equiv \left[k_e + \sum_{i=1}^{N}\left(\frac{n_i}{Y_{P_i}}\right)\right] \tag{6e}$$

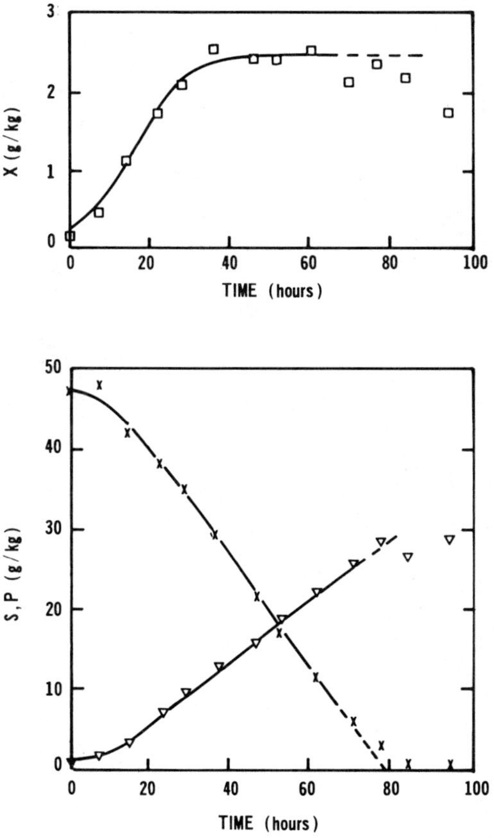

FIGURE 1. Course of *X. campestris* fermentation from Figure 4 of Reference 4. Original data supplied by Silman[5]: (□) biomass, (X) substrate, (▽) products; (—) curves calculated from Equations 2a, b, and c of text using parameters values X_{max} = 2.45 g/kg, $\mu = 0.15\ hr^{-1}$, X_o = .222 g/kg, n = 0.155 gP/gP/gX − hr, m = 1.83 gP/gX, α = 2.0 gS/gX, β = 0.284 gS · gX − hr. (Reprinted from *Biotechnol. Bioeng.* **22:** 859 by permission of John Wiley and Sons.)

As only the number of equations, but not their ultimate form, is changed, parameter evaluation proceeds as before. For a fermentation example[10] leading to formation of ethanol plus acetic, lactic, and formic acids, the corresponding parameter values are given in TABLE 2. The correspondence between data and model is illustrated in FIGURES 2, 3, and 4 for glucose as a substrate, an example with six time-varying concentrations: X, S, P_1 (ethanol), P_2 (acetate), P_3 (formate), P_4 (lactate).

TABLE 1. Parameter Values for Exopolysaccharide Fermentations[1,2]

	Product		Substrate		Biomass
Fermentation	n	m	α	β	μ_{max}
Xanthan gum[1]	.155gP/gX–hr	1.83gP/gX	2.0gS/gX	.284gS/gX–hr	.15hr^{-1}
Pullulan[2]					
(pH) 4.5)	0	89 wt%/(g–day/100ml)	—	—	1.12hr^{-1}
(pH 5.5)	0	135 wt%/(g–day/100ml)	—	—	.89hr^{-1}
(pH 6.5)	0	110 wt%/(g–day/100ml)	—	—	1.12hr^{-1}
Alginic Acid[2]	0	1.60gP/gX	6.6gS/gX	.015gS/gX–hr	.12hr^{-1}
Pseudomonas sp[2]	10^{-3}gP/ODU–hr	0	.165%(w/v)/ODU	$2.8 \cdot 10^{-2}$%(w/v)/ODU	.31hr^{-1}

TABLE 2. Parameter Values for *C. thermocellum* Fermentations[a,b]

	Substrate		
	Glucose	Cellobiose	Fructose
Biomass, X			
X_{max}	.85 g/l	410 Klett units (1.36 g/l)[c]	425 Klett units (1.42 g/l)[c]
$\mu(hr^{-1})$	.174	.250	.191
$X_o(g/l)$[d]	.0065	.0362	.0014
Products			
$n_{ethanol}(g-P/g-X\cdot hr))$	.0814	.0852	.103
$n_{acetic\ acid}$	.029	.0048	.0145
$n_{formic\ acid}$	.0065	NA[e]	NA
$n_{lactic\ acid}$	.062	NA	NA
$m_{ethanol}\ (g-P/g-X)$	2.26	.67	.37
$m_{acetic\ acid}$	.163	.159	.096
$m_{formic\ acid}$	.687	NA	NA
$m_{lactic\ acid}$	.127	NA	NA
Substrate, S			
β(nongrowth assoc., $(g\cdot S/g\cdot X\cdot hr))$	.41	.38	.30
α(growth assoc., $(g\cdot S/g\cdot X))$	6.55	1.55	2.45

[a]Evaluation by procedures in TABLE 1.
[b]Reprinted from *Biotechnol Bioeng.* **23:** 1517 (1981). p. 1521 by permission of John Wiley & Sons, Inc.
[c]Basis: 300 Klett units = 1 g/l.
[d]X_o from Equation 7 is initial *viable* biomass.
[e]NA: products not reported for these fermentations.

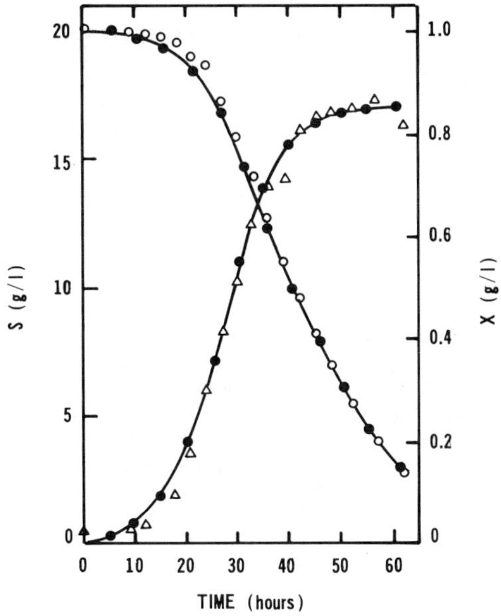

FIGURE 2. Course of *C. thermocellum* fermentation with pH control (6.5) in low iron concentration medium. Data from Figure I.B.1. of reference 10: (□) biomass; (O) substrate. (—) curves calculated from parameter values of TABLE 2. (Reprinted from *Biotechnol. Bioeng.* **23:** 1520 by permission of John Wiley and Sons, Inc.)

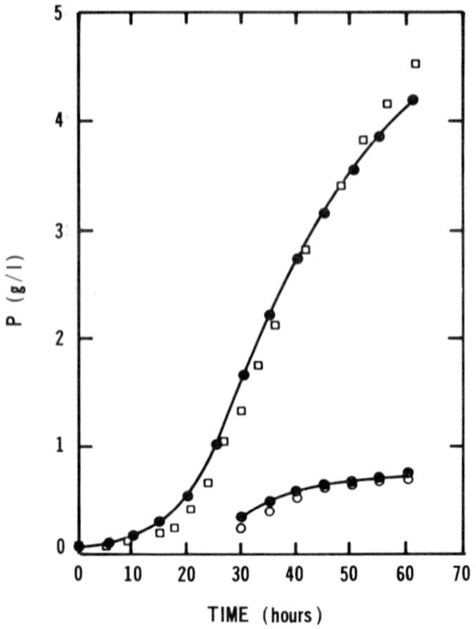

FIGURE 3. Product ethanol and formic acid formation versus time. Data from Figure 1.B.1 of Reference 10: (□) ethanol, (O) formic acid, (—) curves calculated from parameter values in TABLE 2. (Reprinted from *Biotechnol. Bioeng.* **23:** 1522 by permission of John Wiley and Sons, Inc.)

EXAMPLE 3: MICROBIAL EXOTOXIN (INSECTICIDE)[11]

A microbial exotoxin of *B. thuringiensis* is produced under batch aerobic conditions.[12] The biomass is again modeled with the logistic Equation 2a, and the formation of product exotoxin by the Leudeking-Piret Equation 2b. A second product, bacterial spore SP, is formed quite late in the stationary phase and continues into the death phase. This late phenomenon is treated with a lag time τ, where τ is assumed to represent the "finite time interval between initiation of sporulation and appearance of a recognizably refractile spore."[13] Thus,

$$\frac{d(SP)}{dt} = KX(t - \tau) \tag{7a}$$

FIGURE 4. Product acetic acid and lactic acid formation versus time. Data from Figure I.B.1 of Reference 10: (□) lactic acid, (O) acetic acid; (—●—) curve calculated from parameter values in TABLE 2. (Reprinted from *Biotechnol. Bioeng.* **23:** 1522 by permission of John Wiley and Sons, Inc.)

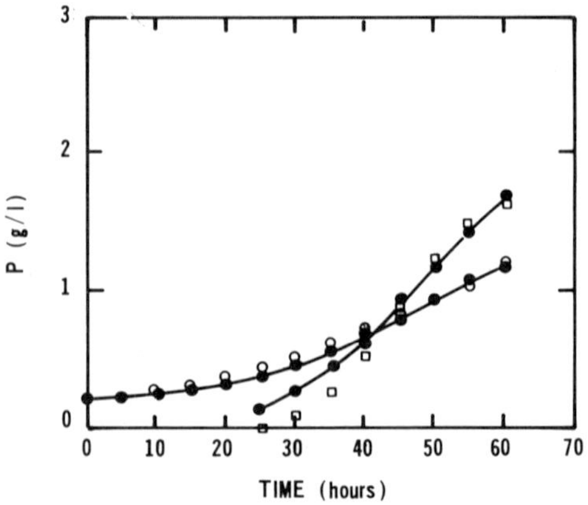

The carbohydrate substrate S was modeled with Equation 2c; as the trace exotoxin presumably consumed a negligible amount of substrate, Equation 2c simplifies to Equation 7b

$$\frac{dS}{dt} = -\frac{1}{Y_x}\frac{dX}{dt} - k_e X \qquad \textbf{(7b)}$$

Dissolved oxygen was monitored in the experiments under consideration. As the dissolved oxygen, C_o, level generally changed slowly over fermentation time compared

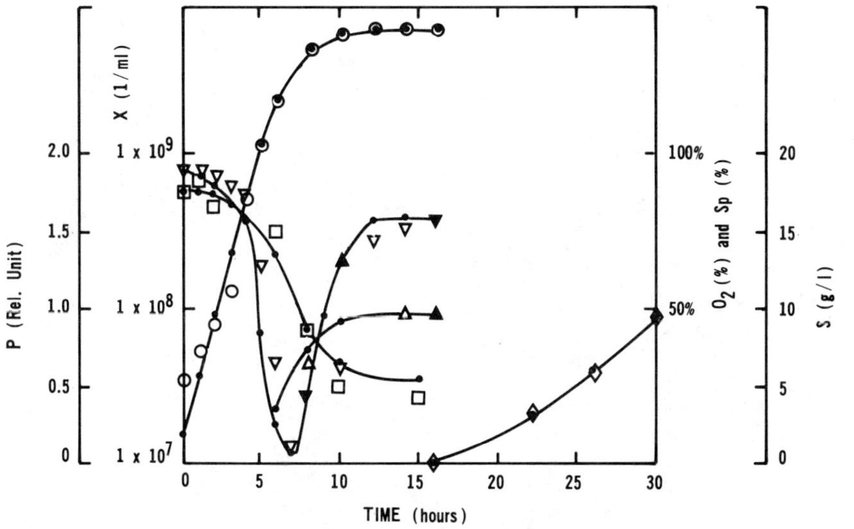

FIGURE 5. *B. thuringiensis* fermentation (data from Fig. 5 of Ref. 12): (○) biomass, (□) carbon substrate, (△) exotoxin product, (▽) dissolved oxygen substrate (◇) spores. Curves (solid symbols and lines) calculated from equations and parameters of text and TABLE 2. (Reprinted from *Biotechnol. Bioeng.*[11] by permission of John Wiley and Sons, Inc.)

with the relative rates of gas transfer and consumption, we may write Equation 7c:

$$\frac{dC_o}{dt} = k_e a\,(C_o^* - C_o) - \frac{1}{Y_o}\frac{dX}{dt} - k_o X \approx 0 \qquad \textbf{(7c)}$$

(mass transfer) (growth) (maintenance)

For the comparison of model and experiment, which is presented in FIGURE 5, parameter evaluation was as follows:

Spore Formation

Integration of Equation 7a gives Equation 7d:

$$SP(t) = K \int_{\tau}^{t} X(t' - \tau)dt' \qquad \textbf{(7d)}$$

Estimation of τ and integration of 7d with Equation 3a allows plotting the data SP(t) versus the calculated integral on the right-hand side of Equation 7d. The lag τ is rechosen until a straight line was obtained, the slope of which gave $K = 6.8 \times 10^{-1}\%$ sporulation/[(cells/ml) − hr] and $\tau = 12$ hours.

Substrate

From stationary phase data, $k_e = (dS/dt)/X_{max} = 10^{-11}$mgS/[(cells/ml) − hr]; from integration of Equation (7b) and a plot of $\Delta S(t) - k_e \int Xdt$ vs. $\Delta X(t)$, $Y_x^{-1} = 1.93 \cdot 10^{-9}$mgS/(cells/ml).

Dissolved Oxygen

As C_o^* and $C_o(t)$ were known, the following ratios were determined:

$$\frac{k_o}{k_\ell a} = 2.65 \times 10^{-9}(\%\ \text{saturation}/(\text{cells/ml}) - \text{hr});$$

and

$$\frac{1}{Y_o k_\ell a} = 6.17 \cdot 10^{-8}\ (\%\ \text{saturation}/(\text{cells/ml}))$$

We note in passing that if the specific oxygen demand, Q_{O_2}, is available from, for example, external mass balancing on oxygen at one or more times, then $k_\ell a$ can be determined, since

$$Q_{O_2} = \frac{k_\ell a\,(C_o^* - C_o)}{X(t)}$$

Given $k_\ell a$, then k_o and Y_o can be calculated.[11]

The Equations 2a and 2b were used to provide the following parameter values:

$$X_{max} = 6 \times 10^9\ \text{cells/ml},\ \mu_{max} = 0.90\ \text{hr}^{-1},\ X_o$$

$$= 1.45 \times 10^7\ \text{cells/ml},\ m = 1.6 \times 10^{-10}\ \text{units P (cells/ml)}$$

EXAMPLE 4: MICROBIAL ENDOPOLYSACCHARIDE

Hydrogen bacteria (*Alcaligenes eutrophus*) produce an intracellular storage produce, poly-β-hydroxybutyric acid, which has received some commercial attention as a possible, potentially biodegradable, replacement for some synthetics such as polypropylene. A recent paper[14] presented batch data for such a system, including biomass, polymer storage product, and a biomass-limiting substrate, ammonium sulfate, $(NH_4)_2SO_4$. A model requiring parameter estimation and numerical solutions was proposed. Here we find that a simpler approach is again useful.[15]

The logistic Equation 2a again fits the data, but the very sharp transition to stationary phase from exponential (FIG. 6) is best fit by the discontinuous autonomous growth Equation 8a, b.

$$\frac{dX}{dt} = \mu_{max}X \qquad \text{(exponential)} \tag{8a}$$

$$\frac{dX}{dt} = 0 \qquad \text{(stationary)} \tag{8b}$$

Here, the time of applicability is Equation 8a until t = 15.5 hrs and Equation 8b for all later times and $\mu = 0.154\ hr^{-1}$ and $X_o = 0.33$ g/l.

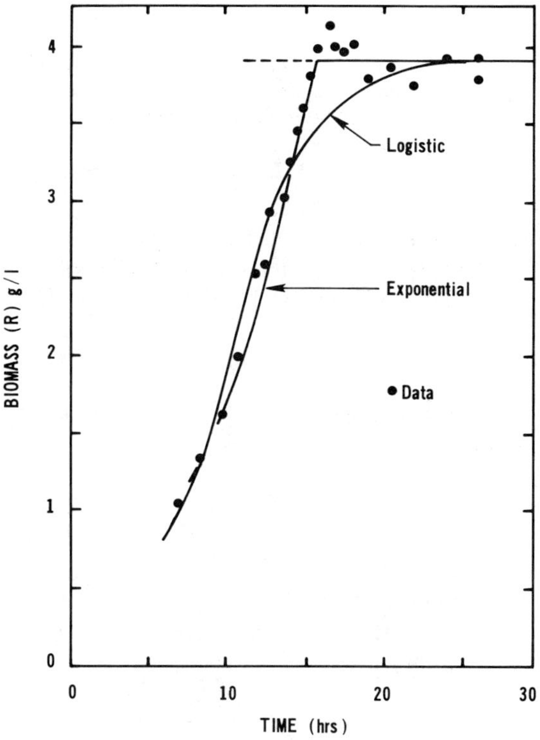

FIGURE 6. Data of Reference 14 for biomass (●) versus calculated best fits for logistic Equation 2a and discontinuous exponential/stationary Equations 8a, b. (Reprinted by permission of *Eur. J. Appl. Microbiol. Biotechnol.* Submitted.)

The substrate measured, ammonium sulfate, does not appear stoichiometrically in the polymer product, thus Equation 7b rather than Equation 2c is appropriate. The data presented in Reference 14 show virtually complete S consumption before the stationary phase. Thus,

$$k_e = \frac{(dS/dt)_{stationary}}{X_{max}} = \frac{0}{X_{max}} = 0$$

and Equation 7b in integrated form then predicts

$$S(t) - S_o = -\left[\frac{1}{Y_x}\right](X(t) - X_o) \tag{8c}$$

a linear relation that represents the data of FIGURE 7 well, giving $Y_x = 1.56$g X/gS. For comparison, we may write a stoichiometric mass balance of Equation 8d:

$$aH_2 + bCO_2 + CO_2 + d[(NH_4)_2SO_4] = \underset{\text{(biomass)}}{C_{4.41}H_{7.3}N_{0.86}O_{1.19}} + e\ (\text{polymer}) \quad \mathbf{(8d)}$$

This typical biomass composition gives an apparent molecular weight of 91.2. With equation 8d requiring d = 0.86/2 = 0.43, the theoretical yield coefficient based on ammonium sulfate is given by

$$Y_x = \frac{1.0 \times 91.2}{.43 \times 132} = 1.61 gX/gS$$

in good agreement with Y_x derived from Equation 8c and FIGURE 7.

The product storage polymer is formed almost only in the stationary phase. Thus, the useful Leudeking-Piret form would predict that

$$\frac{dP}{dt} = nX + m\frac{dX}{dt} \approx nX_{max} = \text{constant}$$

in this phase. The data of Reference 14 show a clearly diminishing value for dP/dt throughout this phase, however, eventually reaching dP/dt = 0 at ~ t = 70 hrs. The alternate form of Equation 8e allows for a decreasing product rate over time:

$$\frac{dP}{dt} = nX_{max}(1 - at') \quad \mathbf{(8e)}$$

where t′ is measured from the onset of stationary phase (t′ = 15.5 hrs). Integration gives Equation 8f,

$$P(t) - P(T' = 0) = nX_{max}\left[t' - \frac{a(t')^2}{2}\right] \quad \mathbf{(8f)}$$

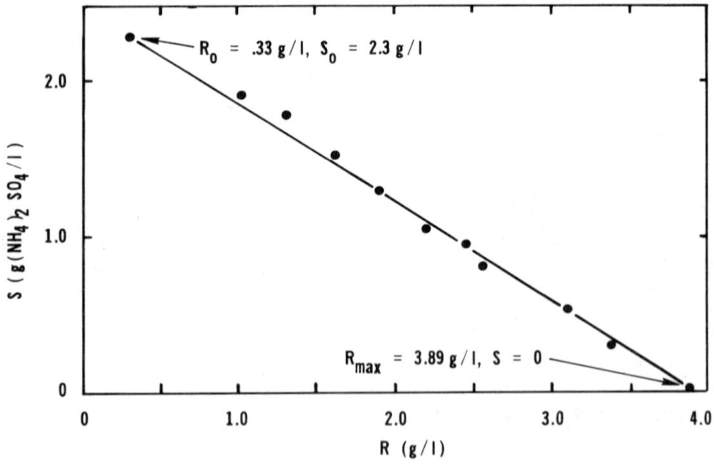

FIGURE 7. Data of Reference 14 plotted as ΔS(t) versus ΔX(t), test of Equation 8c, text. Slope = Y_x^{-1} = 0.64, or Y_x = 1.56 gX/gS. (Reprinted by permission of *Eur. J. Appl. Microbiol. Biotechnol.* Submitted.)

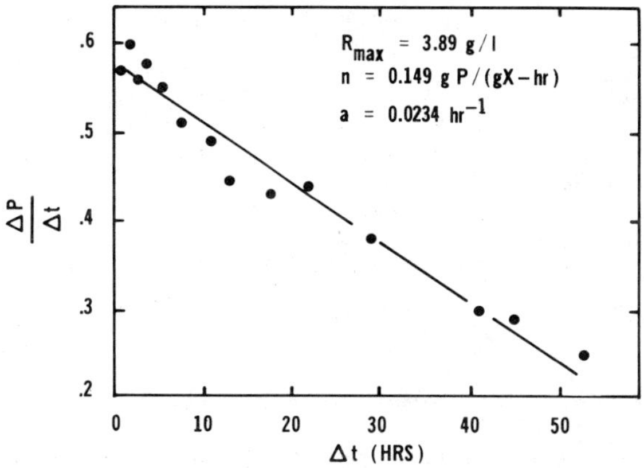

FIGURE 8. Data of Reference 14 replotted as $(\Delta P(t))/t'$) vs. $1 - (a/2)t'$. (Reprinted by permission of *Eur. J. Appl. Microbiol. Biotechnol.* Submitted.)

which describes the stationary phase well (FIG. 8), yielding $n = 0.149$ gP/g − X − hr (with $X_{max} = 3.89$ g/l) and a 0.0234 hr^{-1}. A linear decrease of viable cells with time has been proposed for alcohol fermentations;[16] in the present instance, we have no established cause (nonviability, substrate exhaustion, toxin prodution, etc.) for the cell-decreased, product-forming activity with time.

SUMMARY

The utilization of autonomous growth forms, $dX/dt = f(X)$, allows straightforward analytical representation of a number of *batch* fermentations for such industrially interesting products as exopolysaccharide, endopolysaccharides, exotoxins (insecticide), and ethanol. While autonomous equations are not expected to describe chemostat operation, where μ is necessarily sensitive to S (since μ must change to give $D = \mu$ at each dilution rate D), the very considerable analytical conveniences derived from such an equation for batch fermentations (by virtue of its uncoupling the X, P, S equations) should be considered more often, rather than the too frequently unilluminating resort to numerical schemes and computer time that characterizes some of the recent literature of batch fermentation kinetics.

ACKNOWLEDGMENT

I wish to thank the Department of Energy.

REFERENCES

1. WEISS, R. M. & D. F. OLLIS. 1980. Extracellular microbial polysaccharides I: Substrate, biomass, and product kinetic equations for batch xanthan gum fermentation. Biotechnol. Bioeng. **22:** 859.
2. KLIMEK, J. & D. F. OLLIS. 1980. Extracellular microbial polysaccharides: Kinetics of *Pseudomonas sp., Azotobacter vinelandii* and *Aureobasidium pullulans* batch fermentations. Biotechnol. Bioeng. **22:** 2321.
3. LUEDEKING, R. & E. L. PIRET. 1959. A kinetic study of the lactic acid fermentation. J. Biochem. Microbiol. Technol. Eng. **1:** 393, 431.

4. MORAINE, R. A. & P. ROGOVIN. 1971. Xanthan biopolymer production at increased concentration by pH control. Biotechnol. Bioeng. **13:** 381.
5. SILMAN, R. W. 1978. Private communication.
6. WILLIAMS, A. G. 1974. Ph.D. Thesis. University College. (See Fig. 1 of Reference 7.)
7. DEAVIN, L., T. R. JARMAN, C. J. LAWSON, R. C. RIGHELATO & S. SLOCOMBE. 1977. Extracellular Microbial Polysaccharides. P. A. Sandford & A. Laskin, Eds. ACS. Washington, D.C. ACS Monograph Series. Vol. **45,** p. 14.
8. LEDUY, A., A. A. MARSAN & B. CORPAL. 1974. A study of the rheological properties of a non-Newtonian fermentation broth. Biotechnol. Bioeng. **16:** 61.
9. LAM, J. C. & D. F. OLLIS. 1981. Kinetics of multiproduct fermentations. Biotechnol. Bioeng. **22:** 1517.
10. WANG, D. I. C., C. L. COONEY, A. L. DEMAIN, R. F. GOMEZ & A. J. SINSKEY. 1978. quarterly report (9/1/78–11/30/78). COO-4198-8. DOE under contract No. EG-77-5-02-4198.
11. LAM, J. C. & D. F. OLLIS. 1983. Kinetics of batch exotoxin production. Biotechnol. Bioeng. Submitted.
12. HOLMBERG, A. & R. SIEVANEN. 1980. Fermentation of *Bacillus thuringiensis* for exotoxin production: Process analysis study. Biotechnol. Bioeng. **22:** 1707.
13. DAWES, I. W. & THORNLY, J. H. M. 1970. Sporulation in *Bacillus subtilis.* Theoretical and experimental studies in continuous culture. J. Gen. Microbiol. **62:** 49.
14. HEINZLE, E. & R. M. LAFFERTY. 1980. Kinetic model for growth and synthesis of poly-β-hydroxybutyric acid. Eur. J. Appl. Microbiol. Biotechnol. **11:** 8.
15. OLLIS, D. F. Eur. J. Appl. Microbiol. Biotechnol. Submitted.

Modeling of Cell Viability and Specific Alcohol Productivity

T. CIFTCI[a], A. CONSTANTINIDES, AND S. S. WANG

Department of Chemical and Biochemical Engineering
Rutgers–The State University of New Jersey
Piscataway, New Jersey 08854

INTRODUCTION

Cell structure is maintained by constant replacement of degraded cell biomaterials. This requires an energy input called maintenance energy. Even though cell maintenance is executed, yeasts have limited viability; aging and death could still take place. For instance, cell reproduction cannot continue indefinitely because there is a limit to the number of buds that can be produced by a mother cell. In addition to the physiological limits of viability, extreme environmental effects can cause death.

In the alcohol fermentation condition, although cells use alcohol production to produce a useful form of energy (ATP), the concentration of alcohol reached can also influence the cellular viability.

A correction for changes in the viable fraction of the culture is needed for models describing the behavior of the cells. Common viability tests can be employed to determine this correction. In a recent attempt, a correlation among viability criteria has been proposed by Ciftci *et al.*[1] This correlation involves three viability tests, namely, plate counts, methylene blue staining, and CO_2 production.

Two viable populations are assumed to exist in a culture besides dead cells (i.e., dual viability). One of the viable groups is able to reproduce and show metabolic activity. The other group shows metabolic activity; however, it cannot reproduce (see FIG. 1). The contributions of these two groups to metabolic activity (i.e., CO_2 production) are not equal.

The first viable group, which consists of cells able to reproduce (V_{PC}), has high metabolic activity (A_{PC}). The second viable group, which consists of cells not able to reproduce (V_{NR}), has a low metabolic activity (A_{NR}).

We can introduce a dimensionless term to describe the metabolic activity difference between these two groups. The difference of two metabolic activities ($A_{PC} - A_{NR}$) could be normalized by the higher activity (A_{PC}) to determine the fractional difference. Let us call this ratio the Metabolic Inactivation Index (MII) where,

$$MII = \frac{A_{PC} - A_{NR}}{A_{PC}} \tag{1}$$

This ratio is the fraction of the activity loss due to the loss of reproduction by the killing factor. It could be different for each organism as well as for each condition, since the killing factor is a unique combination of the environmental factors.

When cell production and alcohol production are separated, viable cell concentra-

[a]Present Address: Waksman Institute of Microbiology, Rutgers University, P.O. Box 759, Piscataway, NJ 08854.

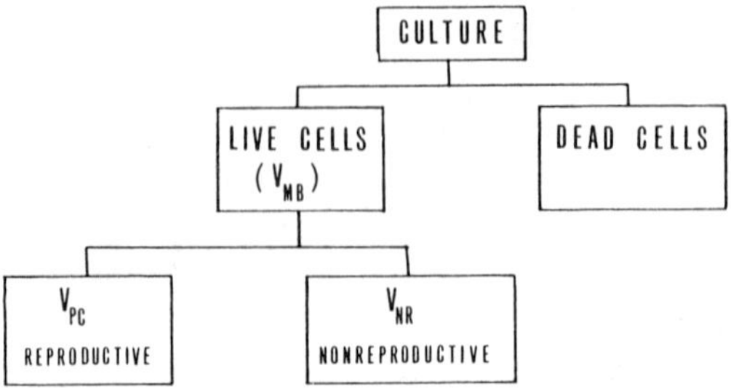

FIGURE 1. Pictorial presentation of the concept of dual viability.

tion can be monitored. It was found that viability variation was very critical, especially in designs where cell feeding into various stages of the fermentation was possible.[2]

In this study, an attempt will be made to model alcohol productivity of a culture on the basis of common viability tests. The objective is not to define viability, but to incorporate the measurements of quality of culture with the variables of the environment.

EXPERIMENTAL PROCEDURES

Materials

Microorganism

A *Saccharomyces cerevisiae* strain (NRRL Y-11572), which is also known as *Saccharomyces sake* (sake yeast), was used for this study.

Cell Growth Medium

The medium composition reported by Margaritis and Wilke[3] was used for cell growth.

Alcohol Fermentation Medium

The medium composition was as presented on TABLE 1 unless otherwise noted.

TABLE 1. Composition of Alcohol Production Medium

Component	Amount[a]
Glucose	200 g/l
Citric Acid	8.6 g/l
Sodium Citrate	2.5 g/l
Silicone Antifoam	200 ppm

[a]Tap water is added to bring the volume to 1 liter.

Methylene Blue Solution

The composition of the methylene blue solution was described by Townsend and Lindegren.[4]

Methods

Inoculum Preparation

One to 5% starter was added to the growth medium. The propagation was carried out in a fermenter (various sizes were used according to the quantity of cells needed) with aeration of 1 vvm at 30°C. After 24 hours, the cells were harvested by centrifugation using a Sorvall centrifuge (RC2-B) with GS-3 rotor at 5000 RPM for 10 minutes.

Dry Weight Measurement

After centrifugation and washing, the cells were dried at about 100°C until no weight change between consecutive measurements was observed.

Optical Density

After centrifugation and washing, the cells were resuspended and diluted to measure optical density at 610 nm using a Bausch and Lomb Spec-20 spectrophotometer. A calibration curve was used to develop the following correlation between the optical density measurements (O.D.) and the dry weight (D.W.) measurements.

$$(\text{D.W.})\ \text{g/lt} = (\text{O.D.}) \cdot \left(\frac{1}{\text{Dilution}}\right) \cdot \left(\frac{1}{1.76}\right) \text{g/lt} \quad (2)$$

Total Cell Count

The total cell number was estimated by using a Petroff-Hausser bacteria counter (C. A. Hausser and Son, Philadelphia). Buds larger than one-half of the mother cell were counted as one.

Plate Counts

Samples were diluted with premeasured, sterilized water and mixed with 15 cc Y. M. Agar at about 50°C. Two to three replicates were made. Colonies of 30 to 300 on one plate were counted and counts of the replicates were averaged. The number of viable cells was divided by the total cell number to determine the fraction of reproducing cells (V_{pc}).

Methylene Blue Counts

After centrifugation and washing, the cells were resuspended and mixed with methylene blue solution (twice as much as the cell solution). After 15 minutes, fractions of stained and unstained cells (V_{MB}) were estimated.

Glucose Concentration

Glucose concentration was measured by using a Lichrosorb NH_2 Sorbent Column on a Spectrophysics SP 8000 High Pressure Liquid Chromatograph with a refractive index detector. Eight percent (v/v) Acetonitrile and twenty percent (v/v) H_2O were used as the mobile phase at a flow rate of 2.0 ml/min.

Alcohol Concentration

A Carle Basic Gas Chromatograph with a thermal conductivity detector is used to detect the concentration of alcohol. Samples were injected into a 6-foot Parapak QS (80/100) column at 110°C.

Standard Specific Alcohol Productivity Test

An arbitrary reference condition for alcohol productivity was assigned and used consistently throughout this study. The purpose was to compare the immediate alcohol productivities of various cell samples at a reference condition. The cells were centrifuged and transferred into a fresh alcohol production medium (30% (w/v) glucose) at 30°C and agitated. Samples were taken at 1, 2, 4, and 8-hour intervals to determine the alcohol concentration reached and initial specific alcohol productivities. Cell concentrations were maintained in the same range for each sample. A correction was introduced if various cell concentrations were used.

This standard condition is not necessarily the condition where the alcohol production rate is the highest. However, it was chosen because of convenience, although ideally it is possible to design a better experiment where substrate, product, and cell concentration effects could be eliminated.

RESULTS AND DISCUSSIONS

To apply the dual viability concept in the case of alcohol fermentation, a modification of the original equation of Ciftci *et al.*[1] could be introduced as follows:

$$\nu = A_{PC} \cdot V_{PC} + A_{NR} \cdot V_{NR} \tag{3}$$

Here V_{PC} and V_{NR} are the fractions of the viable cells in a culture as described above. A_{PC} and A_{NR} are the specific alcohol productivities of the cells of these two viable groups. ν is the specific alcohol productivity of the culture. V_{NR} could be described as

$$V_{NR} = V_{MB} - V_{PC} \tag{4}$$

Substituting the above equation into the specific alcohol productivity relation and

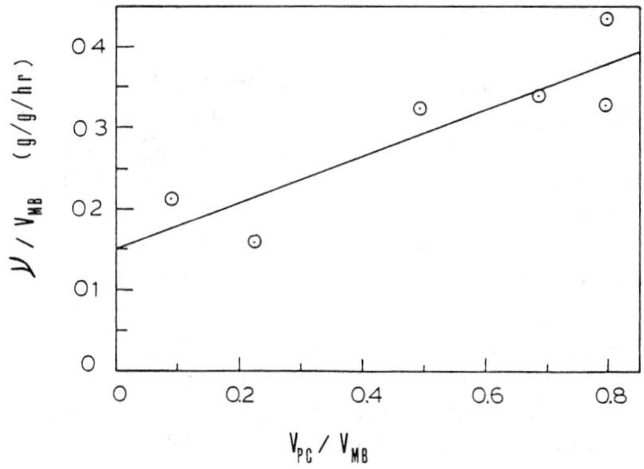

FIGURE 2. Linear regression for viability data.

rearranging, we obtain

$$\frac{\nu}{V_{MB}} = (A_{PC} - A_{NR}) \cdot \frac{V_{PC}}{V_{MB}} + A_{NR} \tag{5}$$

An experiment was conducted to obtain values of fractions of reproductive cells (V_{PC}) and nonreproductive but metabolically active cells ($V_{MB} - V_{PC}$). Specific alcohol productivity of the culture in a fresh medium at standard conditions was also estimated.

The least squares technique was employed to estimate the slope ($A_{PC} - A_{NR}$) and the intercept (A_{NR}) (see FIG. 2). The correlation was 0.87 by which a significance level of 1% was estimated. The slope and intercept were 0.285 and 0.151 g/g/hr, respectively. As a result, the values $A_{PC} = 0.436$ g/g/hr, $A_{NR} = 0.151$ g/g/hr, and MII = 0.66 were found. The implication of this result is that the inactivation in case of alcohol production is somewhat similar to the inactivation in the presence of $HgCl_2$ and heat (see TABLE 2). The reason for this could be the fact that alcohol could denature many proteins as well as other biomaterials of the cell. Therefore, alcohol effects could be on the whole structure and functions of the cell, as opposed to the case of UV or X-rays, where specific biomolecules like DNA or RNA are affected.

The viability of the culture is changed with time and the conditions. V_{PC} and V_{MB} are measured during fermentation for regular intervals. Estimation of V_{MB} is relatively simple and quick, but V_{PC} is rather time consuming since the duration of the experiment is about three days. An alternative to plate count was introduced by Postgate *et al.*[5] An agar film is placed on a slide and cells are put on the surface of this film. The colonies developing are counted under a microscope and the fraction of the

TABLE 2. Effect of Various Factors on Cell Viability

Condition	MII	Level of Significance
Heat	0.83	1
$HgCl_2$	0.89	5
UV	0.44	5
X-ray	0.21	5
Alcohol production	0.65	1

cells that are reproducing is determined. This technique is called slide count and is able to give results in a few hours.

There have been many attempts to model cellular populations and their productivities. In general, models are based on either the size or the age distribution in the population. Productivities are related to the size or the age of individual cells. It should be mentioned here that there is some similarity between age and size of a cell in general.

A well-known example of age distribution was proposed by Shu.[6] An integration is performed to obtain the productivity of the cell population from an equation where age distribution and productivity for each age group is expressed. Size distribution equations are also common. Some utilize bud size or number of bud scars on the membrane as an indication of the age of the individual cell.[7–12]

When only natural aging or death is taking place, the models mentioned above could be applicable. However, in general, death can take place during any part of the aging process. It was observed, during methylene blue staining studies, that in alcohol fermentation, cell death is experienced by groups of all ages (and sizes). As in the case of alcohol fermentation, cell death is generally caused by a combination of factors besides natural age. In the dual viability equation, the differences due to size or age are ignored. Instead, a physiological activity criterion is employed. Cells are divided into three groups. Assuming that the cells of each group have similar size and age distribution, the fractions of two viable populations are multiplied by an average metabolic activity for each group. An attempt can be made to include the size or the age distributions within viable groups. This exercise was not found necessary due to the fact that reasonable adherence was observed by experimental data to the model.

In an alcohol fermentation, we are faced with two distinctly different phenomena. One is the reversible effect of conditions on the alcohol productivity. For instance, by manipulating temperature or substrate concentration, alcohol productivity may be reversibly controlled. The second one is the irreversible effect of conditions where the cell viability is changed by growth or death. These two different effects could be related to each other in some cases. For instance, when there is a strong inhibitory effect of substrate or product on alcohol production, viability could also be drastically affected. Therefore, these two effects should not be considered completely independent.

If all of the cells in a culture were able to reproduce (i.e., $V_{PC} = 1$), at the standard conditions, the productivity of the culture would be:

$$\nu = A^{o}_{PC} = \nu^{o}_{max} \tag{6}$$

But if only some of them are reproductive ($V_{PC} < 1$) and some of them are metabolically active although they are not reproductive, ($V_{NR} > 0$), then:

$$\nu = A^{o}_{PC}\left(V_{PC} + \left(\frac{A^{o}_{NR}}{A^{o}_{PC}}\right) \cdot V_{NR}\right) = \nu^{o} \tag{7}$$

When the culture is at a condition that is different from the standard condition, the effect of the environmental conditions will be as follows:

$$\nu = A^{o}_{PC}\left(V_{PC} + \left(\frac{A^{o}_{NR}}{A^{o}_{PC}}\right) V_{NR}\right) \cdot f\,(S, P, T, C..) \tag{8}$$

or

$$\nu = \nu^{o} \cdot f\,(S, P, T, C..) \tag{9}$$

f (S, P, T, C..) represents the inhibitory (or inducive) effect of the changes in substrate, product concentrations, temperature, cell concentrations, and so forth, respectively.

Similarly, in a culture, if all the cells are able to reproduce ($V_{PC} = 1$), the growth rate at the standard conditions will be

$$\mu = \mu^{o}_{max} \tag{10}$$

But since this is generally not the case, a correction will be necessary ($V_{PC} < 1$):

$$\mu = \mu^{o}_{max} \cdot V_{PC} = \mu^{o} \tag{11}$$

Additionally, when the conditions are different from standard conditions, a general

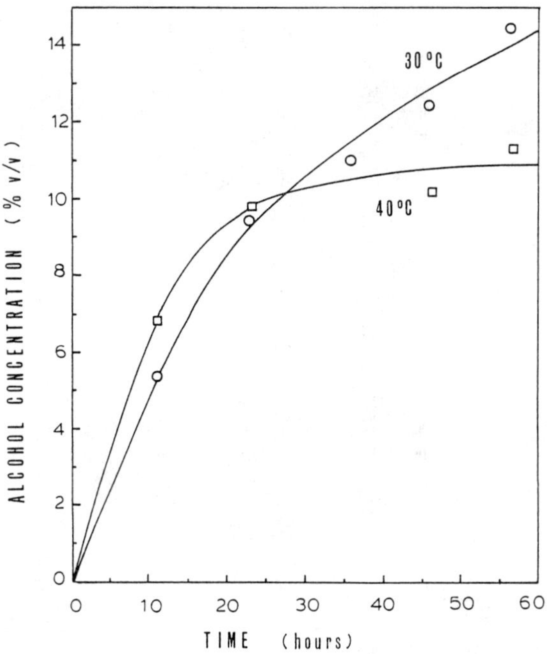

FIGURE 3. Alcohol production at 30° and 40°C.

equation can be introduced with inhibition terms,

$$\mu = \mu^{o}_{max} (V_{PC}) \cdot g\ (S, P, T, C..) \tag{12}$$

where g (S, P, T, C..) represents the effect of the changes from standard conditions.

It should be noted that the growth rate equation includes only V_{PC}, while alcohol production rate equation includes V_{PC} and V_{NR}. If this difference is ignored, an important error may be introduced. In general, equations for growth and alcohol production not only ignore the differences of the populations participating in the processes but also assume the total cell concentration to participate in both of the processes.[13–16] In these equations, inhibition behavior for growth and alcohol fermentation are claimed to be similar (for alcohol and glucose). This point was not investigated in this study, but a significant difference is expected.

An Application of the Model

A study was conducted to observe the number of variables during alcohol fermentation at 30 and 40°C. 12.4 g/l cells were introduced into 30% glucose (alcohol production medium) and the alcohol concentrations were found to be as shown on FIGURE 3. Although the initial alcohol production at 40°C was higher than that at 30°C, the 30°C run was able to do better after about 20 hours, and finally reached higher alcohol concentrations. At first sight, this figure could suggest that inhibitory

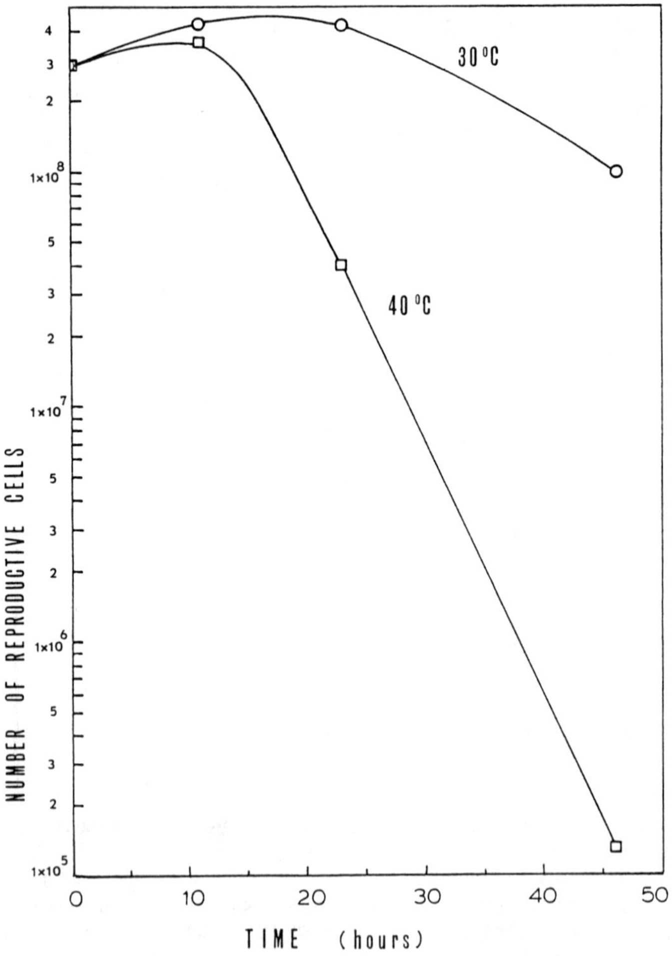

FIGURE 4. Plate counts during alcohol production at 30° and 40°C.

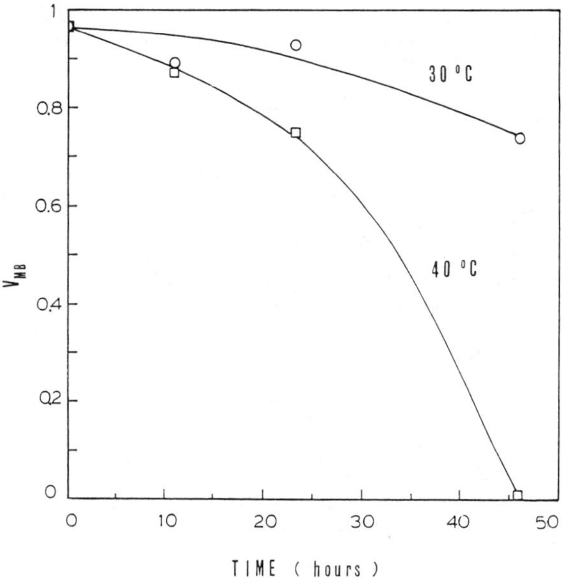

FIGURE 5. Methylene blue counts during alcohol production at 30° and 40°C.

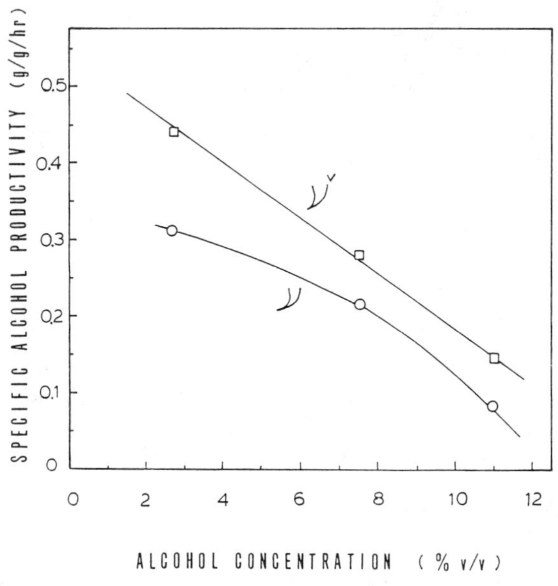

FIGURE 6. Specific alcohol productivity and corrected specific alcohol productivity during alcohol production (30°C).

alcohol concentration at 40°C was much lower than that at 30°C. As it was discussed in the previous section, it is necessary to obtain more information related to viability to analyze this system. Therefore, plate counts and the methylene blue staining test were performed. See FIGURES 4 and 5 for plate counts and M.B. staining tests during fermentations. FIGURES 6 and 7 present the specific alcohol productivities of the cultures at 30 and 40°C in the fermentation conditions as a function of alcohol concentration of the medium. There is a distinct difference between these two systems in terms of alcohol inhibition according to this type of presentation. But if we correct the specific alcohol productivities for the viability of the culture, we obtain linear alcohol inhibition behavior (see FIGS. 6 and 7). This correction was made by utilizing V_{PC} and V_{NR} from the experimental data and incorporating them with A_{NR}/A_{PC}. We can then obtain the inactivation (or activation) term of the system for each specific

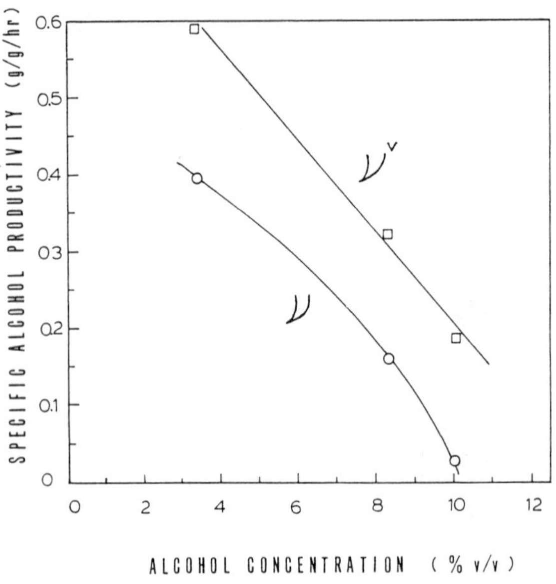

FIGURE 7. Specific alcohol productivity and corrected specific alcohol productivity during alcohol production (40°C).

alcohol productivity value mentioned above.

$$\nu^{V} = \frac{\nu}{V_{PC} + \dfrac{A_{NR}}{A_{PC}} \cdot V_{NR}} \tag{13}$$

Then, values were estimated and plotted. This type of treatment isolated the inhibition effect due to alcohol and induction by temperature,

$$\nu^{V} = \frac{\nu}{V_{PC} + \dfrac{A_{NR}}{A_{PC}} \cdot V_{NR}} = \nu^{o}_{max} \cdot (f(P)\, f(T)) \tag{14}$$

For 30 and 40°C, f (P) was reasonably linear as

$$f\,(P) = \left(1 - \frac{P}{P_{max}}\right) \tag{15}$$

P_{max} values were 14.9 and 13.7% at 30 and 40°C, respectively. These values are reasonably close.

The temperature difference between these two conditions introduced a correction of about 1.45 for 40°C. The effect of change in the substrate concentration was considered negligible in this treatment of the general model, although it could be important.

It can be concluded that incorporation of measurements of quality of the culture (i.e., viability measurements) with the quality of the environment (i.e., environmental factors) improved the modeling of the cellular behavior.

SUMMARY

A model has been proposed to correlate two viability tests and specific alcohol productivity of a yeast culture. Two viable populations with different activity are assumed to coexist. The difference between two populations is the ability to reproduce. Satisfactory agreement between the model and the experimental data has been observed.

REFERENCES

1. Ciftci, T., S. S. Wang & A. Constantinides. 1981. Biotechnol. Bioeng. **23:** 1407.
2. Ciftci, T., A. Constantinides & S. S. Wang. 1983. Biotechnol. Bioeng. In press.
3. Margaritis, A. & C. R. Wilke. 1978. Biotechnol. Bioeng. **20:** 727.
4. Townsend, G. F. & C. C. Lindegren. 1953. Cytologia **18:** 183.
5. Postgate, J. R., J. E. Crumpton & J. R. Hunter. 1961. J. Gen. Microbiol. **24:** 15.
6. Shu, P. 1961, J. Biochem. Microbiol. Technol. Eng. **3:** 95.
7. Jain, V. K. 1972. Biophysik **8:** 133.
8. Vrana, D., J. Lieblova & K. Beran. 1973. *In* Proceedings of Three International Specialized Symposia on Yeasts, Part 2. Otanemi/Helsinki. p. 285.
9. Vrana, D. 1976. Biotechnol. Bioeng. **18:** 297.
10. Votruba, J. & D. Vrana. 1978. Paper presented at the 7th Int. Symp. Cont. Cult. Prague.
11. Vrana, D. 1978. Paper presented at the 7th Int. Symp. Cont. Cult. Prague.
12. Palatt, P. J. & G. M. Seidel. 1979. Ann. Biomed. Eng. **7:** 45.
13. Aiba, S., M. Shoda & M. Nagatani. 1968. Biotechnol. Bioeng. **10:** 845.
14. Bazua, C. D. & C. R. Wilke. 1977. Biotechnol. Bioeng. Symp. No. 7, 105.
15. Ghose, T. K. & R. D. Tyagi. 1979. Biotechnol. Bioeng. **21:** 1401.
16. Jin, C.-K. 1981. Fermentation kinetics and process development for enhanced rate of production of ethanol from carbohydrates. Ph.D. Thesis. Rutgers University. New Brunswick, NJ.

Biofilm Fluidized-Bed Reactors and Their Application to Waste Water Nitrification

IRVING J. DUNN, HIROKI TANAKA, SUHEYLA UZMAN, AND MIRAN DENAC

Chemical Engineering Department (TCL)
Swiss Federal Institute of Technology (ETH)
CH-8092 Zurich, Switzerland

The natural ability of organisms to grow on surfaces can be exploited to provide biomass retention for the design of biological reactors. Reviews on the application of biofilm fluidized beds to wastewater treatment[1] and to the application of fluidization reactors for immobilized enzyme and cell applications[2] have been recently published. Proceedings of meetings on the use of fluidization in wastewater treatment[3] and on the application of anaerobic biofilm systems[4] are available. It is often possible, after a period of selective adaptation, to establish a thin film of organisms on submerged surfaces. There are many mechanisms by which organisms are thought to be able to adhere to surfaces.[5] Many types of solid carrier materials have been used. Some provide simple, open surfaces, such as fine sand or plastic particles,[6,7] others provide a porous, spongelike internal surface such as plastic foam.[3] Depending on the solid carrier, the organisms either adhere under their own power or are assisted by the entrapping characteristics of a solid carrier matrix.

With the biomass adhering to a solid particulate carrier, the reactor can be operated at flow rates that are independent of the maximum specific growth rate. Thus the $D < \mu_m$ restriction, which is applicable to suspended culture, is of no consequence to biofilm reactors, because the residence time of the liquid phase is uncoupled from the residence time of the biomass. The activity of the reactor per unit volume will depend largely on the surface area per unit volume that can be provided by the solid carrier, on the mass transfer limitations that may exist in the bed of particles, and on the diffusional limitations that the biofilm may present.

In principle, a biofilm reactor with particulate carrier could be operated in a number of ways. If the particles are very fine (0.1 mm), their settling velocity will be not much greater than a biological floc. In fact, the mechanism of biofilm formation may not depend on actual adherence to surfaces. Fine carrier material may instead be caught in the floc and provide a heavy floc, which gives higher sedimentation velocities and biomass retention in the reactor. Such flocculating systems may be operated as tower reactors with slow upflow of liquid. Flocculation may be enhanced by slow stirring in a tank geometry with the upflow of liquid through a settler region.

Large particles (0.1–0.5 mm) will probably be associated with the biomass in a film fashion, but perhaps tending also to aggregate into flocs. Their higher density allows higher flow rates to be used. Full fluidization can be expected at a velocity of about 20 m/hr. Below about 10 m/hr, the particles will pack together in a fixed bed and the liquid will channel through it. Fluidization will generally improve mass transfer and prevent channeling or stoppage.

Very large carrier material of a few mm to a few cm in diameter is too heavy for reasonable fluidization and can be operated as a fixed-bed biofilm reactor, in which the liquid and perhaps the gas phase are passed through. Often such reactors are operated as trickle beds with liquid flowing down and gas flowing up.

The retention of biomass in high concentration (10–30 g/l) creates a high activity per unit volume. If the reaction is aerobic, then the volumetric oxygen requirements must be allowed for. A fluidized bed can be operated as a two-phase, liquid-solid system or as a three-phase system including the gas. Aeration or oxygenation of a two-phase fluidized bed would be provided by an external oxygenator located in the liquid recycle loop as shown in FIGURE 1. The liquid flow rate through the reactor ($F + R$) is chosen to give the required fluidization and the oxygen transfer capacity is matched to the size and activity of the reactor. It is important that the entering flow rate and dissolved oxygen concentration are high enough that the reactor is not adversely limited by oxygen transfer. In practice, however, the activity of a biofilm

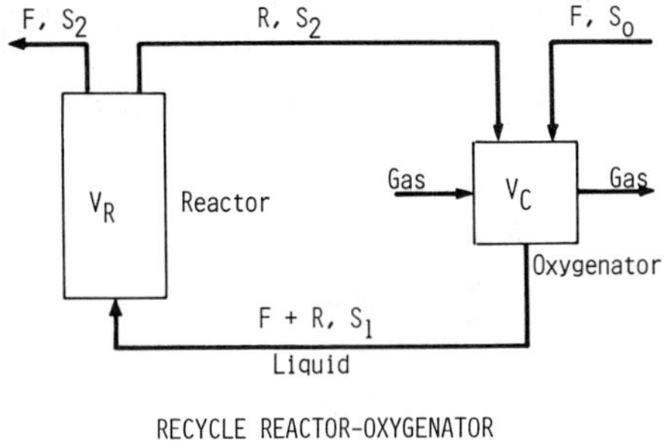

FIGURE 1. Configuration for recycle reactor with external oxygenator and the three-phase fluidized-bed reactor configuration.

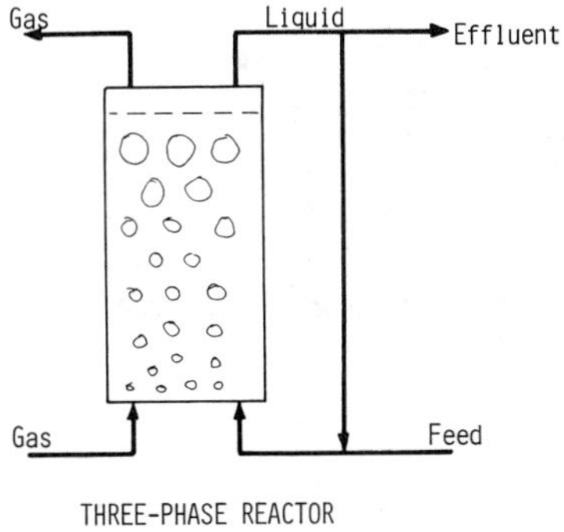

system is such that the reaction is usually oxygen limited. The two-phase system may be preferred because the shearing forces of the gas bubbles passing through the bed can be excessive, leading to loss of biomass and carrier material. The two-phase system also has more flexibility with respect to oxygenation capacity.

A three-phase reactor can be considered to be the simplest as shown in FIGURE 1, but there are problems involving oxygen transfer.[8,9] Very fine solid particles tend to promote bubble coalescence. Thus, in a fluidized bed of 0.5 mm sand, very large gas bubbles of several cm diameter will rise rapidly in the bed. The larger and denser the particles, the better the gas remains dispersed. The problem is one of choosing the surface area of the particle to achieve a volumetric activity that is matched by the oxygen transfer capacity of the particular three-phase system.

Thus, the oxygen transfer rate (OTR) should equal the oxygen uptake rate (OUR). Measurements of the $K_L a$ value for three different solid materials in an air-water fluidization column demonstrate the dependency of $K_L a$ on the particle type and gas velocity; the data are shown in FIGURE 2. The $K_L a$ values are relatively low and indicate that the rates of three-phase aerobic fluidized-bed reactors will be limited by oxygen transfer.

DESIGNING FOR RECYCLE REQUIREMENTS

Both two-phase and three-phase fluidized systems must be operated with high recycle rates to provide adequate fluidization velocities. Axial gradients of oxygen, pH, or other products and substrates can be controlled by varying the recycle ratio. A

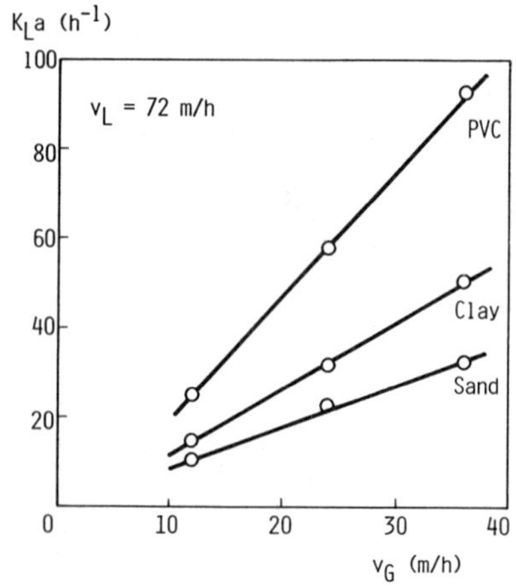

FIGURE 2. $K_L a$ versus gas velocity in a three-phase fluidized bed of plastic (5.7 mm), porous clay (3.5 mm), and sand (1 mm) particles.

substrate balance across the reactor bed gives:

$$S_1 - S_2 = -\int_0^V \frac{r_S \, d\, V_R}{R + F} \tag{1}$$

where r_s will be equal to a constant r_{S_2} when a bed (or section of a bed) is taken to be well mixed. This equation is used to calculate the single pass conversion.

Balancing across the oxygenation tank gives:

$$S_1 = \frac{S_2 + \alpha S_o}{1 + \alpha} \tag{2}$$

where α is the recycle ratio, F/R. This equation can be used to calculate the reactor inlet concentration for known values of the system inflow and effluent concentrations. The single pass concentration difference is given by:

$$S_1 - S_2 = \frac{S_o - S_2}{(1/\alpha) + 1} \tag{3}$$

Equation 1 can be employed with a simple tank, tanks in series, or plug flow description of the mixing. For first-order reactions, the well-mixed reactor system becomes:

$$\frac{S_2}{S_o} = \frac{1}{1 + (kV_R/F)} \tag{4}$$

and for plug flow:

$$\frac{S_2}{S_o} = \frac{\alpha - e^{-k\tau_R}}{1 + \alpha - e^{-k\tau_R}} \tag{5}$$

where $\tau_R = V_R/(F + R)$.

Numerical calculations from the above equations for a particular example indicate that at high rates of recycle (low α), the reactor type makes little difference, and the single pass conversion ($S_1 - S_2$) is primarily dependent on α. The effluent concentration level (S_2) is much more dependent on the reactor type, and if the reactor is well mixed, then recycle rate has no influence, other than, of course, to keep the bed fluidized and to supply oxygen.

The oxygen requirements can be calculated from Equations 1–3. Knowing the yield coefficient $Y_{o/s}$ (or stoichiometry) gives:

$$C_{L_1} = Y_{o/s}\,(S_1 - S_2) + C_{L_2} \tag{6}$$

This equation can be used to calculate the required inlet oxygen concentration for a given single pass conversion.

From this it can conversely be seen that if oxygen limitation is to be avoided:

$$S_1 - S_2 < C_L^*/Y_{o/s} \tag{7}$$

This equation gives the upper limit on $S_1 - S_2$, where $C_L = C_L^*$, the saturation value for a given oxygen concentration in the gas phase.

The oxygen transfer coefficient, K_La, in the well-mixed tank can be obtained, for the case of $C_{L_1} = C_L^*$ and $S_2 = 0$ according to:

$$K_La = \frac{1}{[(C_L^*/YS_o)(1 + \alpha/\alpha) - 1]V_c/R} \tag{8}$$

The importance of oxygenator size, V_c, is clear here.

A further operational limit can be established if $C_{L_2} = 0$ and $S_2 = 0$ and combining Equations 2 and 6:

$$\alpha_{\min} = \frac{1}{1 - (S_oY_{o/s}/C_L^*)} \tag{9}$$

From this equation, the highest rate of recycle (minimum α) necessary for complete conversion with an assumed C_{L_1} can be calculated. The discussion assumes oxygen supply limitation, and the calculated K_La values using air and pure oxygen have been given previously.[10]

BIOFILM KINETICS

The rates of reactions in biofilms are influenced by diffusional transport as they are in porous catalyst particles or immobilized enzyme pellets. Borrowing from the chemical engineering developments 25 years ago, such diffusional-reaction phenomena can be described by:

$$0 = D_S \frac{d^2 S}{d Z^2} + r_S \tag{10}$$

where Z is the film thickness. If the diffusion of a single component is known to be limiting and the intrinsic reaction rate is zero order in dimensionless form:

$$0 = \frac{d^2 \overline{S}}{d \overline{Z}^2} - \frac{kL^2}{S_o D_S} \tag{11}$$

Applying the boundary conditions for incomplete penetration, $\overline{Z} = 0$, $\overline{S} = 1$ and $\overline{Z} = \overline{Z}'$, $\overline{S} = 0$ and $d\overline{S}/d\overline{Z} = 0$, where Z' is the location in the film where S reaches zero. Integrating gives:

$$\overline{S} = \frac{P_{DRO}}{2} \overline{Z}^2 - \sqrt{2 P_{DRO}}\, \overline{Z} + 1 \tag{12}$$

where $P_{DRO} = kL^2/S_o D_S$ (the square of the Thiele Modulus). The gradient at the interface in dimensional form is:

$$\frac{dS}{dZ}\bigg|_{Z=0} = -\sqrt{\frac{2\, kS_o}{D_S}} \tag{13}$$

so that the apparent reaction rate is:

$$r_{app} = \frac{-D_S A_F}{V_{tot}} \frac{dS}{dZ}\bigg|_{Z=0} = -\frac{A_F}{V_{tot}} \sqrt{2\, D_S\, k\, S_o} \tag{14}$$

The important result is that a penetration-limiting substrate with low K_M (effective zero order) follows ½ apparent order reaction kinetics due to the combined reaction and diffusion processes. In practice, however, it may be hard to distinguish between first and ½ order over a limited concentration range.

In the low concentration range ($S < K_M$), first-order kinetics would prevail and the only influence of diffusion would be a reduction of the apparent rate constant. Both zero and first-order reactions would exhibit an apparent activation energy lowered by a factor of ½.

BIOFILM KINETICS OF NITRIFICATION

The modeling of complex reactions occurring within a biofilm requires a mass balance for each reactant and product of the form:

$$\frac{dS}{dt} = D_S \frac{d^2 S}{d Z^2} + r_S \tag{15}$$

Thus for nitrification reactions, considering only the substrate conversion reactions and ignoring the slow organism growth processes, the reactions can be written as:

$$NH_4^+ + \tfrac{3}{2} O_2 \xrightarrow{\textit{Nitrosomonas}} NO_2^- + H_2O + 2\, H^+$$

$$NO_2^- + \tfrac{1}{2} O_2 \xrightarrow{\textit{Nitrobacter}} NO_3^-$$

The oxygen requirements for the first and second steps are 3.5 mg O_2/mg $NH_4^+ - N$ and 1.1 mg O_2/mg NO_2^-/mg $NO_2^- - N$. The low yields and low growth rates make it unnecessary to consider growth requirements and kinetics. Previous work[10,11] has shown the intrinsic substrate uptake kinetics for the two steps to have a double saturation form for the first step:

$$r_{NH_4} = v_{m1}\left(\frac{[NH_4^+]}{K_{NH_4} + [NH_4^+]}\right)\left(\frac{[O_2]}{K_{O_2-1} + [O_2]}\right) \tag{16}$$

and for the second step:

$$r_{NO_2} = v_{m2}\left(\frac{[NO_2^-]}{K_{NO_2} + [NO_2^-]}\right)\left(\frac{[O_2]}{K_{O_2-2} + [O_2]}\right) \tag{17}$$

where v_{m1} and v_{m2} represent the maximum rates for a particular biomass concentration. The four saturation constants have been measured and found to lie below 1 mg/l.

Considering the diffusion phenomena in the biofilm to be represented by one-dimensional diffusion with quasi-homogeneous reaction, differential balance equations can be written for all reactants and products to describe the concentration profile in the film:

for NH_4^+,

$$\frac{d\,[NH_4^+]}{dt} = D_{NH_4}\frac{d^2\,[NH_4^+]}{d\,Z^2} - r_{NH_4} \tag{18}$$

for NO_2^-,

$$\frac{d\,[NO_2^-]}{dt} = D_{NO_2}\frac{d^2\,[NO_2^-]}{d\,Z^2} + r_{NH_4} - r_{NO_2} \tag{19}$$

for NO_3^-,

$$\frac{d\,[NO_3^-]}{dt} = D_{NO_3}\frac{d^2\,[NO_3^-]}{d\,Z^2} + r_{NO_2} \tag{20}$$

for O_2,

$$\frac{d\,[O_2]}{dt} = D_{O_2}\frac{d^2\,[O_2]}{d\,Z^2} - s_1\,r_{NH_4} - s_2\,r_{NO_2} \tag{21}$$

The stoichiometric oxygen requirements for the first and second reaction steps are given by s_1 and s_2. Equations 16 through 21 were used as the model for the biofilm. The boundary conditions used represent the bulk liquid-phase concentration and the zero gradient at the biofilm-solid interface:

$$[NH_4^+]_R = [NH_4^+] \quad \text{at } Z = 0$$

$$\frac{d\,[NH_4^+]}{dZ} = 0 \quad \text{at } Z = L$$

$$[NO_2^-]_R = [NO_2^-] \quad \text{at } Z = 0$$

$$\frac{d\,[NO_2^-]}{dZ} = 0 \quad \text{at } Z = L$$

Simulation techniques can be used to allow the solution of Equations 18–21 to come to steady state. The steady-state profiles obtained using the set of parameters given in TABLE 1 have been given previously.[12] Close agreement with the O_2 profile is obtained by the approximate solution, which assumes a single reaction with zero-order kinetics (Equation 12) and uses a stoichiometry coefficient of 4.6 g O_2/g NH_4^+.[12]

As shown and discussed previously,[11] the differential film balances combined with the kinetics can be written in dimensionless form using the substitutions $[NH_4^+] = [NH_4^+][NH_4^+]_R$:

$$[O_2] = \overline{[O_2]}[O_2]_R,t = \bar{t}\,L^2/D_{NH_4}, Z = \overline{Z}L$$

A comparison of the resulting $[NH_4^+]$ balance with the $[O_2]$ balance revealed that when the second reaction is neglected, the equations are identical if $D_{NH_4} = D_{O_2}$ and if $[O_2]_R = s_1\,[NH_4^+]_R$.

The dimensionless balances are governed by the dimensionless groups: D_{O_2}/D_{NH_4}; $(v_{m1}\,L^2\,s_1/[O_2]_R\,D_{NH_4})$: $(K_{NH_4}/[NH_4]_R)$: and $(K_{O_2-1}/[O_2]_R)$. This analysis indicated that if $[O_2]_R/[NH_4^+]_R = s_1 = 3.5$ then to a good approximation, the penetration distance of O_2 and NH_4^+ will be the same. The ratio $[O_2]_R/[NH_4^+]_R$, which can be varied according to the reactor operating conditions, can thus be used as a criterion to evaluate whether NH_4^+ or O_2 will be penetration limiting, with the pivot value being 3.5. Inspection of the governing dimensionless groups reveals that in a given biofilm situation $[NH_4^+]_R$ and $[O_2]_R$ will determine the penetration distances for NH_4^+ and O_2, respectively. The $[O_2]_R/[NH_4^+]_R$ criterion indicates which component could be penetration limiting, O_2 if less than 3.5, NH_4^+ if greater than 3.5.

TABLE 1. Parametric Values Used in the Simulations

v_{m1}	$= 20$ mg $(NH_4 - N)\ L^{-1}\ min^{-1}$
v_{m2}	$= 10$ mg $(NO_2 - N)\ L^{-1}\ min^{-1}$
K_{NH_4}	$= 0.50$ mg $(NH_4 - N)/L$
K_{NO_2}	$= 0.50$ mg $(NO_2 - N)/L$
K_{O_2-1}	$= 0.50$ mg O_2/L
K_{O_2-2}	$= 0.25$ mg O_2/L
$D_{NH_4}D_{NO_2}D_{NO_3}$	$= 0.096\ mm^2/min$
D_{O_2}	$= 0.126\ mm^2/min$
L	$= 0.5$ mm
s_1	$= 3.5$ mg O_2/mg $(NH_4 - N)$
s_2	$= 1.1$ mg O_2/mg $(NO_2 - N)$
ΔZ	$= 0.056$ mm

Since the reaction cannot proceed without both reactants, a penetration limitation of either O_2 or NH_4^+ may occur, but not of both. Thus if one component falls to zero within the film, the other will penetrate completely. The saturation constants being of order 0.5 mg/l means that for most applications, the intrinsic rate will follow zero-order kinetics, from the film's outer surface to a location near to where either NH_4^+ or O_2 reaches zero. The phenomenon of penetration limitation of either O_2 or NH_4^+ was investigated by simulation. The film profiles are shown in FIGURE 3. The $[NH_4^+]$ outside the film was maintained constant at 2 mg/l and the $[O_2]$ was varied to change the $[O_2]/[NH_4]$ ratio from 5.0 to 1.0. The resulting profiles demonstrated the influence of this ratio on penetration limitation.

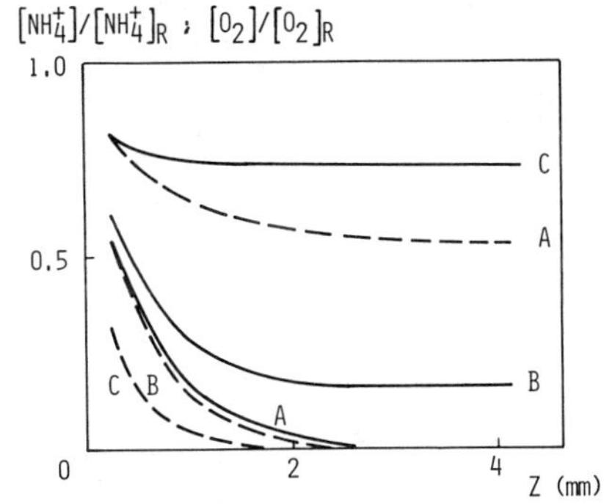

FIGURE 3. Simulated profiles of $[NH_4^+]$ (—) and $[O_2]$ (---) in the biofilm and their dependency of $[O_2]_R/[NH_4^+]_R$. Parameters: $[NH_4^+]_R = 2.0$ mg/l; $[O_2]_R/[NH_4^+]_R$: $A = 5.0$, $B = 3.5$, $C = 1.0$.

The biofilm model, Equations 16–21, can be combined with a well-mixed tank model for the liquid phase written for each component of the form:

$$\frac{d\,S_R}{dt} = \frac{F}{V}(S_F - S_R) + D_S\frac{A}{V}\left(\frac{dS}{dZ}\right)_{Z=0} \tag{22}$$

In this way, it is possible to simulate the performance of a single tank or of a column by using a tanks-in-series approach.[12]

A batch system is obtained by setting the flow terms to zero: the numerical results are given in FIGURE 4.

The importance of oxygen limitation is apparent from the oxygen concentration in the midpoint of the film, which is zero until NH_4^+ is depleted and rises in steps as $[NH_4^+]_R$ and $[NO_2^-]_R$ go to zero.

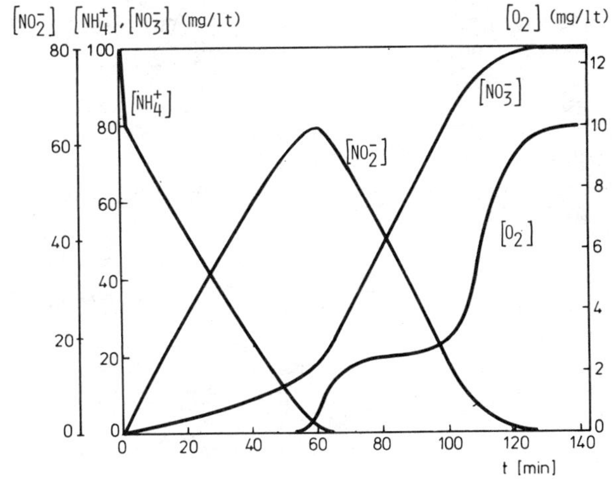

FIGURE 4. Simulated profiles in a batch reactor showing $|NH_4^+|_R$, $[NO_2^-]_R$, and $[NO_3^-]_R$ versus time as well as $[O_2]$ in the middle of the biofilm.

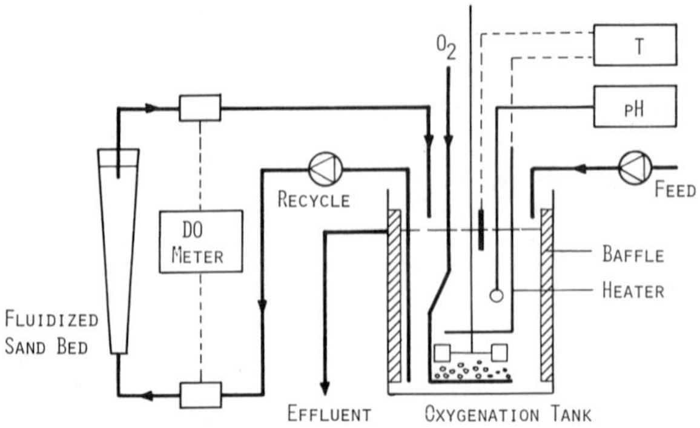

FIGURE 5. Experimental biofilm fluidized bed and equipment.

LABORATORY EXPERIMENTS WITH FLUIDIZED-BED NITRIFICATION

Kinetic experiments were conducted in a laboratory-scale fluidized sand bed reactor with external oxygen supply as shown in FIGURE 5. The bed was cone-shaped, and quartz sand of 0.2–0.3 mm diameter was used. The conditions of the experiments are listed in TABLE 2 and were previously reported in detail.[10,11] The reactor was started up using an inoculum of activated sludge. Approximately 3 months of fed-batch operation was required before appreciable nitrifying activity and biofilm growth were obtained. The substrate contained $(NH_4)_2SO_4$, bicarbonate, and trace minerals, with no organic carbon present. The resulting biofilm consisted of nitrifying bacteria and large amounts of protozoa. A scanning electron microscope photo of the biofilm-covered sand particles is shown in FIGURE 6. TABLE 3 gives the rates of NH_4^+ and NO_2^- oxidation calculated on the basis of bed volume, nominal sand areas, and biomass quantity. Experience with the system running continuously for 3 years indicated that the appearance of the film changed for unexplained reasons. Occasionally the biomass sloughed off the sand and collected in flocs above the bed. After 2–3 weeks this biomass relocated to the sand, progressing downward in the bed from week to week. The nature of fluidization in this tapered bed was stable and the two-phase operation provided gentle, nonturbulent conditions. Vigorous biofilm growth of nitrifying organisms occurred on every plastic, glass, and metal surface of the reactor. Accidents (70°C overnight and pH 11 overnight) caused temporary partial activity losses that recovered within 1–2 weeks.

TABLE 2. Typical Experimental Conditions for Continuous Experiments

Feed rate	= 5.2 l/hr
Recycle rate	= 84 l/hr
Reactor bed volume	= 0.8 liters
Oxygenation tank volume	= 1.9 liters
Total volume	= 2.7 liters
Reactor conditions:	
Settled sand cell volume	= 63 ml
Settled clean sand volume	= 29 ml
Biomass (dry weight)	= 1.7 g
Sand surface area	= 0.5 m^2

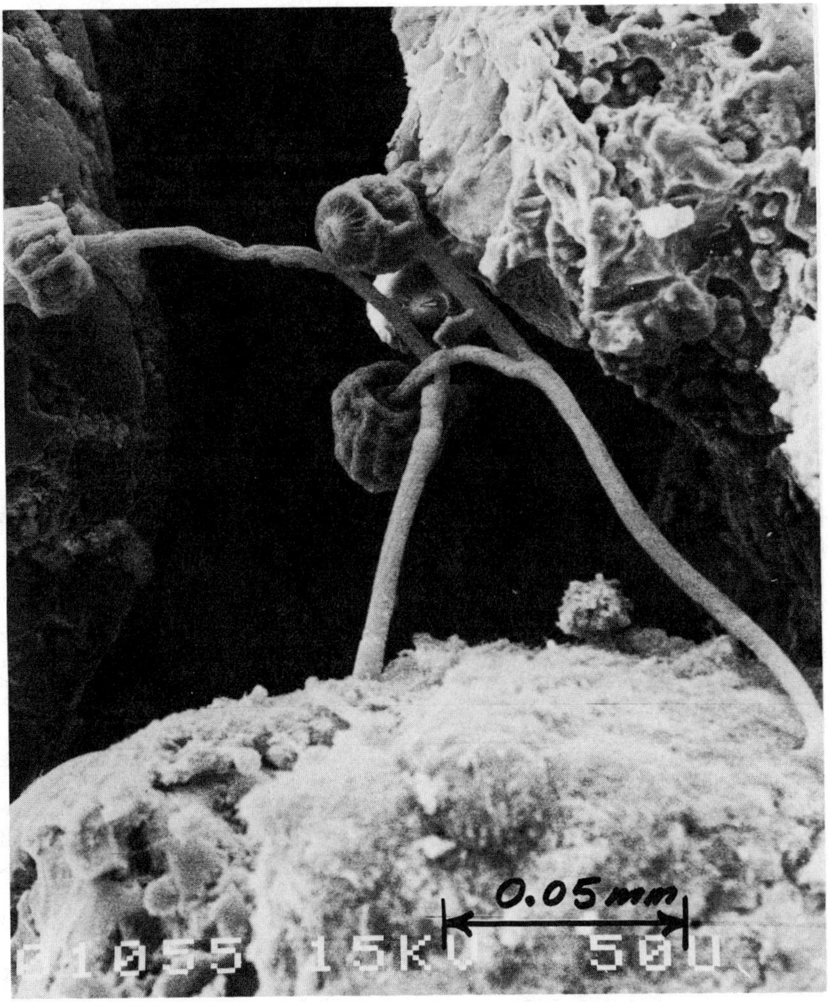

FIGURE 6. SEM photo of sand particles with biofilm.

TABLE 3. Maximum Observed Rates

	Based on Total Volume mg $N\ L^{-1}\ hr^{-1}$	Based on Area mg $N\ m^{-2}\ hr^{-1}$	Based on Biomass mg N/g biomass hr
NH_4	84.5	456	134
NO_2	75.7	406	120

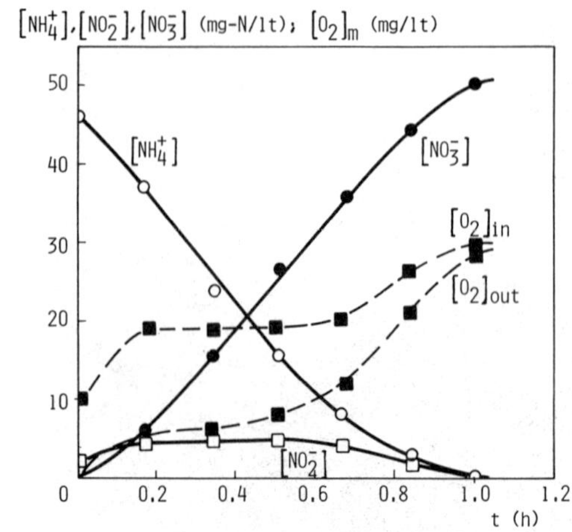

FIGURE 7. Typical experimental batch curves under high $[O_2]_R$ conditions.

The results of experiments with the influence of fluidizing velocity on bed expansion and nitrifying rate showed that the bed expanded linearly with flow rate, and the oxidation rate varied linearly, reaching a maximum at a flow velocity of 20 m/hr taken at the midsection of the bed. At the lowest velocities, channeling occurred; since a high outlet oxygen concentration was maintained, the dependency of reaction rate on flow velocity was due to bed inhomogeneities.

A typical batch concentration–time profile for a case when oxygen was not limiting is shown in FIGURE 7. Under these conditions, $[NO_2^-]_R$ was always relatively low and rose if O_2 was limiting.

Continuous experiments were conducted to investigate the influence of O_2 under controlled conditions, during which the feed concentration, residence time, and recirculation flow rate were kept constant. Only the oxygenation conditions (gassing and P_{O_2}) were changed. Results are shown in FIGURE 8, which indicate that $[NH_4^+]_R$ falls as $[NO_2^-]_R$ and $[NO_3^-]_R$ increase with mean $[O_2]$. Sufficiently long residence times would lead to complete $[NO_2]$ and $[NH_4^-]$ oxidation.

From other continuous data, experimental verification of the change from NH_4^+ to

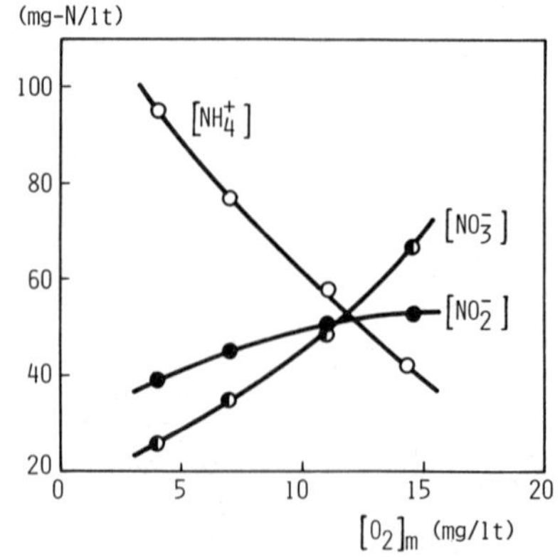

FIGURE 8. Influence of mean $[O_2]_R$ concentration on outlet concentrations for steady-state continuous nitrification experiments.

Conditions: $F = 5.5$ l/hr. $(F + R) = 54$ l/hr. $[NH_4]_F = 100$ mg − N/l. pH = 8.5. $V = 0.8$ l. $T = 30$°C.

O_2 penetration limitation is shown in FIGURES 9a and 9b. In the case of the data in FIGURE 9a, below a value of $[O_2]_R/[NH_4^+]_R = 3.5$ (in brackets) the rates vary linearly with $[O_2]^{0.5}$. Above this value, they are constant.

The data of FIGURE 9b are linearly correlated with $[NH_4^+]^{0.5}$ for $[O_2]_R/[NH_4^+]_R$ above 3.5. In these runs, $[O_2]_R$ was kept nearly constant at 22 mg/l, while in the experiments of FIGURE 9a, both $[NH_4^+]_R$ and $[O_2]_R$ varied. It is seen that in the region of $[NH_4^+]_R < 4$ mg/l with $[O_2]_R/[NH_4^+]_R > 3.5$, NH_4^+ became penetration limiting (FIGURE 9b), and in the region of $[O_2]_R < 12.5$ mg/l with $[O_2]_R/[NH_4^+] < 3.5$, O_2

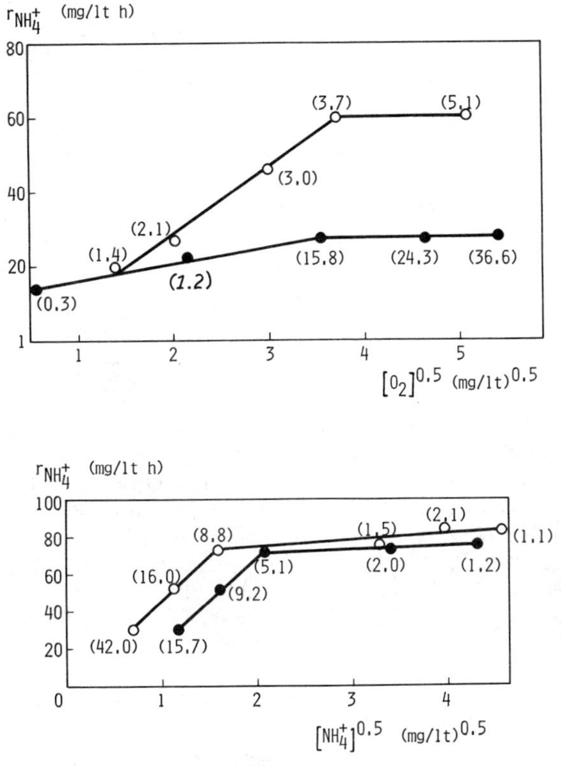

FIGURE 9. Rate dependency on half-order oxygen (9a) and ammonia (9b) kinetics with $[O_2]_R/[NH_4^+]_R$ conditions in brackets.

became penetration limiting (FIGURE 9a). Above these concentration ranges, full penetration was obtained.

Nitrification involves a pH shift to lower values with conversion of NH_4^+ to NO_2^-. It is, therefore, important to design the reactor with sufficient recycle and pH control to avoid extreme axial gradients in pH. Continuous experiments on the pH sensitivity of the biofilm reactor were conducted. The concentration data in FIGURE 10 were obtained from 9 steady-state experiments in the pH 6 to 10 range. The difference in oxygen between bed inlet and outlet (D_{O_2}) is also shown. An ammonia loss to the gas phase above pH 8.5 caused $[NH_4^+]_R$ to be unusually low. From the overall stoichiomet-

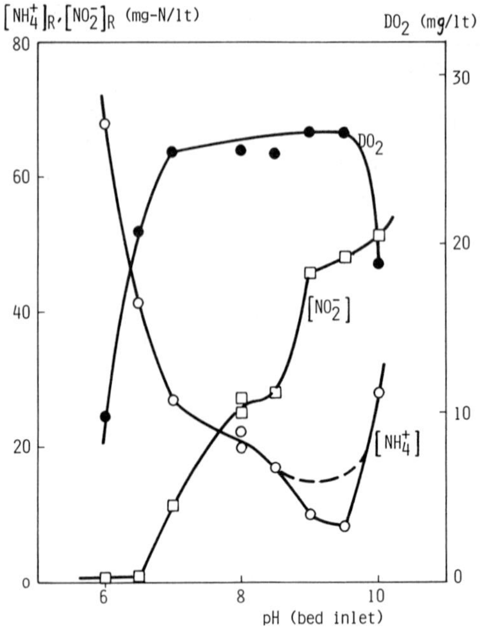

FIGURE 10. Influence of pH on continuous nitrification. Approximate correction for gas-phase NH_4^+ losses. Conditions similar to FIGURE 8.

ric considerations, 2 moles H^+ ion are formed for 2 moles O_2 consumed, but the actual pH is complicated by the buffering capacity of the solution. In practice it appears that a 25 mg/l oxygen difference corresponded to a one-unit pH difference. Experiments with NO_2 only in the feed gave the pH effect on r_{NO_2}, and the data are presented on a rate basis in FIGURE 11. From this the *Nitrobacter* seems to be inhibited at high pH, while the *Nitrosomonas* are inhibited at low pH. This appears contrary to the data of FIGURE 10, which show residual NO_2^- above pH 6.5. This r_{NO_2} decrease is explained by oxygen limitation effects. If *Nitrobacter* were located predominantly deeper in the

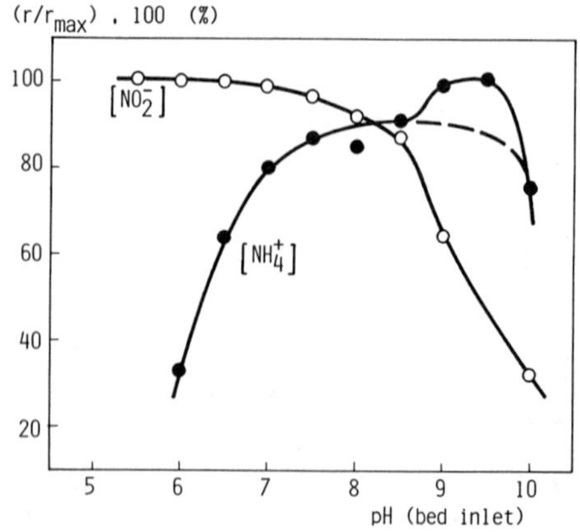

FIGURE 11. Influence of pH on the individual reaction steps. Conditions similar to FIGURE 8 with O_2 excess. Rate data are from separate experiments with NH_4^+ and NO_2^- only in feed.

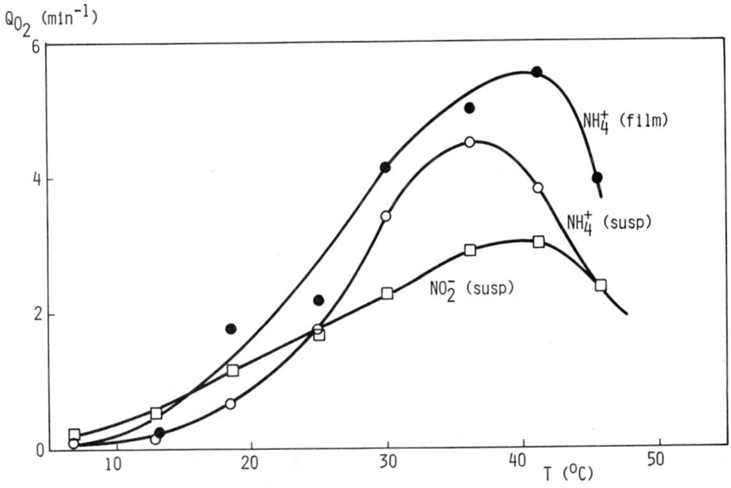

FIGURE 12. Influence of temperature on the individual nitrification reactions as measured in a respirometer from reactor samples of suspended cells (susp) and attached cells (film).

film, then the oxygen profile that developed at $r_{NH_4^+}$ increased above pH 6 would lead to oxygen limitation of the $NO_2^- \rightarrow NO_3^-$ step. A similar effect would occur if the K_M for oxygen were higher for the second step than for the $NH_4^+ \rightarrow NO_2^-$ step.

Temperature effects were measured with a respirometer from the biofilm and suspended culture samples taken from the reactor. The data expressed as Q_{O_2} versus T are shown in FIGURE 12. Activation energies as obtained from Arrhenius plots of this data are compared with those from the literature in TABLE 4.

CONCLUDING DISCUSSION

Biological film reactors provide a means of retaining microorganisms in the reactor, thus obviating biomass separation and recycle. The resulting high biomass concentration and higher volumetric reaction rates lead to greater reactor efficiency. Biomass retention is particularly important in the case of slow-growing organisms because the fluid residence time is then uncoupled from maximum specific growth rates.

Fluidized-bed bioreactors require a minimum upflow liquid velocity to maintain adequate fluidization. The optimal velocity depends on the particle size and density as

TABLE 4. Comparison of Activation Energies

E. (kcal/mol) *Nitrosomonas*	*Nitrobacter*	Data Source
17.3	—	Activ. sludge + pure culture[13]
15.2	—	River water + pure culture[13]
—	10.9	River water + pure culture[13]
—	21.3	Present work
32.0	—	Present work

well as on the biofilm thickness and density. For sand systems, it will lie between 10 and 30 m/hr.

The presence of gas bubbles in a fluidized bed will cause excessive turbulence and undesirable shear effects. Small gas bubbles will rapidly coalesce into large bubbles with high rise velocities. This coalescence phenomenon is detrimental to biofilm growth and leads to low gas-liquid transfer rates. For these reasons, gaseous substrates should usually be supplied in a separate gas-liquid contactor that is located in the recycle loop.

Designing a biological fluidized bed with external gas-liquid transfer requires the use of mass balances, reaction kinetics and gas-liquid transfer quantities. This information permits the calculation of single pass conversion, required recycle rates and gas-liquid transfer coefficients as well as reactor and contactor volumes. Biofilm kinetics describes the influence of diffusion on the apparent rates. Frequently the intrinsic kinetics will be zero order, which will result in an apparent half-order biofilm kinetic expression.

Nitrification is a two-step, oxygen-requiring reaction that involves slow-growing organisms. Diffusion-reaction considerations indicate the important influence of the oxygen/ammonia ratio on the biofilm concentration gradients. Model calculations with appropriate experimentally determined parameters give qualitative agreement with the experimental results, which verify that the stoichiometric ratio is critical in determining the borderline between oxygen or ammonia rate limitation.

The overall nitrification rates achieved indicate the potential of biofilm fluidized beds and point to the difficulty of being able to supply oxygen at a sufficient rate without enrichment or pressure methods.

REFERENCES

1. COOPER, P. F. & D. H. V. WHEELDON. 1980. Water Pollut. Control **79:**286.
2. BAKER, C. G. J., A. MARGARITIS & M. A. BERGOUGNOU. 1981. *In* Advances in Biotechnology I. Pergamon Press. p. 635.
3. COOPER, P. F. & B. ATKINSON. 1981. Biological Fluidized-Bed Treatment of Water and Wastewater. 1980. Manchester, England. Ellis Horwood. London.
4. HENZE C. & P. HARREMOES, Eds. Anaerobic Treatment of Wastewater in Fixed-Film Reactors. 1982. IAWPR Seminar. Copenhagen, Denmark.
5. ASH, S. G. 1979. *In* Adhesion of Microorganisms to Surfaces. D. C. Ellwood, J. Melling & P. Rutter, Eds. Academic Press. New York. p. 57.
6. SHIEH, W. K., P. M. SUTTON & P. KOS. 1981. J. Water Pollut. Control **53:** 1574.
7. SCOTT, C. D., C. W. HANCHER & E. J. ARCURI. 1981. *In* Advances in Biotechnology I. Pergamon Press. p. 651.
8. SHAH, Y. T. 1979. *In* Gas-Liquid-Solid Reactor Design. Chapter 9. McGraw-Hill, New York.
9. OSTERGARD, K. 1968. *In* Advances in Chemical Engineering **7:** 71. Academic Press. New York.
10. TANAKA, H., S. UZMAN & I. J. DUNN. 1981. Biotechnol. Bioeng. **23:** 1683.
11. TANAKA, H. & I. J. DUNN. 1982. Biotechnol. Bioeng. **24:** 669.
12. DENAC, M., H. TANAKA S. UZMAN & I. J. DUNN. 1983. Biotechnol. Bioeng. In press.
13. U.S. Environ. Prot. Agency. 1975. Process Design Manual. pp. 3–9. US Govt. Printing Office.

APPENDIX

Symbols:

C_L Dissolved oxygen concentration ($M\ L^{-3}$)

D Diffusion coefficient ($L^2\ T^{-1}$) and Dilution rate (T^{-1})

DO_2	Difference in $[O_2]$ across bed ($M\ L^{-3}$)
E	Effectiveness factor
F	Feed flow rate ($L^3\ T^{-1}$)
k	Reaction rate constant (T^{-1} and $M\ L^{-3}\ T^{-1}$)
K	Saturation constant ($M\ L^{-3}$)
K_La	Oxygen transfer coefficient (T^{-1})
L	Film thickness (L)
P_{DRO}	Diffusion-reaction (zero order) parameter (dimensionless)
Q	
r	Reaction rate ($M\ L^{-3}\ T^{-1}$)
R	Recycle flow rate ($L^{-3}\ T^{-1}$)
s_1, s_2	Stoichiometric coefficient for step 1 and 2, respectively
S	Substrate concentration ($M\ L^{-3}$)
t	Time (hr)
T	Temperature
V	Volume (m^3)
v_m	Maximum reaction velocity ($M\ L^{-3}\ T^{-1}$)
$Y_{O/S}$	Yield coefficient oxygen/substrate (dimensionless)
Z	Film thickness coordinate (L)
α	Recycle ratio, F/R (dimensionless)
τ	Mean resident time (hr)

Indices:

o	Feed inlet
1	Reactor inlet
2	Reactor outlet and effluent
c	Contactor
F	Film and feed
m	Mean value
R	Reactor
S	Substrate
$\bar{}$	Overbar refers to dimensionless quantities

The Holding Time in Pure and Mixed Culture Fermentations

H. M. TSUCHIYA

Department of Chemical Engineering
and Materials Science
Department of Microbiology
University of Minnesota
Minneapolis, Minnesota 55455

The holding, or residence time, θ, is an essential bit of information when describing a continuous fermentation. It is necessary to know the holding time for the optimal operation of the fermenter.

Continuous fermentations can be run in either of two modes: with a continuous-stirred tank reactor, CSTR, which gives homogeneous growth, or with a plug flow type reactor, which gives heterogeneous growth. With homogeneous growth, there are no stages assumed in the life cycle of the organism; the sample is the same as that which occurs throughout the fermenter. Heterogeneous growth involves an organism that goes through different stages in its life cycle; the sample is terminally the same only with some organism, especially the molds and the actinomycetes. These forms demonstrate various stages in their life cycles.

The difference between pure and mixed culture fermentations is self evident. Mixed cultures are the rule in waste treatment plants where microorganisms are the catalysts that depolymerize the organic compounds. The fact that they do attack a wide variety of compounds is evidence that there are various types of organisms present.

However, there is another aspect of mixed cultures that might be speculated upon. In biological differentiation, there occur daughter cells with different morphological and physiological properties and with different anatomy and function from the parent. Differentiation means the morphological and physiological changes that occur during the temporal and spatial development of an organism, or of a population of protists. In essence, the molds, the actinomycetes, and occasionally the yeasts that undergo differentiation really produce more than one type of cell. One can assign product 1 from the first differentiation stage, product 2 from stage 2, product 3 from stage 3, and product 4 from stage 4. In the orderly differentiation process, as one proceeds from stage 1 to 4, one obtains four different products. These may be called nongrowth-associated products or secondary metabolites. These products are *not* produced in accordance with biomass. They are produced despite the fact that the biomass from which they originate is constant in weight but the cells *differ in stages.*

Now, consider a mold that exists in four stages. Despite the fact that a mold culture is pure from the taxonomist's point of view, the mold culture is functionally a mixture of four different organisms.

We shall discuss both the use of the continuous-stirred tank reactor and the plug flow type of reactor; examples will be discussed.

When two organisms such as *S. cerevisiae* and *L. casei* are placed in a CSTR in a Roberts and Snell type medium with glucose as the limiting substrate, the yeast produces riboflavin, which is required by *L. casei* for growth.[1]

The growth of the yeast is directly proportional to the amount of riboflavin produced; *L. casei,* as the B_2 assay organism, responds to the quantities of riboflavin produced. The θ determines how much of the riboflavin is produced, and *L. casei* responds to it; the more B_2, the higher is the count of *L. casei*. Sugar is the limiting substrate.

In a Roberts and Snell medium with glucose as the limiting substrate, *Propionibacter* and *Lactobacillus* cultures have been grown together.[2] *Lactobacillus,* in pure culture, gives oscillations when grown aerobically at long θ. The oscillations are apparently due to the presence of H_2O_2 produced by lactobacilli when grown aerobically, but not anaerobically.

When the two organisms are grown together, the lactobacilli grow at short θ. The sugar was broken down by the lactobacilli to lactic acid, which was then consumed by the propionibacter; then, the propionibacter grow. The dominance of an organism can be maintained by adjustment of the residence time for a given sugar concentration.

When streptococci, which produce phenylalenine, and lactobacilli, which produce folic acid, are grown together aerobically in a Henderson and Snell medium with glucose as one of the limiting substrates, the ratio of the former to the latter was

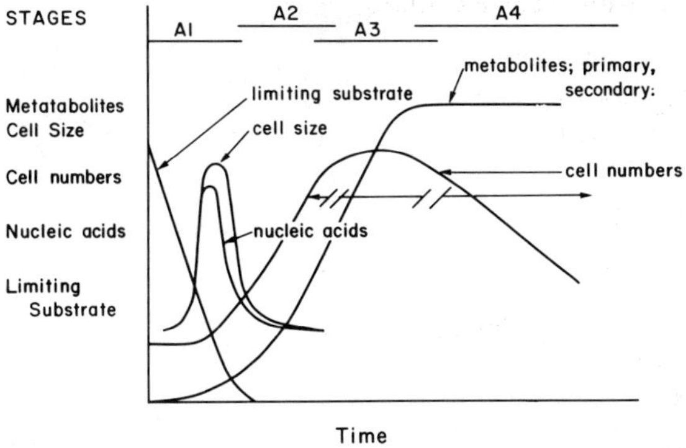

FIGURE 1. Batch growth or differentiation of microorganisms.

dependent upon the θ. When the θ is long, the *Streptococcus* to *Lactobacillus* ratio is on the order of 10.2; when the θ is short, the ratio is on the order of 0.8 to 0.3.

However, when the experiments are run for 25 days, the lactobacilli are essentially the sole survivors; the streptococci were virtually wiped out. In essence, lactobacilli produce H_2O_2 which kills the streptococci. Therefore, when running continuous cultures, it is well to run the CSTR for at least a month before assigning roles to the participants.

All of the foregoing have implications in waste-disposal plants where bacteria occur.

In order to develop a continuous process, it is essential that all details of the batch process be known.

Now, we digress for a moment to review what is known about the "growth curve" or the differentiation of microorganisms (see FIG. 1). All curves are labeled and are self explanatory except for the curve depicting "metabolite" from which two arrows extend outward. If the product is formed coincidentally with the organism, that is, the metabolite curve falls on the "cell numbers" curve, the metabolite is a primary product. If, on the other hand, the metabolite is formed associated with A2, A3, and A4, it is a secondary metabolite.

Shuler, Ng, and Kinoshita have worked upon studies dealing with the lag phase in batch cultures.

Shuler[3] confirmed the lag in numbers, the size distribution of the cells, and the nucleic acid content of azotobacter in the lag phase as compared to the cells in the other phases in a Burk's type medium.

Ng[4] has made careful measurements on lactobacilli with the Coulter channelyzer on Roberts and Snells medium in both batch and continuous cultures with glucose as the limiting substrate. He showed that lactobacilli go through a differentiation process in batch growth in which the cells become larger initially and then smaller in the subsequent phases. Much to our surprise, the cells from the continuous cultures were unlike the cells from batch cultures. In most cases, the continuous culture is an extension of the batch fermentation. We have found that this is *not* necessarily true. What was unexpected was the finding that the peaks of cell size distribution were monopeaked in batch culture, whereas they were multipeaked in continuous culture. This was true at all θ (see FIGS. 2, 3, and 4).

Possibly, the nucleic acid contents should be determined. They might be greater in continuous rather than in batch cultures.

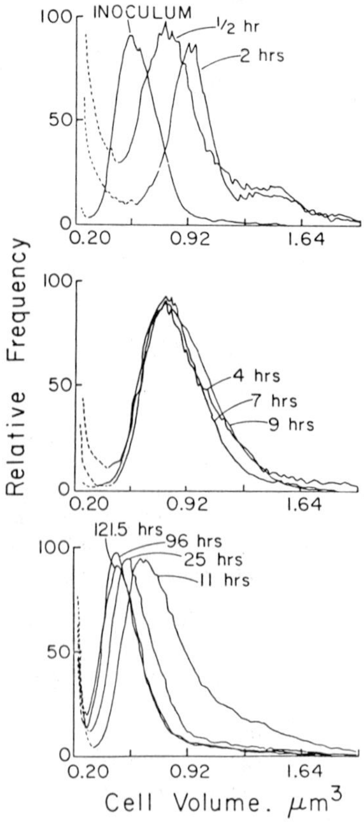

FIGURE 2. Size distribution: (a) lag phase; (b) exponential phase; and (c) stationary and death phases.

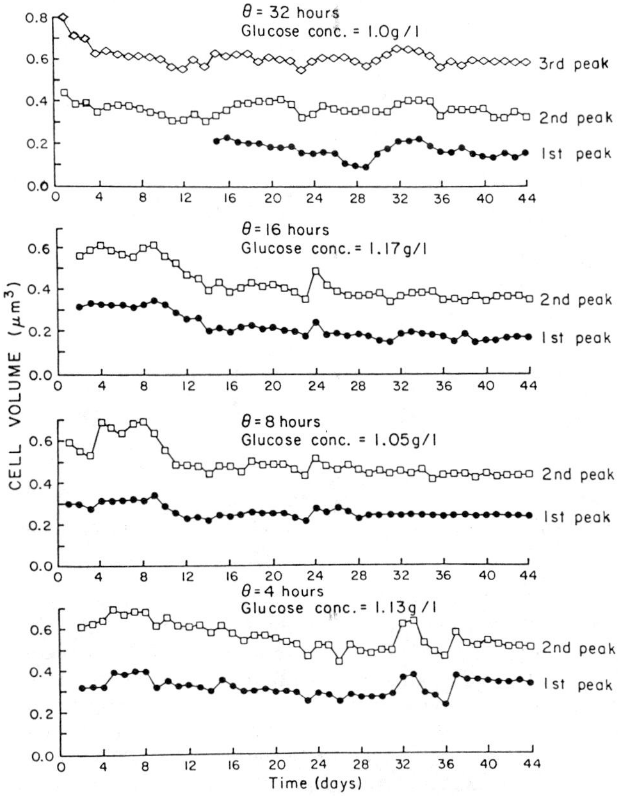

FIGURE 3. Size distribution in continuous cultures. $\theta = 4$ hr., $\theta = 8$ hr., $\theta = 16$ hr., and $\theta = 32$ hr.

Kinoshita[5] grew the mold *Aspergillus awamori* NRRL 3112 in batch cultures upon semisolid medium containing 1.0% glucose, 0.5% yeast extract, and 0.5% peptone. The pH was adjusted to 5.5 to 6.0; he hardened the medium with *1.5% agar*. He grew the mold in a baking dish fitted with a stainless cover. The growth, which spread laterally out towards the sides of the dish, developed in zones. The zones were clearly marked to the eye. He fabricated a knife with razor blades clamped 3.75 mm apart. He then sliced the mycelium *and* the underlying agar lengthwise and extracted each strip with water, the mycelium together with the agar. He ran the biomass, protein, RNA, and DNA determinations on the extracts of these strips. He noted the structures that grew in the zones. The first stage was A1, the second stage was A2, the third stage A3, and the fourth stage A4.

A1 is the hyaline zone and is the zone of vegetative hyphae. A2 is the yellowish zone and is the stage where conidiophores appear profusely with their vesicles plus hyphae. A3 is the black stage and is the stage where metulae and phialides appear on the vesicles of the conidiophore plus hypae. A4 is the dense, brownish black zone and is the zone where conidiation occurs in addition to the metulae, the phialides, and the

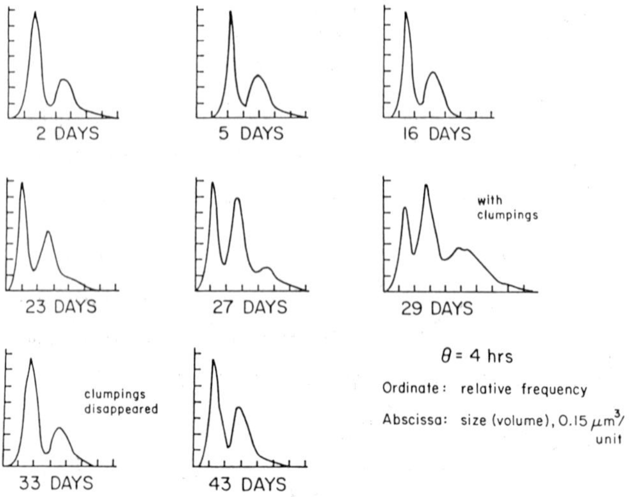

FIGURE 4. Size distribution at $\theta = 4$ hr. Similar data were obtained at 8, 16, and 32 hr.

conidiophores with their vesicles and their hyphae (see FIG. 5). If one were to see these diverse forms, he might consider them four different organisms.

Yanagita and Kogane[6] have also noted four zones based on their histochemical studies. We based our zones on visual observation of the structures of the molds.

Semisolid medium is recommended for seeing these four zones. The states are not discrete and clear cut; there is some overlapping between the four zones.

The mycelial biomass, the protein, the RNA, and the DNA of the mold and the agar were measured. This was done for each of the strips and associations between the figures and these structures were made. These are the *nonspecific* indicators because all organisms go through these stages (see FIG. 6). They, together with the *specific* indicators, define the physiological stages of this organism. They can be determined by chemical measurements (see FIG. 7).

Henrici[7] has noted the initial enlargement in size of bacteria during the early phase of growth. He summarized much of the early literature beginning with Muller[3] up to his time and added observations of his own on several bacteria. Then Malmgren and Heden[9] and Hinshelwood[10] showed that there was an initial increase in the amount of nucleic acid content of cells taken from the early stages of growth. They also noted the lag in numbers of cells and the initial rise in cell size. They found that the nucleic acid content increased substantially during the initial phase and then subsequently diminished. Herbert[11] and Maaloe and Kjeldgaard[12] (1966) have also seen this with their bacteria. Malek[13] has been calling for a better definition of physiological stages; he has long called for it. Bacteria and fungi go through different stages of physiological age as do the higher animals and plants. Dawson[14] is one of the applied microbiologists who is cognizant of the importance of the physiological stages.

We also noted the difference in the appearance of the enzymes: α-amylase, protease, acid phosphatase, glucamylase, and alkaline phosphatase. These are the *specific* indicators of physiological age. They are called specific because they are unique for this organism. The high α-amylase and glucamylase content make *A. awamori* valuable industrially for the amylolytic breakdown of starch to glucose. The α-amylase is associated with A1. The glucamylase is associated with A3 and A4. The occurrence of protease in A2 and A3 may be the reason why α-amylase decreases slightly during the growth of the mold. Alkaline phosphatase may be associated with A4 and may be involved in the death process, the lytic phenomenon. It may also be involved in the liberation of glucamylase.

Jeffreys[15] first described a continuous method for the production of mold enzymes. For some reason or other, the tray method did not catch the fancy of chemical engineers in this country. Suffice it to say that the growth of mold in a plug-flow type reactor was not followed up.

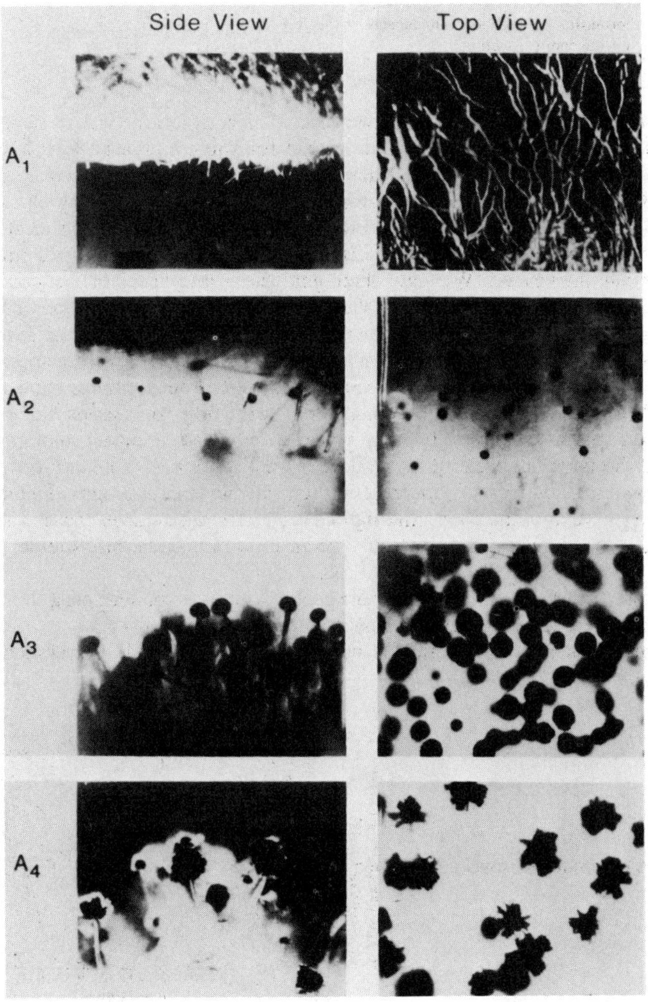

FIGURE 5. Mold structures at stages A1, A2, A3, and A4.

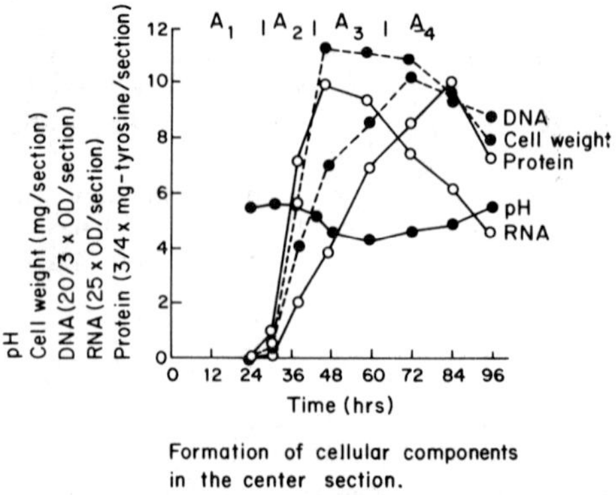

FIGURE 6. Occurrence of *nonspecific* indicators.

Yokotsuka and his associates[16] have described a toroidal type of reactor for the preparation of koji. Koji is the Japanese word for the enzymatic hydrolyzate of soy bean and wheat which is also the starter for the moromi from which soy sauce is produced. Koji is the aerobic fermentation of soybeans and wheat by *A. oryzae*. The moromi is the anaerobic fermentation and extraction of koji with yeasts and pediococci of the brine. The fermentation can be allowed to go a certain allotted time and the semisolid mash harvested. We can then get the α-amylase, or we can allow the fermentation to go longer and harvest the glucamylase also. It is suggested that we in this country have not paid sufficient attention to the plug-flow type of reactor which allows A1, A2, A3, and A4 to develop in an orderly manner. The Chinese and Japanese have long been familiar with the preparation of shoyu. The Japanese have automated, mechanized, and optimized the tray method. According to Yokotsuka, the θ of the process is less than 48 hours. In this process, they use a cutter and mixer at the initiation of the process; it could as well be a CSTR for self-reinoculation. Then, the mold is allowed to develop without relative shear in the various stages. The mash bed is approximately 2 feet deep. Some mixing occurs as the disc moves around; the mash is aerated and cooled. The economics of the process are such that shoyu, a common condiment, can be prepared from it.

It is a pity that Jeffrey's patents were never followed up. The plug-flow reactor is still worth using in certain cases; we have the technology to do so.

The continuous reactor should be run initially as a CSTR to reinoculate itself.

FIGURE 7. Occurrence of *specific* indicators.

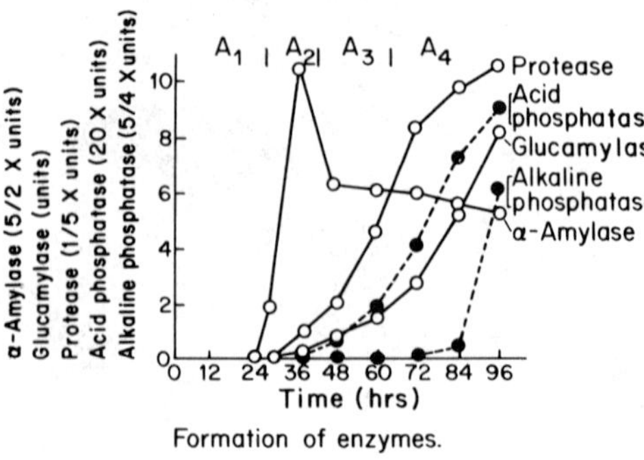

Then, it should be run as a plug flow reactor to continuously develop stages and harvest the products. Examination of FIGURE 8 shows that α-amylase is produced in zones A1 and A2 and the glucamylase in zones A3 and A4; a product is specific for the zone in which it occurs, it can neither be formed before nor after a particular stage.

SUMMARY

Continuous mixed culture fermentations have been studied in the continuous-stirred tank reactor. The residence or holding time, θ, is important in determining which of two mixed organisms shall dominate in numbers.

Continuous ethanol and acetic-acid fermentations are known to the brewing industry. The continuous production of ethanol and acetic acid are contingent upon the cells of *Saccharomyces* and *Acetobacter* being alive and growing. These are known as growth-associated products.

On the other hand, α-amylase and glucamylase, or fungal amylase, are known as nongrowth-associated products or secondary metabolites.

The organisms that produce the secondary metabolites, for example, penicillin,

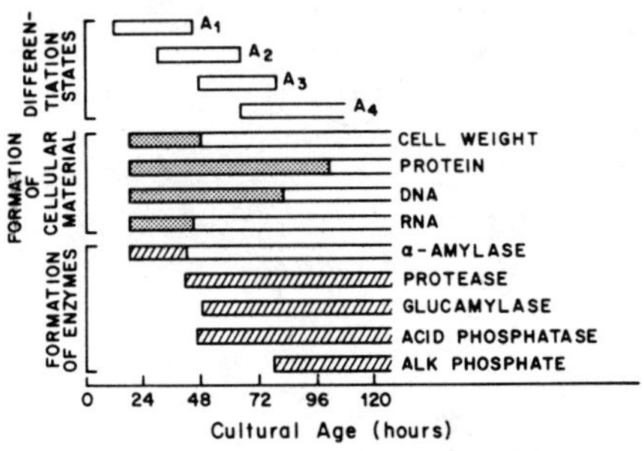

FIGURE 8. Association of *nonspecific* and specific indicators with stages.

cephalosporin, streptomycin, and aureomycin, undergo differentiation and growth. These are the higher microbial forms and are produced by batchwise fermentation of molds and actinomycetes.

It is submitted that these higher microbial forms can be grown continuously and produce secondary metabolites in amounts so as to make the processes economically viable. It is possible to grow the organisms continuously in plug flow reactors so that the secondary metabolites that have hitherto not been considered for continuous production be so considered.

REFERENCES

1. MEGEE, R. D., J. F. DRAKE, A. G. FREDRICKSON & H. M. TSUCHIYA. 1972. Can. J. Microbiol. **18:** 1733.
2. LEE, I. H., A. G. FREDRICKSON & H. M. TSUCHIYA. 1976. Biotechnol Bioeng. **18:** 513.
3. SHULER, M. L. & H. M. TSUCHIYA. 1975. Can. J. Microbiol. **21:** 927.
4. NG, K. & H. M. TSUCHIYA. 1982. In press.

5. KINOSHITA, S. 1971. Private communication.
6. YANAGITA & KOGANE. 1962. J. Gen. Appl. Microbiol. **8:** 201.
7. HENRICI, A. T. 1982. *In* Morphological Variation and Rate of Growth of Bacteria. C. C. Thomas. Baltimore, MD.
8. MULLER, M. 1895. Z. J. Hyg. **20:**245.
9. MALMGREN, B. & C. J. HEDEN. (1947) Acta Pathol. Microbiol. Scand. **24:** 448.
10. HINSHELWOOD, C. M. 1946. *In* The Chemical Kinetics of the Bacterial Cell. Clarendon, Oxford.
11. HERBERT, D. O. 1956. J. Gen. Microbiol. **14:** 60.
12. MAALOE, L. & N. O. KJELDGAARD. 1966. *In* Control of Macromolecular Synthesis. W. Benjamin, Inc. New York.
13. MALEK, I. 1958. *In* Continuous Cultivation of Microorganisms, Czechoslovak Acad. Sci., Prague.
14. DAWSON, P. S. S. & K. L. PHILLIPS. 1974. Adv. Appl. Microbiol. **17:** 195.
15. JEFFREYS, G. A. 1948. Food Inds. **20:** 688.
16. AKAO, T., T. SAKASAI, M. SAKAI, A. TAKANO & T. YOKOTSUKA. 1972. First Pacific Chem. Eng. Congr. Part II Soc. Chem. Eng. Japan (SCEF), Am. Inst. of Chem. Eng. (AIChE), 127.

Computer Control of Fermentations with Biosensors

B. MATTIASSON,[a] C. F. MANDENIUS,[a] J. P. AXELSSON,[b]
B. DANIELSSON,[a] AND P. HAGANDER[b]

[a]Department of Pure and Applied Biochemistry
[b]Department of Automatic Control
Lund Institute of Technology
S-220 07 Lund, Sweden

INTRODUCTION

Ethanol production from biomass has during recent years been one of many ways considered to better utilize the world's energy resources. The use of microorganisms for this purpose is historical. However, optimization and control of the processes are rare, mainly because of insufficient knowledge about process dynamics and especially because of lack of reliable on-line sensors for substrates and products. Today, advanced methods for computer estimation and optimization of productivity are usually based on oxygen and carbon dioxide on-line analysis. These measurements are then used in mass and energy balances together with substrate feed-rate data.[1]

Recently it has been shown that enzyme thermistor sensors can be used for controlling the carbohydrate concentrations in fermentations.[2,3] This report is on computer control of ethanol concentration based on measurements obtained using a semiconductor gas sensor. The substrate (sucrose) concentration is monitored by an enzyme thermistor sensor.

THE PROCESS

The experiments are performed with a continuous stirred-tank reactor of 5-liter capacity (FIG. 1) containing *Saccharomyces cerevisiae* immobilized by entrapment in calcium alginate beads (d = 2–3 mm). The tank is fed with 0.5 mol/l sucrose solution by a peristaltic pump equipped with voltage control. The sucrose content in a small flow continuously withdrawn from the reactor is calorimetrically measured using an enzyme thermistor containing immobilized invertase. The flow is diluted with buffer and sterilized by addition of cyanide. Using a similar on-line flow system, broth is pumped to a piece of silicone tubing to get the vapor for the ethanol sensor.

THE MODEL

The process and the sensors are described by the following equations:

$$dC_s/dt = -\,k_s\,C_s - v\,(C_s - C_{s_0})/V$$

$$dC_e/dt = k_e\,C_e - v\,C_e/V$$

$$Y_s(t) = C_c(t - \tau_{ds})$$

$$dY_e/dt = (C_e(t - \tau_{de}) - Y_e(t))/\tau_e$$

where

C_s = sucrose concentration in the reactor (mol/l)

C_{s_0} = sucrose concentration in the feed (mol/l)

C_e = ethanol concentration in the reactor (mol/l)

v = flow rate through the reactor (l/min)

V = volume of the reactor (l)

k_s = rate of sucrose hydrolysis (1/min)

k_e = rate of ethanol formation (1/min)

Y_s = measurement signal for sucrose concentration

Y_e = measurement signal for ethanol concentration

τ_{ds} = delay time of sucrose signal (min)

τ_{de} = delay time of ethanol signal (min)

τ_e = time constant of ethanol sensor (min)

Simulation is used to tune the parameters of the model to get a reasonable fit to experimental data. The model is rather crude and further research is needed.

AIM WITH CONTROL

The objective of the control is to maintain a constant ethanol concentration despite variations in incoming feed and in cell condition. Rapid response to set-point changes

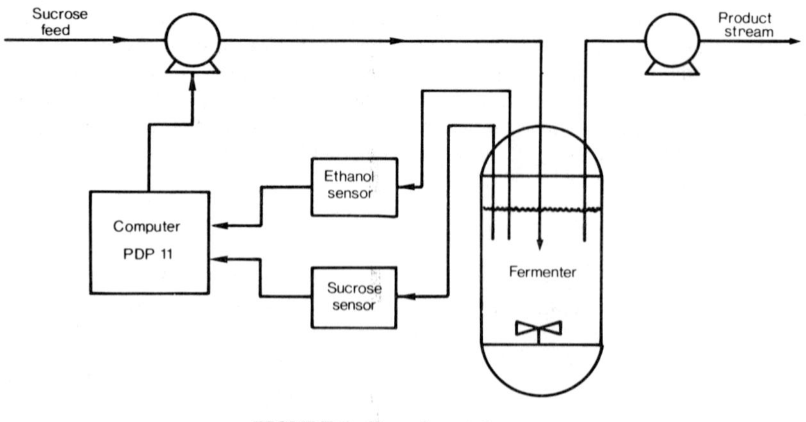

FIGURE 1. Experimental setup.

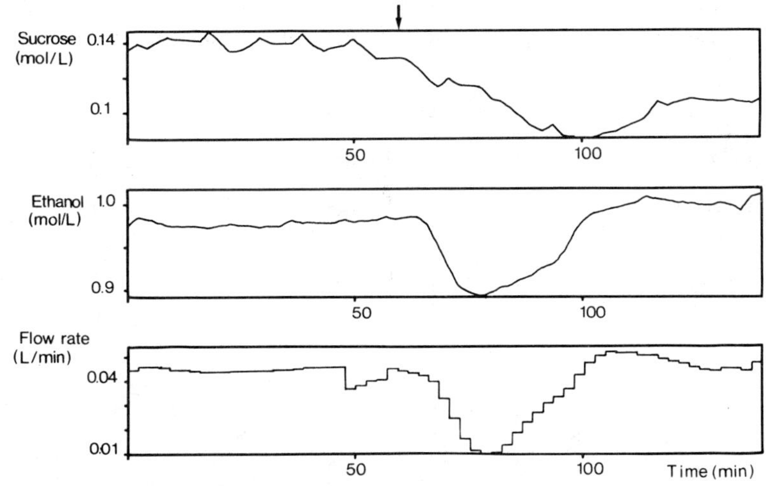

FIGURE 2. Experiment with addition of water (details in the text).

from outer loops is also desirable. This is obtained by measuring the ethanol concentration and controlling the feed rate of sucrose. A broad flow rate range is required for the feed pump.

By simultaneous monitoring of sucrose concentration, it is possible to estimate the activity of the cells.

REGULATION

The main control difficulties of the process are due to the pump characteristics and time delays in the measurement signals. These are compensated for by using gain scheduling and state feedback from a third-order bilinear Kalman filter. The regulator is tested using simulation and experiments on the real process.

FIGURE 2 shows an experiment where water is added to the system at $t = 60$ min. The measured sucrose and ethanol concentrations thus decrease after the sensor delay times. The regulator adjusts for the disturbance by changing the substrate flow rate, and at $t = 110$ min, the ethanol concentration is back at its set-point value in this case. Without control, this would have required several hours.

CONCLUSION

Reliable sensors for on-line measurement of substrate, intermediates and products are very useful for control of fermentation. If we accept time delays in the signals, reliable analysis can be done. However, time delays impose limitations on control, and conventional regulators are of less value. Knowledge of process dynamics should be utilized in the design. See, for example Reference 3. In this study, it is shown that a rather crude process model is of great help.

This type of regulator is easily implemented on a microcomputer. During experimental work, it is, however, convenient to have a more powerful program system,[3] so that different aids for start up and tuning of the regulator can be incorporated.

REFERENCES

1. WANG, H. Y., C. L. COONEY & D. I. C. WANG. 1979. Biotechnical. Bioeng. Symp. No. 9. pp. 13–24.
2. MATTIASSON, B., B. DANIELSSON, C. F. MANDENIUS & F. WINQUIST. 1981. Ann. N.Y. Acad. Sci. **369:** 295–306.
3. AXELSSON, J. P., P. HAGANDER, C. F. MANDENIUS & B. MATTIASSON. 1983. Proc. 1st IFAC workshop on Modelling and Control of Biotechnical Processes. Helsinki, Finland, August 17–19, 1982. A. Halme, Ed.: 273–280. Pergamon Press.

The Influence of Dilution Rate, Temperature, and Influent Substrate Concentration on the Efficiency of Steady-State Biomass Production in Continuous Microbial Culture

ALICIAN V. QUINLAN[a]

Department of Mechanical Engineering
Massachusetts Institute of Technology
Cambridge, Massachusetts 02139

INTRODUCTION

The maximum achievable biomass concentration in continuous culture equals the influent substrate concentration S_0 times a yield coefficient Y. This is the biomass concentration that would be produced were all the substrate entering the culture vessel permanently incorporated into biomass. However, microbial biomass production in continous culture is inherently inefficient. Some of the influent substrate is flushed unchanged out of the culture vessel, and some is converted into extracellular metabolites within the culture vessel. As a result, only a fraction of the maximum possible biomass concentration is ever actually achieved. This fraction is a direct measure of the efficiency of biomass production. Hence, whatever factors affect this fraction also directly affect the efficiency of biomass production.

The purpose of this paper is to identify these factors and determine how they can be manipulated to increase the efficiency of steady-state biomass production in continuous microbial culture. The results obtained apply to nonattached microbial populations whose growth rate saturates in substrate concentration according to a Monod rate law.

MONOD BIOMASS PRODUCTION KINETICS

Reaction Mechanism

A simple Monod reaction mechanism for microbial growth in continuous culture is schematized in FIGURE 1. Its rationale is given in Reference 1. According to it, the total biomass concentration B may be partitioned into two portions: B_c, capable of consuming substrate; B_n, not capable. Substrate at a concentration S is consumed at a rate cSB_c/Y. For each unit of substrate consumed, Y units of biomass are produced. Substrate consumption causes B_c to become saturated and converted to B_n at a rate

[a]Present address: Department of Mechanical Engineering and Materials Science, Duke University School of Engineering, Durham, NC 27706.

sSB_c. Eventually, B_n may become unsaturated and converted back to B_c at a rate uB_n. Production of extracellular metabolites causes biomass to waste away at rates wB_c, wB_n, wB. In a constant-volume culture vessel, inflow of substrate at a concentration S_0 and a dilution rate d causes substrate and biomass to be flushed out at rates dS, dB_c, dB_n, dB.

Rate Laws

If B_c is assumed to respond to changes in S much more slowly than B_n does, then the familiar differential equations for substrate depletion and biomass production may be derived from the reaction mechanism shown in FIGURE 1.[1]

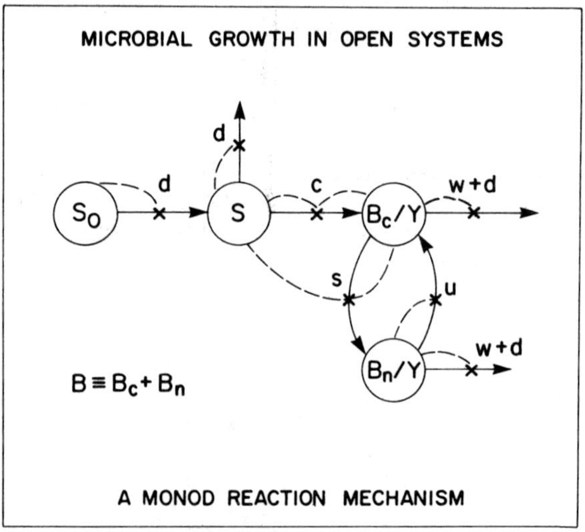

FIGURE 1. Reaction mechanism assumed to govern the rate of microbial biomass production in batch (d = 0) and continuous (d > 0) culture. Symbols defined in text and legend of symbols.

The differential equation obtained for substrate depletion is:

$$\dot{S} \simeq d(S_0 - S) - \mu_m(B/Y)S/(K + S) \tag{1}$$

where

$$\mu_m \equiv (c/s)(u + w + d) \tag{2}$$

$$K \equiv (1/c)\mu_m \tag{3}$$

The second term of Equation 1 represents the rate of substrate consumption. It saturates in substrate concentration according to a Monod rate law. The macroscopic coefficient μ_m represents the maximum specific growth rate; K, the substrate concentration at which half this rate is achieved. The ratio μ_m/Y gives the maximum specific rate of substrate consumption.

The differential equation found for biomass production is:

$$\dot{B} \simeq \mu_m BS/(K + S) - (w + d)B \tag{4a}$$

$$\simeq V_B B(S - S_T)/(S + K) \tag{4b}$$

where

$$V_B \equiv \mu_m - (w + d) \tag{5a}$$

$$\equiv (c - s)K + u \tag{5b}$$

$$S_T \equiv (w + d)K/V_B \tag{6}$$

The *net* rate of biomass production is given by Equations 4. From Equation 4b, the net rate can be seen to saturate in substrate concentration with a maximum net specific growth rate V_B and a threshold substrate concentration S_T that must be exceeded for net growth.[1] The first term in Equation 4a is the Monod rate law for *gross* biomass production.

Steady-State Conditions

At steady state,

$$\dot{S} \equiv \dot{B} \equiv 0 \tag{7}$$

Under these conditions, the steady-state substrate concentrations $\hat{S}$ may be found from Equation 4b to be:

$$\hat{S} = S_T \tag{8}$$

Similarly the steady-state biomass $\hat{B}$ may be found from Equations 1, 4a, and 8 to be:

$$\hat{B} = YS_0[1 - (S_T/S_0)]/[1 + (w/d)] \tag{9}$$

As S_T/S_0 and w/d both become small relative to unity, the steady-state biomass approaches the maximum achievable biomass in value.

Efficiency of Steady-State Biomass Production

The efficiency of steady-state biomass production η_{ss} may be defined as the ratio of the biomass actually produced at steady state $\hat{B}$ to the maximum achievable biomass YS_0. Thus,

$$\eta_{ss} \equiv \hat{B}/YS_0 \tag{10}$$

A simple rearrangement of Equation 9 shows the efficiency of steady-state biomass production depends on the ratios S_T/S_0 and w/d as follows:

$$\hat{B}/YS_0 = [1 - (S_T/S_0)]/[1 + (w/d)] \tag{11}$$

So, whatever makes S_T/S_0 and w/d decrease also makes the efficiency increase. Elevating S_0 will obviously improve the efficiency until $S_0 \gg S_T$. It is not at all obvious,

TABLE 1. Experimental Data[2] Used to Estimate Y, w, s, u as Functions of Temperature

T (°C)	d (Khr^{-1})	S_0 (ppm)	$\hat{S}$ (ppm)	$S_0 - \hat{S}$ (ppm)	$\hat{B}$ (ppm)
10	42	943	72	871	380
	56	954	(55)[b]	899	390
	83	973	(92)	881	402
	167	1109	818	(291)	(106)
	333	[a]	[a]	[a]	[a]
20	42	970	19	951	346
	56	938	28	910	327
	83	992	53	939	361
	167	979	(54)	925	484
	333	921	375	546	301
30	42	909	(16)	893	300
	56	931	14	917	346
	83	939	23	916	395
	167	996	(97)	(899)	(380)
	333	948	77	871	428
40	42			(921)	(416)
	56			854	225
	83			910	291
	167			911	324
	333			956	388

[a]Washout occurred.
[b]Data in parentheses not used to estimate coefficient values.

though, from theory alone how changes in w and d will actually affect the efficiency because S_T is a complicated function of w and d (see Equations 6, 5, 3, 2).

EVALUATION OF COEFFICIENTS

To shed light on how changes in w and d can affect S_T and thereby the efficiency of steady-state biomass production, experimental data were analyzed.

Experimental Data

TABLE 1 gives the experimental data[2] used to evaluate s, u, w, Y, and from them c, μ_m, K, S_T. The data describe the growth of a nonattached community of microorganisms obtained from the primary clarifier effluent of a municipal sewage treatment plant. The experiments were performed under carbon-limited conditions with glucose as the sole carbon source in a mineral salts medium. The culture vessel was an aerated,

TABLE 2. Procedure Used to Estimate Values of Y and w

(1) Linearize Equation 9:
$$[(S_0 - \hat{S})/\hat{B}] = (1/Y) + (w/Y)[1/d]$$
(2) Evaluate $[(S_0 - \hat{S})/\hat{B}]$ and $[1/d]$ from TABLE 1
(3) Regress $[(S_0 - \hat{S})/\hat{B}]$ on $[1/d]$:
*Obtain Y from intercept;
*Obtain w from slope.

TABLE 3. Procedure Used to Estimate Values of s and u

(1) Assume c ~ s.
(2) Substitute Equations 6, 5b, 3, and 2 into 8 and linearize the result:
$$[\hat{S}/(w+d)] = (1/s) + (1/su)\,[w+d]$$
(3) Evaluate $[\hat{S}/(w+d)]$ and $[w+d]$ from TABLES 1 and 4.
(4) Regress $[\hat{S}/(w+d)]$ on $[w+d]$:
*Obtain s from intercept.
*Obtain u from slope.

baffled, single-pass cylindrical, one-liter continuous stirred-tank reactor (CSTR). The values of influent and steady-state substrate concentration, respectively S_0 and $\hat{S}$, represent biologically available carbon measured in ppm COD; they were calculated from TABLE 2 of Reference 2 by subtracting "nondegradable COD" from respectively "influent COD" and "reactor soluble COD." The steady-state biomass values $\hat{B}$ were measured in ppm dry cell weight; they were taken directly from the "reactor suspended solids" column in Table 2 of Reference 2.

Microscopic Rate Coefficients

If each substrate consumption site becomes saturated when one substrate molecule occupies it, then the microscopic saturation and consumption rate coefficients are equal.[1] This case was assumed to apply here, and c was set equal to s.

The values of the microscopic rate coefficients s, u, and w were then estimated as functions of temperature by applying the procedures outlined in TABLES 2 and 3 to the data given in TABLE 1. The values thus obtained are presented in TABLE 4. The temperature dependencies of these coefficients are shown in FIGURE 2. Each coefficient rises exponentially with temperature and can be shown[3] to have an Arrhenius temperature dependence with a constant apparent activation energy, namely, E_s, E_u, and E_w.

The values of w obtained here are close to those given in Table 3 (5th column) of Reference 2. The remaining microscopic rate coefficients were not evaluated in Reference 2.

Macroscopic Coefficients

The yield coefficient Y was estimated by applying the procedure outlined in TABLE 2 to the data given in TABLE 1. Its values (units: ppm dry cell weight/ppm COD) are reported as a function of temperature in TABLE 4. These values average out to roughly

TABLE 4. Coefficient Values Estimated by Applying Procedures Outlined in Tables 2 and 3 to Data Given in Table 1

T (°C)	Y (ppm/ppm)	w (Khr^{-1})	s (Khr^{-1}ppm^{-1})	u (Khr^{-1})
10	.473	3.94	2.61	14.9
20	.574	(28.7)[a]	3.44	122.
30	.539	24.2	5.39	2,050.
40	.449	37.8	[b]	[b]

[a]Omitted from linear regression of ln w on T in FIGURE 2.
[b]Washout occurred.

0.5 and show a maximum of 0.574 at 20°C. They agree well with the yield coefficient values listed in Table 3 of Reference 2.

With the assumption $c \simeq s$, the maximum specific growth rate coefficient μ_m (Equation 2) reduces to:

$$\mu_m \simeq u + w + d \tag{12}$$

Its dependence on both temperature and dilution rate is illustrated in FIGURE 3. As

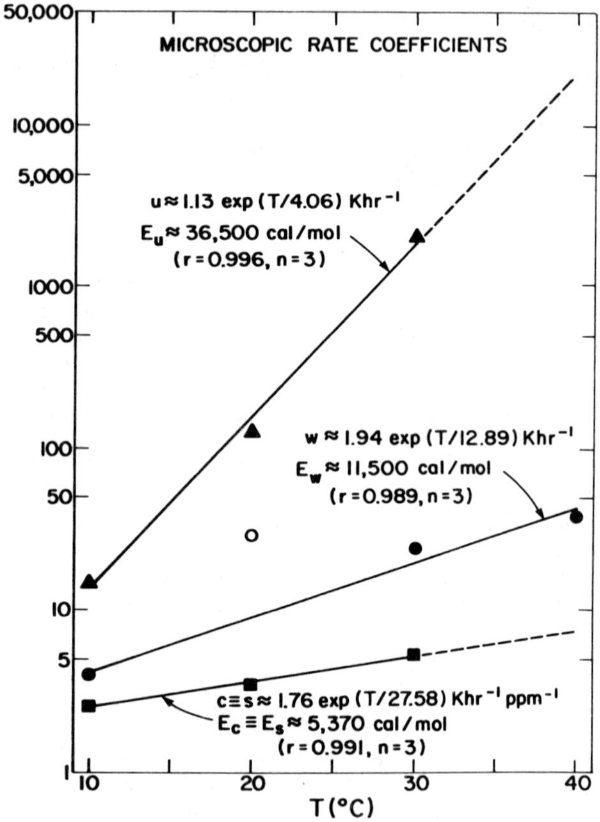

FIGURE 2. Log-linear relationships between microscopic rate coefficients and temperature. Solid lines represent interpolations of data; dashed lines, extrapolations. Data points tabulated in TABLE 4.

temperature rises or dilution rate falls, μ_m approaches a pure Arrhenius temperature dependence. The influence of dilution rate is strongest at low temperatures where d dominates u and w. The values of μ_m obtained here agree well with those given in TABLE 3 of Reference 2, which did not consider the influence of d on μ_m.

The influence of temperature and dilution rate on the half-saturation coefficient K (Equations 3, 2) is displayed in FIGURE 4. Again, dilution rate exerts its strongest

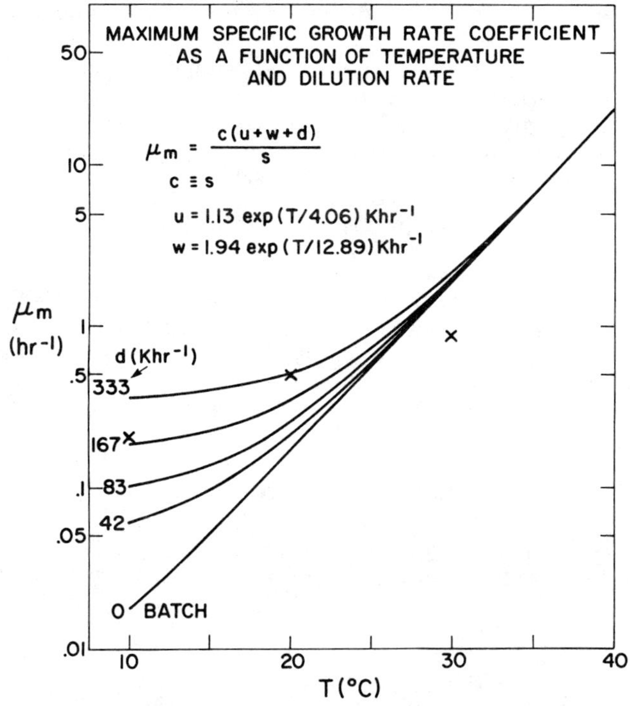

FIGURE 3. Thermal sensitivity of the maximum specific growth rate coefficient in batch (d = 0) and continuous (d > 0) culture with the microscopic coefficients specified in FIGURE 2. Crosses mark values reported in Table 3 of Reference 2.

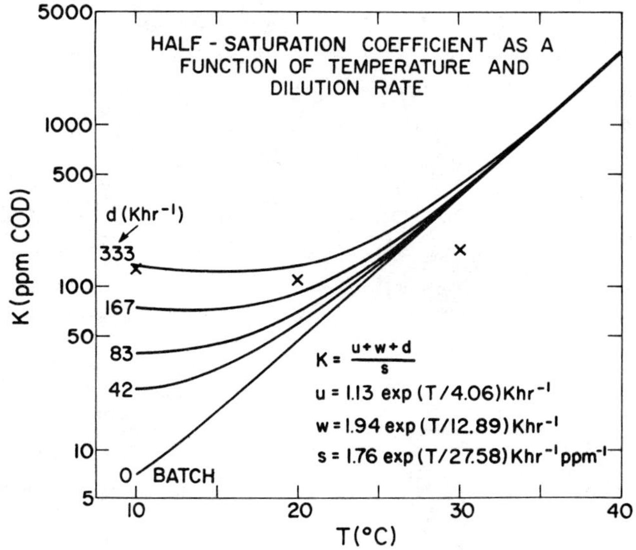

FIGURE 4. Thermal sensitivity of the half-saturation coefficient in batch (d = 0) and continuous (d > 0) culture with the microscopic coefficients specified in FIGURE 2. Crosses mark values reported in Table 3 of Reference 2.

effects at low temperature. As temperature rises or dilution rate falls, K also tends to approach a classic Arrhenius temperature dependence. As was found for μ_m, the values of K obtained here agree well with those listed in Table 3 of Reference 2.

The threshold substrate concentration coefficient S_T was evaluated by setting $c \simeq s$ and combining Equations 6, 5b, 3, 2, with the result:

$$S_T \simeq (w + d)(u + w + d)/us \tag{13}$$

The dependence of S_T on temperature and dilution rate is illustrated in FIGURE 5. Both temperature and dilution rate strongly affect S_T across their ranges of variation. In continuous culture ($d > 0$), S_T increases as dilution rate rises, but falls as temperature rises. Over the temperature range studied, S_T does not follow an Arrhenius temperature dependence. For the most part, the values of S_T calculated from Equation 13 and FIGURE 2 agree with the experimental data presented in Reference 2.

SENSITIVITY ANALYSIS

As shown by Equation 11, decreasing w/d and S_T/S_0 should improve the efficiency of steady-state biomass production. To decrease w/d, the dilution rate should be raised

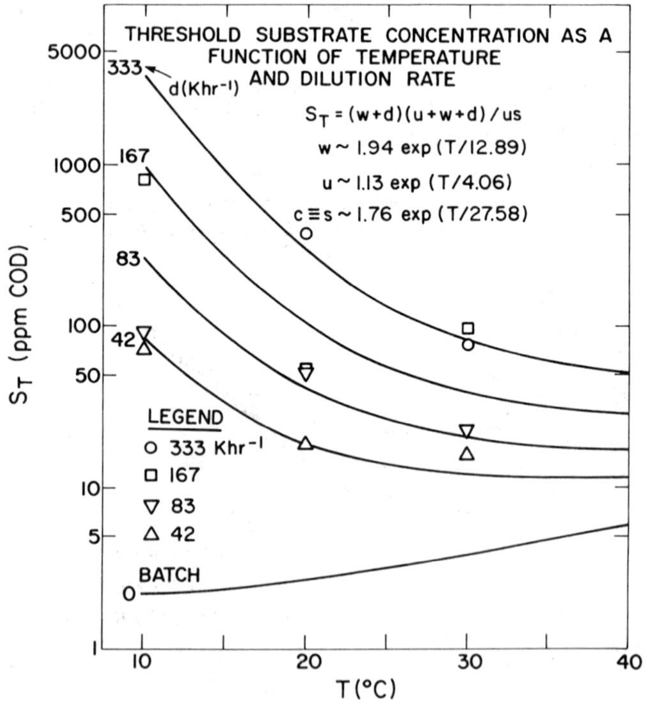

FIGURE 5. Thermal sensitivity of the threshold, or steady-state, substrate concentration in batch (d = 0) and continuous (d > 0) culture with the microscopic coefficients specified in FIGURE 2. Data points represent $\hat{S}$ ($\equiv S_T$) values given in TABLE 1.

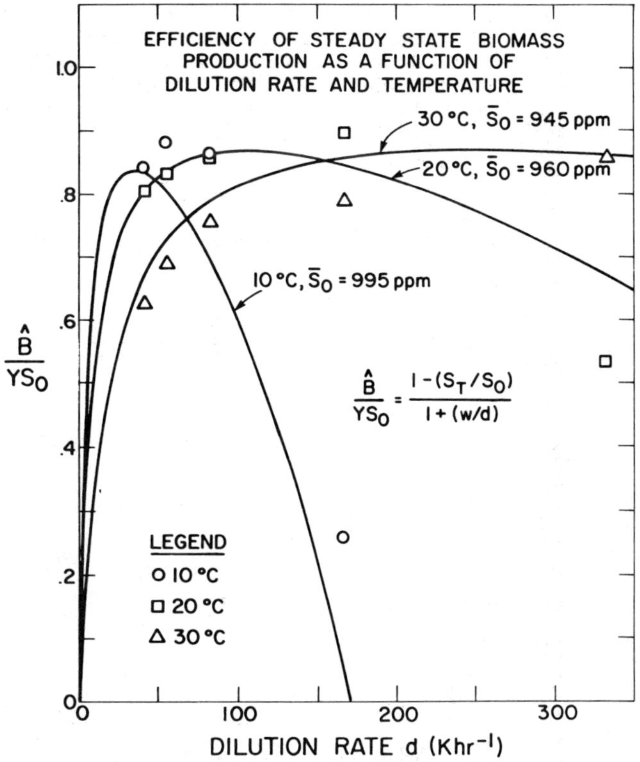

FIGURE 6. Influence of dilution rate on the efficiency of steady-state biomass production with temperature as a parameter. Curves were generated from equations for w and S_T shown in FIGURE 5. Points were calculated from values of $\hat{S}$ ($\equiv S_T$) given in TABLE 1 and w given in TABLE 4 and FIGURE 2. Note shift of optimum dilution rate to higher values with rising temperature.

and/or the temperature should be lowered (see FIGURE 2). Unfortunately, for microbial cultures with coefficient values similar to those specified in FIGURE 5, either action would at the same time elevate S_T, which is counterproductive. As a consequence, the efficiency of steady-state biomass production in such microbial cultures cannot be a monotonic rising function of dilution rate or temperature. Indeed, FIGURES 6–9 show the efficiency passes through a single peak as either dilution rate or temperature increases.

When efficiency is plotted versus dilution rate, the peak can be seen to shift toward higher dilution rates as either the temperature or influent substrate concentration rises (FIGS. 6 and 7, respectively). In other words, the optimum dilution rate for steady-state biomass production rises with both temperature and influent substrate concentration. Furthermore, as the peak shifts toward higher dilution rates, it broadens. The peak efficiency is fairly insensitive to temperature changes (FIG. 6), but improves considerably as influent substrate concentration increases (FIG. 7).

In contrast, plots of efficiency versus temperature show that raising the dilution rate shifts the peak toward higher temperatures (FIG. 8), but elevating influent substrate concentration shifts it toward lower temperatures (FIG. 9). This means the

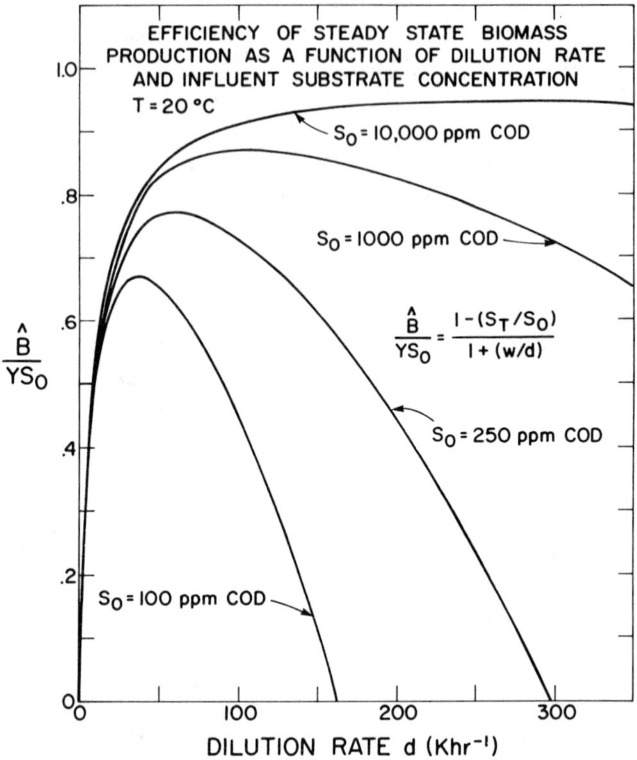

FIGURE 7. Influence of dilution rate on the efficiency of steady-state biomass production with influent substrate concentration as a parameter. Curves generated from equations for w and S_T shown in FIGURE 5. Note how increasing influent substrate concentration both raises and broadens the range of optimum dilution rates, and also how it improves the peak efficiency.

optimum temperature for steady-state biomass production rises with dilution rate, but falls as influent substrate concentration increases. The breadth of the peak is not strongly affected by changes in either dilution rate or influent substrate concentration. However, the peak efficiency can be significantly improved by elevating the influent substrate concentration (FIG. 9), but it is relatively insensitive to dilution rate.

DISCUSSION AND CONCLUSIONS

With the information presented in FIGURES 6–9, a strategy may be devised for manipulating d, T, and S_0 to optimize the efficiency of steady-state biomass production by microbial cultures obeying Monod rate laws with coefficients that behave as those shown in FIGURES 2–5.

Suppose, for example, S_0 is 1000 ppm. Then, for a moderate dilution rate of, say, 100 Khr^{-1} (i.e., a hydraulic retention time of 10 hr), the optimum temperature would be close to 20°C (FIG. 9). Now, suppose S_0 were raised to 10,000 ppm to increase the peak efficiency. The optimum temperature could be kept close to 20°C by increasing the dilution rate to a value within the range 220–320 Khr^{-1} (cf. FIGS. 6 and 7). Conversely, the optimum dilution rate could be kept near 100 Khr^{-1} by lowering the temperature roughly 7°C to about 13°C (cf. FIGS. 8 and 9).

Therefore, for the process under consideration, relatively high influent substrate concentration, low temperature, and moderate dilution rate would be needed to make the steady-state biomass approach the maximum achievable biomass YS_0.

The applicability of this finding is restricted by the thermal sensitivity of S_T. According to Equation 11, the efficiency of steady-state biomass production is a function of S_T, S_0, w, and d. Of these four variables, only the thermal sensitivity of S_T may vary significantly from one microbial culture to another.[1] For the experimental culture considered in this paper, S_T was a monotonic decreasing function of temperature under continuous culture conditions ($d > 0$, FIG. 5). However, for other microbial cultures, S_T may also show a minimum as temperature rises, or it may appear to be a monotonic increasing function of temperature.[1] The influence of these other thermal sensitivity patterns is the subject of continuing research on the topic of this paper.

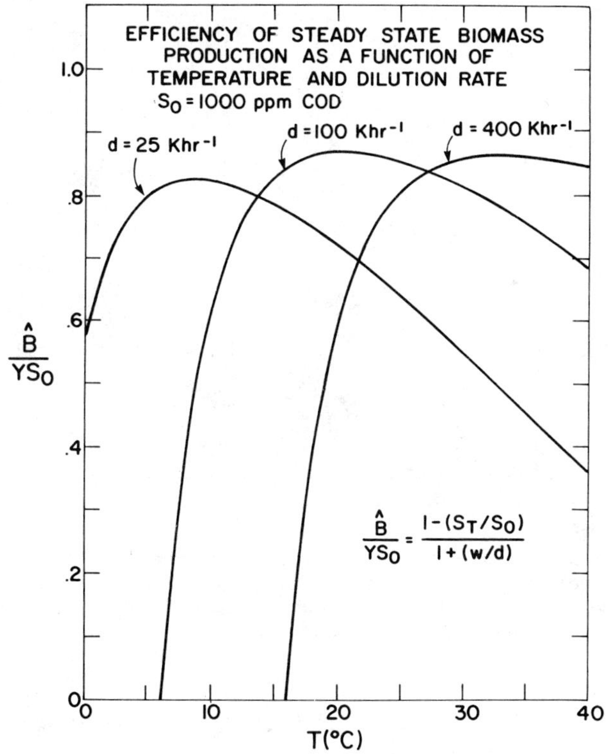

FIGURE 8. Influence of temperature on the efficiency of steady-state biomass production with dilution rate as a parameter. Curves generated from equations for w and S_T shown in FIGURE 5. Note how the optimum temperature rises with dilution rate.

SUMMARY

The efficiency of steady-state biomass production was defined as the ratio of the biomass produced at steady state to the biomass that would be produced if influent substrate were completely and permanently incorporated. The scope of analysis was confined to microbial growth processes described by a Monod reaction mechanism. A thermo-kinetic analysis of this mechanism with coefficient values estimated from experimental data showed:

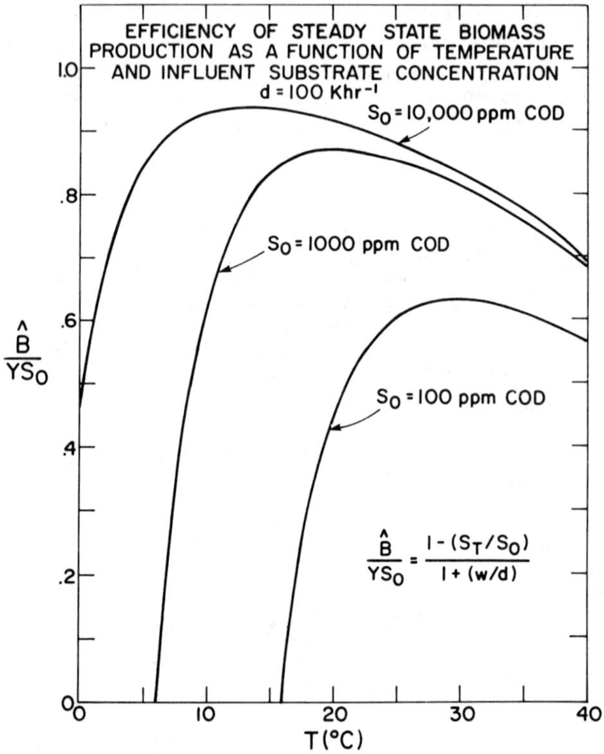

FIGURE 9. Influence of temperature on the efficiency of steady-state biomass production with influent substrate concentration as a parameter. Curves generated from equations for w and S_T shown in FIGURE 5. Note how the peak efficiency is improved and the optimum temperature lowered by increasing influent substrate concentration.

(1) As the dilution rate was increased, the efficiency passed through a single peak. Raising the temperature did not markedly change the peak efficiency, but did broaden and shift the peak toward higher dilution rates. In contrast, elevating the influent substrate concentration significantly improved the peak efficiency while still broadening and shifting the peak toward higher dilution rates.

(2) As the temperature was raised, the efficiency also showed a single peak.

Increasing the dilution rate did not significantly alter the peak value or breadth, but did shift the peak to higher temperatures. In contrast, elevating the influent substrate concentration substantially improved the peak efficiency and caused the peak to shift toward lower temperatures without broadening.

(3) As the influent substrate concentration was elevated, the peak efficiency improved and asymptotically approached unity.

These results suggested that relatively high influent substrate concentrations, low temperatures, and moderate dilution rates would be needed to optimize the efficiency of steady-state biomass production by Monod processes with coefficient values similar to those used in this paper.

ACKNOWLEDGMENTS

The research reported in this paper was supported by the Department of Mechanical Engineering and the School of Engineering at M.I.T. The figures were drawn by Andy Poynor.

REFERENCES

1. QUINLAN, A. V. 1983. Thermochemical optimization of microbial biomass-production and metabolite-excretion rates. *In* Foundations of Biochemical Engineering: Kinetics and Thermodynamics in Biological Systems. H. W. Blanch, E. T. Papoutsakis & G. Stephanopoulos, Eds. ACS Symposium Series 207. Washington, DC. pp. 463–488.
2. MUCK R. E. & C. P. L. GRADY, JR. 1974. Temperature effects on microbial growth in CSTR's. ASCE J. Environ. Eng. Div. **100:** 1147–1163.
3. QUINLAN, A. V. 1981. The thermal sensitivity of generic Michaelis-Menten processes without catalyst denaturation or inhibition. J. Thermal Biol. **6:** 103–114.

APPENDIX

SYMBOLS

B ≡ total biomass concentration in culture vessel; equals sum of B_c plus B_n; mg dry cell weight per liter.

B_c ≡ biomass concentration associated with substrate consumption sites capable of consuming substrate; same units as B.

B_n ≡ biomass concentration associated with substrate consumption sites not capable of consuming substrate; same units as B.

$\hat{B}$ ≡ steady-state biomass concentration in culture vessel; same units as B.

E ≡ apparent activation energy; cal/mol.

K ≡ half-saturation coefficient; same units as S.

S ≡ substrate concentration in culture vessel; ppm COD.

S_0 ≡ influent substrate concentration; same units as S.

S_T ≡ threshold substrate concentration coefficient; steady-state substrate concentration that must be exceeded for a positive net biomass production rate; same units as S.

$\hat{S}$ ≡ steady-state substrate concentration in culture vessel; same units as S.

T ≡ temperature; °C.

V_B ≡ maximum possible specific rate of net biomass production in culture vessel; Khr^{-1}.
Y ≡ yield coefficient; ppm dry cell weight produced per ppm COD consumed.
c ≡ microscopic substrate consumption rate coefficient; ppm COD^{-1} Khr^{-1}.
d ≡ dilution rate (flow/volume); rate at which substrate and biomass are displaced from culture vessel; rate at which substrate is added to culture vessel; Khr^{-1}.
s ≡ microscopic substrate saturation rate coefficient; ppm COD^{-1} Khr^{-1}.
u ≡ microscopic unsaturation rate coefficient; Khr^{-1}.
w ≡ microscopic wastage rate coefficient; Khr^{-1}.
η_{ss} ≡ efficiency of steady-state biomass production; dimensionless.
μ_m ≡ maximum specific growth rate coefficient; Khr^{-1}.

Primary Metabolite or Microbial Protein from Cellulose: Conditions, Kinetics, and Modeling of the Simultaneous Saccharification and Fermentation to Citric Acid

JUAN A. ASENJO AND CHUCK JEW

Biochemical Engineering Laboratory
Department of Chemical Engineering
and Applied Chemistry
Columbia University
New York, New York 10027

INTRODUCTION

The utilization of cellulose and cellulose hydrolysates as raw materials for fermentations is a subject of growing interest. One alternative is the hydrolysis of cellulose substrates into sugars with the subsequent fermentation of these sugars. Another is to carry out a direct bioconversion as a simultaneous saccharification of cellulose and fermentation into products (SSF). This can be done by obtaining expression of the cellulolytic enzymes in the microorganism or by using a compatible mixed enzyme-microbe system.

In the enzymatic hydrolysis of cellulose, rates are greatly lowered due to a high end-product inhibition caused by cellobiose and glucose. This inhibition would be overcome in a simultaneous process where the accumulation of sugars is minimized.

In order to evaluate the potential advantages of a simultaneous saccharification and fermentation, an appropriate kinetic model has been established that describes the rates of cellulose degradation, formation of sugars, microbial growth, and product formation. The use of this model to describe the simultaneous bioconversion of cellulose and for maximizing production of microbial protein or fermentation product (citric acid) is shown.

CONDITIONS FOR THE SIMULTANEOUS BIOCONVERSION (SSF)

As an intermediate of the TCA cycle, citric acid is considered to be a primary metabolite. However, the production of this acid, by appropriate deregulated mutants able to accumulate high titers, follows the trophophase-idiophase relationship that is characteristic of secondary metabolism.[1] Studies of the fermentation of glucose to citric acid show that nitrogen starvation and the resulting limitation of growth are fundamental for the accumulation of citric acid.[1–3] Such conditions are also valid for the fermentation of cellulose hydrolysates and for the direct bioconversion using a mixed enzyme-microbe system.[4] Therefore, it has been concluded that the simultaneous bioconversion of cellulose to citric acid should be carried out in two phases:

(1) Trophophase (growth), in which conditions for simultaneous saccharification and rapid cell growth (SSG) have to be maintained; and

(2) Idiophase (production), where conditions for simultaneous saccharification and accumulation of citric acid are maintained (SSA).

If the aim is to maximize the production of cell mass (microbial protein), it is evident that only trophophase conditions are needed. The main difference between the two phases is an adequate supply of nitrogen for the growth phase and conditions of nitrogen starvation for the production phase.

In any cellulose utilization scheme, it is expected that the conversion of cellulose into products will be maximized. A simultaneous saccharification and fermentation will have a strong effect on increasing the rates of hydrolysis by overcoming product inhibition. However, there are other factors that are crucial for maximizing cellulose utilization. One is a good pretreatment such as ball milling,[7,8] which can dramatically increase the amount of exposed substrate available for attack. Another is an appropriate ratio of cellobiase/cellulases (exo- and endocellulases). This will increase the rate of formation of fermentable glucose[9,10] since most natural cellulases are low in cellobiase.

MATHEMATICAL DESCRIPTION OF THE KINETICS

The kinetic model developed for simultaneous saccharification and fermentation considers rate equations for microbial mass (X), citric acid product (P), cellobiose (Cb), glucose present (G), and total glucose formed (Gf). For microbial growth, the logistic rate equation was used:

$$\frac{dx}{dt} = \left(\frac{\mu_m G}{K_{sx} + G}\right)\left(1 - \frac{x}{x_{max}}\right)x \tag{1}$$

where:

$$\mu = \frac{\mu_m G}{K_{sx} + G} \tag{2}$$

takes into account the dependence of microbial growth on the concentration of glucose present (Monod kinetics).

The product formation rate equation was considered as having a growth-associated term and a nongrowth-associated term (Eq. 3). For the production of citric acid, the growth-associated term was found to be negligible in most cases.

$$\frac{dP}{dt} = \left(m\frac{dx}{dt} + nX\right)\frac{G}{K_{sp} + G} \tag{3}$$

It also takes into account its dependence on the concentration of substrate ($G/(K_{sp} + G)$).

As has been discussed in the previous section, in the trophophase before nitrogen starvation begins, there is a lag time for acid accumulation. This period has been referred to as maturation time, t_m. Therefore, for $0 \leq t \leq t_m$; $dP/dt \simeq 0$, Equation 3 being valid for $t > t_m$.

In the simultaneous model, the degradation of cellulose into glucose has been simplified into two equations that adequately represent the limiting effect of cellobiase on the reaction rate. Equation 4 is for the hydrolysis of cellulose into cellobiose with competitive cellobiose and glucose inhibition. Equation 5 describes the hydrolysis of cellobiose into glucose with competitive product inhibition.

$$\frac{dC_b}{dt} = \frac{V_{m2}(\alpha S_0 - C_b)}{K_{m2} + (\alpha S_0 - C_b) + \frac{K_{m2}}{K_{I2}}(C_b - G_f) + \frac{K_{m2}}{K_{I3}} G} \tag{4}$$

$$\frac{dG_f}{dt} = \frac{V_{m1}(C_b - G_f)}{K_{m1} + (C_b - G_f) + \frac{K_{m1}}{K_{I1}} G} \tag{5}$$

Temperature denaturation of the enzymes was not taken into account. It was assumed to be unimportant at the temperature used (35°C).

Recent work has shown that initial rate kinetics of cellulose hydrolysis does not describe adequately long-term results, since the structure of untreated cellulose is highly crystalline and the total amount of insoluble substrate present is not exposed and available for attack. In Equation 4, the term αS_0 takes this into account, α being the fraction of substrate susceptible to attack by the enzymes.

The final rate equation is a mass balance for the amount of glucose present in the reaction:

$$\frac{dG}{dt} = \frac{dG_f}{dt} - \frac{1}{Y_{x/s}}\frac{dx}{dt} - \frac{1}{Y_{p/s}}\frac{dP}{dt} - m'x \tag{6}$$

where $m' \simeq 0$ if a global yield of product over substrate ($Y_{p/s}$) is considered. Such a yield would take into account all glucose consumed in the production phase apart from that used for growth.

VERIFICATION OF MATHEMATICAL MODEL

Cellulose Hydrolysis

A large amount of data on cellulose hydrolysis has been reported in the literature. Our model considers the conversion of cellulose into cellobiose over extended periods of time and the conversion of cellobiose into glucose in a parallel reaction. In both cases, the use of a crude fungal enzyme complex has been assumed. Therefore, the Michaelis and inhibition constants, determined by Howell and Mangat,[11] were chosen as appropriate for the conversion of cellulose into cellobiose. For the conversion of cellobiose into glucose, the constants determined by Lee and Fan,[12,13] using a crude fungal preparation of cellulases were used (TABLE 1). The effect of temperature on the Michaelis and inhibition constants was considered to be negligible between 50°C and 35°C. Parameters V_{m1}, V_{m2}, and α were evaluated for two different cases (see FIG. 1, a and b). FIGURE 1a shows the experimental data obtained by Lee[13] as well as the curves corresponding to Equations 4, 5, and for total accumulated cellobiose [(Eq. 4) – (Eq. 5)]. α, V_{m1}, and V_{m2} were found by nonlinear regression analysis. FIGURE 1b shows the data from our previous work.[4] The addition of cellobiase is reflected in a larger V_{m1}/V_{m2} ratio and the smaller particle size of the substrate in a slightly larger value of α (see TABLE 1). The conditions used in this experiment were later maintained in a direct bioconversion (SSF) experiment.

TABLE 1. Cellulose Hydrolysis Parameters Used in Equations 4 and 5

Parameter	Value[a]	Comments
K_{m2} K_{12}	42.8 2.05	As determined by Howell and Mangat[11] for the conversion of cellulose to cellobiose over extended periods of time using Solka Floc BW100 and a crude enzyme preparation from *Trichoderma.*
K_{13}	4.01	Value determined using experimental results of Lee[13] for glucose and cellobiose inhibition of cellulose hydrolysis.
K_{m1} K_{11}	0.86 0.1	As determined by Lee and Fan[12,13] for the conversion of cellobiose to glucose using a crude enzyme preparation from *Trichoderma.*
V_{m2} V_{m1} α	18.4 1.3 0.56	Values obtained for the hydrolysis of Solka Floc SW40 using a crude enzyme preparation from *Trichoderma* (FIG. 1a).
V_{m2} V_{m1} α	7.5 1.3 0.65	Values obtained for the hydrolysis of Solka Floc BW300 using a commercial enzyme preparation from *Trichoderma* supplemented with cellobiose. α was fixed as 0.65 since the total amount of substrate hydrolyzed in the 50 g/l simultaneous bioconversion was estimated as 32.5 g/l (Asenjo *et al.*[4]). (FIG. 1b)

[a]Units are gl^{-1} and $gl^{-1}hr^{-1}$.

Growth and Production of Citric Acid

Growth and production parameters were estimated from citric acid fermentations carried out using three different yeast strains and a glucose substrate (pure glucose as well as glucose present in a cellulose hydrolysate). Results obtained in our previous work[4,14] as well as those of Briffaud and Engasser[3] were used. FIGURES 2 and 3a show such experimental results as well as the curves that were fitted to this data by nonlinear regression analysis. The values of μ_m, n, and t_m obtained are shown in TABLE 2. Initially we used a strain of *Candida guilliermondii* that could only accumulate modest

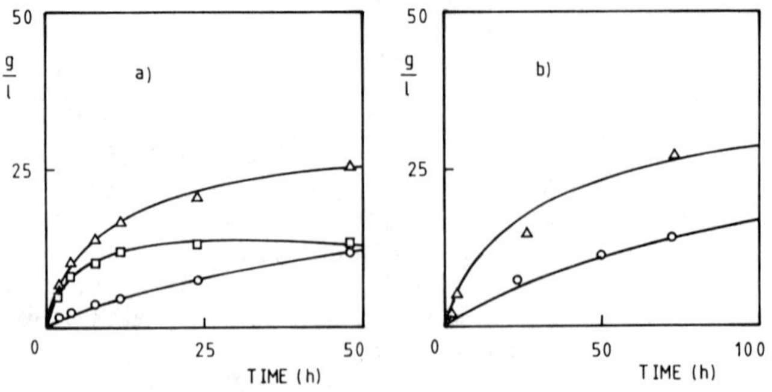

FIGURE 1. Comparison of Equations 4 and 5 with experimental data of cellulose hydrolysis. Parameters in TABLE 1. (O) glucose; (Δ) reducing sugars; (□) cellobiose.

(a) Solka Floc SW40 using crude cellulase from *Trichoderma.* Based on data from Ref. 13 (temp = 50°C, pH = 4.8).

(b) Solka Floc BW300 using commercial cellulase (*Trichoderma*) supplemented with cellobiase. Based on data from Ref. 4 (temp = 35°C, pH = 4.9)

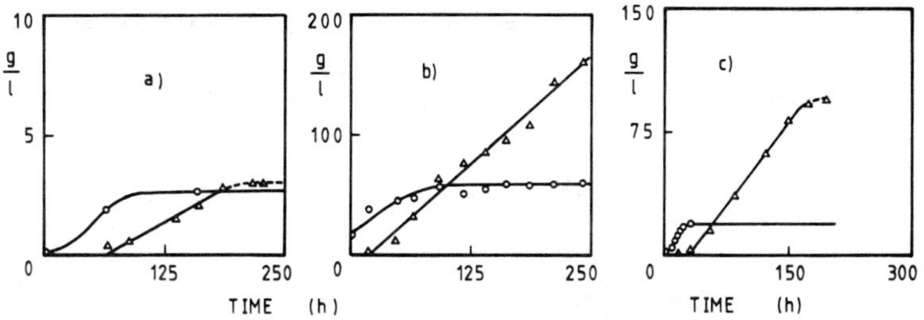

FIGURE 2. Growth of yeast and production of citric acid on a glucose substrate. Comparison with Equations 1 and 3. Parameters in TABLE 2. (O) microbial cells; (Δ) citric acid.

(a) *Candida guilliermondii* NRRL Y488. Experimental data from Ref. 2 (temp = 35°C, pH = 4.9).

(b) *Candida* (*Saccharomycopsis*) *lipolytica* NRRL Y1095 at high cell concentrations. Experimental data from Ref. 14 (temp = 35°C, pH = 5.0).

(c) *Saccharomycopsis lipolytica* D1805. Experimental data from Ref. 3 (temp = 30°C, pH = 4.7).

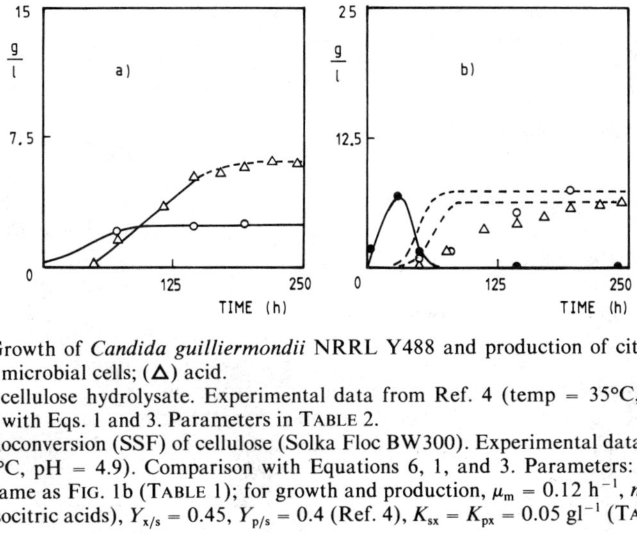

FIGURE 3. Growth of *Candida guilliermondii* NRRL Y488 and production of citric acid. (●) glucose; (O) microbial cells; (Δ) acid.

(a) On a cellulose hydrolysate. Experimental data from Ref. 4 (temp = 35°C, pH = 4.9). Comparison with Eqs. 1 and 3. Parameters in TABLE 2.

(b) In a bioconversion (SSF) of cellulose (Solka Floc BW300). Experimental data from Ref. 4 (temp = 35°C, pH = 4.9). Comparison with Equations 6, 1, and 3. Parameters: for cellulose hydrolysis, same as FIG. 1b (TABLE 1); for growth and production, $\mu_m = 0.12\ h^{-1}$, $n = 0.03\ hr^{-1}$ (citric and isocitric acids), $Y_{x/s} = 0.45$, $Y_{p/s} = 0.4$ (Ref. 4), $K_{sx} = K_{px} = 0.05\ gl^{-1}$ (TABLE 3).

TABLE 2. Citric Acid Fermentation Parameters (From FIGS. 2a–c and 3a)[a]

Strain	Substrate	$x(0)$	μ_m	x_{max}	$n \times 10^{-2}$	t_m
Y488	Glucose	0.13	0.06(0.22)[b]	2.7	0.85[c]	65
Y1095	Glucose	17.5	0.05(0.23)[b]	60	1.2	20
D1805	Glucose	0.55	0.20–0.21	19.1	3.5[d]	30
Y488	Cellulose hydrolysate	0.12	0.06–0.08	2.5	2.1[c]	45

[a]Units are gl^{-1}, hr^{-1}, hr.

[b]In parentheses, maximum values that have been obtained under optimum growth conditions.[14]

[c]Values of citric acid only for strain Y488. The ratio of citric to isocitric acid is 1:1; hence, $n = 1.7 \times 10^{-2}$ and 4.2×10^{-2}, respectively.

[d]3.5 corresponds to citric acid only. $n = 4.1 \times 10^{-2}$ was reported for citric and isocitric acids.[3]

TABLE 3. Parameters Used in the Simulation Studies (FIG. 4)[a]

K_{m2} =	43.0
K_{m1} =	0.86
K_{I2} =	2.0
K_{I3} =	4.0
K_{I1} =	0.1
V_{m2} =	18.4
V_{m1} =	1.3
α =	0.90[b]
K_{sx} =	0.05[c]
K_{sp} =	0.05[c]
μ_m =	0.2
m =	0.0
n =	0.04 (citric and isocitric acids)
$Y_{x/s}$[d] =	0.45
$Y_{p/s}$[d,e] =	0.95 (80% of theoretical for cellulose ($Y_{p/s}$ = 1.18))

[a]Units are gl^{-1}, $gl^{-1}hr^{-1}$, hr^{-1}.
[b]Is strongly dependent on pretreatment; with appropriate ball milling values close to 1.0 can be obtained.[13]
[c]Value of K_{sx} for yeast from Wang *et al.*[15] $K_{sp} = K_{sx}$ was assumed.
[d]Yields on cellulose ($Y_{x/s}$ on glucose = 0.41).
[e]$Y_{p/s}$ takes into account $m' = 0$.

quantities of citric acid (up to 6.5 g/l, FIG. 3a) with a low ratio of citric to isocitric acid (1:1). At present, we are using a strain of *Candida* (*Saccharomycopsis*) *lipolytica* that can accumulate considerably higher quantities (up to 160 g/l, FIG. 2b) with a much higher ratio of citric to isocitric acid (4:1).[14] In all the fermentations shown, the length of the trophophase was controlled by the amount of nitrogen supplied for growth. This evidently determined the point at which the production of citric acid was started (t_m). In all cases, the accumulation of acid appears to be well interpreted by a nongrowth-associated model without the need for a growth-associated term ($m = 0$).

In the simultaneous saccharification and fermentation (FIG. 3b), rapid use of the glucose formed was observed between 30 and 50 hours. This could only be attributed to

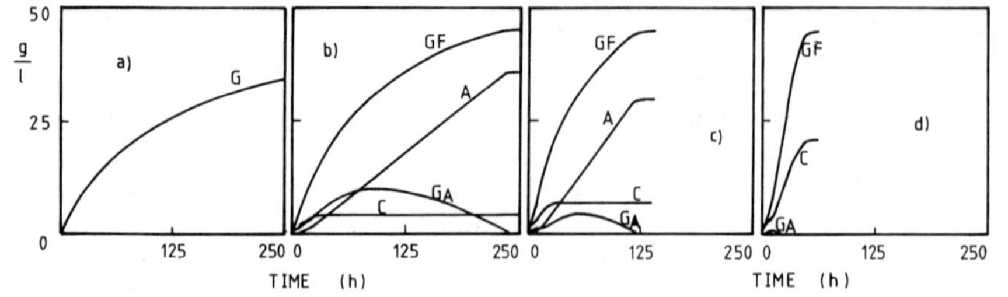

FIGURE 4. Use of mathematical model for simulation of the direct bioconversion (SSF) of cellulose into citric acid and microbial protein. Parameters in TABLE 3. (Temp = 35°C, pH ≈ 5.0). C = microbial cells; A = acid; G = glucose; GF = glucose formed; GA = glucose accumulated.
(a) No cells ($x_{max} = 0$).
(b) $x_{max} = 4$.
(c) $x_{max} = 7$.
(d) Cells only (no product).

growth and the associated product formation. However, the experimental results of growth (viable cell count) and production of acid (citric + isocitric) show a longer lag or "retarding" effect. We chose to fit the glucose accumulation; this is well described by the parameters obtained for cellulose hydrolysis and by values of μ_m within the range observed for this strain.

The individual kinetic parameters determined in the cellulose hydrolysis and citric acid fermentations were appropriate for the description of glucose accumulation in the simultaneous process. The low cell growth rates that have been observed experimentally should be investigated in more detail.

USE OF THE MODEL FOR MAXIMIZATION OF PRODUCTION OF CITRIC ACID OR MICROBIAL PROTEIN

TABLE 3 shows the values of the parameters chosen for the simulation of the direct bioconversion of cellulose into citric acid and microbial protein. The results obtained are shown in FIGURE 4. A substantial increase in the total amount of cellulose utilized in the simultaneous bioconversion is observed (FIG. 4, b–d). For citric acid production (FIG. 4, b and c), a larger amount of cells (7 g/l) gives a higher productivity but a lower concentration. The production of microbial protein only, without acid accumulation (FIG. 4d), gives the fastest cellulose utilization.

REFERENCES

1. ROEHR, M., O. ZEHENTGRUBER & C. KUBICEK. 1981. Biotechnol. Bioeng. **23:** 2433–2445.
2. SZUHAY, J. 1982. M.S. Thesis. Columbia University.
3. BRIFFAUD, J. & J. M. ENGASSER. 1979. Biotechnol. Bioeng. **21:** 2083–2092.
4. ASENJO, J. A., J. SZUHAY & D. CHIU. 1982. Biotechnol. Bioeng. Symp. **12:** 111–120.
5. DEMAIN, A. L. 1971. *In* Advances in Biochemical Engineering, Vol 1. T. Ghose & A. Fiechter, Eds. Springer. pp. 113–142.
6. AIBA, S. & M. MATSUOKA. 1979. Biotechnol. Bioeng. **21:** 1373–1386.
7. RYU, D. D. Y., S. B. LEE, T. TASSINARI & C. MACY. 1982. Biotechnol. Bioeng. **24:** 1047–1067.
8. FAN, L. T., Y. H. LEE & D. H. BEARDMORE. 1980. Biotechnol. Bioeng. **22:** 177–199.
9. CONTRERAS, I., R. GONZALEZ, A. M. RONCO & J. A. ASENJO. 1982. Biotechnol. Lett. **4:** 51–56.
10. ASENJO, J. A. Maximizing the Formation of Glucose in the Enzymatic Hydrolysis of Insoluble Cellulose. Biotechnol. Bioeng. In press.
11. HOWELL, J. A. & M. MANGAT. 1978. Biotechnol. Bioeng. **20:** 847–863.
12. LEE, Y. H., M. M. GHARPURAY & L. T. FAN. 1982. Biotechnol. Bioeng. Symp. **12:** 121–138.
13. LEE, Y. H. 1981. Ph.D. Dissertation. Kansas State University.
14. LEE, M. 1982. M.S. Thesis. Columbia University.
15. WANG, D. I. C., C. L. COONEY, A. L. DEMAIN, P. DUNNILL, A. E. HUMPHREY & M. D. LILLY. 1979. Fermentation and Enzyme Technology. John Wiley & Sons, Inc. New York.

Measurement of Gas-Phase Oxygen Concentrations with an Oxygen Electrode

ROBERT L. FIREOVED,[a] R. MUTHARASAN, AND
Y. H. LEE

Department of Chemical Engineering
Drexel University
Philadelphia, Pennsylvania 19104
and
[a]Wyeth Laboratories, Inc.
West Chester, Pennsylvania 19380

Very low oxygen uptake rates can adversely affect the growth of obligate anaerobes. On the other hand, yeasts and other facultative anaerobes require small amounts of oxygen for lipid synthesis in order to grow and ferment.[1] Consequently, it is desirable to have a means of accurately measuring low oxygen uptake rates in laboratory fermentation systems. Fermenter oxygen uptake rates are determined by measuring the difference in oxygen concentration of the fermenter feed and off-gases. A galvanic probe is used to measure gas-phase oxygen concentrations because of its exceptional accuracy.[2]

The technique was applied to a continuous culture of *Saccharomyces cerevisiae* NRRL Y-132. The continuous stirred-tank fermenter consisted of a 1-liter Erlenmeyer flask fed with a semisynthetic medium by a fixed-drive peristaltic pump. The working volume of the fermenter was between 900 and 1000 ml and the feed rate was adjusted in the range of 0.6–3.0 ml/min. In addition, the fermenter was sparged with standard gas mixtures (MG Scientific) of 1,157 and 990 ppm oxygen in nitrogen. Flow rates of gas were measured with ball-type rotameters (Brooks Instruments). From measurements of oxygen concentration of fermenter inlet and outlet gas streams, the oxygen uptake rate was computed. The dissolved oxygen (DO) concentration of the liquid feed stream was made negligible by enclosing both the polypropylene feed tank and the silicon-rubber pump tubing inside a glove box (Instruments for Research and Industry) filled with commercial-grade nitrogen. Experiments in which the chemostat was operated without inoculation showed no difference in the oxygen content between the feed and outlet gas streams. Consequently, the amount of oxygen leaving the fermenter in the liquid effluent stream was no more than 0.4% of the amount of oxygen that entered in the feed gas.

FIGURE 1 is a schematic of the apparatus used to sample the gas streams for oxygen determination. By utilizing three-way ball valves, either the fermenter off-gas or a sample of the fermenter feed gas could by analyzed. The temperature of the measured gases and oxygen electrode was controlled by immersion inside a water bath (Precision Scientific). The feed gas was passed through a water aspirator to reduce the electrode drift due to electrolyte-water loss through the electrode membrane. The load resistance was either 97.6 kohm or 1000 kohm (precision, metal film-type), depending on the desired range and sensitivity. The voltage generated over the load resistance was measured to the nearest microvolt by a Keithly multimeter (model 177). Readings were taken of the gas-phase oxygen concentrations by passing the sample streams over the oxygen electrode and comparing the reading to a calibration curve made with standard gases. A sample was passed over the electrode for 15 min in order to allow the electrode output to reach equilibrium before a reading was taken.

RESULTS AND DISCUSSION

The precision of the oxygen concentration measurement was determined by taking 14 successive, 15-min readings of a 1 ppm oxygen feed gas and the resulting fermenter off-gases. The percent deviation (standard deviation/mean) of the feed gas reading was 0.11%. This represented a measurement error of 0.27% or 3 ppb for a confidence level of 95%. The electrode drift was found to be 0.05%/hr.

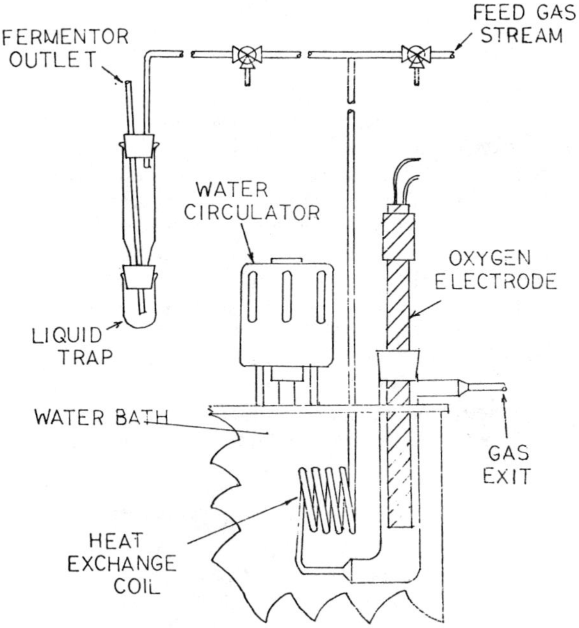

FIGURE 1. Oxygen uptake measurement apparatus.

With a load resistance of 1000 kohm and a step change in oxygen concentration from 1000 to 970 ppb, the probe response had a time constant of 8.2 min. The time constant of the probe response when used in conjunction with the 97.6 kohm load resistor was 2 min. The long response time of the probe when connected to the high load resistance is due to the large reservoir of electrolyte characteristic of the New Brunswick electrode.[3] The high load resistance is necessary when the oxygen concentration is low so that a detectable voltage drop can be obtained from the minute current of the probe.

FIGURE 2 gives the calibration curves for the electrode current as a function of the gas-phase oxygen concentration. Three standard gases of 1,157 and 990 ppm oxygen in nitrogen were used to obtain the data. The electrode current signal was also found to be

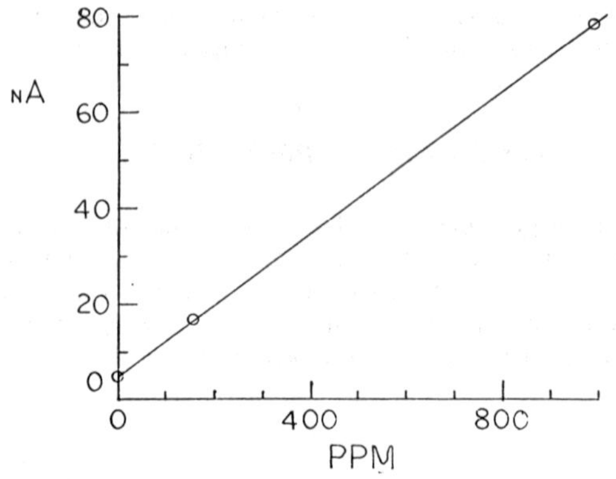

FIGURE 2. Oxygen electrode calibration curve; electrode curve vs. oxygen concentration.

proportional to the oxygen concentration when readings of 22 mol% and 990 ppm oxygen in nitrogen over a 1-kohm load resistance were compared. It is also reasonable to assume that the calibration curve is linear when the oxygen is between 0 and 1 ppm.[4] After a five-month period, the slope of the calibration curve decreased by 18%.

Oxygen uptake rates by the inoculated fermenter were calculated by measuring the output of the oxygen probe and using the calibration curves in FIGURE 2. The results are presented in TABLE 1. A load resistance of 97.6 kohm was used for feed-gas oxygen concentrations of 999 and 127 ppm while a 1000-kohm resistor was used for 1-ppm feed-gas oxygen concentration. The errors in the gas concentrations are given for a 95% level of confidence and the error in the fermenter oxygen uptake rate was calculated by the sum of the squares method.[5] The lowest previously reported oxygen uptake rates were paramagnetic measurements for continuous cultures of yeast and were 1320[6] and 1560 μg/hr.[7]

CONCLUSIONS

A hydro- and thermal-stabilized galvanic oxygen probe can measure gas-phase oxygen concentrations of less than 1 ppm with a measurement error of 0.27% for a 95% level of confidence. Applications of this gas-phase galvanic probe include the determination of low oxygen uptake rates by microorganisms. Measurement of oxygen uptake rates as low as 0.23 ± 0.03 μg/hr was demonstrated.

TABLE 1. Oxygen Uptake Rates with Low Feed-Gas Oxygen Concentration

Incoming Gas (ppm)	Outgoing Gas (ppm)	Oxygen Uptake Rate (μg/hr)
999.0 ± 2.7	922.0 ± 2.5	600.0 ± 28.8
127.000 ± 0.35	123.00 ± 0.33	31.20 ± 3.78
1.000 ± 0.003	0.970 ± 0.003	0.234 ± 0.029

REFERENCES

1. ANDREASEN, A. A. & T. J. B. STIER. 1953. J. Cell. Comp. Physiol. **41:** 23.
2. WENNBERG, L. A. 1975. J. Appl. Physiol. **38:** 540.
3. LEE, Y. H. & T. T. TSAO. 1979. Adv. Biochem. Eng., **13:** 35.
4. EVANGELISTA, T., N. H. GARIN & W. J. HAMILTON. 1977. Power. June: 83.
5. HOLMAN, J. P. 1971. Experimental Methods for Engineers, 2nd ed. McGraw-Hill. New York.
6. COWLAND, T. W., D. R. MAULE. 1966. J. Inst. Brew. **72:** 480.
7. ROGERS, P. J. & P. R. STEWART. 1974. Arch. Microbiol. **99:** 25.

Sugar (Glucose, Fructose, Sucrose) Sensor for Fermentation Control

HARUO OBANA, MOTOHIKO HIKUMA,
AND TAKEO YASUDA

Central Research Laboratories
Ajinomoto Co., Inc.
Suzuki-Cho, Kawasaki-Ku
Kawasaki 210, Japan

ISAO KARUBE AND SCHUICHI SUZUKI

Research Laboratory of Resources Utilization
Tokyo Institute of Technology
Nagatsuta, Midori-Ku
Yokohama 227, Japan

Cane and beet molasseses, which are used as the raw materials of fermentation processes, mainly contain three kinds of sugars, glucose, fructose and sucrose. Accordingly, on-line measurements of those sugars would be required for advanced fermentation control.[1] A glucose sensor[2,3] consisting of immobilized glucose oxidase and an oxygen electrode can be used for the determination of glucose. However, the determination of fructose is quite tedious.

On the other hand, three sugars can be specifically determined by a glucose sensor system consisting of chemical isomerization and chemical hydrolyzation reactors. The schematic diagram of a sugar sensor system is shown in FIGURE 1. The principles of each sugar measurement are as follows:

(1) Glucose measurement—Glucose + $O_2 \xrightarrow{\text{glucose oxidase}}$ Glucono-lactone + H_2O_2. Oxygen consumed can be determined by a dissolved-oxygen probe.

(2) Fructose measurement—Fructose $\xrightarrow[\text{chemical isomerization}]{\text{NaOH (pH 14)}}$ Glucose. Glucose produced can be measured by the glucose sensor.

(3) Sucrose measurement—Sucrose $\xrightarrow[\text{chemical hydrolyzation}]{\text{HCl (pH 1)}}$ Fructose + Glucose. Glucose produced can also be measured by glucose sensor.

Chemical isomerization and hydrolyzation were continuously performed at 75°C by addition of 2.0 *N* NaOH and 2.0 *N* HCl, respectively. Untreated broth is transferred to the sugar sensor system within a certain time interval. As a result, pulse currents corresponding to the concentrations of glucose, fructose plus glucose, and sucrose plus glucose in the broth were obtained. Concentrations of the individual sugars could be calculated from these pulse currents. Sensitivity ratios of fructose and sucrose (weight basis) to glucose were 0.62 and 0.53, respectively. Good correlation between concentrations determined by conventional methods and this proposed method could be obtained as shown in FIGURE 2. Total time required for the measurement of these sugars was about 8 min per sample. The selectivity of this sensor system is satisfactory as shown in TABLE 1, and this system can be used for at least a month.

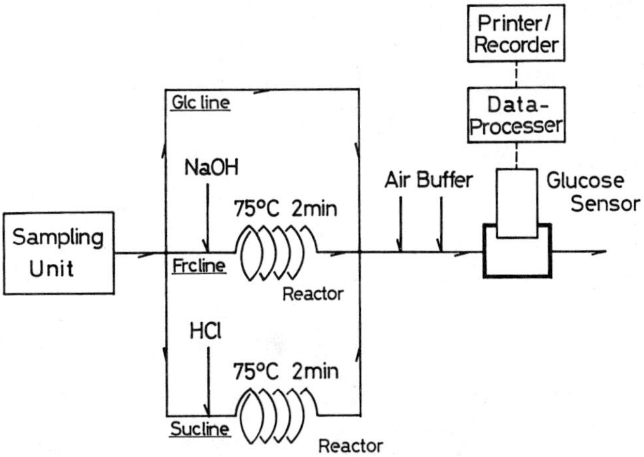

FIGURE 1. Schematic diagram of the sugar sensor system.

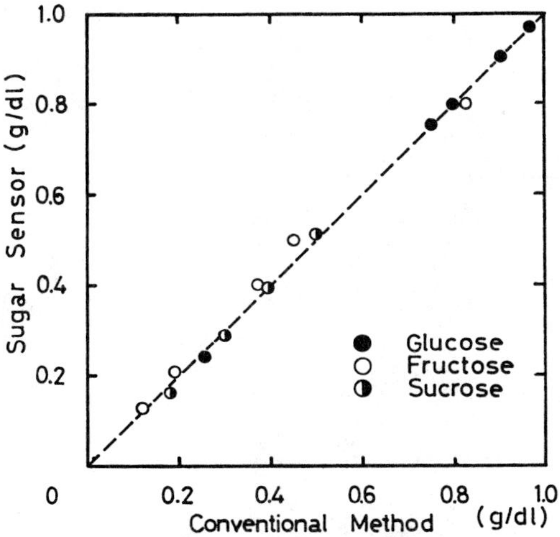

FIGURE 2. Comparison between concentrations determined by the sensor system and those determined by the liquid chromatography method. Fermentation broths (glutamic acid) were employed for experiments. Diluted sample solutions (2–5 times) were applied to the system. (●) Glucose, (○) fructose, and (◑) sucrose concentrations, respectively.

TABLE 1. Selectivity of the Sensor System to Various Sugars

	Measurement of		
Kind of Sugar	Glucose	Fructose	Sucrose
Glucose	1.00	—	—
Fructose	0	1.00	0
Galactose	0.05	0	0.15
Sorbose	0	0	0
Mannose	0	0.15	0
Sucrose	0	0	1.00
Maltose	0.01	0.63	0.31
Lactose	0	0	0.09
Trehalose	0	0	0
Raffinose	0	0	0

REFERENCES

1. HIKUMA, M., H. OBANA, T. YASUDA, I. KARUBE & S. SUZUKI. 1980. Enzyme Microb. Techn. **2:** 234–238.
2. KARUBE, I., S. MITSUDA & S. SUZUKI. 1979. Eur. J. Appl. Microbiol. Biotechnol. **7:** 343–350.
3. HIROSE, S., M. HAYASHI, N. TAMURA, S. SUZUKI & I. KARUBE. 1979. J. Mol. Catal. **6:** 251–260.

Dynamic Behavior of the Glucose Sensor Using the Glucose Oxidase/Glucose Isomerase Membrane

SHINICHIRO GONDO, MICHIO MORISHITA, AND HIDEKAZU KOYA

Fukuoka Institute of Technology
483, Shimowajiro, Higashiku
Fukuoka, 811-02, Japan

In order to obtain reliable information on the dynamic behavior of the glucose sensor through experimental studies, it is essential that we use a sensor of longer stability. We have found that the immobilized glucose oxidase/glucose isomerase can improve the sensitivity and stability of the glucose sensor where the enzymic oxidation of glucose by immobilized glucose oxidase was sensed by a dissolved-oxygen sensor of the galvanic type.[1,2] Especially the bilayered membrane, where each of two enzymes were immobilized individually in different layers of a two-layered membrane, improved the sensor performance remarkably. With a glucose sensor equipped with an immobilized glucose oxidase/glucose isomerase membrane, we studied the sensor response to glucose concentration changes and found that the first-order system equation can be successfully applied to express the dynamic behavior of the response. Collagen fibrils prepared from cowhide powder[3] were used to prepare the immobilized enzyme membrane. "B(10/10)" appearing in TABLE 1 and FIGURE 1 designates that 10 mg of glucose oxidase (Sigma, Type II) and 10 mg of glucose isomerase (industrial grade, Toyobo Co., Japan) were used per 50 mg of cowhide powder for the preparation of each layer of the bilayered membrane, respectively.

The initial response of the glucose sensor was measured experimentally by removing the glucose sensor from the vessel of glucose concentration C_A (sensor response R_A) and putting it quickly into another vessel of glucose concentration C_B (sensor response R_B). The transient response of the sensor, R, was successfully expressed by the first-order system equation:

$$\left\langle \ln \left| \frac{R - R_A}{R_B - R_A} \right| = -\frac{t - D}{\tau} \right\rangle \tag{1}$$

where D is the time lag and τ is the time constant. Typical values of these parameters are listed in TABLE 1.

The frequency response of the sensor was also measured by the equipment where the inlet flow of glucose concentration C_A and that of C_B were alternately fed (switching interval = T) at the flow rate of 30 ml/min into the measuring cell, mixed by a magnetic stirrer at 1200 rpm, of volume 6.8 cubic cm, in which the glucose sensor was set. The theoretical analysis for this experimental system gives the following equations:

$$R = R_1 e^{-t/\tau} + C_B(1 - e^{-t/\tau}) - \frac{t_R(C_B - C_{\max})}{t_R - \tau}(e^{-t/t_R} - e^{-t/\tau}) \tag{2}$$

$$\text{for } 2nT \leqq t \leqq (2n+1)T, \ (n = 0, 1, 2, \ldots)$$

TABLE 1. Parameters of the First-Order System Equation for the Membrane B(10/10)

C_A, mg/100 ml	C_B, mg/100 ml	Temp, °C	rpm	T, sec	D, sec
0	35	25	800	8.3	4.6
35	0	25	800	8.4	4.3
0	35	25	500	8.2	5.8
35	0	25	500	8.6	4.5
0	25	25	100	9.7	7.2
25	0	25	100	9.8	4.1
0	35	15	800	11.6	6.6
35	0	15	800	12.1	7.4
0	35	5	800	16.8	9.4
35	0	5	800	17.1	10.5

or

$$R = R_2 e^{-t/\tau} + C_A(1 - e^{-t/\tau}) - \frac{t_R(C_A - C_{\min})}{t_R - \tau}(e^{-t/t_R} - e^{-t/\tau}) \tag{3}$$

$$\text{for } (2n + 1)T \leqq t \leqq (2n + 1)T, \ (n = 0, 1, 2, \ldots)$$

where t_R is the average residence time of the cell,

$$R_1 = \frac{C_A + C_B}{2} + \frac{1}{2} \cdot Q \cdot (C_A - C_B)$$

$$R_2 = \frac{C_A + C_B}{2} - \frac{1}{2} \cdot Q \cdot (C_A - C_B) \tag{4}$$

$$Q = \frac{1 - e^{-T/\tau}}{1 + e^{-T/\tau}} - \frac{2t_R}{t_R - \tau} \frac{e^{-T/t_R} - e^{-T/\tau}}{(1 + e^{-T/t_R})(1 + e^{-T/\tau})}$$

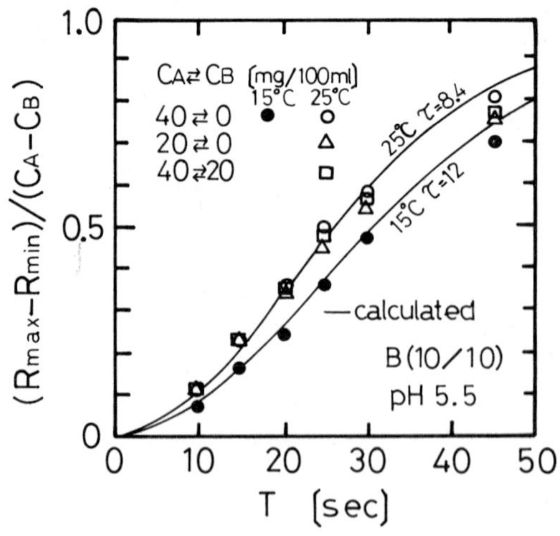

FIGURE 1. Comparisons of the amplitude ratio with theoretical calculations.

and, C_{max} and C_{min} are the maximum and minimum glucose concentrations in the cell, respectively. FIGURE 1 shows the relationship between T and the amplitude ratio. The real lines in FIGURE 1 were calculated with values of τ from TABLE 1 by Equations 2 and 3. Good coincidence of the experimental plots with the theoretical lines confirms the adequacy of the application of the first-order system equation for studies of glucose-sensor dynamics.

REFERENCES

1. S. GONDO, M. MORISHITA & T. OSAKI. 1980. Biotechnol. Bioeng. **22:** 1287.
2. S. GONDO, T. OSAKI & M. MORISHITA. 1981. J. Molec. Catal. **12:** 365.
3. S. GONDO & H. KOYA. 1978. Biotechnol. Bioeng. **20:** 2007.

Application of Kalman Filter to Automatic Monitoring System of Microbial Physiological Activities

ISAO ENDO AND TERUYUKI NAGAMUNE

The Institute of Physical and Chemical Research
Hirosawa 2-1, Wako-shi
Saitama 351, Japan

ICHIRO INOUE

Tokyo Institute of Technology
Department of Chemical Engineering
Ookayama 2-12-1, Meguro-Ku
Tokyo 152, Japan

INTRODUCTION

Various sensors and measuring devices have been invented and developed recently. These sensors and devices made possible the realization of many kinds of automatic monitoring and/or control systems for the fermentation processes with the aid of a microcomputer. However, measuring noises are inherent to these devices and hinder correct estimations of the microbial physiological activities such as the specific rate of substrate consumption, that of oxygen respiration, that of cellular growth, that of metabolites production, and so on. In order to avoid the problems originating in the noises, the authors have applied a Kalman filter in the design of an automatic monitoring system and have shown the effectiveness of the filter for correct estimation of the physiological activities of yeast in a batch alcoholic fermentation process.

KALMAN FILTER

Let signals of the process and its noises at the time (k) be expressed in terms of vectors $\boldsymbol{x}_k$ and $\boldsymbol{u}_k$, respectively. Then the signals at the time of (k + 1) are given by

$$\boldsymbol{x}_{k+1} = A_k \boldsymbol{x}_k + B_k \boldsymbol{u}_k \tag{1}$$

These signals can not be measured directly, but they are observable through the following measuring system;

$$\boldsymbol{y}_k = C_k \boldsymbol{x}_k + \boldsymbol{w}_k \tag{2}$$

where $\boldsymbol{y}_k$ and $\boldsymbol{w}_k$ are vectors of the output signals from the measuring system and that of the noises, respectively.

If we describe the least square estimation of the signals $\boldsymbol{x}_k$ as $\hat{\boldsymbol{x}}_k$, we have the following Equations 3 and 4 under the initial conditions written in Equation 5.

$$\hat{x}_k = \tilde{x}_k + K_k \{y_k - (C_k \tilde{x}_k + \overline{w}_k)\} \tag{3}$$

$$\left.\begin{aligned} \tilde{x}_k &= A_{k-1} \hat{x}_{k-1} + B_{k-1} \overline{u}_{k-1} \\ K_k &= P_k C'_k W_k^{-1} \\ P_k &= (M_k^{-1} + C'_k W_k^{-1} C_k)^{-1} \\ M_k &= A_{k-1} P_{k-1} A'_{k-1} + B_{k-1} U_{k-1} B'_{k-1} \end{aligned}\right\} \tag{4}$$

$$\hat{x}_o = \overline{x}_0, \; M_o = X_o \tag{5}$$

where $\overline{w}_k$, $\overline{u}_{k-1}$ and $\overline{x}_o$ are the mean values of w_k, u_k and x_o, respectively, while W_k, U_k, and X_o are variance matrices of w_k, u_k, and x_o. Actually, when we have designed the monitoring system, we neglected the term u_k.

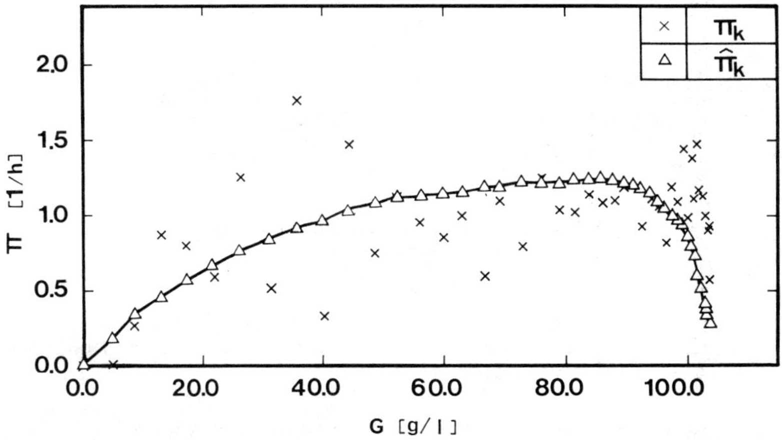

FIGURE 1. Effectiveness of the Kalman filter for the estimation of the specific production rate of alccohol.

DISCUSSION

FIGURE 1 shows the effectiveness of Kalman filter for the estimation of the specific production rate of alcohol, π. The keys of (X) in the figure show the ordinal calculational results of π without any filtration. They are scattered in the figure, and from the features of this scatter, we cannot estimate correctly the specific production rate of alcohol.

Many attempts at selecting the input signals for the Kalman filter led us to the conclusion that we should select quantities in each category that are the same as those of the output signals. That is, we have selected y_k as the derivatives of alcohol concentration and x_k as the specific production rate of alcohol, π. Then, we have succeeded in obtaining the correct estimation of π as shown in the symbol (Δ) in FIGURE 1. By this method, we could correctly estimate the time course of the other activities of the organism.

FIGURE 2 shows the total results of the monitoring of a batch fermentation by brewer's yeast that was operated under the following conditions: the initial concentra-

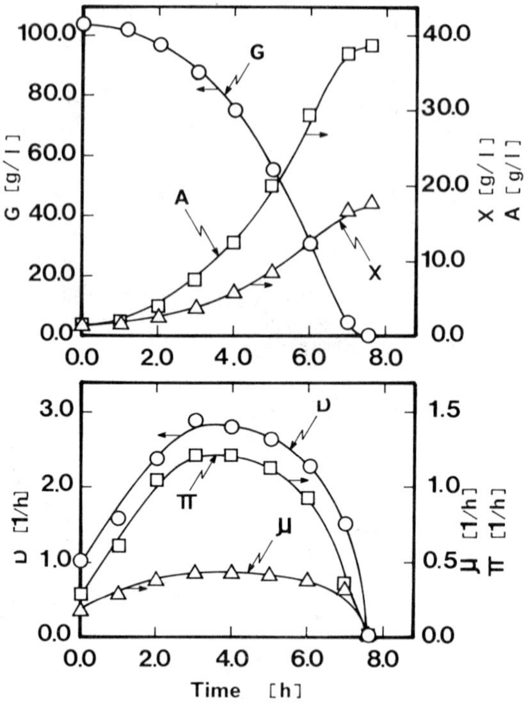

FIGURE 2. Total results of the automatic monitoring system. The solid lines in the figure show the smoothing data of the experimental results while the symbols express the output from the Kalman filter, which was marked at one-hour intervals.

tion of glucose was 104 (g/l) and the inoculum size of the cells was 1.2 (g/l). It is obvious from FIGURE 2 that the filtering results agreed well with the smoothed lines.

CONCLUSIONS

A microcomputer was used to assemble available data from a 2-liter jar fermenter in which brewer's yeast was cultivated. By applying a Kalman filter, the authors have constructed a fully automated monitoring system for the correct estimation of the physiological activities of yeast.

Large-scale Isolation and Purification of Proteins from Recombinant *E. coli*

W. C. McGREGOR

Biopolymer Research Department
Hoffmann-La Roche Inc.
Nutley, New Jersey 07110

INTRODUCTION

Three medically important proteins have already been produced in clinically significant quantities by means of recombinant DNA technology. These are summarized in TABLE 1. Although the human hormone somatostatin has also been expressed in *E. coli,*[1] to this author's knowledge it is not yet being produced for clinical study. The high expectations for genetic engineering to produce other therapeutically interesting proteins or polypeptides have been widely publicized. This paper, however, will concentrate on the three products from *E. coli* that are currently being produced for clinical study and that have been reported on.

ESCHERICHIA COLI (E. COLI) AS HOST

The evolution of molecular biology including the breakthroughs collectively termed "recombinant DNA technology" has been intimately linked to the study of *E. coli* and its viruses. This gram-negative organism has been a favorite for at least three reasons. It has relatively simple genetics, a rapid growth rate, and is well characterized generally. It was natural, therefore, that *E. coli* be the common host for initial large-scale work with the products in TABLE 1.

It has been suggested that other microbial hosts may be more advantageous for production of commercial quantities of proteins.[2,3] Yeast, for example, is nonpathogenic and has the enzymatic capability for adding sugar side chains. Glycosylation may or may not have therapeutic importance even if the native protein is a glycoprotein.

Another alternative host, *Bacillus,* has been used for producing antibiotics and enzymes. Its secretory capability, as in that of yeast, is thought to make purification easier. This has yet to be demonstrated, however. Meanwhile, we face the real challenges with *E. coli,* which has been used commercially mainly for the production of amino acids.[4] L-Asparaginase has been the only medically important protein produced by *E. coli* until now.[5]

There are some inherent problems associated with this organism as it relates to purification. *E. coli* has been shown to contain at least eight soluble proteolytic activities.[6] These or other proteases may have been responsible for product degradation as suspected by those investigating recombinant somatostatin[1] and recombinant growth hormone.[7] It is known, for example, that *E. coli* selectively degrades polypeptides with abnormal conformations.[8] Recombinant proteins would no doubt be "seen" by the cell as being abnormal. It is also known that normal *E. coli* proteins are degraded more slowly, but when cells are starved for essential amino acids, energy source, or certain inorganic ions, the rate is greatly increased.[8–10]

TABLE 1. Recombinant Proteins Produced for Clinical Study

Protein	Number of Amino Acids	Approximate Molecular Weight	Reference/s
Human insulin	51	6,000	22,23,24
Human growth hormone	191	22,000	7,25
Human leukocyte interferon	165	19,000	28,12

In many purification protocols, the protease inhibitor, phenyl methyl sulfonyl fluoride is included in the buffers, but it is only effective against serine proteases. Of the eight *E. coli* protease activities reported by Goldberg,[6] six were of the serine type and two were metalloenzymes.

When an *E. coli* fermentation is depleted of iron, an enzyme is derepressed that catalyzes the formation of an iron chelator, 2,3-dihydroxy benzoyl serine.[11] This compound, when complexed with iron, is bright red and may cause unexpected problems during purification, such as binding to agarose gel (McGregor, unpublished observations).

Another potential purification problem arising from the use of *E. coli* is the presence of pyrogens. Lipopolysaccharide from the cell envelope is a potent pyrogen, causing fever response at levels $\geq$ 0.5 ng/kg body weight. Such contamination has been experienced in the purification of L-asparaginase from *E. coli.*[5]

ISOLATION AND PURIFICATION

Flow charts for the isolation and purification of human insulin, human growth hormone, and human leukocyte interferon are shown in FIGURE 1. They are derived from the earliest publications (TABLE 1) describing breakthroughs in the new technology.

For the most part, the steps as shown are conventional laboratory techniques and have no doubt been improved upon by workers in the respective companies. It is expected that newer or additional purification steps will be included to yield product of higher purity than the naturally derived counterparts.

Conventional wisdom dictates that more steps will give higher purity but lower overall yield. Newer techniques, however, such as immunosorbent columns may serve to eliminate some steps as suggested in FIGURE 1 and yet give relatively high purity. This was shown to be the case in the purification of interferon.[12]

Immunosorbent separation, a form of affinity separation, has held great promise both analytically and preparatively for the last decade. Now, hybridoma technology has opened the way for producing monoclonal antibodies that are both uniform and highly selective ligands.

Preparation of interferon specific hybridomas and the subsequent steps of purification and immobilization of monoclonal antibodies are not trivial steps.[13] Once available, however, immunosorbent columns are relatively simple to use. It is basically adsorption-desorption chromatography. J. W. Eveleigh, in an appeal for wider use of the technique, recently pointed to some of the practical considerations and design strategies.[14] He worked with N. G. Anderson some years ago in the development of the "Cyclum." The "Cyclum" approach allowed a column chromatographic separation to be repeated precisely a large number of times.[15] Purified, immobilized antibodies or antigens are expensive, and it becomes important to keep the column of immobilized species small but in continuous use. Newer developments in microprocessor control afford a much simpler array of hardware than was available with the original "Cyclum."

Some of the practical considerations of immunosorbent chromatography first pointed out by Eveleigh[14] and supported by our experience are as follows:

(1) Design strategy is largely determined by the resources (mainly antibody) available and the need for the product. Recombinant *E. coli* cells would not be expected as a limiting factor, that is, fermentation processing is better developed and improvements generally easier to obtain than is the case with the purification steps.

(2) A relatively small column with repetitive operation is almost mandatory when monoclonal antibodies are the adsorbent.

(3) Agarose is the most common and probably the safest support matrix. Surface density of the immobilized antibody must be optimized.

(4) Inclusion of precolumn purification steps must be weighed against the required purity of the final product and the long-term operation of the column.

(5) The first cycle on a virgin immunoadsorbent column invariably gives lower recovery than subsequent runs because of an apparent subset of irreversible binding sites.

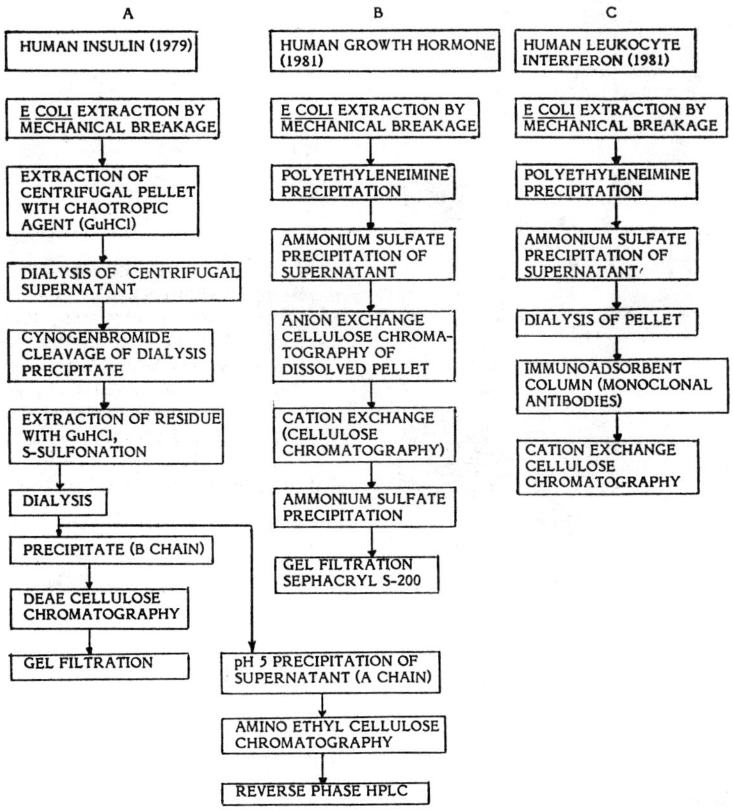

FIGURE 1. Flowcharts for the purification of proteins from recombinant *E. coli*.

(6) Determining the number and volume of buffer washes is a major part of the strategy. Product purity requirement, affordable time for overall cycling, and the yield per cycle are all critical factors in the decisions. They are particularly important for scale-up.

Monoclonal antibodies have been used not only for the purification of recombinant leukocyte interferon but also for development of a rapid, quantitative enzyme immunoassay of the same interferon.[16]

PROTEIN PURITY

The conventional method for monitoring protein purity is by electrophoresis in polyacrylamide gel, usually in presence of the denaturing detergent, sodium dodecyl sulfate (SDS). Protein bands are detected with Coomassie blue stain, which has a

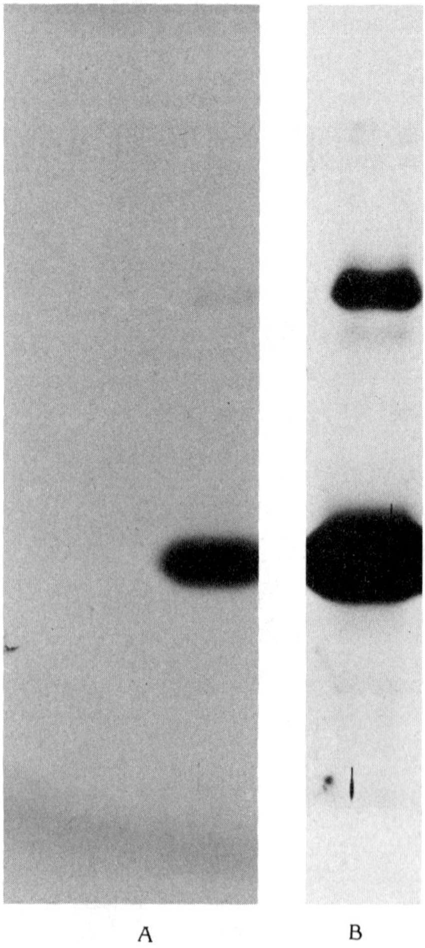

FIGURE 2. Silver stain of recombinant leukocyte interferon following SDS-PAGE: 10 μg recombinant leukocyte interferon alpha was run in SDS-PAGE, then stained with Coomassie blue essentially as described by Laemmle.[27] The gel after Coomassie blue staining as shown in A was re-equilibrated in a solution of methanol/37% w/w formaldehyde/water (50:2:50 v/v) for 18 hr, then stained with silver as described by Wray *et al.*[19] and shown in B.

TABLE 2. A Comparison of Integration Units per μg Protein Stained by the Silver Method[a]

Protein	Molecular Weight	Integration Units/μg Protein: Method of Ref. 19 Aver. of 2 Runs	Method of Ref. 17
Ribonuclease A	13,700	46 ± 8	187
Recombinant leukocyte interferon	19,000	186 ± 16	170
Chymotrypsinogen A	25,000	236 ± 66	271
Ovalbumin	43,000	172 ± 2	228
Bovine serum albumin	68,000	60 ± 34	107

[a]0.5 μg of the above proteins were electrophoresed in the presence of SDS, stained by the methods referenced, and scanned with a densitometer to determine the integration units/μg protein.

detection limit of about 0.1 μg protein or 1% of a 10 μg load. Newer staining techniques with silver[17] have afforded sensitivities as much as 100 times greater than Coomassie blue.[18,19]

FIGURE 2 demonstrates the increased sensitivity of the silver stain by the appearance of additional stained bands in a recombinant leukocyte interferon sample that was first stained with Coomassie blue. The procedure clearly allows visualization of minor protein impurities that would otherwise remain undetected.

The silver staining method allows visualization of interferon on sodium dodecyl sulfate polyacrylamide gel electrophoresis (SDS-PAGE) in amounts below detection by Coomassie blue. In our hands, 0.005 μg of recombinant leukocyte interferon could be detected with silver stain whereas a minimum of 0.05 μg was needed for detection with Coomassie blue.

The mechanism of the silver stain is not known. However, it has been speculated that proteins in the alkaline environment produced by the ammonium hydroxide are complexed with silver cations via amino groups, and the sulphur in cysteine and methionine residues.[20] We have observed variability by almost an order of magnitude in the silver stain intensity of several proteins (TABLE 2), but this depends much on the method used as well as destaining techniques. Impurities could be actually higher or lower than estimated depending on their staining intensity relative to the protein of interest.

Another emerging analytical technique in protein purification is high performance liquid chromatography (HPLC).[21] Reverse-phase HPLC columns have been used to monitor purity of recombinant insulin[22–24] and recombinant interferon.[12]

This technique was also used to "fingerprint" tryptic digests of the recombinant products.[12,23,24,29] The resolving power of reverse-phase columns permits analysis of subnanomole quantities of complex peptide mixtures in a short time with high reproducibility.[26] Fingerprinting or peptide mapping of recombinant proteins is an important tool for confirming structure especially when making scale-up changes in the purification protocol.

HPLC has also played an indirect but important preparative role in the production of recombinant leukocyte interferon. To obtain specific hybridomas (used in production of monoclonal antibodies), it was necessary to immunize the mice with natural interferon that was greater than 10% pure. This relative high purity (for that time) was obtained by reverse-phase HPLC.[13] The use of reverse-phase columns is emerging for preparative purification of interferons and other proteins. (Pestka, S. and M. Rubinstein, 1981. Protein purification process and product. U.S. Patent 4,289,690; Friesen, H.-J., and S. Pestka, 1981. Preparation of homogeneous human fibroblast interferon. U.S. Patent 4,289,689). Isoelectric focusing is yet another technique used to monitor purity of the recombinant proteins.[12,25]

It is important to emphasize the role of analytical techniques in the development of large-scale purification techniques. While recombinant DNA technology now opens the way for preparing large amounts of heretofore rare proteins, the simultaneous development of more sensitive analytical tools will mean higher purity requirements. In other words, the challenge is not only to produce large quantities of protein but also to approach 100% in purity. Further developments in techniques such as immunoaffinity separations and HPLC are expected to bring this about.

ACKNOWLEDGMENTS

I acknowledge the dedicated efforts of many colleagues in the Biopolymer Research Department of Hoffmann-La Roche Inc. The assistance of Lyn Nelson in preparing and typing the manuscript is gratefully acknowledged.

REFERENCES

1. Itakura, K., T. Hirose, R. Crea, A. D. Riggs, H. L. Heynecker, F. Boliver & H. W. Boyer. 1977. Expression in *Escherichia coli* of a chemically synthesized gene for the hormone somatostatin. Science **198:** 1056–1063.
2. Hitzeman, R. A., F. E. Hagie, H. L. Levine, D. V. Goeddel, G. Ammever & B. D. Hall. 1981. Expression of a human gene for interferon in yeast. Nature **293:** 717–722.
3. Dean, D. H. & M. J. Kaelbling. 1981. A genetic engineering manifesto for the genus *Bacillus*. Ann. N.Y. Acad. Sci. **369:** 23–32.
4. Hirose, Y. & H. Okada. 1979. Microbial production of amino acids. *In* Microbial Technology. H. J. Peppler & D. Perlman, Eds. Academic Press. New York. pp. 211–240.
5. Grinnan, E. L. 1981. L-Asparaginase: A case study of an *E. coli* fermentation product. *In* Insulins, Growth Hormone & Recombinant DNA Technology. J. L. Gueriguian, Ed. Raven Press. New York.
6. Swamy, K. H. S. & A. L. Goldberg. 1981 *E. coli* contains eight soluble proteolytic activities, one being ATP dependent. Nature **292:** 652–654.
7. Goeddel, D. V., H. L. Heynecker, T. Hozumi, R. Arentzen, K. Itakura, D. G. Yansura, M. J. Ross, G. Miozzari, R. Crea & P. H. Seeburg. 1979. Direct expression in *Escherichia coli* of a DNA sequence coding for human growth hormone. Nature **281:** 544–548.
8. Goldberg, A. L. & A. C. St. John. 1976. Intracellular protein degradation in mammalian and bacterial cells. Part 2. *In* Annual Review of Biochemistry **45:** 747–803. E. E. Snell, Ed. Annual Reviews, Inc. Palo Alto, CA.
9. Miller, C. G. 1975. Peptidases and proteases of *Escherichia coli* and *Salmonella Typimurium*. *In* Annual Review of Microbiology **29:** 485–504. Annual Reviews, Inc. Palo Alto, CA.
10. Voellmy, R. & A. L. Goldberg. 1980. Guanosine-5′-diphosphate (ppGpp) and the regulation of protein breakdown in *Escherichia coli*. J. Biol. Chem. **255:** 1008–1014.
11. Brot, N., J. Goodwin & H. Fales. 1966. *In vivo* and *in vitro* formation of 2,3-dihydroxybenzoyl serine by *Escherichia coli* K_{12}. Biochem. Biophys. Res. Commun. **25:** 454–461.
12. Staehelin, T., D. S. Hobbs, H. -F. Kung, C.-Y. Lai & S. Pestka. 1981. Purification and characterization of recombinant human leukocyte interferon (IFLrA) with monoclonal antibodies. J. Biol. Chem. **256:** 9750–9754.

13. STAEHELIN, T., B. DURRER, J. SCHMIDT, B. TAKACS, J. STOCKER, V. MIGGIANO, C. STAHLI, M. RUBINSTEIN, W. P. LEVY, R. HERSHBERG & S. PESTKA. 1981. Production of hybridoma secreting monoclonal antibodies to the human leukocyte interferons. Proc. Natl. Acad. Sci. USA **78:** 1848–1852.
14. EVELEIGH, J. W. 1982. Practical considerations in the use of immunosorbents and associated instrumentation. *In* Affinity Chromatography & Related Techniques. T. C. J. Gribnau, J. Visser & R. J. F. Nivard, Eds. Elsevier, Amsterdam.
15. ANDERSON, N. G., D. D. WILLIS, D. W. HOLLADAY, J. E. CATON, J. W. HOLLAMAN, J. W. EVELEIGH, J. E. ATTRILL, F. L. BALL & N. L. ANDERSON. 1975. Analytical techniques for cell fractions. XIX. The cyclum: An automatic system for cyclic chromatography. Anal. Biochem. **66:** 159–174.
16. STAEHELIN, T., C. STEHLI, D. S. HOBBS & S. PESTKA. 1981. A rapid quantitaive assay of high sensitivity for human leukocyte interferon with monoclonal antibodies. Meth. Enzymol. **74:** 589–595.
17. MORRISSEY, J. H. 1981. Silver stain for proteins in polyacrylamide gels: A modified procedure with enhanced uniform sensitivity. Anal. Biochem. **117:** 307–310.
18. OAKLEY, B. R., D. R. KIRSCH & N. R. MORRIS. 1980. A simplified ultrasensitive silver stain for detecting proteins in polyacrylamide gels. Anal. Biochem. **105:** 361–363.
19. WRAY, W., T. BOULIKAS, V. P. WRAY & R. HANCOCK. 1981. Silver staining of proteins in polyacrylamide gels. Anal. Biochem. **118:** 197–203.
20. FREEMAN, H. C. 1973. Metal complexes of amino acids and peptides. *In* Inorganic Biochemistry. G. L. Eichhorn, Ed. Elsevier. Amsterdam. **1:** 121–166.
21. REGNIER, F. E. & K. M. GOODING. 1980. High-performance liquid chromatography of proteins. Anal. Biochem. **103:** 1–25.
22. GOEDDEL, D. V., D. G. KLEID, F. BOLIVAR, H. L. HEYNECKER, D. G. YANSURA, R. CREA, T. HIROSE, A. KRASZEWSKI, K. ITAKURA & A. D. RIGGS. 1979. Expression in *Escherichia coli* of chemically synthesized genes for human insulin. Proc. Natl. Acad. Sci. USA **76:** 106–110.
23. CHANCE, R. E., J. A. HOFFMANN, E. P. KROEFF, M. G. JOHNSON, E. W. SCHIRMER & W. W. BROMER. 1981. The production of human insulin using recombinant DNA technology and a new chain combination procedure. *In* Peptides: Synthesis, Structure and Function; Proceedings of the Seventh American Peptide Symposium. D. G. Rich & E. Gross, Eds. Pierce Chemical Co. Rockford, IL. pp. 729–738.
24. FRANK, B. H., J. M. PETTEE, R. E. ZIMMERMAN & P. J. BURCK, 1981. The production of human proinsulin and its transformation to human insulin and C-peptide. *In* Peptides: Synthesis, Structure and Function; Proceedings of the Seventh American Peptide Symposium. D. H. Rich & E. Gross, Eds. Pierce Chemical Co. Rockford, IL. pp. 729–738.
25. OLSON, K. C., J. FENNO, N. LIN, R. N. HARKINS, C. SNIDER, W. H. KOHR, M. J. ROSS, D. FODGE, G. PRENDER & N. STEBBING. 1981. Purified human growth hormone from *E. coli* is biologically active. Nature **293:** 408–411.
26. RUBINSTEIN, M., S. STEIN & S. UDENFRIEND. 1980. Fluorometric methods for analysis of proteins and peptides: Principles and applications. Horm. Proteins Peptides **9:** 1–24.
27. LAEMMLI, U. K. 1970. Cleavage of structural proteins during the assembly of the head of bacteriophage T4. Nature (London) **277:** 680–685.
28. GOEDDEL, D. V., E. YELVERTON, A. ULLRICH, H. L. HEYNEKER, G. MIOZZARI, W. HOLMES, P. H. SEEBURG, T. DULL, L. MAY, N. STEBBING, R. CREA, S. MAEDA, R. MCCANDLISS, A. SLOMA, J. M. TABOR, M. GROSS, P. C. FAMILLETTI & S. PESTKA. 1980. Human leukocyte interferon produced by *E. coli* is biologically active. Nature **287:** 411–416.
29. KOHR, W. J., R. KECK & R. N. HARKINS. 1982. Characterization of intact and trypsin digested biosynthetic human growth hormone by high-pressure liquid chromatography. Anal. Biochem. **122:** 348–359.

Ultrafiltration Processes in Biotechnology

ROBERT S. TUTUNJIAN

Scientific Systems Division
Amicon Corporation
17 Cherry Hill Drive
Danvers, Massachusetts 01923

INTRODUCTION

During the past decade, ultrafiltration has become established as a useful separation process for the pharmaceutical and biological industries. Present applications are found in the areas of plasma products, vaccines, diagnostic enzymes, fermentation products, and depyrogenation of parenterals. Ultrafiltration has been used to replace and/or supplement many conventional separation techniques, such as centrifugation, dialysis, distillation, salt precipitation, lyophilization, evaporation, and chromatography. The results have shown dramatic savings in cost and time as well as improvements in product purity.

The recent industrial activity in the field of biotechnology has also utilized ultrafiltration as an integral part of purification processes. Present production of recombinant DNA products and of monoclonal antibodies generally require extensive isolation techniques. Products are typically very dilute and highly impure directly after cellular extraction. Consequently, ultrafiltration has allowed economical, efficient concentration and purification of these biologicals and simple scale-up of production processes. The advantages of ultrafiltration should have increased impact on this emerging industry as process research continues in the future.

BACKGROUND

Ultrafiltration is a technique for separation of dissolved and suspended materials on the basis of size and molecular scale. Substances smaller than the pore size of the filter are driven through with the solvent while larger solutes are retained. This is a very simple procedure and requires no phase change, no chemical addition, and little energy.

The separation involved in ultrafiltration can deal with species ranging in molecular weight from 500 to 1,000,000 daltons, or approximately 10 to 100 ångströms. This distinguishes the process from reverse osmosis, where even salts are retained, and from microporous filtration, where pore size ranges from 0.01 to 10 microns (see FIG. 1). It should also be noted that ultrafiltration membranes are produced in a wide variety of polymers that exhibit high resistance to acids, bases, alcohols, and temperature allowing very effective membrane cleaning. In this manner, ultrafilters can be reused without deterioration of flow rates and without cross contamination. In most cases, filter lifetimes last one to four years. UF membranes are also differentiated from microporous filters by their unique "skinned" or anisotropic surface structure (see FIG. 2). This allows retention of retained substances to take place on the membrane surface rather than within the filter structure.

The concentration of dissolved species can create some processing problems. If the

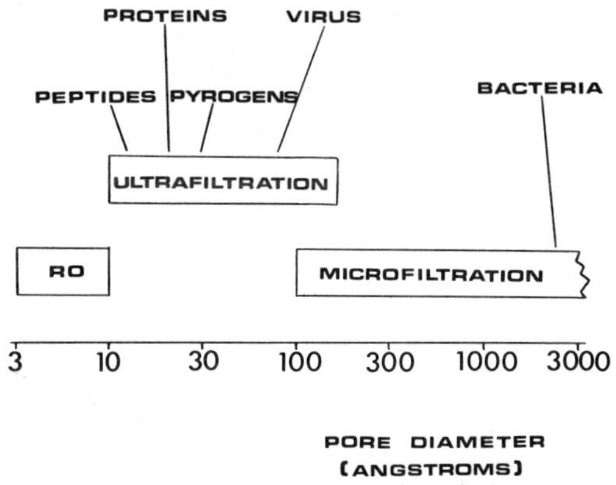

FIGURE 1. Membrane processes as a function of pore size.

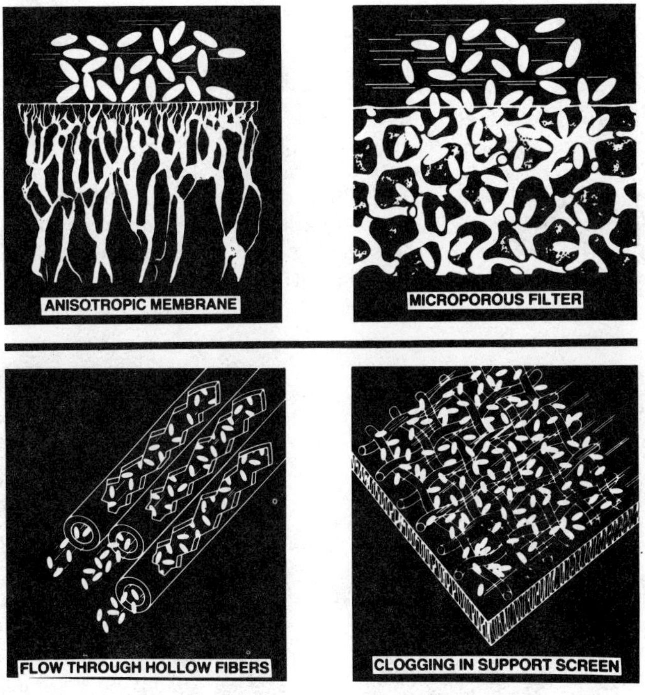

FIGURE 2. Comparison of cross-flow filtration designs. Anistropic membrane compared to microporous filter (*above*); flow-through hollow fibers compared to clogging in support screen device (*below*).

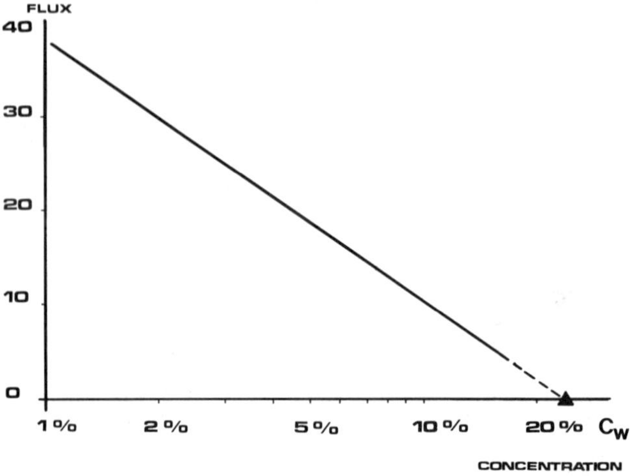

FIGURE 3. Typical filtration flux rate as a function of protein concentration.

membrane retains a solute, the solute concentration adjacent to the membrane will increase, resulting in a phenomenon known as concentration polarization. Polarization will cause the filtration flux rate to decline logarithmically with increasing concentration of bulk solute (see FIG. 3).

The problems created by concentration polarization can be controlled by various methods. The principle behind these centers on maximizing movement of solute away from the membrane. This is usually accomplished by cross-flow filtration where retained fluid is recirculated over the membrane surface (see FIG. 4). Other approaches involving mechanical agitation, with stir bars and vibrators, are limited to small-scale equipment. The rate of filtration flux will increase with the efficiency of polarization control. Where cross-flow methods are used, flux will generally improve with increased velocity across the membrane surface.

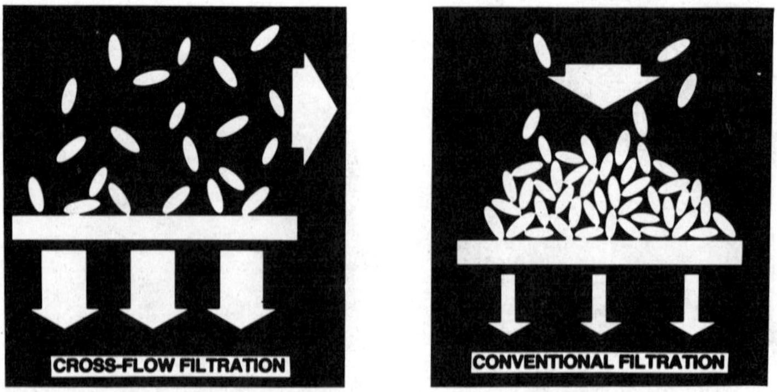

FIGURE 4. Conventional filtration compared to cross-flow filtration.

MEMBRANE CONFIGURATIONS

There are two basic configurations of ultrafiltration membranes—hollow fibers and flat sheets. Filtration equipment with these membranes utilize a variety of operational modules (see FIG. 5).

Hollow fibers are generally supplied in self-contained cartridge housings. Fibers are easy to clean and allow good product recovery; they are somewhat limited in

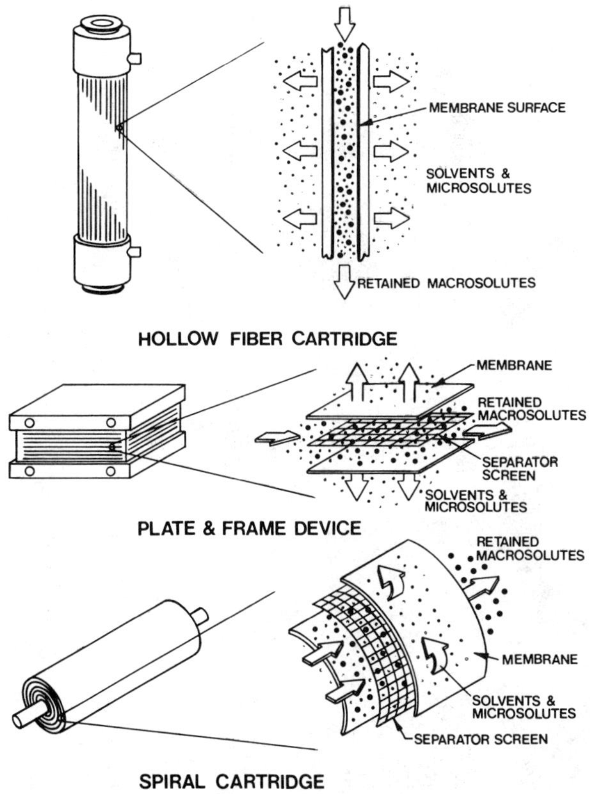

FIGURE 5. Design of hollow-fiber, plate and frame, and spiral cartridge ultrafiltration modules.

pressure capability. Large diameter (¼–1") membrane tubes are basically a variant of hollow fibers. Due to their large hold-up volume and high space requirements, tubular membranes are not well suited for biological purification.

Flat sheet membranes have been fabricated in several types of modules. Spiral wound cartridges use a rolled membrane design with spacer screens separating the filters. Certain plate-and-frame-type devices also use separator screens except that the membranes are stacked vertically. These devices allow high-pressure operation but can

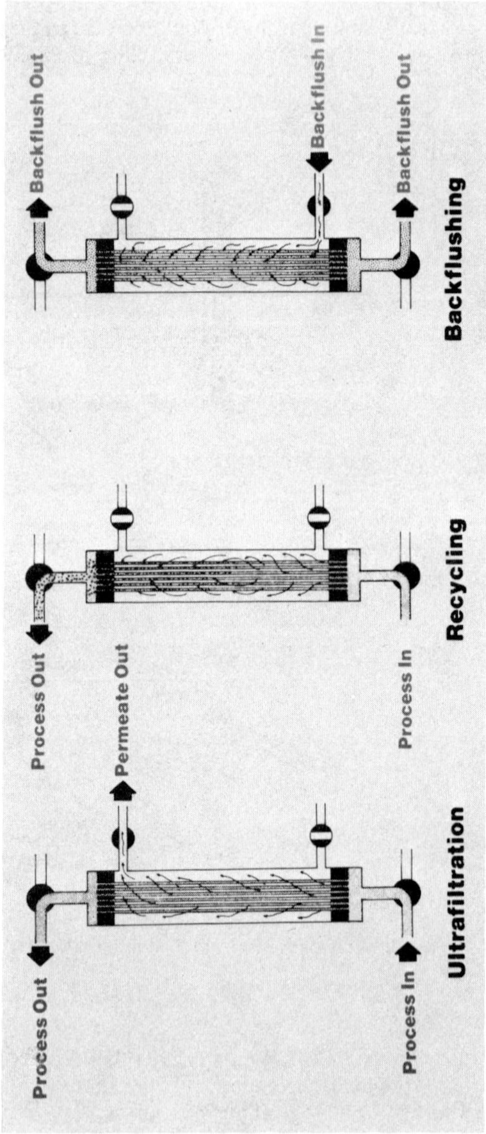

FIGURE 6. Operating modes for hollow-fiber ultrafiltration.

have cleaning and product recovery problems due to clogging of the screens. There are some true plate-and-frame devices available that are easier to clean, but these tend to require more space.

SYSTEM OPERATION

As discussed previously, ultrafiltration systems use cross-flow filtration to control concentration polarization and associated membrane fouling. This recirculating flow creates a pressure differential between the inlet and outlet of the system on the retentate side of the filter (see FIG. 6). The relative recirculation velocity can be measured by the pressure drop from cartridge inlet to outlet. The recirculation rate is directly proportional to ΔP in laminar flow and to the square root of ΔP in turbulent flow.

The filtration driving force through the membrane is determined by difference between the retentate pressure and that of the permeate. Since the retentate pressure decreases from inlet to outlet, an average value is used to determine the transmembrane pressure (driving force). In most cases, the permeate pressure is negligible.

TABLE 1. Filtration Flux Rate as a Function of Recirculation Rate and Pressure Drop

P_i	P_o	ΔP	ΔPTM	Recirculation Rate (LPM)	Flux Rate (LPM)
25	0	25	12.5	50	2.3
25	5	20	15	40	2.8
25	10	15	17.5	30	2.9
25	15	10	20	20	2.6
25	20	5	22.5	10	1.7

In system operation, flux rates are optimized by variation of the recirculation rate and outlet pressure; inlet pressure is generally maintained at the maximum allowed by the system. Typical data from a test are shown in TABLE 1.

In this case, the optimal flux rate is reached with a ΔP of 15 lb/in^2. Further reduction in velocity caused polarization and a drop in filtration rate. Higher velocity resulted in lower driving force, ΔPTM, due to the given inlet pressure.

It should also be noted that the optimum operating conditions are a strong function of product concentration; a higher ΔP is required at elevated solute concentrations (see FIG. 7). Above a certain point, in fact, outlet pressures of zero are generally used.

Another consideration in system operation is that of maintaining flux rates and cleaning. Most systems are cleaned following operation by flushing with sodium hydroxide, sodium hypochlorite, or other suitable chemical agent. This is normally done in the standard ultrafiltration mode of operation, but in cases where severe membrane fouling has occurred, membranes can be backflushed with cleaning solution (see FIG. 6).

A method useful for cleaning during product processing is that of recycling (see FIG. 6). In this case, the permeate outlets are closed while the cartridge is filled with ultrafiltrate. This causes backflushing on the outlet end of the cartridge, where the permeate pressure now exceeds the retentate pressure. By then reversing the direction

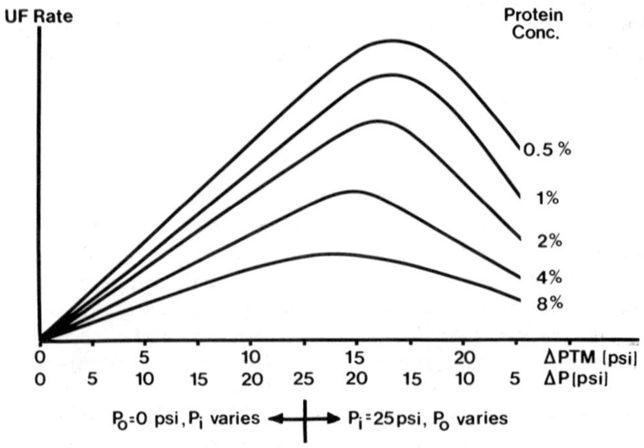

FIGURE 7. Ultrafiltration rate as a function of recirculation pressure drop (ΔP), average transmembrane pressure (ΔPTM) for varying protein concentrations.

of the process flow, the other end of the filter can be similarly cleaned. This procedure requires only a few minutes to perform and has proven very useful in regenerating flux rates in process. In fact, the process can be automated for easy operation. Recycling is particularly effective in dealing with process streams with high-suspended solids, such as bacterial cells and protein precipitates.

PROCESS CONSIDERATIONS

The applications of ultrafiltration can be broadly categorized as follows: concentration, diafiltration, or purification.

(1) Concentration of suspended particles or macromolecules is where the retained stream is the product (see FIG. 8).

In a batch system, the so-called concentration factor is merely the ratio of the starting volume to the final volume. Thus if the feed stream was a 1,000-gallon batch

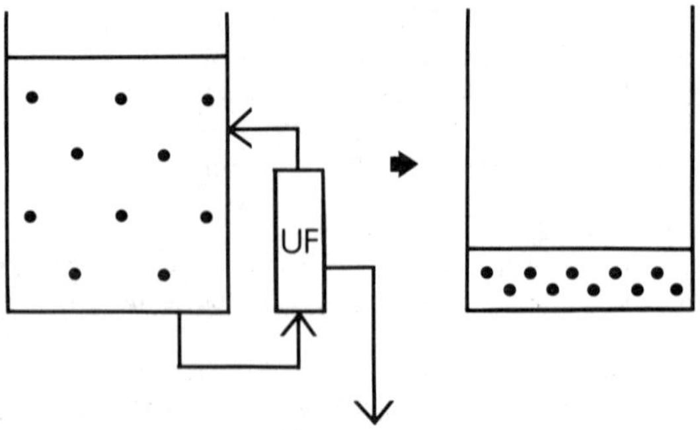

FIGURE 8. Schematic of batch concentration by ultrafiltration.

and the system was allowed to run until the final volume dropped to 100 gallons, the concentration factor would be 10. A macromolecule that was completely retained by the membrane would have been increased by a factor of 10 in concentration. The concentration of a completely permeable species such as a salt will not be affected by ultrafiltration.

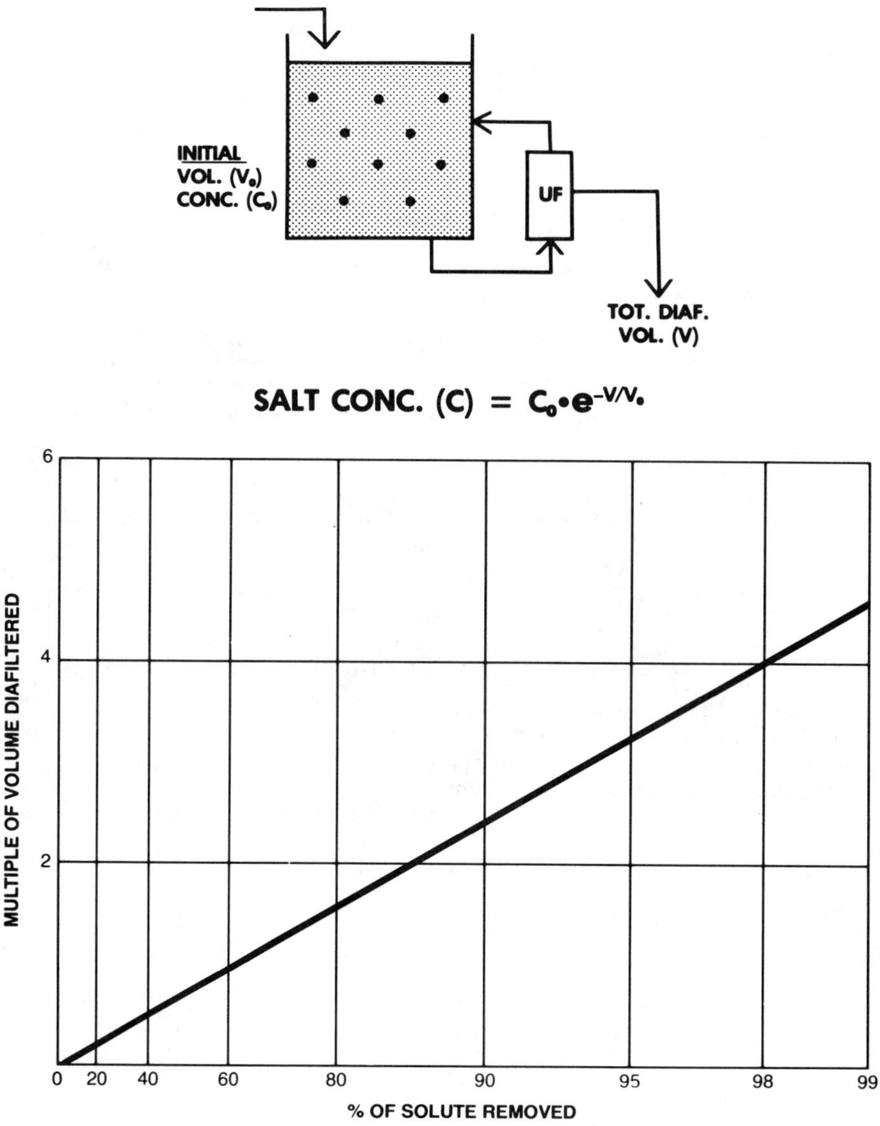

FIGURE 9. Schematic of diafiltration operation showing salt removal as a function of diafiltrate volume.

TABLE 2. Diafiltration time as a Function of Protein Concentration and Batch Volume

Protein Concentration	Batch Volume (l)	Diafiltration Volume (l)	Filtration Rate (l/hr)	Diafiltration Time (hr)
2%	1000	4600	1300	3.5
5%	400	1840	840	2.2
10%	200	920	500	1.8
20%	100	460	150	3.1

(2) Diafiltration of suspended particles or macromolecules is where the retained stream contains the product and low molecular weight solutes such as salts, sugars, and alcohols, pass through the membrane by the addition and removal of water. This operation is shown schematically in FIGURE 9. The permeate that leaves the system is replaced with deionized water, ideally through a level controller at the same rate that permeate is removed. A buffer may be used instead of water if salt exchange is desired.

The equation governing the concentration of a microsolute during such a diafiltration operation is also shown here. The key parameter is the ratio of volume added to the sample volume during diafiltration (v/v_o). The theoretical decrease in concentration of the permeating species with water addition may also be shown graphically (see FIG. 9).

In processes where diafiltration is combined with concentration, the desalting can be performed before or after concentration. After concentration, smaller diafiltrate volumes are needed but flux rates are slower due to higher protein levels. In one actual example (see TABLE 2), the optimum point for diafiltration is between the desired initial and final protein concentrations.

(3) The third type of UF process involves the purification of solvents and solutions of low molecular solutes where the permeate stream contains the product; the retained species may also contain a product of interest (see FIG. 10).

Processes where the low molecular weight solute is the product of interest generally combine concentration and diafiltration. The retained stream is first concentrated to near its maximum level, while allowing purified solutes to pass through the membrane. A diafiltration step then washes out additional solute from the retentate. In this way, maximum recovery of product is achieved with minimal dilution. The retentate is also

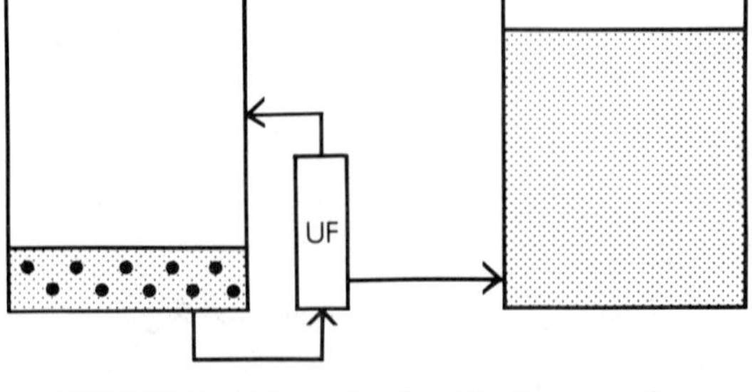

FIGURE 10. Schematic of purification operation.

highly purified; this is important in some fractionation processes where the macromolecules may also be a product of interest.

BACTERIA CONCENTRATION/CELL HARVESTING

The recovery of bacteria and cells after fermentation and culture can be tedious, inefficient, and costly when using centrifugation processes. Other problems include aerosol formation, poor product recovery, and slow processing. These difficulties are more pronounced when dealing with large batch volumes.

Ultrafiltration has been demonstrated to be an attractive alternative to centrifugation for cell harvesting. The unique membrane structure rejects bacteria at the filter surface and prevents clogging. Highly porous microfilters permit penetration of bacteria into the filter matrix[1,2]; this can lead to irreversible clogging even when used in a cross-flow arrangement. Clogging problems can also be experienced in devices with separator screens,[3] particularly at high solids levels (see FIG. 2).

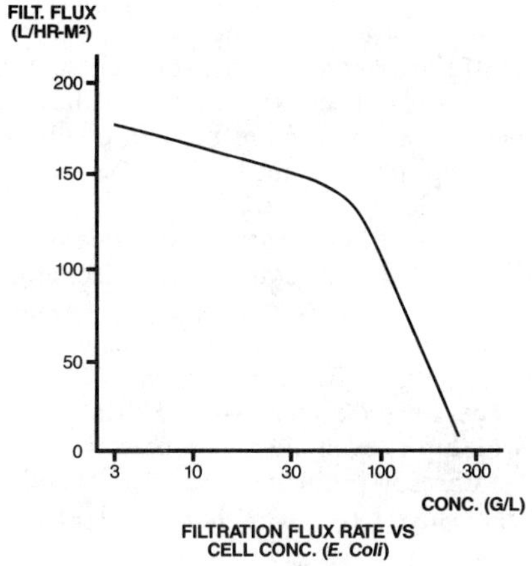

FIGURE 11. Filtration flux rate as a function of cell concentration (*E. coli*).

Hollow fiber devices have been particularly effective in concentration of bacteria and cells due to their high resistance to clogging. *E. coli* suspensions have been concentrated to small volumes, achieving very high cell densities—over 80–90% by volume or 200–250 g/l.[4] Other microorganisms have included: *B. subtilus, Brucella, Pseudomonas, Streptomyces,* yeasts, molds, and algae.[5]

Ultrafiltration of such cells behaves differently than a typical protein solution in terms of flux/concentration profiles. Most bacteria suspensions will exhibit relatively little decline in flux at lower cell densities (see FIG. 11). However, when the cells approach a near-packed condition, filtration rates will drop sharply. This behavior results from two factors: (1) suspension viscosity rises markedly at higher cell densities, and (2) the cells are discrete particles, not dissolved molecules. The membrane does not see a buildup of macrosolute on the surface that causes true polarization.

Cells may also be easily washed without a number of pelleting and suspension steps. Broth, medium, and extracellular products can be recovered while washing solution is used in a diafiltration mode.[6] Hollow-fiber systems have been used

TABLE 3. Concentration and Diafiltration of *E. Coli* with Initial Concentration

	Cell Volume (l)	Filtration Rate (l/hr)	Average Rate (l/hr)	Permeate Volume (l)	Filtration Time (hr)
1) Initial	1000	920			
			490	600	1.2
2) Concentration 2.5×	400	210			
			250	800	3.2
3) Diafiltration 2×	400	290			
			Totals:	1400	4.4

successfully in several such applications, including the recovery of interferon from cell suspensions.[7]

In cell washing processes, the choice of concentration at which the diafiltration is performed is critical. If cell density exceeds the critical point where rapid flux decay takes place, the overall process time can be much longer. TABLES 3 and 4 show the impact of such process variations. TABLE 3 gives data where cell washing is carried out at the final cell concentration; TABLE 4 shows that initial diafiltration allows processing in about 60% of the time.

If the product is extracellular, such as an antibiotic, minimizing process time may not be the only consideration. As shown in the tables, initial washing is the more rapid method but uses more permeate volume. This would further dilute an extracellular product and may not be desirable; this choice would depend on the cost of recovering this product.

Use of ultrafiltration can solve many of the problems associated with centrifugation. First of all, equivalent processing rates can be achieved with a fraction of the capital expense. In one case (see TABLE 5), a UF system gave equal rates for less than one-third the cost.[8] Systems have been easily scaled up to allow the processing of thousands of liters at rates of 20 liters/minute. Such systems have an initial cost of about $35,000 compared to almost $100,000 for a comparably sized centrifuge. Scale-up to even higher process rates is currently being carried out.

Aerosol problems can also be eliminated with ultrafiltration.[9,10] Biohazardous organisms have been harvested and contained in reservoirs with sterile air vent filters.[8]

High recovery of cells is also simple with hollow-fiber membranes. Bacterial aggregation can cause high losses with continuous centrifuges; in one case, over 30% cell loss was observed (see TABLE 6). A hollow-fiber system allowed recovery of over 98% with a 5-fold increase in process rates.[11] Other cases have been cited where 100% recovery is achieved routinely.[12]

TABLE 4. Concentration and Diafiltration of *E. coli* with Initial Diafiltration

	Cell Volume (l)	Filtration Rate (l/hr)	Average Rate (l/hr)	Permeate Volume (l)	Filtration Time (hr)
1) Initial	1000	920			
			1090	2000	1.8
2) Diafiltration 2×	1000	1250			
			690	600	0.9
3) Concentration 2.5×	400	310			
			Totals:	2600	2.7

TABLE 5. Comparison of Ultrafiltration and Continuous Centrifuge Showing Reduction of System Cost[a]

	UF	Centrifuge
Initial Concentration	← 1×10^7 cells/ml →	
Concentration Factor	100×	500×
Recovery	98%	90%
Process Rate	40 l/hr	30 l/hr
System Cost	$6,000	$20,000

[a]See Reference 8.

PRODUCT CONCENTRATION/DESALTING

Aside from bacteria concentration, ultrafiltration has been important in concentration and diafiltration of many other biological products, such as: interferon, insulin, lymphokines, peptides, plasma proteins, growth hormones, monoclonal antibodies, and enzymes.

Although most data are proprietary, some information is available, particularly for interferon purification. In one such case, a small hollow-fiber system was used to concentrate 10–20 liter batches to 100 mls, or 100–200×. The average rate of filtration was 1–1.2 liters/hour. After removing the concentrated product, a backflush of 100 mls of filtrate was carried out and the resulting fluid was pooled with the concentrate. In this fashion, over 98% recovery of interferon activity was achieved. A cleaning with sodium hydroxide gave effective restoration of flow rates and removal of residual protein. In fact, the system was used for several different types of interferon with no detectable cross-contamination.[13]

Protein adsorption to the membrane can present a problem in product recovery, particularly with very dilute solutions. In most cases, adsorbed product can be recovered by backflushing with buffers or similar washing techniques. Pretreatment of the membrane with a protein such as albumin has also been effective. In these situations, process trials on a small scale are critical in determining optimum membrane type and procedure.

PURIFICATION

Major product purification applications of ultrafiltration are: pyrogen removal, deproteinization of growth medium, and cell debris removal. As mentioned previously, in purification the filtrate contains the product of interest.

TABLE 6. Comparison of Ultrafiltration and Continuous Centrifuge Showing Improved Product Recovery[a]

	UF	Centrifuge
Initial Concentration	← 4×10^9 cells/ml (0.7 vol %) →	
Concentration Factor	10×	10×
Recovery	98%	70%
Process Rate	30 l/hr	6 l/hr
System Cost	$9,000	$8,000

[a]See Reference 11.

Products of fermentation, such as antibiotics, often contain pyrogen levels unacceptably high for injectable drugs. Pyrogens are generally lipopolysaccharides derived from fragments of bacterial cell walls.

Removal of pyrogens has been effectively carried out by ultrafiltration with successful purification of water as well as low-molecular-weight parenteral products.[14,15] Pyrogens have been filtered to below detectable levels from concentrations as high as 10,000 ng/ml.[16]

Intracellular products of cell culture must also be free of cellular debris for further purification. The same diafiltration techniques used in cell washing may be applied to solutions containing lysed cell fragments. Cell debris has been successfully removed from such products as interferon with hollow-fiber filters; yields were shown to be 80–100%.[17]

Fermentation broths frequently incorporate hydrolysates of soy, casein, and other proteins. Such broth contains high-molecular-weight contaminants before fermentation.[18] This simplifies later product recovery steps. The filtration step also removes bacteria from the broth before use.[19]

CONTINUOUS FERMENTATION

Batch fermentation has been the traditional means for growth of bacteria. Research on the use of membranes in small-scale dialysis fermentation systems shows substantial increases in cell mass and total product yield.[20,21] Continuous removal of inhibitory by-products allows growth to continue for weeks.

Hollow-fiber membranes offer the same advantage as dialysis fermentation but allow easy use in an industrial enviroment, (see FIG. 12). They have proven successful in increasing total cell densities to substantially higher levels than those found in normal batch fermentations—up to 10^{15} cells/liter in certain cases.[4] These bacteria can be withdrawn continuously to extract intracellular products or to test cell viability. Where extracellular products are of interest, filtrates are removed for further purification.

The use of hollow-fiber ultrafiltration cartridges in a high-solid fermentation system has been reported.[22] The membrane cartridge was connected to a fermenter to remove waste effluent, while fresh substrate was added to the broth. Production efficiency was increased by 100%. It was also determined that at high cell density, lactic acid biomass production was controlled by substrate limitations rather than by product inhibition.

CONTINUOUS CELL CULTURE

Large-scale growth of animal or plant cells in culture can be difficult with conventional techniques. Approaches examined for suspension cultures as well as anchorage-dependant cell growth include roller bottles, microcarrier beads, stacked-plate systems, and multiple tubes.

Hollow-fiber cartridges have been evaluated extensively for growth of animal cell lines. Cells grow to tissue-like densities as opposed to monolayers in roller bottles and on microcarriers[23] (see FIG. 12). Typical densities are 10^6–10^7 cell/cm^2 of fiber area, about 10 times higher than those achieved in roller bottles. Cultures can be maintained for as long as months at a time and have been scaled up for production purposes with specialized cartridges and systems.

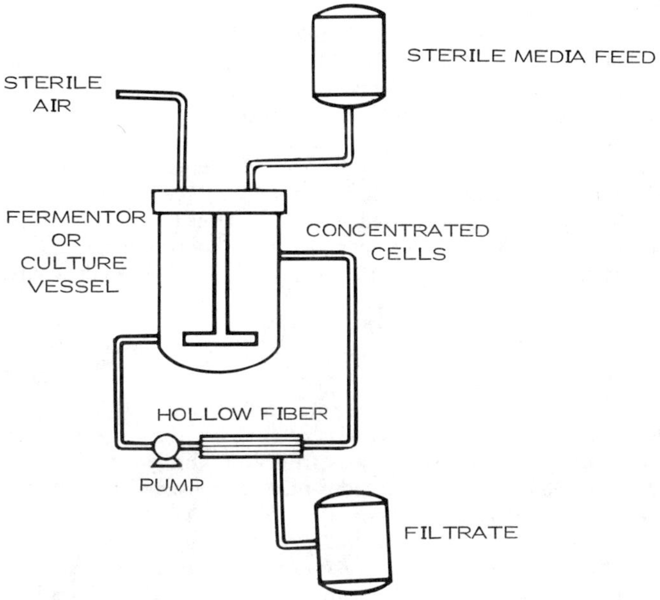

CONTINUOUS CULTURE/FERMENTATION

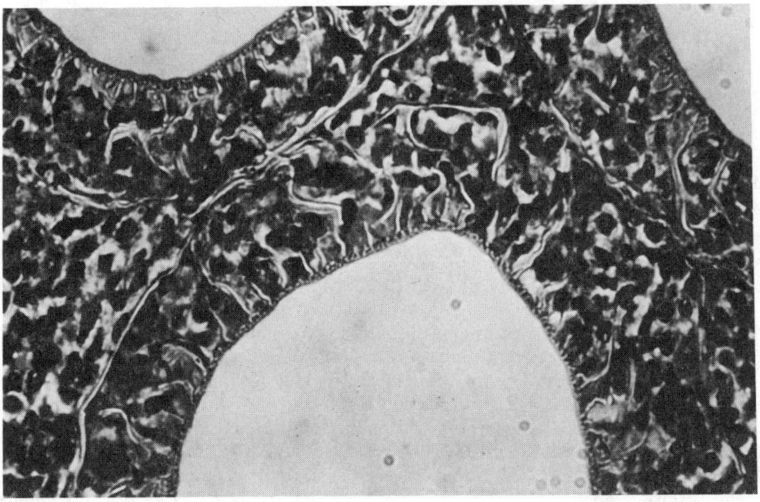

CELLS GROWING ON HOLLOW FIBERS

FIGURE 12. Continuous culture/fermentation operation (*above*) and photograph of cells growing on hollow fibers (*below*).

Monoclonal Antibodies

Researchers report the use of hollow fibers to grow hybridomas producing monoclonal antibody.[24] Mouse hybridomas were constructed by fusing myeloma cells with spleen cells from mice immunized with β-galactosidase. Hollow fibers with 50,000 MW cutoff were inoculated with the hybridoma cells. Solid masses of cells were observed to fill the interstices of the fibers.

Daily output of monoclonal antibody was equivalent to a suspension culture 50 times larger in volume and containing about 5×10^5 cells/ml. Weekly analysis showed that antibody production increased with time. After two days of culture, the antibody concentration was 33 times higher than the control medium. After 29 days of continuous culture, the concentration increased to 136 times the control level.

Interferon Production

A novel method for the production of interferon has also been reported.[25] Fibroblast cells were first attached to microcarrier beads. The cells and beads were then placed in a hollow-fiber system. The fibroblasts tripled in number after 18 days and were still growing after 30 days. Cells formed a three-dimensional matrix in the hollow fibers with the beads interspersed throughout.

Interferon production was induced after nine days of culture. The output per cell was three times greater than that of monolayer cultures. This production level was also maintained ten times longer. It resulted in a total yield 30 times higher than that of monolayer cells. In addition, the cells could be restimulated to produce interferon again at the same high rate. It was concluded that interferon production was extended in the hollow-fiber system due to removal of inhibitors through the membrane.

SUMMARY

The uses of ultrafiltration membranes have been reviewed for a variety of processes used in the emerging field of biotechnology. These uses range from concentration and harvesting of cells to product isolation to production by continuous fermentation and cell culture. As biotechnology expands into production use, the importance of ultrafiltration will continue to grow in the years to come.

REFERENCES

1. Howard, G. & R. Duberstein. 1980. A case of penetration of 0.2 μm rated membrane filters by bacteria. J. Parent. Drug Assoc. **34**(2): 95.
2. Reti, A. R., *et al.* 1979. The retention mechanism of sterilizing and other submicron high efficiency filter structures. Proc. Second World Filt. Congr. pp 427.
3. Lukaszewicz, R. C., *et al.* 1981. Functionality and economics of tangential flow microfiltration. J. Parent. Sci. Tech. **35**(5): 231.
4. Michaels, A. S., *et al.* 1980. Hollow fiber bioreactor: A novel approach to continuous, immobilized whole cell biochemical synthesis. Paper presented at Second Chem. Congr. North Am. Cont. San Francisco, CA.
5. Hollow Fibers for fermentation and cell culture. 1982. Amicon Publication. **402:** 1,4.
6. Maizell, A. L., *et al.* 1981. Human T lymphocyte/monocyte interaction in response to lectin. J. Immunol. **127**(3): 1058.
7. Van Reis, R., *et al.* 1981. Production and purification of interferon from human

leukocytes/sendai virus using filtration technology. Abstr. First Eng. Fdn. Conf. on Advances in Fermentation Recovery Process Technology. Banff, Alberta, Canada.
8. D'AGOSTINO, R. 1982. Private communication. University Micro Reference Lab. Ann Arbor, MI.
9. TAMURA, T. & T. TAKANO. 1978. A new, rapid procedure for the concentration of C-type viruses from large quantities of culture media. J. Gen. Virol. **41:** 135.
10. TRUDEL, M. & P. PAYMENT. 1980. Concentration and purification of rubella virus hemagglutinin by hollow fiber ultrafiltration. Can. J. Microbiol. **26:** 1334.
11. LAWRIE, J. 1982. Private communication. Codon, Inc. Brisbane, CA.
12. HANISCH, W. H., *et al.* 1982. Separation and purification techniques applicable to the biological processes. Aust. Biotech. Conf. pp. 153–170.
13. TAYLOR, F. 1981. Private communication. Lee Biomolecular, Inc. La Jolla, CA.
14. MCGREGOR, W. C., *et al.* 1978. Stabilized thymosin composition and method. U.S. Patent 4,082,737.
15. HENDERSON, L. W. & E. BEANS. 1978. Successful production of sterile pyrogen-free electrolyte solution by ultrafiltration. Kidney Int. **14:** 522.
16. TUTUNJIAN, R. S. 1982. Pyrogen Removal by Ultrafiltration, Endotoxins and Their Detection with the LAL Test. Alan R. Liss Publishers. New York. pp. 319.
17. LEUTHARD, P. & A. R. SCHUERCH. 1980. A simple and rapid method for concentration of interferon and removal of concentrated inducing virus. Experimentia. **36**(12): 1447.
18. DEESLIE, W. D. & M. CHERYAN. 1981. A CSTR-hollow fiber system for continuous hydrolysis of proteins. Biotechnol. Bioeng. **23:** 2257.
19. COX, J. C. 1975. New method for the large-scale preparation of diphtheria toxoid. Appl. Microbiol. **29**(4): 464.
20. ABBOTT, B. S. & P. GERHARDT. 1970. Dialysis fermentation I. enhanced production of salicylic acid from naphthalene by *P. fluorescens*. Biotechnol. Bioeng. **12:** 577.
21. FRIEDMAN, M. R. & E. L. GADEN. 1970. Growth and acid production by *L. delbrueckii* in a dialysis culture system. Biotechnol. Bioeng. **12:** 961.
22. CHANG, W. T. H. & J. J. FURJANIC. 1981. Production of biomass with hollow fiber ultrafiltration fermentation system. SIM News. July. pp. 24.
23. KU, K., *et al.* 1981. Development of a hollow fiber system for large-scale culture of mammalian cells. Biotechnol. Bioeng. **23:** 79.
24. CALABRESI, P., *et al.* 1981. Monoclonal antibody production in artificial capillary cultures. Proc. AACR ASCO. pp. 302.
25. STRAND, J. M., *et al.* 1982. Human fibroblast interferon production in a matrix perfusion microcarrier bead system. Proc. ASM 82nd Annual Meeting. Paper No. T20.

Reactor Design for Protein Precipitation and Its Effect on Centrifugal Separation

M. HOARE AND P. DUNNILL

Department of Chemical and Biochemical Engineering
University College
London, England

D. J. BELL

DSIR
Private Bag
Petone, New Zealand

INTRODUCTION

The systems used in the past for protein precipitation and centrifugal recovery of precipitates have relied heavily on technological know-how with little engineering science input. Increased emphasis on efficient processes for the recovery of high-value or large-volume protein products demands a more fundamental understanding of the processes involved.[1]

The range of important proteins is wide, embracing vegetable, animal, and microbial enzymes, food proteins, human and animal blood plasma proteins, and bacterial and viral protein antigens. Recently, mammalian proteins have been prepared from microbial sources after recombinant DNA manipulations and must be recovered. Each material will pose particular practical problems especially with regard to the presence of nonprotein contaminants such as nucleic acids, fats, and cellular debris. However, it seems likely that this will give rise to variations of detail rather than radical differences in the character of the precipitation process.

A general flowsheet, (FIG. 1), can be drawn for protein precipitation. The first stage involves the addition of a precipitating agent under conditions that do not damage the protein. At the same time, particles of protein precipitate should be produced that can be readily formed into large aggregates by control of the growth processes in the second stage. The other objective in this aging stage is to produce aggregates that are resistant to shear disruption. This process of aging is important in that significant improvements in separation can be achieved by design to aggregate fine particles that are difficult to recover and to optimize the density and strength of the aggregates. The final stage involves separation of the precipitated protein phase and this is usually achieved by centrifugation.

In most cases the sticky, gelatinous nature of protein solids precludes the use of filtration mechanisms and contamination by filter aids is normally unacceptable. The design of the process used to transfer the suspension from the precipitation and aging vessels can be critical if the aggregates are susceptible to shear disruption. Similarly the design of the centrifuge feed zone may have an overriding influence on the size distribution of the feed, and so affect separation efficiency.

Apart from the mechanical operating problems associated with the processing of food-quality or potentially hazardous biological materials, the main difficulties in the separation step are connected with the physicochemical properties of the protein. The amphipathic nature of proteins can lead to foaming and air entrapment resulting in low effective settling velocities, as well as protein denaturation due to the presence of large air/liquid interfaces. Denaturation may also occur as a result of effects associated with solid/liquid interfaces. Such effects are limited to the soluble protein and depend on the concentration and sensitivity of the particular protein.[2,3]

The present study considers the influence of the physical characteristics of protein precipitates on the centrifugal separation performance. This requires a protein source for which fundamental biochemical information is available. Another important consideration is the ability to readily prepare pilot-plant-scale quantities of protein solution. Soy protein has been chosen as meeting these requirements and also is of major importance as a food protein.

The operations of protein precipitation, aging, and centrifugal recovery are considered in order to define the interrelated factors that influence the overall process efficiency. Physicochemical aspects of protein precipitation are examined first, leading to the choice of precipitation agent studied. Factors governing the properties of protein precipitates that affect centrifugal recovery are then analyzed before a description of results with industrial centrifuges.

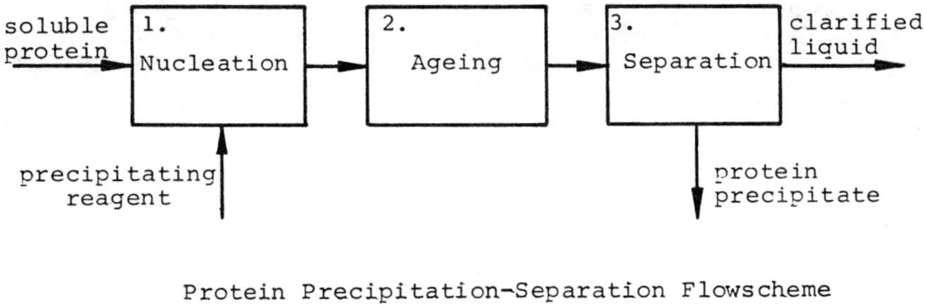

FIGURE 1. Protein precipitation–separation flow scheme.

PHYSICOCHEMICAL ASPECTS OF PROTEIN PRECIPITATION

Protein precipitation involves a reduction in protein solubility by either changing the nature of the solvent environment or by direct interaction of a reagent with the protein. Methods that affect the solvent environment include:

(a) "salting-out" by the addition of high concentrations of neutral salts, e.g. ammonium sulfate; and
(b) reduction in the solvent dielectric constant by the addition of miscible organic solvents, e.g., ethanol, acetone.

Methods involving direct interaction with the protein include:

(a) isoelectric point precipitation by pH adjustment. Solubility is a function of the net charge on the weakly acidic and basic amino-acid side chains of proteins. The ionization of these groups is strongly influenced by pH;
(b) addition of charged polyelectrolytes that act as flocculating agents under appropriate pH conditions, e.g. carboxymethyl-cellulose; and
(c) polyvalent metal ions, e.g., Ca^{2+}, Ba^{2+}.

In some instances, reagents used to precipitate proteins can damage them. This can be the case for alcohols,[4] acids,[5] and ionic polyelectrolytes.[6] It is then necessary to establish the extent to which the mixing conditions must be geared to reducing damage rather than favoring precipitate particle recovery characteristics. To be able to examine this problem, and because mineral acids are a cheap and widely used reagent for soy protein recovery, isoelectric precipitation from a total water extract was chosen for detailed study. It was found that the extent of protein damage at low levels of mixing was dependent on the acid anion. For hydrochloric acid, considerable denaturation occurs (FIG. 2). The degree of damage varies inversely with the Hofmeister series,[5] that is, sulphate < nitrate < phosphate < chloride and derives directly from the destabilizing effect of some anions. No damage was observed with the use of sulfuric acid.

FORMATION AND GROWTH OF PROTEIN PRECIPITATES

The processes of protein precipitate nucleation and the subsequent aggregate growth are distinct and controlled by different mechanisms. Nucleation involves removal or reduction of any barrier that will allow molecular association to proceed. The rate of molecular association will be controlled by the diffusivity of the colliding species. Smoluchowski[7] described the rate of decrease of particle number concentration, N, due to diffusion-controlled collisions for a monosized dispersion, as a second-order process:

$$-\frac{dN}{dt} = KN^2 \tag{1}$$

where

$$K = 8\pi Dd \tag{2}$$

and D is the particle diffusivity. Parker and Dalgleish[8] found that once the stabilizing forces had been removed, the growth of calcium-precipitated bovine α_s-casein followed this theory.

Any remaining protective electrical barrier around the molecules will cause a reduction in the rate of aggregation. Corrections must also be made for particle

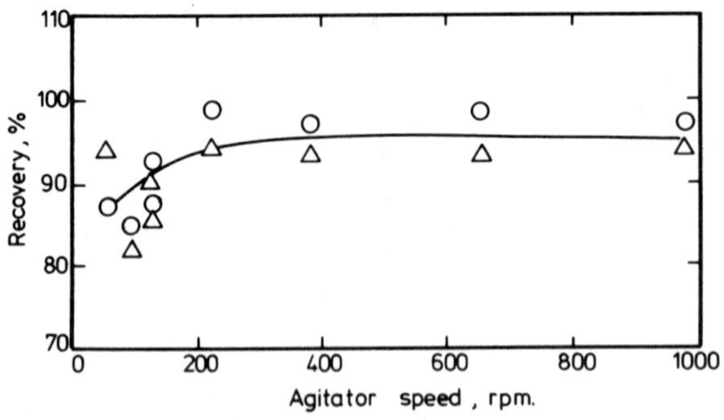

FIGURE 2. The effect of mixing speed during isoelectric precipitation of soy protein with hydrochloric acid upon the percentage recovery of unchanged protein.[5] Measurements made using polyacrylamide gel electrophoresis in the presence of mercaptoethanol. Soy protein constituents studied are: O, glycinin; △, conglycinin.

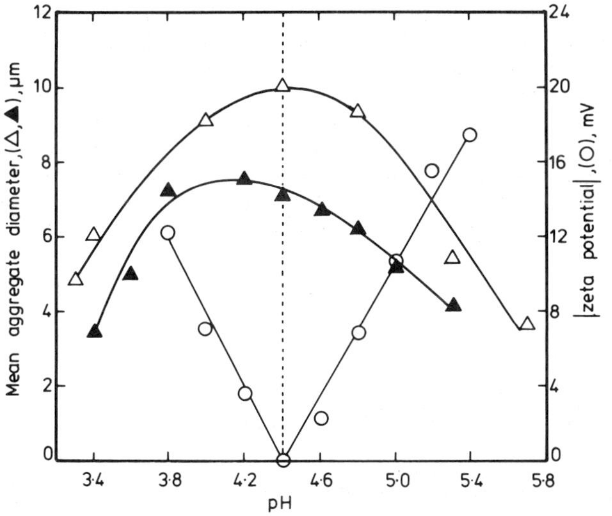

FIGURE 3. The effect of the pH of precipitation of soy protein on the mean aggregate diameter.[14] Preparation conditions: △, continuous tubular reactor, aging time 26 sec; ▲, batch turbine stirred tank reactor, aging time = 1800 sec. Absolute value of zeta potential, O; construction line for zero zeta potential, --------. Positive values of zeta potential at pH <4.4, negative values at pH >4.4.

nonsphericity and for the hydrodynamic removal of free water as the particles approach. However, application of this theory and the modifications of it do not account for the presence of a hydration barrier of associated water nor for the effect of a varying surface potential as the molecules approach. The rate of association may depend on the rate of removal of hydration and electrical barriers, and the interaction of protein particles may also be orientation specific.

The time for first appearance of precipitate particles will depend on the rates of nucleation and diffusion-controlled growth. This can vary from a few seconds for the isoelectric precipitation of casein,[9] to as high as several minutes for the ammonium sulfate salting-out of fumerase[10] and the ethanol precipitation of catalase.[4]

Following nucleation and an initial stage of diffusion-controlled growth, the influence of hydrodynamic shear in the aging vessel will cause particles larger than a given size to collide. These orthokinetic collisions promote further growth as well as aggregate breakup. The growth process can be described for a suspension of uniformly sized, spherical particles of diameter, *d*, in a uniform shear field by:

$$\frac{-dN}{dt} = \frac{-2}{3} \alpha N^2 G d^3 \tag{3}$$

where α is a collision effectiveness factor, and G is shear rate (or mean velocity gradient in nonuniform shear).[11] As with perikinetic growth, the validity of Equation 3 depends on the influence of factors such as orientation effects and interaction barriers. Van de Ven and Mason[12] have shown the effectiveness of collisions to be dependent on the hydrodynamics of flow around the approaching particles with α proportional to $G^{-0.18}$ and $d^{-0.73}$.

The aggregate size of a protein precipitate at a given shear rate will be dependent on the interparticle forces, which are related to the surface charge of the particle. Riddick[13] has reported that for colloids, the maximum aggregate size occurs either at the isoelectric point or at small positive values of the zeta potential. This is observed[14] for soy protein precipitated in the region of its isoelectric point by acid (FIG. 3). Values of the pH corresponding to the maximum aggregate size differ by approximately 0.3 pH units for the precipitates prepared in batch and continuous reactors. The shapes of the pH/size profile are also different. These variations are probably due to the method of acid addition and the local pH environment of the protein. A continuous reactor is characterized by the addition of protein solution to a region of constant pH. In this

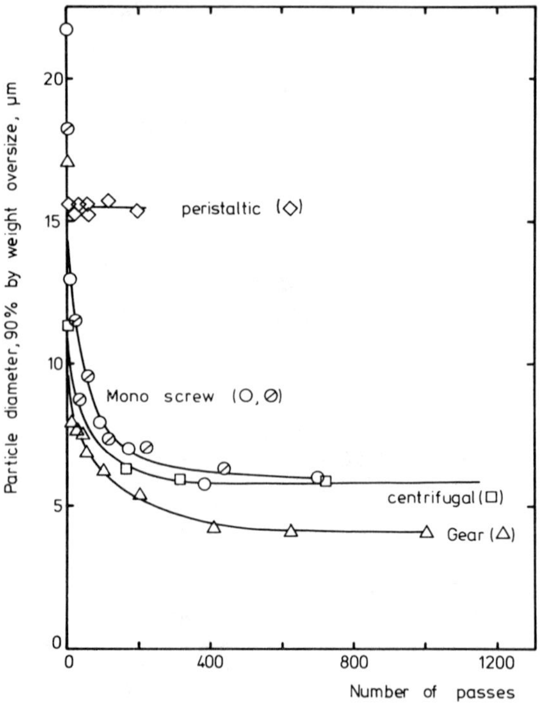

FIGURE 4. Particle size change of isoelectric soy protein precipitate after flow through various pumps under total recycle.[16] Total protein concentration = 2.5% by weight.

case, the protein moves immediately to the final condition of pH so that all the particles making up the aggregates are exposed to the same precipitation conditions. Protein prepared in a batch vessel is exposed to a gradual change in pH. This means that on adjusting the pH to below the isoelectric point, the aggregates have already experienced the conditions of maximum attraction associated with the isoelectric point.

SHEAR BREAKUP OF PROTEIN PRECIPITATES

Breakup of flocs or aggregates during exposure to shear may proceed by several mechanisms. These include bulgy deformation due to pressure gradients across the aggregate and fragmentation or primary particle erosion due to hydrodynamic shear or particle–particle or particle–surface collisions.[15] Isoelectric soy precipitate aggregates exposed to a laminar Couette shear rate of $2000 s^{-1}$ give an initially rapid decrease in aggregate size due to the removal of large fragments.[16] The fragment size decreases during exposure to shear, and it is probable that this process proceeds until almost spherical aggregates are obtained, further breakup being due to the erosion of primary particles of 1 μm diameter. The fragment size is a very weak function of the shear rate, whereas the rate of aggregate breakdown is strongly dependent on shear rate.[17] Continued exposure to shear for up to 50 hr does not show attainment of an equilibrium value expected from a balance between orthokinetic aggregation and shear breakup, indicating the partly irreversible nature of precipitate breakup.

Shear-controlled breakup of precipitate aggregates proceeds by particle–surface or particle–particle collisions.[18] The latter is determined by Smoluchowski's theory for orthokinetic collisions (Equation 3), α now being an effectiveness factor for collisions promoting shear breakup. Precipitate exposed for short times to very high shear rates in laminar capillary flow shows the concentration and shear-rate dependence expected from Equation 3.[17]

The transport of protein precipitate using pumps can cause considerable aggregate breakup (FIG. 4).[16] However, as is emphasized by the peristaltic pump data, it is possible in some cases to design transport systems that do not cause significant shear damage of the protein precipitates.

A most important factor determining protein precipitate aggregate strength is the history of preparation. Precipitate exposed to sufficient deformation forces will develop a compact structure that is more resistant to shear breakup. This improved aggregate strength is attributed to an infilling/rearrangement mechanism resulting in a closer, more stable packing.[17] Optimum strength of soy protein precipitate aggregates is achieved when the product of mean velocity gradient and aging time in a batch, turbine-stirred vessel exceeds 10^5.

REACTOR DESIGN FOR PRECIPITATION AND AGING

The basic reactor configurations used for protein precipitation are the batch tank and, for continuous operation, the stirred-tank reactor and the tubular or plug-flow reactor. The initial design problem is to establish the requirements of contacting of reactants for nucleation and whether these conflict with the optimum conditions of particle growth. In the case of fast reactions, the major design consideration for the nucleation process is to achieve good mixing to ensure a uniform distribution of precipitating reagent and for efficient heat transfer in the case of exothermic reactions. Although few data exist to define the relationship between the reagent mixing conditions and the subsequent particle-size characteristics, it might be expected that, for very fast protein precipitation reactions, the rate of intermixing of reactants will affect the primary particle size and so affect the growth characteristics. Under turbulent mixing conditions, the protein solution may be thought of as being divided into minute packets of fluid defined by the turbulent microscale of mixing. The lifetime of these packets depends on the diffusivity of the protein species and the recirculation rates of the fluid through regions of high turbulence. A low-molecular-weight precipitating agent, such as an acid, salt or organic solvent, will rapidly diffuse into these packets. The relative rates of protein precipitation and dissipation of protein from the packets will determine the size of primary protein precipitate particles formed by this mechanism. If the rate of precipitation is very fast with respect to the lifetime of the turbulent eddies, then the basic particle size will be determined by the eddy size and the protein concentration. On the other hand, the rate of diffusion of the precipitating agent to the protein surface or the rate of precipitation may be sufficiently slow that individual groups of protein molecules are insolubilized. These then aggregate readily at a rate controlled by perikinetic and subsequently orthokinetic collisions. The isoelectric point precipitation of soy protein[19] in turbulent mixing vessels, shows the formation of basic particles of approximately 1 μm diameter (FIG. 5).

Having formed the primary particles, the objectives of any subsequent treatment are to form large, dense, stable aggregates that can be readily recovered by centrifugation. The growth of aggregates (larger than about 0.1–1 μm) is a shear-controlled process with the history of preparation influencing their size, overall shape, and degree

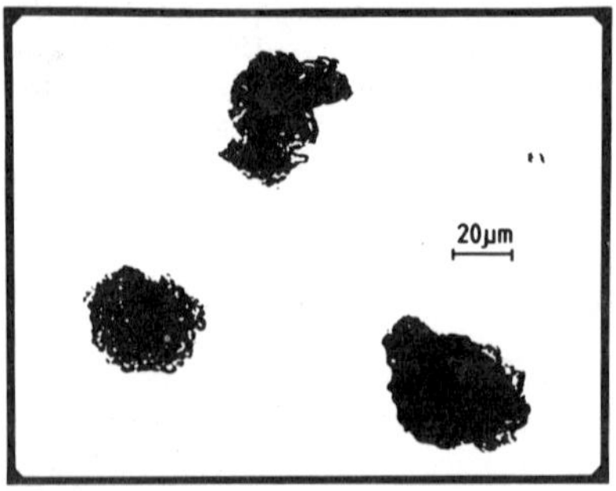

a) Batch stirred tank precipitate

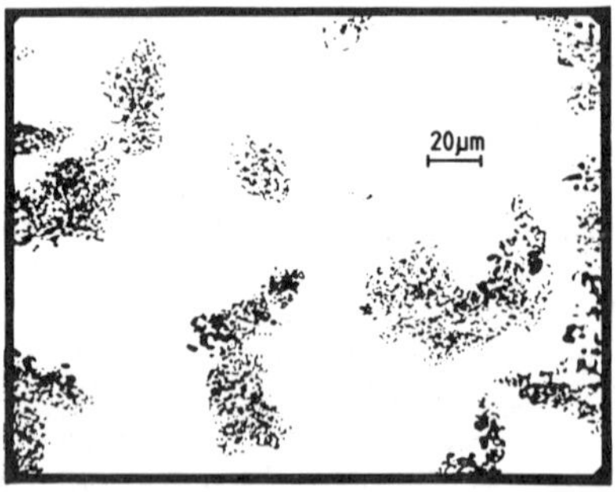

b) Tubular reactor precipitate

FIGURE 5. Photomicrographs of isoelectric soy protein precipitate prepared in (a) batch turbine stirred tank reactor, and (b) continuous tubular reactor.[21]

of compactness. Even though the precise growth kinetics have not been defined for protein systems, the manipulation of concentration and shear rate has been shown to result in an advantageous variation in the particle size distribution for isoelectric soy protein precipitates in batch and continuous tubular reactors.[20] The growth of soy protein isoelectric precipitate in a tubular reactor shows increased rates of growth at higher mean velocity gradients and greater protein concentrations (FIG. 6). Both of these results would be expected from orthokinetic-controlled aggregation. However, precipitate prepared at a higher mean velocity gradient tends to a smaller final particle size due to the greater rates of shear-induced breakup. Turbulence promotors incorporated at intervals along the continuous tubular reactor increase the rate of collision between aggregates and give greater precipitate growth rates (FIG. 7).

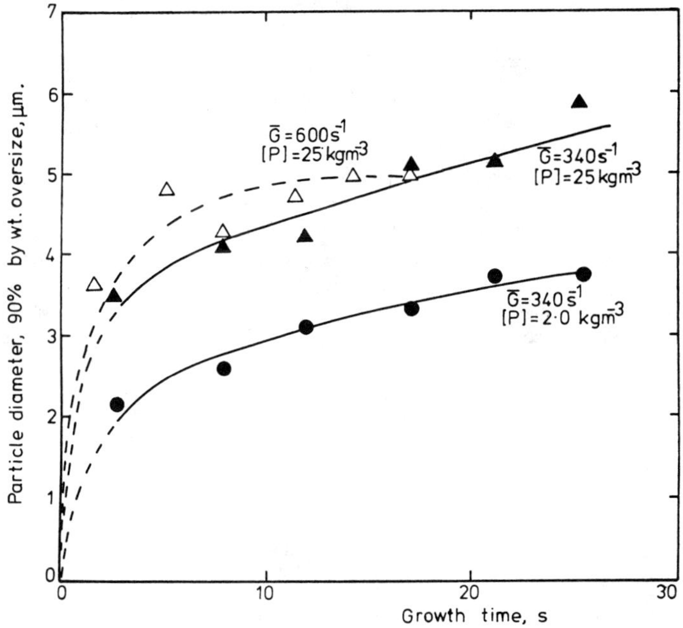

FIGURE 6. Preparation of isoelectric soy protein precipitate in a continuous tubular reactor.[20] Operating conditions as shown in figure; $\overline{G}$ is the mean velocity gradient; $[P]$ is total protein concentration.

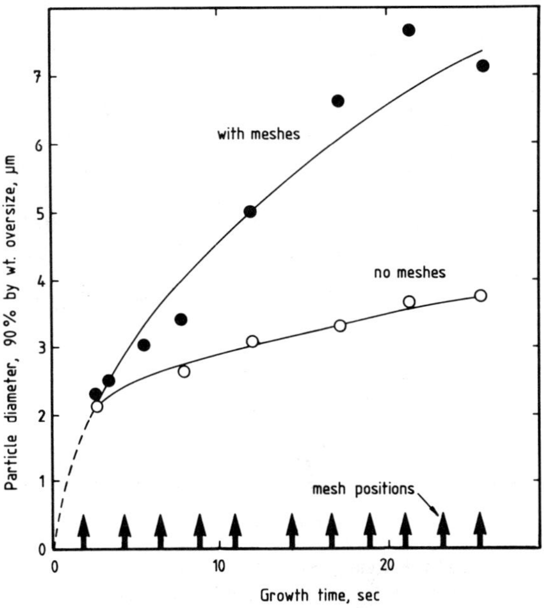

FIGURE 7. Effect of turbulence promotors on isoelectric soy protein precipitation in a continuous tubular reactor.[20] Protein concentration = 2.0 kg m^{-3}; average fluid velocity = 0.47 ms^{-1}.

The effect of shear-controlled breakup on the size of precipitate aggregate is most clearly shown for the long-residence-time batch reactor (FIG. 8).[16] Batch reactors may be scaled up on the basis of the same mean velocity gradient to produce precipitates with similar d_{90} values (FIG. 8).

Comparison of the strength of particles prepared in a batch reactor and a continuous tubular reactor shows significant differences.[21] FIGURE 9 gives the volume of fine particles produced as a result of subjecting a sample to a capillary rate of shear of 9×10^4 sec^{-1} for varying times. The continuous tubular reactor precipitate exhibits significantly less resistance to disruption with the volume of fine particles increasing from about 2.5% to over 25% for shear exposure times greater than about 0.04 sec. Subsequent aging of the tubular-reactor-prepared precipitate in a batch tank only

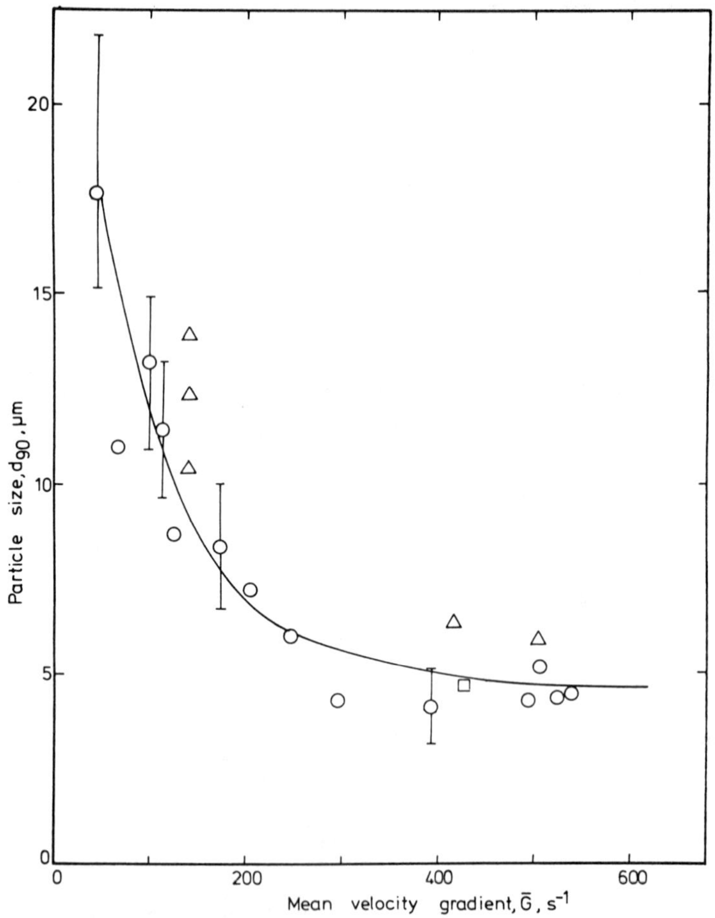

FIGURE 8. The relationship between aggregate size and mean velocity gradient (based on power dissipation per unit volume of reactor) for the preparation of isoelectric soy protein precipitate in batch-stirred tanks.[16] Bars indicate range of experimental results. Protein concentration: 30 kg m^{-3}; vessel volumes: △, 0.27 l; ○, 0.67 l; □, 200 l. Aging time: 600 sec.

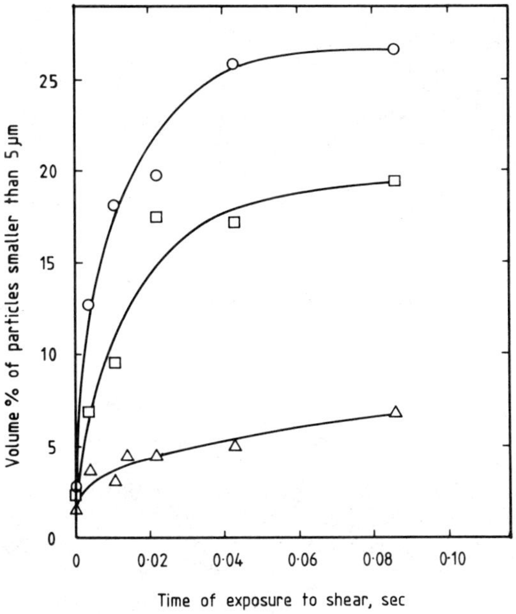

FIGURE 9. The effect of preparation conditions on the stability of precipitate particles to short-duration shear.[21] Total protein concentration: 25 kg m^{-3}; average rate of capillary shear: 9×10^4 sec^{-1}. Preparation conditions of precipitate: ○, tubular reactor, $\overline{G} = 340$ sec^{-1}, $t = 25$ sec; □, tubular reactor plus 2400 sec in batch-stirred vessel, $\overline{G} = 200$ sec^{-1}; △, batch-stirred tank reactor, $\overline{G} = 200$ sec^{-1}, $t = 1800$ sec.

resulted in a small improvement in the aggregate strength. This shows that the precise conditions of mixing during precipitate preparation also strongly influence the final aggregate strength.

DENSITY OF AGGREGATED PROTEIN PRECIPITATES

The design of protein precipitate recovery processes requires a knowledge of the effective settling density of the solid phase. Since the growth of protein precipitates involves aggregation, the density of an aggregate will be determined by its fluid voidage and the density of the protein solid phase.

The measured density of isoelectric soy protein precipitate was found to depend on the method of analysis. Gradient centrifugation in cesium chloride and sucrose/cesium chloride solutions indicated no variation of buoyant density with increasing aggregate size. This is consistent with an aggregate structure consisting of primary particles. Methods involving determination of the aggregate volume by Coulter counter and comparison of Coulter counter and centrifugal photosedimentometer size distributions indicated the existence of a density–size relationship of the form:[19]

$$\rho_a = 1004 + 246\, d^{-0.408} \qquad (4)$$

where ρ_a is the aggregate density (kg m^{-3}) and d its diameter (μm). The basic particle buoyant density is approximately 1296 kg m^{-3}.

The existence of a density–size relationship is important in the consideration of solid–liquid separation by centrifugation since the calculation of separation limits requires a knowledge of the density difference between the aggregate and the bulk fluid. For isoelectric soy protein precipitate, the use of a constant density based on the buoyant density overestimates the Stokes settling velocity by up to five times. Using a constant density difference based on the density of large aggregates will give a safe design value for predicting centrifuge performance.[19]

RHEOLOGY OF PRECIPITATE SUSPENSIONS AND SEDIMENTS[21]

The flow behavior or rheology of a concentrated precipitate suspension will be indicative of the aggregate structural properties and the interaction between aggregate particles. Concentrated suspensions of isoelectric soy precipitate (118 and 202 kg m^{-3}) exhibit thixotropic and rheopectic behavior. At low shear rates (up to about 50 sec^{-1}), there is a sharp decrease in apparent viscosity with increasing shear rates. During shear cycling, a hysterisis effect is noted with lower apparent viscosities on the decreasing shear cycle until at shear rates below about 8 sec^{-1}, higher apparent viscosities are observed. This rheological behavior indicates the formation of a floc-type network at low shear rates that readily disperses on increasing shear rate. This can be physically observed in gravity sedimentation and low-shear tube flow where flocs of the order of 1 mm diameter are temporarily formed.

These rheological characteristics are significant with respect to the recovery of precipitate suspensions in continuous centrifuges. The high viscosity and rheopectic behavior at low rates of shear could create difficulties in the dewatering of the sediment and its flow properties will be important in relation to its discharge from the centrifuge. For example, the sediment is removed from intermittent-discharge disc centrifuges by opening ports at the bowl periphery. The discharging force is the centrifugal force acting on the solids, which is typically between 10^5 and 10^7 Nm^{-2}. Sediment discharge from decanter centrifuges is continuous and results from the action of the scroll, which rotates at a small differential speed with respect to the centrifuge bowl.

An investigation of the creep flow properties of sedimented precipitate can provide information on the influence of the centrifugal force and residence time on the discharge characteristics. These properties describe the solid behavior of the sediment in the initial stages of shear and subsequently the viscous or fluid behavior at longer times. The influence of the precipitation reactor configuration on the creep flow characteristics is shown in FIGURE 10 for 135 kg m^{-3} precipitate sediments prepared in a batch and a continuous tubular reactor. Both sediments show nonlinear viscoelastic properties at low applied shear stresses with a twofold increase in shear stress resulting in more than twice the deformation. Also, flow reversal is observed on removal of the applied shear stress. The sediment prepared from protein precipitated in the continuous tubular reactor required a stress approximately two orders of magnitude greater than that required to give an equivalent deformation for sediment prepared in the batch-stirred tank. This result suggests significantly greater aggregate–aggregate interactions are occurring for the tubular reactor precipitate. These interactions can be attributed to the more nonuniform shape of the tubular reactor precipitate (FIG. 5).

The relationships between creep rheology and the parameters of force, time, pH, and shear history suggest that creep-flow measurements may be used as an empirical method of predicting the performance of, for example, scroll-discharge centrifuges.

The relationship between applied shear stress and resulting deformation, defined as the sediment shear modulus, will determine if the sediment will flow freely in a particular centrifuge. Measurement of shear modulus with respect to the time of centrifugation will define the residence time limit of the centrifuge, which should not be exceeded to ensure continuous discharge of the sediment.

CENTRIFUGAL SEPARATION

The rate of particle sedimentation is defined by the balance between the centrifugal force, the Brownian diffusional force and the kinetic drag of the fluid. For an

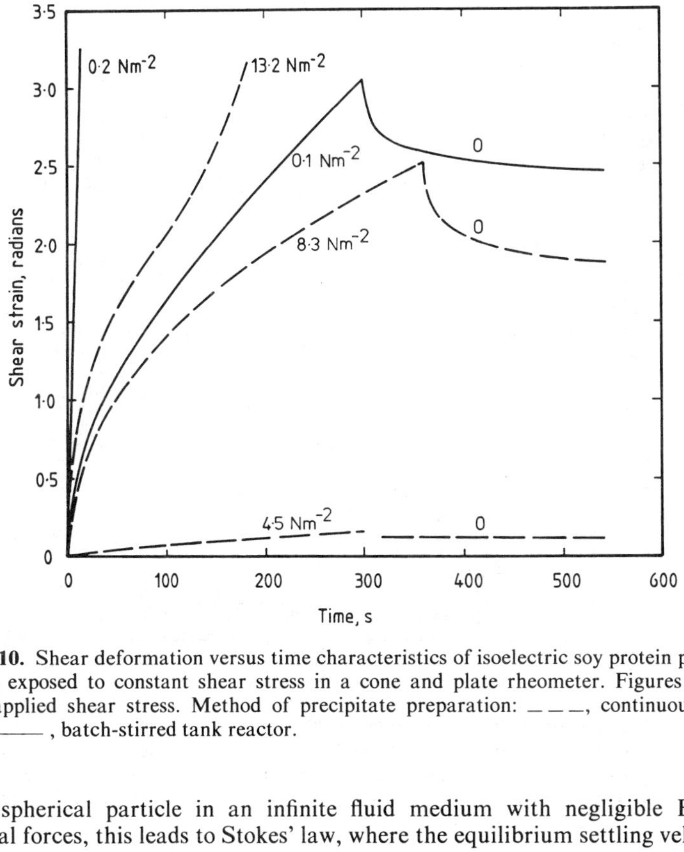

FIGURE 10. Shear deformation versus time characteristics of isoelectric soy protein precipitate sediments exposed to constant shear stress in a cone and plate rheometer. Figures on graph indicate applied shear stress. Method of precipitate preparation: — — —, continuous tubular reactor; ———, batch-stirred tank reactor.

isolated spherical particle in an infinite fluid medium with negligible Brownian diffusional forces, this leads to Stokes' law, where the equilibrium settling velocity, v_o, is given by:

$$v_o = \frac{d^2 \Delta\rho \omega^2 r}{18\mu} \quad (5)$$

where d is particle diameter, $\Delta\rho$ is density difference, μ is viscosity, $\omega^2 r$ is relative centrifugal acceleration. This description is, of course, simplistic and gives an estimate for particles with low settling Reynolds numbers, $Re_p < 0.2$. Particle nonsphericity can have a significant effect resulting in a reduced sedimentation velocity even for slight surface roughness. Another significant factor is the effect of hindered settling. Using Stokes' law as a basis for machine design, it can be seen that improved recovery can be achieved by either reducing the settling distance or increasing the sedimentation velocity.

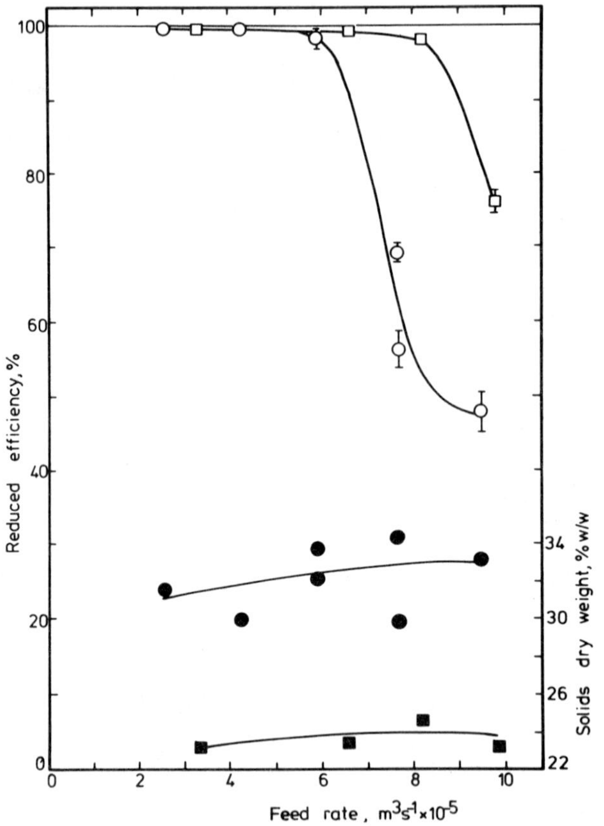

FIGURE 11. Comparison of the clarification and sludge solids content of isoelectric soy protein precipitate prepared in continuous tubular and batch-stirred tank reactors.[21] Centrifuge: scroll discharge Sharples P 600. Total protein concentration in feed, 35 kg m^{-3}; □, ■, batch-stirred tank reactor; ○, ●, continuous tubular reactor. Open symbols: efficiency; closed symbols: solids dry weight.

A reduction in settling distance is achieved by flowing in thin channels such as in disc machines, although this can result in deviations from Stokes' law due to particle–wall interactions. Increasing the sedimentation velocity can be achieved by either increasing the particle diameter or density, reducing the fluid viscosity, or by increasing the centrifugal force. This last method can be carried out by increasing the speed of rotation or the radius of settling, but is limited by the materials of construction and the requirements of mechanical strength, chemical resistance, and sanitary design.

The maximum conditions of shear in a centrifuge occur in the feed zone where the precipitate suspension is subjected to a sudden increase in its tangential and radial velocity components. In tubular bowl and scroll decanter centrifuges, the feed suspension is usually jetted into a chamber that contains axially mounted spinning vanes, and in multichamber and disc centrifuges, the suspension is accelerated down a conical annulus with vanes. Order-of-magnitude calculations indicate shear rates of 10^4 to 10^5 sec^{-1}. Studies with aged soy isoelectric precipitates, exposed to average capillary shear rates ranging from 9×10^3–9×10^4 sec^{-1} for times between 0.004 and 0.22 sec, indicate a bimodal distribution of fragment sizes caused by erosion and fragmentation. With an 0.086 sec exposure to the most severe conditions of shear, a 23% reduction in mean size was observed, which could result in a significant decrease in centrifugal efficiency.[17]

Isoelectric soy protein precipitate prepared in a continuous tubular reactor and a batch-stirred tank reactor showed no difference with respect to recovery in an intermittent-discharge disc centrifuge.[21] The relatively high density difference of approximately 120 kg m^{-3} for 5 μm particles rising to about 250 kg m^{-3} at 1 μm for this protein precipitate[19] and the large size of the feed particles (d_m = 8–18 μm) suggests that even if breakup is occurring in the feed zone, the resulting fine particles are still readily recovered. The critical particle size for this centrifuge was approximately 1 μm at the maximum flow rates studied. Operation of the discharge mechanism of the intermittent desludging centrifuge was unsatisfactory due to the long residence times. The extent of sludge dewatering was indicated by the high volume (9×10^{-2} m^3) of 7.6% v/v dispersion that could be clarified in the centrifuge (nominal sludge volume: 3.9×10^{-3} m^3) before the separation efficiency dropped below 98.5%. These results indicate that the sludge-handling characteristics are more important in separating this precipitate than the clarification, which creates no problem in this type of centrifuge.

The effect of preparation conditions on the separation efficiency and sludge dry solids content for a scroll discharge (decanter) centrifuge is shown in FIGURE 11. The batch-prepared precipitate gave approximately 30% greater throughput before the efficiency dropped below 98%. However, the solids content of the scroll-recovered material for the continuous tubular-prepared precipitate was approximately 30% higher than that for the batch precipitate. Both sludges showed a small increase in concentration with increased throughput, which is consistent with a slightly compressible sludge being dewatered by the pressure of a deeper sludge layer. The higher sludge concentration for the tubular reactor precipitate material, at the same feed concentration, indicates either a more compressive sludge or better drainage. The creep rheology experiments (FIG. 10) indicated the tubular precipitate to be more resistant to deformation compared to the batch-prepared material. This suggests a shear-induced dewatering mechanism in which liquid is effectively drained from the sedimented phase into the bulk fluid by the action of the scroll.

CONCLUSION

A number of points may now be summarized with respect to the design of reactor systems and related solid-liquid recovery equipment:

(a) The precipitating reagent should be added under mixing conditions that minimize denaturation. This may dominate all other considerations and design may need to be based on factors such as mixing efficiency, heat removal and avoidance of gas-liquid interfaces.

(b) The shear rate should be reduced as soon as possible after nucleation to promote orthokinetic growth.
(c) The aggregated precipitate should be aged at moderate shear levels for a sufficient time to maximize its strength. The rate of aggregation is improved by incorporating a distribution of shear rates such as that which exists in a stirred tank. This can be achieved also in a continuous tubular reactor by using turbulence promoters.
(d) Exposure to high rates of shear after the precipitation stage should be minimized by careful selection of pumps and design of the piping system and centrifuge feed zone.
(e) A critical diameter that corresponds to the smallest particle size that is just recovered can be used to define centrifuge performance. The mixing and aging process will then be designed to ensure, say, 95% or 99% of particles are larger than this initial size allowing also for any subsequent shear-induced breakup.
(f) The flow characteristics of the sedimented precipitate will be influenced by the aggregate size and shape. Greater aggregate–aggregate interactions will result in increasing resistance to flow and will affect the extent of solids dewatering. The dewatering process must be designed to achieve maximum solids content, while at the same time ensuring that the sediment will discharge from the centrifuge.

The design of protein precipitation and precipitate recovery processes requires information from a number of disciplines including biochemistry, colloid chemistry, and biochemical engineering. Some of the key mechanisms must still be inferred but a scientific foundation is beginning to emerge for industrial protein precipitation. It will increasingly allow laboratory tests to be translated into efficiently engineered processes. The further development of a fundamental design basis is not only important for conventional protein systems such as food proteins and microbial enzymes but will also form a foundation for the demanding purifications of mammalian proteins produced by recombinant DNA techniques in bacteria.

REFERENCES

1. Bell, D. J., M. Hoare & P. Dunnill. 1983. Advances in Biochemical Engineering. A. Fiechter, Ed. Springer Verlag. Berlin, Heidelberg. Vol. **26:** 1–72.
2. Virkar, P. D., T. J. Narendranathan, M. Hoare & P. Dunnill. 1981. Biotechnol. Bioeng. **24:** 425–429.
3. Charm, S. & B. L. Wong. 1981. Enzyme Microb. Technol. **3** (Suppl. 2): 111–118.
4. Schubert, P. F. & R. K. Finn. 1981. Biotechnol. Bioeng. **23:** 2569–2590.
5. Salt, D., P. Dunnill, R. B. Leslie & P. J. Lillford. 1982. Eur. J. Appl. Microbiol. Biotechnol. **14:** 144–148.
6. Imeson, A. P., D. A. Ledward & J. R. Mitchell. 1977. J. Sci. Food Agric. **28:** 661–668.
7. Smoluchowski, M. 1917. Z. Phys. Chem. **92:** 129–168.
8. Parker, T. G. & D. G. Dalgleish. 1977. Biopolymers. **16:** 2533–2547.
9. Southward, C. R. & R. M. Aird. 1978. N. Z. J. Dairy Sci. **13:** 77–96.
10. Foster, P. R., P. Dunnill & M. D. Lilly. 1976. Biotechnol. Bioeng. **18:** 545–580.
11. Ives, K. J. 1978. *In* The Scientific Basis of Flocculation. K. J. Ives, Ed. Sijthoff and Noordhoff. Alpen aan den Rijn. pp. 37–61.
12. Van de Ven, T. G. & S. G. Mason. 1977. Colloid and Polym. Sci. **255:** 468–479.
13. Riddick, T. M. 1968. Control of Colloid Stability Through Zeta Potential. Livingstone Publ. Co. Pennsylvania.
14. Chan, M. Y. Y., D. J. Bell & P. Dunnill. 1982. Biotechnol. Bioeng. **24:** 1897–1900.
15. Glasgow, L. A. & R. H. Luecke. 1980. Ind. Eng. Chem. Fundam. **19:** 148–156.

16. HOARE, M., T. J. NARENDRANATHAN, J. R. FLINT, D. HEYWOOD-WADDINGTON, D. J. BELL & P. DUNNILL. 1982. Ind. Eng. Chem. Fundam. **21:** 402–406.
17. BELL, D. J. & P. DUNNILL. 1982. Biotechnol. Bioeng. **24:** 1271–1285.
18. TWINEHAM, M., M. HOARE, D. J. BELL & P. DUNNILL. 1982. To be published.
19. BELL, D. J., D. HEYWOOD-WADDINGTON, M. HOARE & P. DUNNILL. 1982. Biotechnol. Bioeng. **24:** 127–141.
20. VIRKAR, P. D., M. HOARE, M. Y. Y. CHAN & P. DUNNILL. 1982. Biotechnol. Bioeng. **24:** 871–882.
21. BELL, D. J. & P. DUNNILL. 1982. Biotechnol. Bioeng. **24:** 2319–2336.

APPENDIX A

Nomenclature

d	particle diameter (m or μm)
d_m	mean particle diameter (m or μm)
d_{90}	particle diameter at 90% by weight oversize (m or μm)
D	diffusivity ($m^2\ sec^{-1}$)
G	shear rate (sec^{-1})
$\overline{G}$	mean velocity gradient (sec^{-1})
K	rate constant (sec^{-1})
N	particle number concentration (m^{-3})
$[P]$	total protein concentration (kg m^{-3})
r	radius (m)
Re_p	particle Reynolds number $= \rho v_o d/\mu$
t	time (sec)
v_o	equilibrium settling velocity (m sec^{-1})
α	collision effectiveness factor, Equation 3
$\Delta\rho$	density difference (kg m^{-3})
μ	dynamic viscosity (N sec m^{-2})
ρ	suspension density (kg m^{-3})
ρ_a	aggregate density (kg m^{-3})
ω	angular velocity (rad sec^{-1})

Recent Developments in Separation and Purification of Biomolecules

H. SCHÜTTE, K. H. KRONER, W. HUMMEL,
AND M.-R. KULA

Gesellschaft für Biotechnologische Forschung mbH.
Mascheroder Weg 1
D-3300 Braunschweig-Stöckheim,
Federal Republic of Germany

INTRODUCTION

Proteins—linear polymers with a defined sequence of amino acids—can be produced effectively only in living cells. Molecular biology and recombinant-DNA technology have provided methods and examples to improve the biosynthesis of a protein of interest in easily grown microorganisms, for example *E. coli,* to very high levels. To develop an industrial process, however, more is needed, in particular biochemical engineering for the large-scale cultivation of different microorganisms and an adequate recovery technology. Recovery processes may require 50% or more of the floor space and the initial investment of a plant. It is this that will be discussed in the following paper with emphasis on the integration of the different steps outlined in FIGURE 1. Leucine dehydrogenase from *Bacillus* strains is used as an example for the isolation and purification of an intracellular, biologically active protein. The enzyme was chosen because we needed larger amounts in a joint program to investigate coenzyme-dependent, enzyme-catalyzed reactions.[1] Engineering aspects discussed will also be valid for other proteins with biological activity, for example, interferon or proteohormones. In this case, the purity of the final product has to meet more stringent criteria than an enzyme used as an industrial catalyst. Even if purification has to proceed to protein homogeneity, the initial steps are similar and have to cope with the same difficulties.

HARVESTING

When the culture leaves the bioreactor, either continuously or, most often, at the end of a batch process, separation of the cells and culture broth is usually carried out. The necessity of this step when working with bacteria should be carefully considered as it is an energy and/or labor-intensive operation.

In principle, any insoluble material could be removed together with the cell debris in step III, FIGURE 1. Harvesting may therefore be avoided by processing the whole broth provided there are no components in the broth that interfere with the subsequent operations or use and cannot be separated during the purification of the protein. The following step of cell disintegration requires a certain cell concentration for optimal performance, for example, approximately 40% packed volume for wet milling.[2] If the cells cannot be grown to such high densities, a concentration step becomes necessary. We investigated three different approaches for concentrating a *Bacillus cereus* culture. The strain used was a constitutive mutant obtained in a screening program for

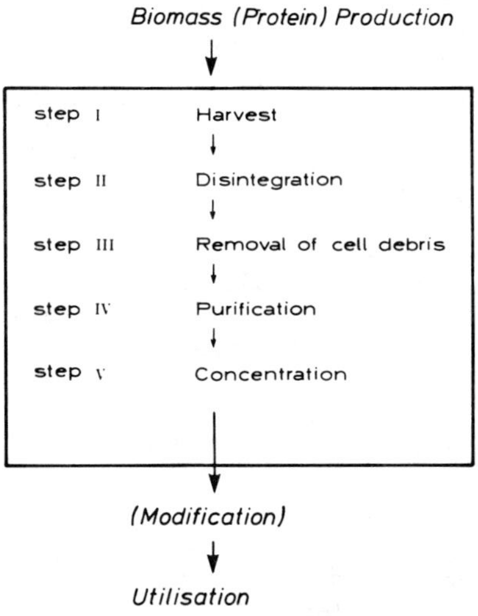

FIGURE 1. Schematic flow sheet for the isolation of intracellular proteins.

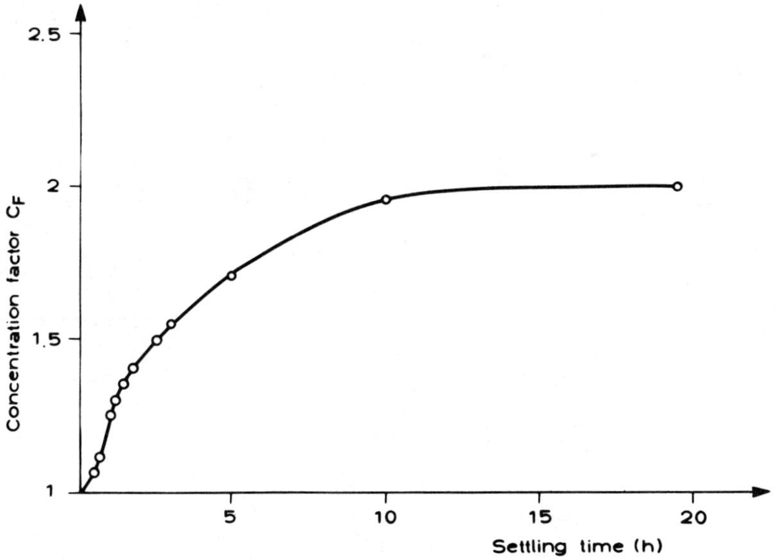

FIGURE 2. Concentration of *Bacillus cereus* from fermentation broth (100 l) by settling under gravity at room temperature. Dimensions of settling tank:

volume: 150 l
height of liquid: 1.20 m
diameter: 0.45 m
height/diameter ratio: 2.67

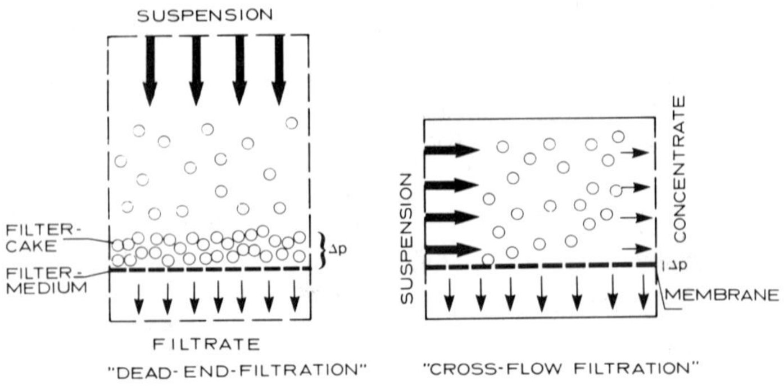

FIGURE 3. Comparison of filtration principles.

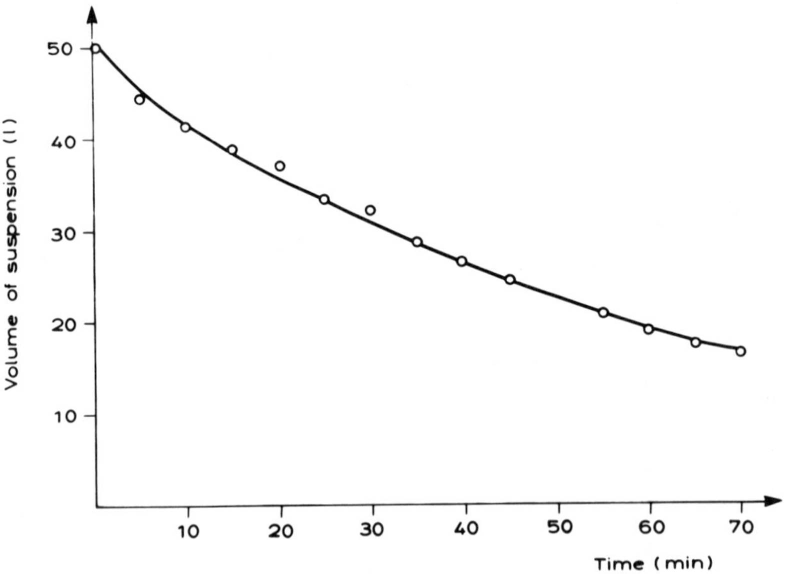

FIGURE 4. Concentration of *Bacillus cereus* cell suspension by cross-flow filtration in a hollow-fiber module. The volume of the recirculating suspension is plotted as a function of time. Operating parameters:

temperature: 20°C
pressure-drop Δp: 0.25 bar
mean transmembrane pressure Δp_M: 0.9 bar
recirculation rate: 2.05 m^3/hr

Hollow-fiber module:

Membrana NW 40/0.5/1.8-K
material: Accurel® Polypropylen
$d_i = 1.8$ mm $d_a = 2.6$ mm
length: 50 cm pore size: 0.3 μm
inlet area: 2.5 cm^2 filtration area: 0.23 m^2
Membrana GmbH (Wuppertal, W. Germany)

leucine dehydrogenase and was grown on yeast extract and glucose to cell densities of approximately 12% packed volume (12,000 × *g*, 20 min).

By chance, the strain turned out to be self flocculating; therefore, settling in a tank was analyzed, and this is described in FIGURE 2. A twofold concentration could be obtained within 10 hours, but even on prolonged storage, no further compaction of the cell slurry occurred. This behavior is related to the rheological properties of the suspension, which will be discussed below. Addition of flocculating agents, such as ionogenic acrylamides, will give higher cell concentrations in the settled flocs in shorter times, but the consequences of such additions on the subsequent extraction and purification of leucine dehydrogenase have not yet been determined. Conventional filtration techniques are difficult to perform when harvesting bacteria due to the high

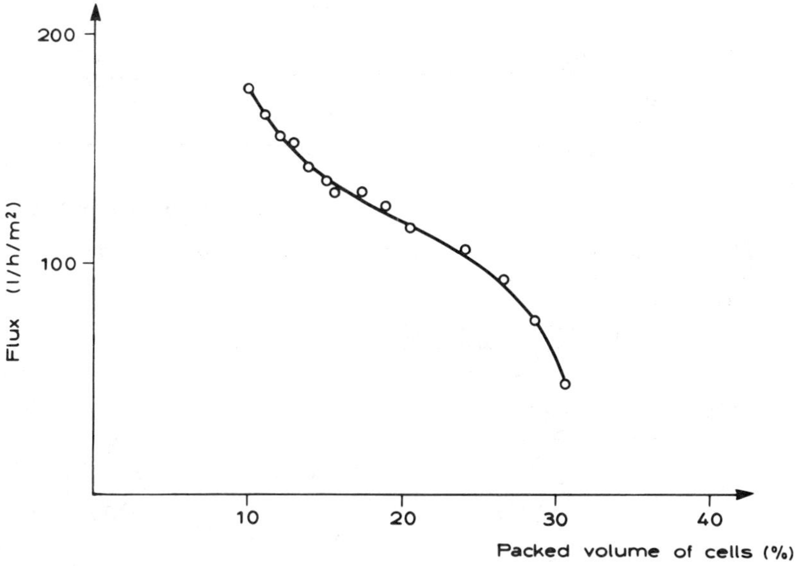

FIGURE 5. Concentration of *Bacillus cereus* cell suspension by cross-flow filtration. The flux is plotted during prolonged operation against the packed volume of cells. For operational variables, see the legend of FIGURE 4.

resistance of the filter cake. Addition of filter aid or the use of a precoat cannot be recommended for the recovery of intracellular products since the presence of such materials will lead to severe problems in the disintegration and clarification steps. However, the formation of a filter cake may be prevented by suitable hydrodynamic measures by the so-called cross-flow filtration technique leading to considerable increases in cell concentration.

FIGURE 3 presents a schematic illustration of the various principles involved. Cross-flow filtration was developed during recent years through the application of ultrafiltration principles and improved membrane technology.[3] The cell concentrate is pumped along the surface of a membrane and recirculated with a sufficient velocity to prevent cake formation while a stream of filtrate passes through the membrane with a rate dependent on the differential pressure applied.

FIGURE 4 illustrates the concentration of *Bacillus cereus* starting with 50 liters

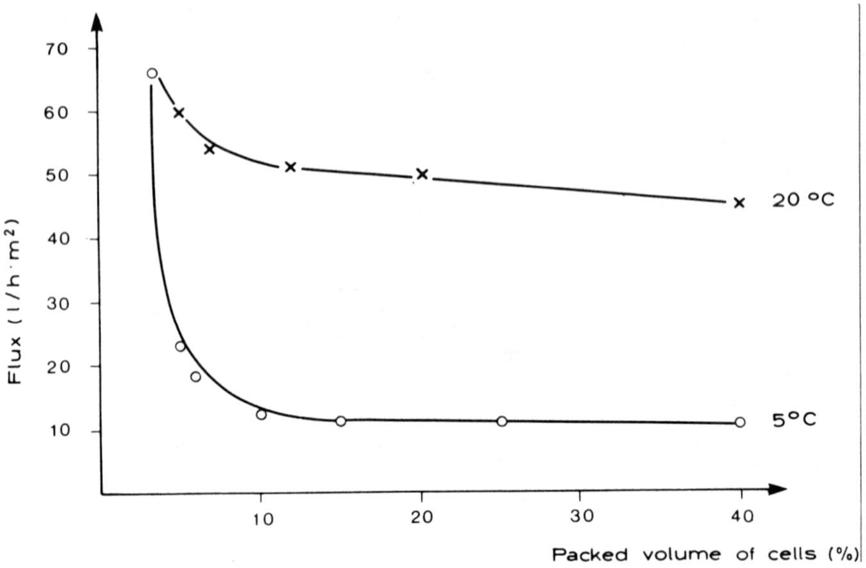

FIGURE 6. Concentration of *E. coli* K5 cell suspension by cross-flow filtration. For type of hollow fiber see the legend of FIGURE 4.
Operating parameters:

pressure-drop Δp: 0.4 bar
mean transmembrane pressure Δp_M: 1.0 bar
recirculation rate: 1.4 m^3/hr

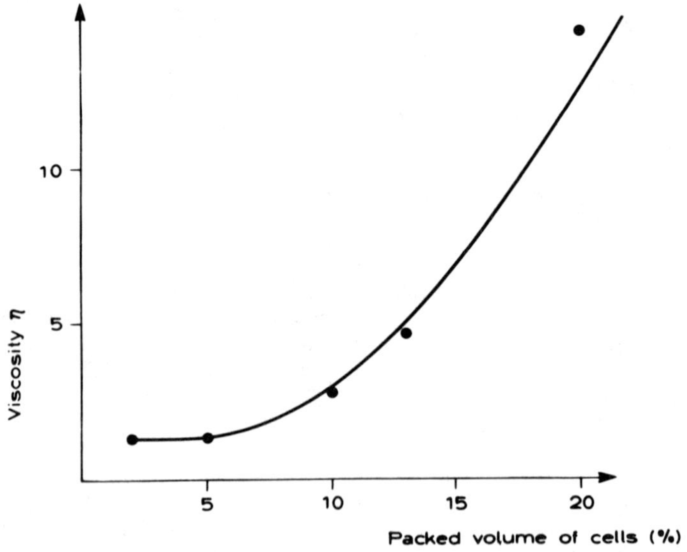

FIGURE 7. Viscosity of *Bacillus cereus* cell suspensions at a temperature of 20°C (Cuette Viscosimeter: D = 124 sec^{-1})

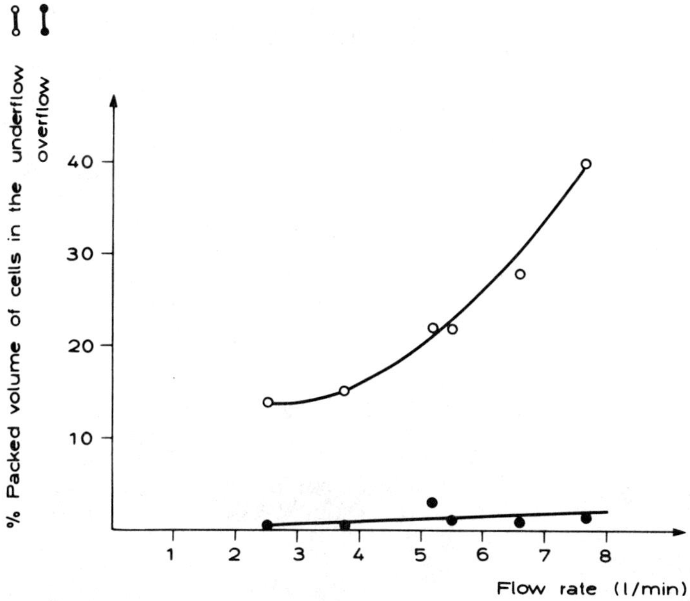

FIGURE 8. Harvest of *Bacillus cereus* cells from fermentation broth with a nozzle separator, model YEB 1334 from Alfa-Laval. Nozzles: 4 × 0.5 mm.

The packed volume of cells were determined in the underflow O—O and overflow ●—● at various flow rates.

culture using a microporous membrane with 0.3 μm pore size and a surface area of 0.23 m^2. Within 1 hour, a 2.5-fold concentration was achieved. However, analyzing the transmembrane flux as a function of increasing cell concentration (FIGURE 5) shows that the flux is rapidly decreasing. An initial drop in the flux is expected due to the buildup of a concentration polarization layer, but the steep decline around and above 25% is surprising. It was also unexpected since cultures of *E. coli,* for example, could be concentrated with no apparent difficulty (see FIGURE 6) to over 40% packed volume of cells. However, in contrast to *E. coli* suspensions, which exhibit rather low viscosity

TABLE 1. Comparison of Three Methods for Cell Concentration of *Bacillus cereus*

Method/Apparatus	Separation Area (m^2)	Volume Reduction (1/hr)	Concentration Factor	Performance Factor (1/hr · m^2)	Energy Demand (kW · hr/1)
Nozzle separator (YEB 1334)	1 200[a]	264	3	0.22	$2.1 \cdot 10^{-3}$
Settling tank	0.16[b]	5	2	31.3	$1 \cdot 10^{-4}$
Cross-flow module (Membrana NW 40/0.15/1.8-K)	0.23[c]	28.5	3	124.0	$13.0 \cdot 10^{-3}$

[a]Area equivalent (Σ-value).
[b]Settling area of the tank (height/diameter ratio 2.67).
[c]Area of the filter element.

(2–3 mPas) and apparent Newtonian behavior during flow, the viscosity in *B. cereus* suspensions is much higher and strongly dependent on cell concentrations as demonstrated in FIGURE 7. The flow behavior of *B. cereus* suspensions is also different and can be modeled as a Bingham plastic fluid. The rheological properties of the fluid are of profound influence in cross-flow filtration.[3]

The magnitude of the viscosity effect on flux may be seen from FIGURE 6, which demonstrates that the flux increased 3.9-fold while raising the temperature from 5°C to 20°C and keeping all other experimental parameters constant. To achieve the same degree of concentration, the operation time and therefore exposure to all possibly adverse conditions are much shorter at the higher temperatures. Concentration of non-Newtonian fluids are more difficult to achieve, especially under conditions of laminar flow. We are presently exploring larger diameter, tubular modules allowing higher Reynolds numbers to overcome some of the limitations observed when handling *B. cereus* cultures.

Since the first tests of cross-flow filtration in hollow-fiber devices did not yield the desired cell concentration, we turned to centrifugation. Our previous experience with nozzle separators[4] prompted us to try such a centrifuge for the concentration of *B. cereus* from culture broth. As shown in FIGURE 8, the cell concentration rose to 40% in the nozzle effluent with a loss of 1.5% in the overflow operating with a feed rate for the suspension of 7.7 l/min (≅460 l/hr) and with 4 nozzles of 0.5 mm bore. TABLE 1 summarizes and compares the different methods investigated for harvesting of *B. cereus*. For moderate cell concentrations (as needed for subsequent cell disintegration) and favorable rheological properties of bacterial suspensions, cross-flow filtration may be a useful alternative to centrifugation and provides sufficient speed in volume reduction. It can be operated in a recycle and bleed arrangement and integrated into continuous processing. However, the high superficial velocities needed in the recirculating suspension also lead to a rather high specific energy demand as shown in TABLE 1. It can be seen that backflushing or other measures to improve flux during prolonged operations and thereby to shorten the operation time are vital for the economy of the technique.[5] This was not taken into account in the experiments described here. Certainly more work is needed to arrive at an optimized harvesting procedure.

CELL DISINTEGRATION

The solubilization of leucine dehydrogenase was analyzed during mechanical treatment of *Bacillus* strains in a high-pressure homogenizer and high-speed bead mills. Wet milling appeared to be most effective after optimization of the operating variables.[2] FIGURE 9 illustrates the results. The capacity of a 22.6-liter industrial bead mill for the disintegration of *B. cereus* was estimated at 16 kg/hr with a degree of enzyme release of 85%. High-pressure homogenizers as well as bead mills can be integrated into continuous processing.

HEAT TREATMENT

Ohshima *et al.*[6] and Hummel *et al.*[7] have noted the unusual heat stability of leucine dehydrogenase from *B. sphaericus*. We investigated the heat stability of the enzyme in cell homogenates in detail with the intention of removing any precipitated material together with the cell debris. FIGURE 10 shows that the loss in activity is small up to 64°C for heating times between 10 and 30 min. We selected 63°C as the operating

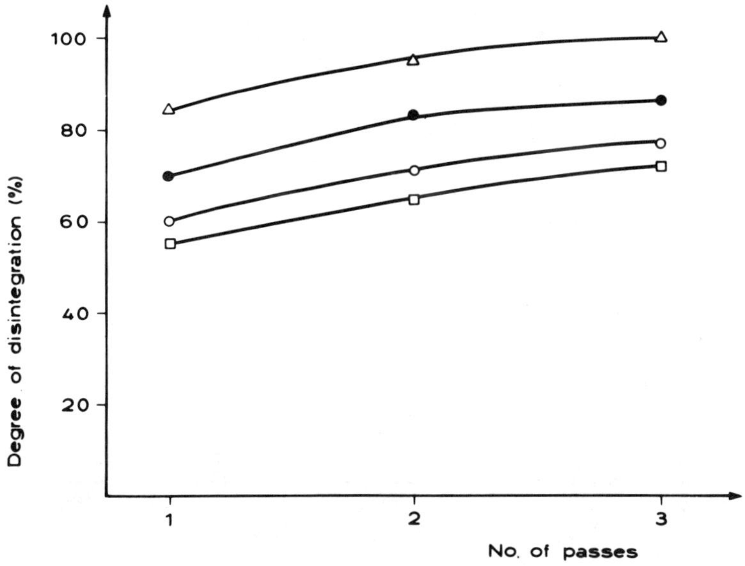

FIGURE 9. Disruption of *Bacillus* strains using various types of homogenizers. The degree of disintegration was followed for three passes in the Dyno-Mill Type KDL (Δ—Δ, *B. sphaericus*); Netzsch LME 20-Mill (○—○, *B. sphaericus*), (●—●, *B. cereus*); and a Manton-Gaulin high pressure homogenizer, Type 15 M-8TA (□—□, *B. sphaericus*).

Operating parameters:
- Dyno-Mill:
  - agitator tip speed: 10 m/sec.
  - glass bead diameter: 0.25–0.50 mm
  - flow rate: 5 l/hr
  - bead load volume: 85%
  - cell suspension: 40%
- Netzsch-Mill:
  - agitator tip speed: 9.5 m/sec
  - glass bead diameter: 0.55–0.85 mm
  - flow rate: 100 l/hr
  - bead load volume: 85%
  - cell suspension: 40%
- Manton-Gaulin:
  - pressure: 600 Kp/cm^2
  - flow rate: 54 l/hr
  - cell suspension: 40%

temperature and analyzed the content of soluble protein for different residence times. As demonstrated in FIGURE 11, heat coagulation of protein is fast in the first five minutes and continues at a slower rate for at least another 25 min. This way, the specific activity is improved 6-fold. 80–90% of the soluble protein can be removed while maintaining >90% yield of the desired enzyme. Comparable results were obtained in pilot-scale studies conducted at 63°C and with a residence time of 10 min. These conditions were chosen so as to avoid attaining the critical temperature.

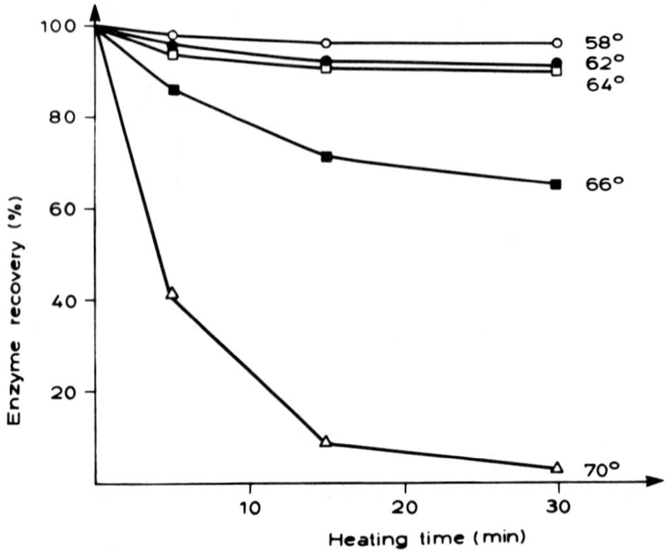

FIGURE 10. Denaturation of Leucine dehydrogenase from *Bacillus sphaericus* at various temperatures. Enzyme recovery was measured after 5, 15, and 30 minutes heating time by standard assay procedures.

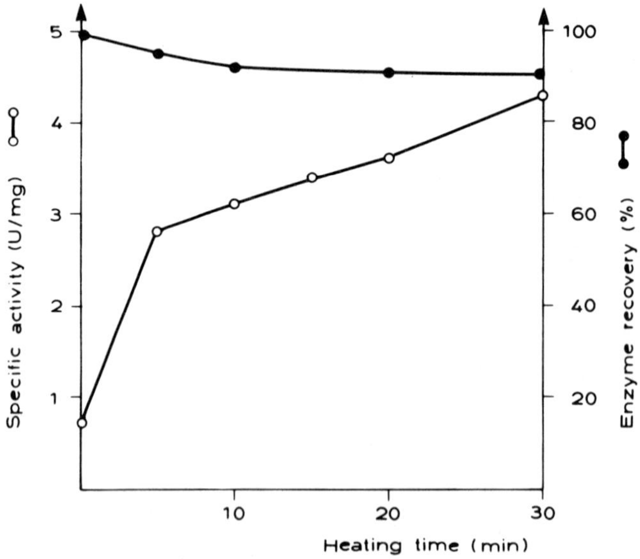

FIGURE 11. Heat denaturation of Leucine dehydrogenase from *Bacillus cereus* at 63°C. Enzyme recovery ●—● and specific activity ○—○ were measured after various heating times using standard assay procedures.

TABLE 2. Isolation of Leucine Dehydrogenase from *Bacillus* Strains Using Various Purification Procedures[a]

Crude Extract (NETZSCH LME 20-mil)

Heat Denaturation (10 min. 63°C)

A		B		C	
Centrifugation	2.1 U/mg	PEG-dextran-system	5.2 U/mg	PEG-salt-system I	6.8 U/mg
Diafiltration	3.7 U/mg	PEG-salt-system	6.5 U/mg	PEG-salt-system II	9.4 U/mg
DEAE-Cellulose Chromatography	10 U/mg	Diafiltration	11 U/mg	Diafiltration	16 U/mg

[a]A, B *Bacillus sphaericus*, specific activity of crude extract: 0.4 U/mg. C *Bacillus cereus*, specific activity of crude extract: 0.6 U/mg.

TABLE 3. Comparison of Different Leucine Dehydrogenase Preparations

Method	Total Cell Mass (kg)	Initial Units (μmol min^{-1})	Purity (U mg^{-1})	Yield (%)	Net Time (hr)	Performance Factor (U hr^{-1}kg^{-1})	Cost Index[a] (DM U^{-1})
A	25	390 000	10	72	80	140	0.008
B	32	346 000	11	69	27	276	0.0037
C	30	434 000	16	83	27	444	0.0031

[a]Rough calculation based on labor, material, and energy costs (without cultivation).

REMOVAL OF CELL DEBRIS AND COAGULATED PROTEINS

Insoluble material could be sedimented quite well after heat treatment using a tubular centrifuge as shown earlier.[7] To shorten the operating time for this step, and at the same time to improve the specific activity of the enzyme, we developed an extraction process for leucine dehydrogenase in aqueous two-phase systems.[8] Partition coefficients of 9.5 and 24 could be obtained in poly(ethylene glycol)/crude dextran or poly(ethylene glycol)/salt systems, respectively, removing 97% of the enzyme into the poly(ethylene glycol)-rich upper phase while the insoluble material as well as some soluble proteins collected in the dense lower phase. Liquid–liquid separation was carried out as described earlier[8,9] employing a disc stack separator with feed rates of 60 l/hr. Performance of the poly(ethylene glycol)/crude dextran system and the poly(ethylene glycol)/salt system were quite similar, but the latter was slightly more selective and gave higher yields in the subsequent extraction step (see TABLE 2). The specific activity of leucine dehydrogenase in the first clarified extract was 2.5 to 3-fold higher after extraction compared to the conventional removal of insoluble material by solid–liquid separation techniques.

PURIFICATION

For the intended use of leucine dehydrogenase, a specific activity of 10–20 U/mg was desired. A purification from *B. sphaericus* by conventional techniques employing diafiltration and DEAE-Chromatography has been described[7] yielding 10 U/mg in the final product. To avoid the chromatographic step, we investigated the reextraction of the enzyme from the poly(ethylene glycol)-rich top phases obtained in the previous step. By adding salt to this phase, conditions could be established that induced the formation of two liquid phases and partitioned the enzyme this time into the salt-rich

TABLE 4. Performance of Scale-up for the Partition Steps

Method	Step	Yield of Leu-DH in 10 cm^3 Scale (%)	Yield of Leu-DH in Process Scale (%)	Scale-up Factor
B	PEG-dextran	99	97 (160 l)	16 000
	PEG-salt	88	85 (130 l)	13 000
	Overall yield	87	82	~1.5 × 10^4
C	PEG-salt I	99	98 (127 l)	12 700
	PEG-salt II	95	91 (172 l)	17 200
	Overall yield	94	89	~1.5 × 10^4

lower phase. Phase separation could be carried out in a settling tank within 30 min. The final product was obtained by diafiltration of the lower phase using a Romicon Hollow Fiber cartridge (Type HF 30-20-GM80) followed by a concentration step by ultrafiltration using an Amicon Hollow Fiber cartridge (H1X50). The choice of UF membranes with high cutoff values resulted in relatively high fluxes. At the same time, a rather large portion of the residual protein was removed with the filtrate, improving the specific activity almost twofold to 11 U/mg or 16 U/mg, respectively. The three different preparations are summarized and compared in TABLES 2 and 3. It can be seen that route C gives the highest yield and lowest cost index for the best specific activity of the final product. The overall purification of leucine dehydrogenase achieved was 27-fold. The homogeneous enzyme has a specific activity of about 40 U/mg. From this value, it can be estimated that the final product is 25 to 40% pure, depending on the route followed.

It is noteworthy that the cost index is reduced to 50%, replacing a rather straightforward, conventional purification scheme by an enzyme extraction process. The cost index, however, is relatively insensitive with regard to the choice of the extraction system, that is, PEG/crude dextran or PEG/salt. The better performance of route C is predominantly due to the higher yield. The higher specific activity obtained is not decisive but, of course, valuable. For the separation of the phases, commercially available equipment can be used as discussed previously[8,9] and further scale-up appears simple. TABLE 4 illustrates this for each of the extraction systems included in routes B and C. The large-process-scale yield can be calculated with high confidence from the partition coefficient of the enzyme and the volume ratio of the corresponding phase system from laboratory data. The close correspondence between calculated and experimental values demonstrates that equilibrium is reached for the partition in large scale and that phase separation was adequate using a continuously operated liquid–liquid separator.

If necessary, further purification steps can follow immediately after diafiltration. No problems are encountered if conventional chromatographic techniques are employed for purification of the diafiltered extract. It appears possible to devise an integrated system for continuous operation from harvesting to the last diafiltration. This would shorten the recovery process considerably and lower the investment cost necessary for large-scale production. These aspects are currently under investigation and first results appear very promising.[10]

ACKNOWLEDGMENTS

We thank Mr. W. Hahn for the supply of *E. coli* K5 and Dipl.-Ing. W. Wania and the staff of the pilot plant for their support during the cultivation of the bacteria. The skillful technical assistance of Miss A. Schulz, Mr. R. Kraume-Flügel, and Mr. W. Stach is gratefully acknowledged.

REFERENCES

1. WICHMANN, R., C. WANDREY, A. F. BUECKMANN & M.-R. KULA. 1981. Continuous enzymatic transformation in an enzyme membrane reactor with simultaneous NAD(H) regeneration. Biotechnol. Bioeng. **23:** 2789–2802.
2. SCHÜTTE, H., K. H. KRONER, H. HUSTEDT & M.-R. KULA. 1983. Experiences with a 20-l industrial bead mill for the disruption of microorganisms. Enzyme Microb. Technol. **5:** 143–148.

3. HENRY, J. D. 1972. Cross-flow filtration. *In* Recent Developments in Separation Science. **2:** 205–225. N. N. Li, Ed. CRC Press. Cleveland, OH.
4. KRONER, K. H., H. HUSTEDT & M.-R. KULA. 1982. Evaluation of crude dextran as phase-forming polymer for the extraction of enzymes in aqueous two-phase systems in large scale. Biotechnol. Bioeng. **24:** 1015–1045.
5. KLEIN, W. 1981. Cross-flow microfiltration. Verfahrenstechnik **15** No. 7: 490–492.
6. OHSHIMA, T., H. MISONO & K. SODA. 1977. Properties of crystalline leucine dehydrogenase from *Bacillus sphaericus*. J. Biol. Chem. **253:** 5719–5725.
7. HUMMEL, W., H. SCHÜTTE & M.-R. KULA. 1981. Leucine dehydrogenase from *Bacillus sphaericus*. Optimized production conditions and an efficient method for its large-scale purification. Eur. J. Appl. Microb. Biotechnol. **12:** 22–27.
8. KULA, M.-R., K. H. KRONER & H. HUSTEDT. 1982. Purification of enzymes by liquid–liquid extraction. Adv. Biochem. Eng. **24:** 73–118.
9. KULA, M.-R., K. H. KRONER, H. HUSTEDT & H. SCHÜTTE. 1981. Technical aspects of extractive enzyme purification. Ann. N. Y. Acad. Sci. **369:** 341–354.
10. HUSTEDT, K. H. KRONER, H. SCHÜTTE & M.-R. KULA. Extractive purification of enzymes. W. Braun Melsungen AG., 3. Rotenburger Fermentations Symposium "Enzyme Technology." R. M. Lafferty, Ed. In press.

The Microcomputer Control of Lyophilization

RAYMOND P. JEFFERIS III

Widener University
Chester, Pennsylvania 19013

INTRODUCTION

The number of biochemical products dried by lyophilization techniques has been rapidly increasing in recent years for, although costly, it is the most effective method for preserving labile biochemicals for long periods. This is especially true of products that must be dried in their intended dosage forms without loss of biological activity. For this reason, lyophilization has proven to be especially valuable for high-value-added pharmaceuticals, special enzymes, vaccines, and diagnostic reagents that must maintain accurately specified activity levels in dosage units over long storage intervals.

A number of authors have proposed control strategies for reducing the time, and thus the cost, of lyophilization while retaining desired product qualities. Rey[1–3] has discussed control of the shelf temperature based upon measurements of the electrical resistivity of the product, a sensitive indication of eutectic state. Rieutord[4] has described the controlled introduction of inert gases into the drying chamber in proportion to changes in this same electrical property. Jefferis[5] elaborated a cascade control algorithm in which drier pressure (vacuum) is made a function of product resistivity and condenser temperature. All of these control algorithms act to balance the heat input rate to the product against the sublimation heat removal rate, and all are based upon the well-known change in electrical resistivity of a product with temperature-dependent phase changes. There are significant differences in control performance, however. The control of heat transfer by variation of shelf temperature, mentioned first above, is very slow and unstable because the shelf heating and cooling lag is large compared with the very rapid consequential effect upon resistivity. The second method described, direct control of inert gas flow into the drying chamber in proportion to resistivity measurements, is more rapid but does not inherently balance the heat transfer rates. It can also become unstable during condenser saturation. The third of the aforementioned algorithms, which cascades resistivity control onto pressure control, is very stable early in the drying cycle, but the published equations fail to adequately balance the heat transfer rates later in the drying cycle and are overly complex in implementation. This paper will describe an algorithm similar to the latter, but which is simple to implement and gives improved control response.

MEASUREMENT

Reliable and accurate instrumentation is the key to obtaining reproducible and optimal drying conditions. Certain process variables are critical to drier operation. The typical instrumentation for monitoring and control purposes is shown in FIGURE 1. The product, shelf, and condenser temperatures are typically measured by means of either

Type T thermocouples or 100-Ohm Resistance-Temperature Detector (RTD) elements. Both sensor types are inherently nonlinear, and require additional signal processing and linearization. The drying chamber pressure (vacuum) is best measured by means of a capacitance sensor, which is linear and produces a high-level signal requiring no further processing. Chamber pressure should be closely controlled for most reproducible results. In addition, many investigators have shown that the bulk electrical resistivity of a product undergoing freeze drying is an important indicator of the changes in physical state that take place as collapse points and eutectic points are approached by product heating. Such phase changes are not easily detected by thermometric means during drying, due to the temperature-regulating action of the sublimation process. Resistivity, however, is readily measured to high accuracy by detecting the current that results when a sinusoidal 1000-Hz alternating potential is impressed across a pair of inert electrodes of fixed geometry. The corresponding resistance measurement must be amplified and logarithmically scaled in order to permit control over the entire dynamic range of 10^3 to 10^8 Ohms (3 to 8 log-Ohms) found in drying operations.

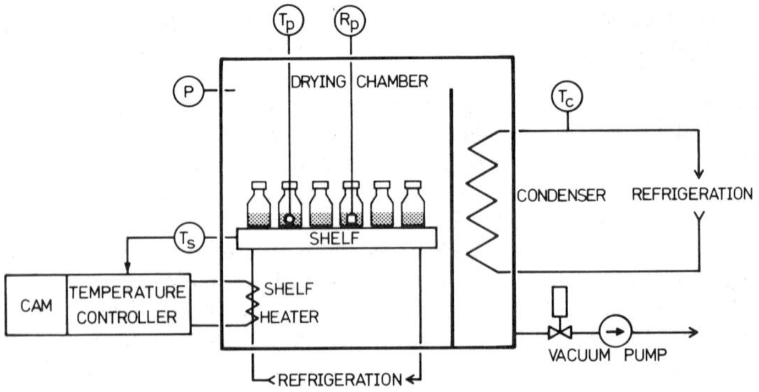

FIGURE 1. Typical instrumentation for lyophilization.

With these measurements in place, a typical primary drying cycle with pressure control will appear as shown in FIGURE 2. Notice that, as the temperature initially rises from −50°C to −20°C (233°K to 253°K), the measured resistance falls by a factor of 1000 (3 log-Ohms). This illustrates the very great sensitivity of resistivity as an indication of physical state. Both temperature and resistivity gradually rise, in subsequent hours, as the product becomes dry. The final abrupt rise in product temperature, to 0°C at 7 hours drying time, signals the transition to secondary drying. Only the most sensitive instruments can follow the corresponding four-decade rise in resistivity, but such sensitivity is necessary for stable and optimal control during primary drying.

CONTROL

To understand the design of a proper control system for primary drying, consider the net heat transfer on a vial of product at temperature, T. For a given difference

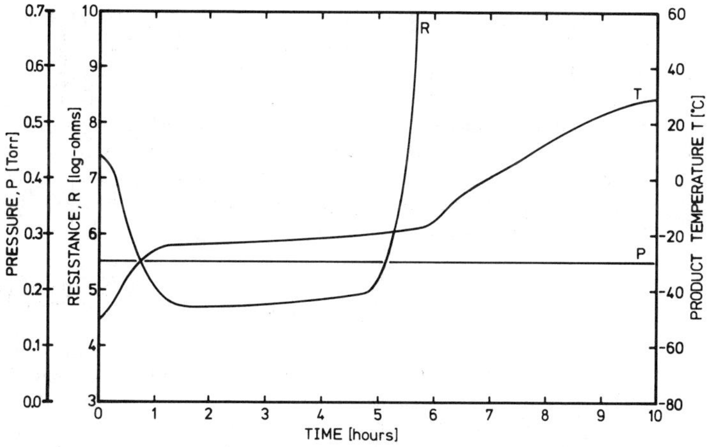

FIGURE 2. Typical primary drying cycle at constant pressure.

between product and condenser temperatures, there will be a given driving force for mass transfer derived from the resulting vapor pressure difference, as illustrated in FIGURE 3. This will result in a water removal rate (drying rate), which will vary with the mass transfer conditions in the vial, and a correspondingly varying heat removal rate by sublimation. If this heat removal rate is not exactly balanced by a heat input rate, through the shelf heat transfer system, then a net heat flux will result in a corresponding rate of increase or decrease in product temperature. The temperature will continue to change until thermal equilibrium is reached at a new product temperature, possibly at an undesirable operating state. The objective of control, then, is to attain this equilibrium *at* the desired operating temperature. This is achieved by letting the heat removal rate vary with the mass transfer conditions, and producing an exactly corresponding heat transfer rate. This will drive the net heat flux, and the rate

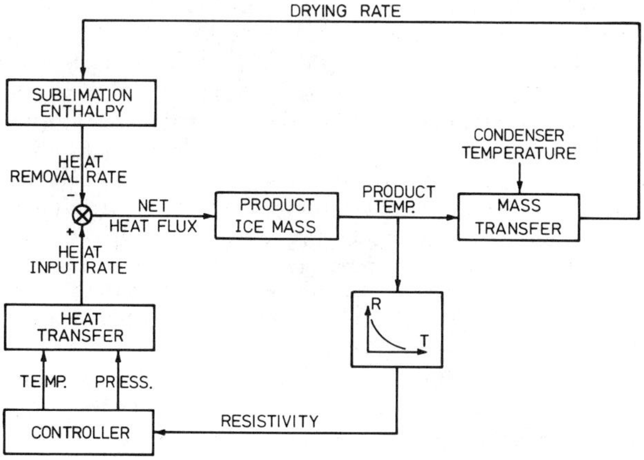

FIGURE 3. Model of heat and mass flow during lyophilization.

of temperature change, to zero at the desired operating temperature. This is a classical application for cascade control. It is implemented, as shown in FIGURE 4, by cascading a Proportional-Plus-Integral (PI) resistivity controller onto a Proportional Only (P) pressure controller. The integral action of the PI controller forces the heat transfer rate to balance the heat removal rate by integrating deviations of the resistivity-derived temperature from the setpoint, R_s. By cascading this integration onto a proportional pressure controller, smooth and stable control action is obtained, as will be shown. However, this control strategy must be implemented accurately and reliably to insure its success in production lyophilization.

CONTROLLER DESIGN

The principal design criteria for a lyophilization controller are reliability and accuracy. Reliability is difficult to attain in any measurement system that depends upon electrodes as sensing elements. Open-circuit and short-circuit failures of such electrodes can easily occur under the extreme stress of repeated sterilization, freezing, and drying. The failure of an electrode, upon which a control system depends, could result in the loss of very valuable product. Means must therefore be devised for operating reliably in spite of this hazard. This can be accomplished by introducing

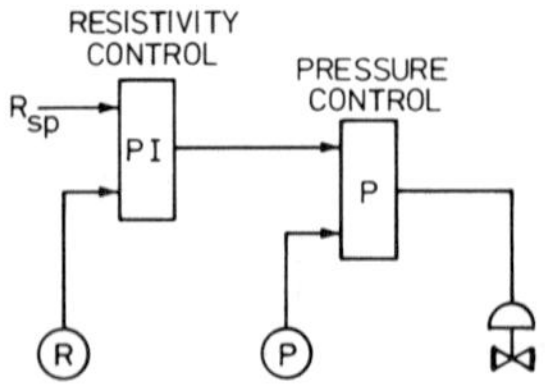

FIGURE 4. Cascade resistivity and pressure control loops.

multiple probes, with voting logic to eliminate the effects of a probe failure. Accuracy is readily obtainable with the logarithmically scaled resistivity instruments available today, as the logarithmic scaling gives constant accuracy (typically 1%) over the entire dynamic range. This accuracy can be maintained in the cascade control configuration shown by the use of microcomputer technology and is the key factor in dealing with the rapid changes in resistivity that characterize the terminal phase of primary drying.

A microcomputer controller appears to be an ideal means to implement the proposed cascade control algorithm, since it offers the accuracy and computing power needed for both control loops. Microcomputer technology also offers a means for obtaining high reliability from electrode sensors by means of triple redundancy. The application of microcomputer components to the design of one such lyophilization controller is illustrated in FIGURE 5. In this design, the median of three resistivity signals R1, R2, and R3 is selected as input to a Proportional-Plus-Integral (PI) control loop. This voting scheme minimizes the effects of open-circuit and short-circuit probe failures. The control loop, in turn, provides the setpoint to a proportional pressure control loop that uses the simultaneously acquired pressure measurement, P. The controller could provide an on/off signal to operate the inert gas valve directly. It could also provide the pressure setpoint, through a properly isolated Digital-to-Analog (D/A) converter, as a continuous output for use by an external pressure controller. For a resistivity setpoint, S, and a measured resistivity, M, the discrete-time equations that

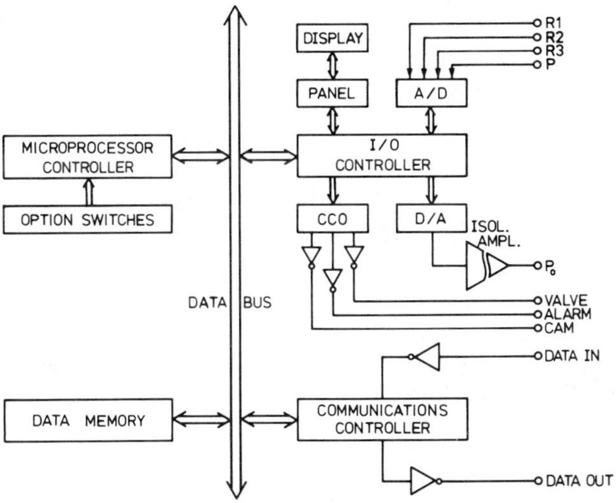

FIGURE 5. Microcomputer design for lyophilization control.

yield the pressure setpoint, P, are

$$I_n = K_p (S_n - M_n) \tag{1}$$

$$J_n = J_{n-1} + K_i I_n \tag{2}$$

$$P_n = J_n + I_n \tag{3}$$

where K_p and K_i are the proportional and integral gains, respectively. Other signals for operating alarm circuits and the shelf temperature cam can also be provided by the

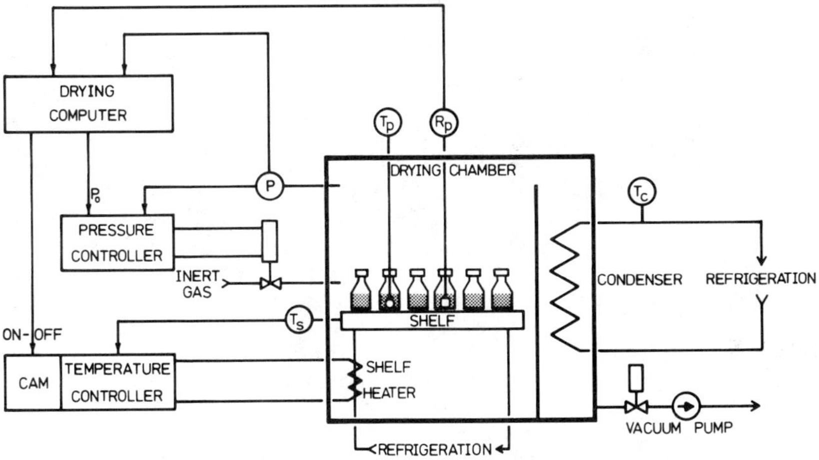

FIGURE 6. Lyophilizer instrumented for microcomputer control by resistivity and pressure measurement.

Contact Closure Outputs (CCO) of the controller. With modern technology, this entire controller can be integrated onto a single circuit card.

The controller as described could be readily applied to existing drying equipment. The installation would appear as suggested in FIGURE 6. Resistivity and pressure signals are connected directly to the controller. If an external pressure controller is provided, it receives the pressure setpoint, P_o. Otherwise the self-contained pressure controller in the microcomputer design would be able to operate the gas bleed valve directly. A controller design based upon this configuration has now been tested for hundreds of operating hours with success.

EXPERIMENTAL RESULTS

Vials of a proprietary reagent were prepared and frozen in a pilot-scale lyophilizer (Hull Model 8FS12) equipped with eutectic resistivity monitoring instruments (Hieronetics, Inc. Model RL-100 with PT-314 probes). Chamber pressure (vacuum) was measured by means of a capacitance manometer (MKS Model 222A, 10 Torr full scale). The shelf temperature was measured by resistance bulb (PT-100) and controlled by means of a cam-programmed recorder-controller (Leeds & Northrup Trendtrak programmer and Electromax III controller). Product and condenser temperatures were measured by means of Type T thermocouples. Resistivity and pressure control were performed by the microcomputer controller described (Hieronetics, Inc. Model LC-100). The controller was placed in its "Automatic" mode when chamber pressure had fallen to 0.8 Torr, and it controlled resistivity to a setpoint of 4.05 log-Ohms thereafter, as shown in FIGURE 7, until the onset of secondary drying at 4 hours. The rapid fall of resistance to its setpoint value early in the drying cycle, and the exponentially increasing pressure required to hold this setpoint are especially noteworthy. This control action applies the maximum heat transfer possible to the product and shortened the primary drying cycle by more than 40%.

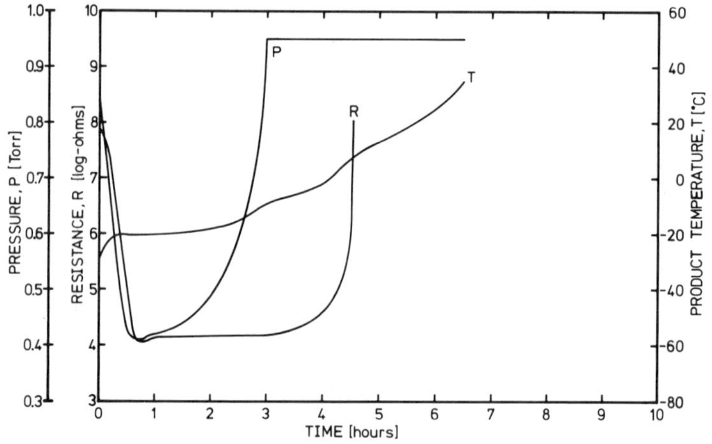

FIGURE 7. Resultant drying cycle using resistivity control of chamber pressure.

CONCLUSIONS

A new and simple algorithm for the control of lyophilization has been proposed. A microcomputer implementation of this algorithm, which uses redundant electrode inputs for high reliability, has been tested and found to markedly decrease primary drying time over constant-pressure drying.

SUMMARY

Lyophilization is a critical process in the production of many biochemicals that can be greatly improved by proper instrumentation and control. This paper describes the application of a microcomputer controller to lyophilization. Design criteria and the control algorithm are explained. Proportional-plus-integral control was found to be effective and stable while maintaining accurate heat balance during primary drying. A controller that implements this algorithm is described in detail. Experimental data using this controller illustrate its effectiveness in reducing the cycle time of primary drying.

REFERENCES

1. REY, L. 1960. Ann. N.Y. Acad. Sci. **85:** 510.
2. REY, L. 1961. Biodynamica **8:** 241–260.
3. REY, L. 1963. U.S. Patent No. 3,078,586.
4. RIEUTORD, L. M. A. 1965. U.S. Patent No. 3,192,643.
5. JEFFERIS, R. P. 1981. Ann. N.Y. Acad. Sci. **369:** 275–284.

Electric Membrane Processes for Protein Recovery

SURENDAR M. JAIN

Ionics, Incorporated
65 Grove Street
Watertown, Massachusetts, 02172

Application of electrodialysis (ED) and recirculating isoelectric focusing (RIEF) for the recovery and purification of genetically engineered proteins or selective separation of proteins for human and animal plasma is discussed, with examples showing the potential of these techniques.

ELECTRODIALYSIS

In electrodialysis, ions are electrically transported through semipermeable anion and cation selective membranes. The ED process is essentially independent of diffusion and hence can be used either to selectively remove or add salts. These two characteristics of ED, (a) reducing ionic concentration (desalting) and (b) increasing ionic concentration (salting out), form the basis of novel fractionation techniques for proteins and can serve as a tool either for a broad-range purification method of removing impurities that become insoluble at low or high ionic strengths or simply to recover the protein of interest selectively. The separations can be further enhanced by combining a pH adjustment step along with ED to approach the pI of the proteins to be precipitated.

FIGURE 1 illustrates fractionation of plasma proteins by ED desalting. As the salt level of the plasma is decreased, most of the nonalbumin proteins precipitate out and the solution is enriched in albumin and globulin such as IgA. By continuing desalting, it is possible to obtain an IgA-rich precipitate and an albumin-rich solution. This technique is also applicable for broad purification or impurity removal or concentration of genetically engineered protein solutions.

ED desalting for protein fractionation has many advantages over conventional methods. It effects a fast and controlled removal of salts, without dilution of the product. Because of low membrane area requirements, there is negligible nonspecific adsorption of protein resulting in high yields of proteins. Pore size of membranes being very small (10–100Å), low-molecular-weight proteins and peptides can be rapidly desalted by this technique. The process is easy to scale up.

FIGURE 2 illustrates separation of proteins by ED salting out. The example chosen to illustrate the principle is separation of plasma protein albumin from IgG. ED effects about 80% removal of IgG at a bulk concentration of 1.1 *N* Na_2SO_4, whereas the same separation is obtained at about 1.8 *N* Na_2SO_4 by direct addition. Another advantage is the ability to recycle the salting-out agent—ED can salt out one stream and simultaneously desalt the other. This obviates the necessity of waste dispoal or recovery of the salting-out agent. FIGURE 3 illustrates how the salting-out agent can be recycled.

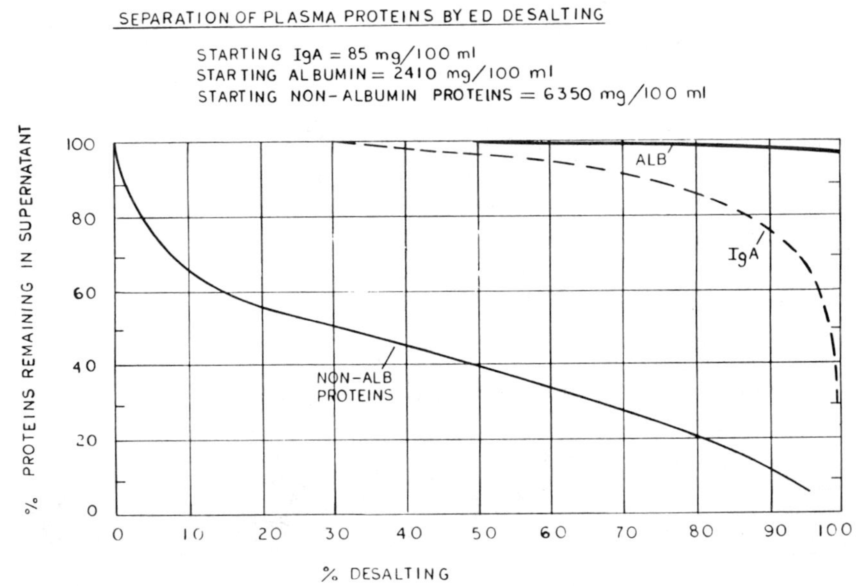

FIGURE 1. Removal of plasma proteins as a function of desalting.

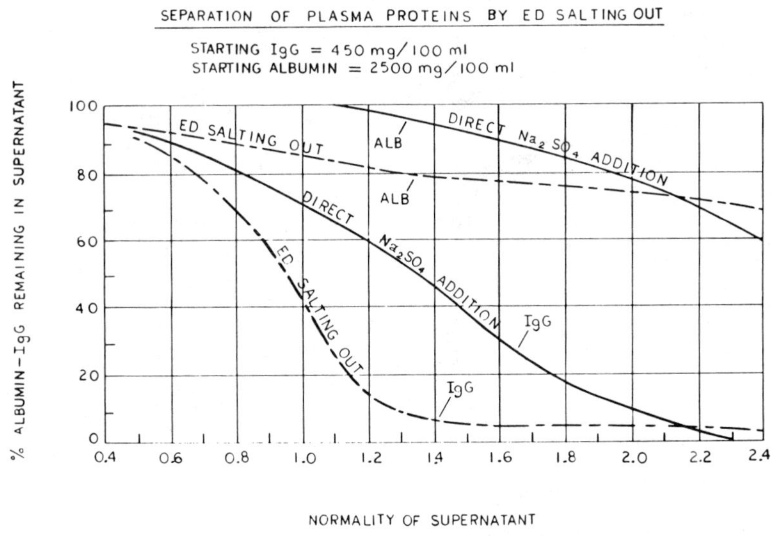

FIGURE 2. Removal of IgG as a function of bulk Na_2SO_4 concentration.

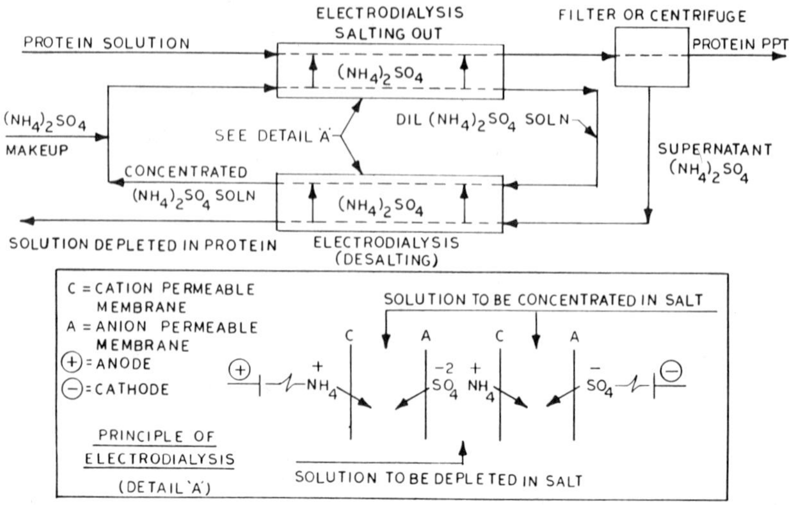

FIGURE 3. Schematic for ED salting out and subsequent recovery of salting-out agent.

RECIRCULATING ISOELECTRIC FOCUSING (RIEF)

RIEF separates proteins of differing pI's on a preparative scale under the influence of an applied electrical potential. A pH gradient is usually established by carrier ampholytes in the multiple channels formed by a parallel array of filter elements in a

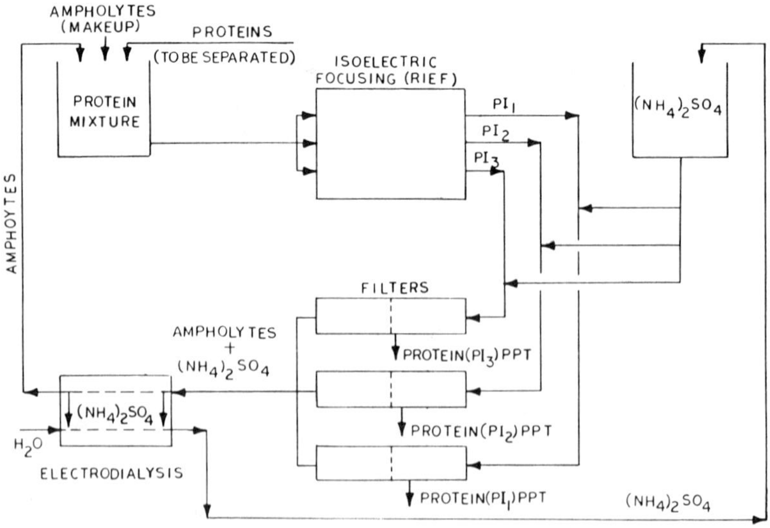

FIGURE 4. Schematic for recovery of ampholytes used in RIEF.

focusing cell stack. This arrangement streamlines fluid flows and eliminates the problems of boundary distortion commonly observed in conventional electrophoresis apparatus. Joule heat is dissipated by recirculating the solution externally through a heat exchange reservoir eliminating throughput constraints. Proteins migrate until they reach a channel with a pH equal to their respective isoelectric points, concentrate there, and then can be separated.

One of the obstacles to industrial-scale use of the process is the high cost of ampholytes. The scheme of ampholyte recovery suggested in FIGURE 4 has the potential of overcoming this obstacle. After the protein constituents have been separated, the proteins of interest are then salted out by adding appropriate amounts of a salting-out agent such as $(NH_4)_2SO_4$; the precipitated proteins of various pI's are recovered by filtration and the supernatant consisting of ampholytes with the salting-out agent is electrodialyzed to separate the salting-out agent and the ampholytes for recycle.

The Isolation of Proteins from Complex Mixtures by Immobilized Monoclonal Antibodies

DAVID A. VETTERLEIN AND GARY J. CALTON

Purification Engineering, Inc.
9505 Benger Road
Columbia, Maryland 21046

INTRODUCTION

Immobilized monoclonal antibodies (MAB) offer a streamlined purification system for the commercial-scale isolation of specific proteins found in complex mixtures. MAB to the light chain of urokinase was immobilized and used as a model system to demonstrate the utility of affinity purification. Urokinase was isolated directly from human urine, cell culture harvest fluids, and *E. coli* broth. The urokinase isolated was substantially free of contaminants and contained most of its original enzymatic activity.

RESULTS

Purification of Urokinase From Commercial Sources and Complex Mixtures

The specific activity (S.A.) of commercial urokinase was 60,000–70,000 CTA units/mg protein. After a one-step affinity purification, the S.A. was 130,000 to 140,000 CTA units/mg protein, with over 85% recovery of enzymatic activity. A SDS-gel (silver stain) of purified commercial urokinase is shown in FIGURE 1.

TABLE 1 provides examples of urokinase enzyme activities before and after separation and MAB affinity purification. Purified urokinase preparations were substantially pure when analyzed by silver-staining methods (similar to FIG. 1) and recoveries varied from 60–100% depending on the complex mixture being tested. No pretreatment was required before separation. The estimated urokinase concentration in urine was 1 to 10×10^{-11} *M*.

CONCLUSION

MAB affinity purification offers a number of advantages over conventional techniques. For example, many protein molecules are partially degraded by proteases before they can be purified. As a result, the protein isolated by conventional purification schemes is a mixture of degradation products and the molecule of interest. The native molecule can be isolated by affinity chromatography as demonstrated in the case of urokinase. Native urokinase has a different mode of action than commercially available products isolated by conventional methods. Native urokinase binds tightly to

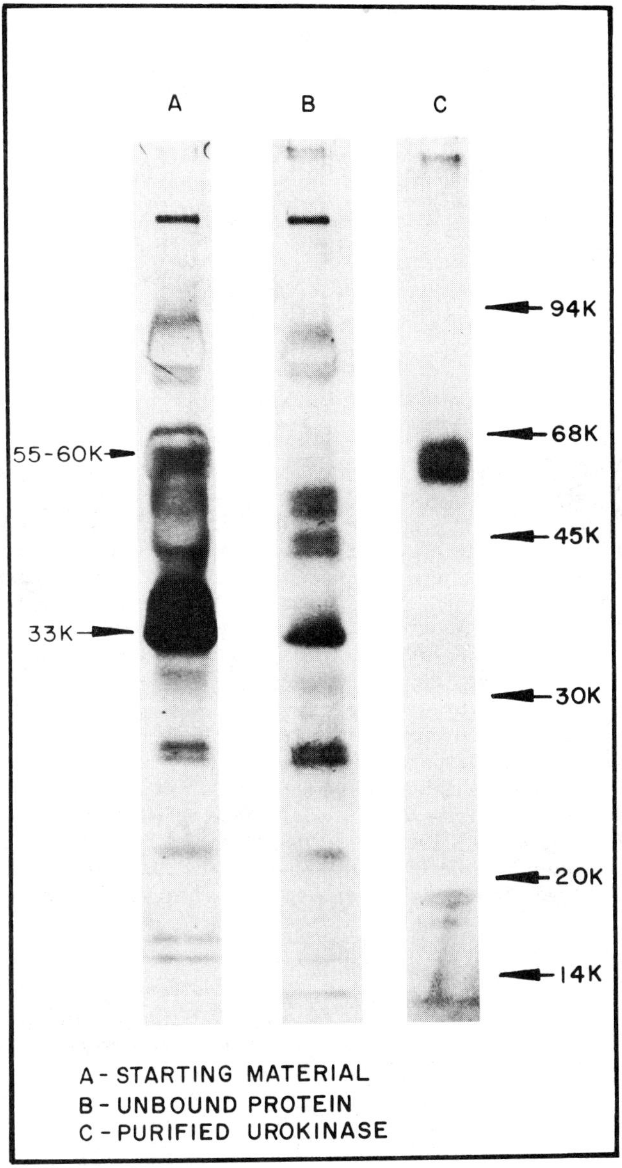

FIGURE 1. Hybridoma antibody affinity purification of urokinase.

TABLE 1. Urokinase Purification from Different Sources

Starting Material	Starting Activity (CTA Units)	Nonretained Fraction	Wash	Acetic Acid Eluant (CTA Units)
Urine (150 ml)	76.6	0.1	0	62.9
Calu-3 cell HF (100 ml)	68	0	0	86.8
Bowes cell HF (140 ml)[a]	924	841	0	9
E. coli broth (4×10^8 cells/ml, 500 ml)[b]	12.5×10^4	Not detected	0	7.8×10^4

[a]Positive control to test the specificity of the monoclonal antibody system. Bowes cell harvest fluid above contains tissue activator (TA) exclusively. TA is a different plasminogen activator gene product.

[b]Protease inhibitor added to prevent degradation of urokinase.

fibrin[1] and therefore may be more useful for the treatment of cardiovascular disorders such as heart attack or stroke.

The MAB purification procedures described here demonstrate that molecular recognition by MABs can be used for: (1) selection of a dilute pharmacologically important compound from several complex mixtures, and (2) rapid isolation of that compound without significant losses of enzymatic activity. MAB purification techniques should lower purification costs by improving yield and eliminating many of the unit operations utilized conventionally. Because of the rapidity of the technique, isolation of native (nondegraded) protein molecules with improved pharmacological properties may be possible. Contaminants not easily removable by other commercial purification processes can be removed using MAB affinity techniques.

The use of MAB as a feasible commercial purification technique depends on a number of factors. Immobilization support materials capable of withstanding high flow rates and operating pressures of several atmospheres will be advantageous in process engineering. The stability of the immobilized MAB is also an important economic consideration. Evidence to date suggests that over 100 cycles (conservatively) can be achieved. Suitable polymer supports are being developed.

REFERENCE

1. SUMI, H., M. MARUAMA, O. MATSUO, H. MIHARA & N. TOKI. 1982. Higher fibrin-binding and thrombolytic properties of single polypeptide chain-high molecular weight urokinase. Thromb. Haemostas **47:** 297.

Nylon Tubing as an Affinity Matrix in the Purification of Acetylcholine Receptors and Immunosorption Studies

P. V. SUNDARAM

Abteilung Klinische Chemie
Medizinische Klinik Universität Göttingen
3400 Göttingen, West Germany

A new approach to affinity chromatography has been adopted in a method used for the isolation of acetylcholine receptors (Ach receptor) from electric eel in that a chemically modified, polymeric nylon tube is employed as an affinity support. BSA, α-cobratoxin, and the Ach receptors were coupled to the activated tubular matrix as biospecific ligands. The affinity tube made with the Ach receptor as ligand binds receptor-specific ligands such as immunoglobulins and can thus be used in affinity chromatography and immunoassays.

Although methods of affinity purification of Ach receptors using agarose-type gels as matrices are known,[1] isolation of receptor-binding proteins such as neurotoxins and antibodies using affinity adsorbents carrying Ach receptor as the ligand has not been satisfactory because of the large loss of binding capacity of the receptors and noncovalent adsorption to the gel matrix. Thus the need for a better affinity matrix that would permit a simple procedure of affinity trapping and purification of compounds that interact biospecifically with the receptor. The trapping device may not only selectively account for the antibodies and other unknown compounds but also facilitate in early detection of myasthenia gravis through immunoassays.

METHODS

Ligands were attached to nylon tubes activated by partial hydrolysis[2] by three different methods: by cross-linking with glutaraldehyde or bismidates or through carbodiimide-activated COOH groups of the hydrolyzed nylon.[3] Cross-linking with glutaraldehyde, 5% (v/v) was done at pH 9.5 in borate buffer (0.05 *M*) and with bisimidates (4 mg/ml) in pH 9.5–10.0 buffer. Activation of COOH groups of hydrolyzed nylon with 100 m*M* EDAC at pH 5 for 2 hr was followed by coupling of BSA and neurotoxins at pH 5 and Ach receptor at pH 6.

Coupling yields were assayed with ^{125}I-labeled toxin, DEAE filter disc assay for Ach receptor, and UV absorption for BSA. However, the toxin-trapping capacity of receptor-linked nylon tubing was estimated by using ^{3}H-labeled toxin.

RESULTS

Among the three methods of coupling tried, the best suited in terms of coupling yield and retention of properties of the coupled ligands was that using EDAC. With BSA, a typical protein couples 420 pmol/cm, toxin couples least (41 pmol/cm), and

TABLE 1. Efficiency of Protein-Coupled Nylon Tubes in Trapping and Reversible Binding of Biospecific Ligands[a]

Ligand Linked to Nylon Tube					Capacities of:			
					Trapping		Reversible Binding	
Type	Concentration (pmol/cm)	Density (pmol/cm²)	Coupling Reagent	Bound Ligand	(pmol/cm)	(% of concentration of linked ligand)	(pmol/cm)	(% of trapped ligand)
Acetylcholine receptor	137	436	3-(3-dimethylaminopropyl)-1-ethylcarbodi-imide hydrochloride	α-Neurotoxin	36	26	32	89
α-Cobratoxin	0.6	2	Glutaraldehyde	Acetylcholine receptor	0.42	70	0.35	83
	15	48	Glutaraldehyde	Acetylcholine receptor	3.8	26	0.4	11
	41	131	3-(3-dimethylaminopropyl)-1-ethylcarbodi-imide hydrochloride	Acetylcholine receptor	7	17	0.3	4

[a]Coupling density is expressed in pmol of linked ligand per cm^2 of nylon surface. Trapping capacity of toxin-linked tubes was determined by binding of acetylcholine receptor. Reversible binding refers to the amount of receptor competitively detached from these tubes after perfusion with 0.1 *M*-hexamethonium for 3 hr. Trapping capacity of receptor-linked tubes was determined with ^{125}I-labeled α-cobratoxin (see the METHODS section). Reversible binding refers to the amount of toxin competitively detached after perfusion of the tubes with 0.1 *M*-hexamethonium for 3 hr.

the receptor couples 137 pmol/cm. The results for the glutaraldehyde method are 121, 15, and 0.6 pmol/cm, respectively, in the last case receptor being diluted 1:25 with BSA.

Data in TABLE 1 show that the trapping and reversible binding capacities of affinity tubes depend only on the density of ligands (binding sites) and not on the total concentration of nylon-linked proteins. The low-density tubes (results of toxin coupled with a 25-fold excess of BSA) were the most efficient in trapping and reversible binding because the high density of toxin in toxin-linked tubes bind the receptor molecules too tightly to be reversible.

It is shown in this study that (1) receptor coupling to nylon is efficient, (2) these affinity surfaces bind ligands biospecifically, (3) nylon-linked receptor retains its native ligand-binding properties even after repeated application, and (4) anti-(receptor) immunoglobulins and anti-(toxin) immunoglobulins can be trapped by affinity chromatography using the proper toxin-bound and receptor-bound affinity tubes. The sensitivity of detection matched that of radioimmunoassays.[4]

REFERENCES

1. MAELICKE, A., B. W. FULPIUS & E. REICH. 1977. The nervous system. Handb. Physiol. Sect. 1. pp. 493–519.
2. SUNDARAM, P. V. & W. E. HORNBY. 1970. FEBS Lett. **10:** 325–327.
3. SUNDARAM, P. V. 1977. *In* Biomedical Applications of Immobilized Enzymes and Proteins. T. M. S. Chang, Ed. Plenum Press. New York. pp. 317–340.
4. YANG, B. -H., P. V. SUNDARAM & A. MAELICKE. 1981. Biochem. J. **199:** 317–322.

Metal-Chelate Affinity Chromatography as a Separation Tool

L. FANOU-AYI AND M. VIJAYALAKSHMI

Université de Technologie de Compiègne
B.P. 233
60206 Compiegne, France

INTRODUCTION

The transition metals are known to form complexes with compounds rich in electrons. Apart from the aromatic and heterocyclic compounds including the nucleotides, the proteins due to their contents in Cys, His, Trp, and Tyr can show affinities for the transition metal chelates. The differences in the exposed groups can be made use of in designing liquid chromatographic separation techniques. Such adsorbents can be prepared by coupling the chelator chemically and binding the metal onto them. Porath *et al.*[1] have shown the possibility of separating serum proteins by this technique with Cu and Zn immobilized onto an agarose matrix.

For this technique to be useful in large-scale purifications, we have to take into consideration the following criteria:

(1) the adsorbent should have excellent mechanical properties and low compressibility;
(2) the kinetics of association and dissociation should be rapid in order to use high flow rates on a preparative scale; and
(3) the chelation of the metal should be stable enough to avoid the metal leakage during elution and at the same time, there should be free cordination sites available for the association of solute molecules.[2]

The work reported so far makes use of only agarose-based adsorbents. However, agarose has limitations related to its mechanical properties. In this paper, we compare the agarose-based metal chelate adsorbents with those of silica-based ones under similar aqueous elution conditions, in order to design scaled-up metal-chelate chromatographic procedures for active proteins and peptide purification. Synthetic nucleotides and dipeptides are used as model substances to better elucidate the mechanism involved.

METHODS

Preparation of Adsorbents

A typical Sepharose-based metal-chelate adsorbent is prepared by first epoxy-activating the commercial Sepharose 6B and then coupling the amino-carboxyl ligand (e.g., iminodiacetate, carboxymethylated tetraethylene pentamine, carboxymethylated polyethylene imine) in an alkaline medium according to Porath *et al.*[1]

The silica-based adsorbent is prepared after introducing oxiran groups by reacting

TABLE 1. Retention Behavior of Model Substances on the Organic and Inorganic Matrices Coupled with Different Chelating Ligands[a]

	Sepharose-based Gels				Silica-based Gels								
	Chelating Ligand				Chelating Ligand								
	IDA Cu = 900 μmole/g				CM PEI Cu = 100 μmole/g			CM TEPA Cu = 87 μmole/g			IDA Cu = 90 μmole/g		
Test Substance:	pH 5.0	pH 6.0	pH 7.0	pH 8.0	pH 5.0	pH 6.0	pH 7.0	pH 5.0	pH 6.0	pH 7.0	pH 5.0	pH 6.0	pH 7.0
Nucleotides:													
AMP	—	1.20	5.75	1.35	2.55	1.23	1.13	1.59	0.98	1.18	5.0	7.01	2.24
GMP	—	1.20	4.10	1.45	4.34	2.07	1.06	1.69	1.25	1.18	2.69	2.85	1.67
CMP	—	1.05	1.40	1.08	2.55	1.23	1.06	1.35	0.98	1.27	1.4	1.41	1.38
TMP	—	1.20	1.35	1.10	1.89	1.23	—	1.35	1.0	—	1.4	1.10	1.10
Dipeptides:													
His–Tyr	≫	≫	≫	—	—	3.97	—	1.42	5.23	4.59	2.21	1.14	4.3
Tyr–His	21.4*	6.45*	13.7*	—	—	1.35	—	1.62	2.04	2.30	3.93	4.70	3.1

[a]Note: The V_E/V_T values are calculated by taking the volumes at the maximum of the peaks. However, a large spreading of the peaks is observed in the case of results marked with (*). So, the values are not significant.

the Spherosil XOB030 (Rhône Poulenc, France) with 3-glycidoxy-trimethoxysilane in aqueous medium and the same chelating ligands as mentioned before are coupled to this suspension in 50% DMF as described elsewhere.[3]

The metal coupling is done by percolating a 50 m*M* solution of the metal salt ($CuSO_4$ or $FeSO_4$ or $FeCl_3$) in water through the columns containing the chelator adsorbent. After complete loading with metal, as judged visually, the bed is thoroughly washed first with water and then with the buffers used in the elution, in order to remove all the loosely bound metal.

The ligand contents in terms of nitrogen were on the order of 1000 μM for the Sepharose gel and 300 μM for the silica gel. The metal contents (e.g., Cu) were 200 μM and 90 μM, respectively.

Chromatography

1 × 10 cm columns are used in the preliminary investigations for studying the influence of pH, ionic strength, and flow rates on the retention of model substances like mono- and dinucleotides and the dipeptides as shown in TABLES 1 and 2.

For the separation of the peptide hormones and the protein, columns of about 1 × 30 cm are used with flow rates of about 30 $ml/cm^2/hr$.

Hormone Peptides Elution

A crude acetone extract of the bovine pituitary glands reconstituted in 0.1 *M* ammonium acetate buffer, pH 5.0 + 0.5 *M* NaCl is applied to a Cu IDA Sepharose column and the elution is continued with the same buffer. One-ml fractions are collected and the elution monitored by a UV cord LKB. The peak fractions (2 well-separated ones) are pooled and first analyzed by electrophoresis and TLC for identifying the peptides and proteins. The second peak containing the peptides plus the receptor protein is chromatographed on a Sephadex G25 column with 0.1 *M* formic acid and the peaks are analyzed as before. The biological activity of oxytocin is assayed on a fresh rat uterus according to the standard technique.[4]

Protein Separation

The separation of the iron transport protein lactoferrin present in milk whey is demonstrated as an example. The whey, which contains the lactoferrin, SIgA,

TABLE 2. Influence of Ionic Strength on the Retention Behavior of Model Substances[a]

	Cu—IDA—Silica				Cu—IDA—Sepharose			
	Dipeptide:		Nucleotide:		Dipeptide:		Nucleotide:	
NaCl Conc	His–Tyr	Tyr–His	AMP	CMP	His–Tyr	Tyr–His	AMP	CMP
0 *M*	1.14	4.97	7.01	1.41	≫	13.70*	3.0	1.30
0.2 *M*	1.88	—	7.01	1.57	≫	27.36*	5.65	1.15
0.5 *M*	2.62	2.99	6.85	1.57	≫	24.90*	5.50	1.15
1.0 *M*	2.62	1.88	8.81	1.89	≫	9.01	5.75	1.20

[a] Note: The V_E/V_T values are calculated by taking the volumes at the maximum of the peaks. However, a large spreading of the peaks is observed in the case of results marked with (*). So, the values are not significant.

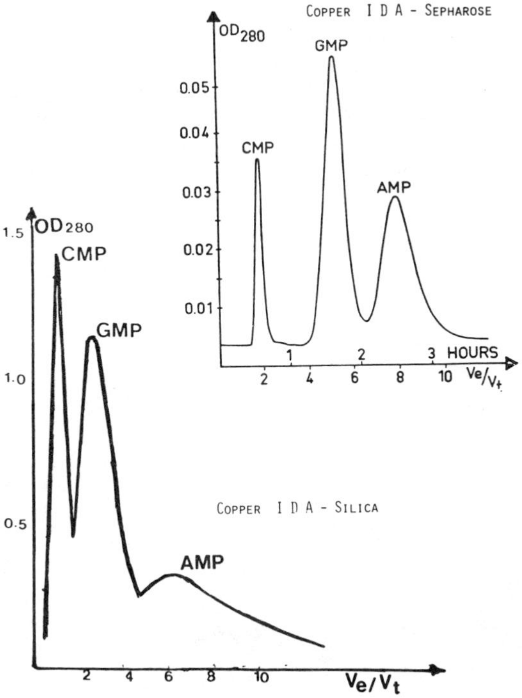

FIGURE 1. Group separation of mononucleotides on Cu iminodiacetate silica (elution with 0.05 *M* ethylene morpholine acetate buffer at pH 6.0 and 20°C), and Cu iminodiacetate Sepharose (elution with 0.05 *M* Tris-HCl buffer at pH 7.0 and 20°C).

lactabumin, lactoglobulin, and so forth after removing the fat and casein, is applied to a Cu-IDA-silica column with 0.05 *M* Tris-HCl buffer at pH 7.5.

The elution program is: (1) 0.05 *M* Tris-HCl pH 7.5; (2) + 0.5 *M* NaCl; (3) + 0.5 *M* NaCl + 20 m*M* glycine; and (4) + 0.5 *M* NacL + 10 m*M* His. The elution is monitored by UV adsorption. Fractions of ~1 ml are collected and the protein contents are determined after dialysis according to Lowry *et al.*[5] The peak fractions are characterized by electrophoresis, immunoelectrophoresis, and gel filtration.

RESULTS AND DISCUSSION

Influence of Elution Parameters and the Nature of the Chelating Ligand

Of the different chelating ligands tested in our experiment, only IDA shows selectivity for the purine nucleotides (TABLE 1), inspite of similar amounts of bound copper in all the cases. The nature of the support matrix does not seem to play an important role as already showed by us.[3] The elution pH seems to play a complex role. At pH 6.0 for Sepharose-based gel and at pH 7.0 for silica-based gel, a clear separation of purine and pyrimidine mononucleotide is seen (FIG. 1 and TABLE 1). The added NaCl concentration does not seem to affect the retention much, thereby excluding the possibility that only electrostatic interactions are responsible for the separation (TABLE 2).

The kinetics of association/dissociation seem to be rapid as we could not see any significant change in the retention patterns for the flow rates between 20 ml/hr/cm^2 and 120 ml/hr/cm^2.

A clear cooperative effect is shown as purine dinucleotides are more strongly retained than the pyrimidine dinucleotides. Moreover, the position of the purine base plays an important role. This is very useful in the separation of pairs in a mixture as shown in FIGURE 2.

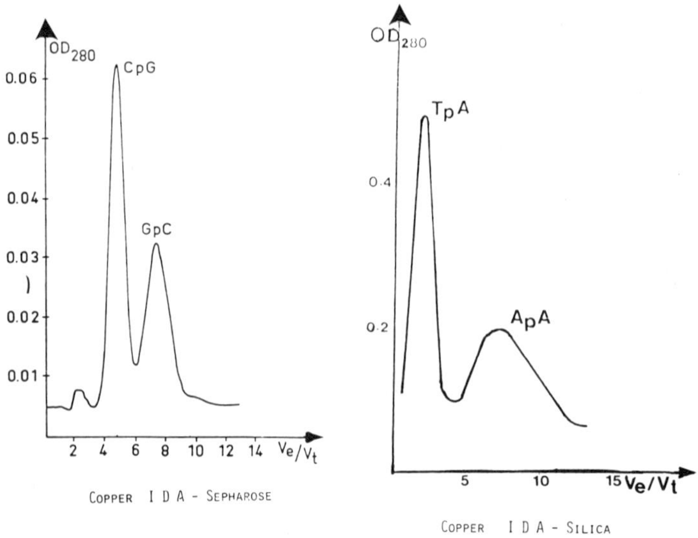

FIGURE 2. Dinucleotide pair separation on Cu iminodiacetate silica and Cu iminodiacetate Sepharose. The elution conditions are the same as in FIGURE 1.

Hormone Peptides Purification

The elution pattern of the crude gland extract on a Cu-IDA Sepharose column is shown in FIGURES 3, and 4. The second peak contains hormones, oxytocin, and vasopressin plus the receptor neurophysins. The second elution of this peak on a Sephadex G25 column with 0.1 *M* ammonium acetate buffer shows the resolution seen in FIGURE 4. The first peak contains the neurophysin and the second corresponds to oxytocin and the third to vasopressin as revealed by TLC analysis. The oxytocin thus obtained is about 80-fold purified and has a biological activity of 400 units/mg as compared with 5 units/mg in the initial extract. The peptide yield is 80%. Moreover, the neurophysin, a high value by-product, can be recovered in fact. The vasopressin fractions do not show any contamination with oxytocin.

Protein Separation

The elution pattern of the milk whey proteins on a Cu-IDA-silica column is shown in FIGURE 5. The adsorption of lactoferrin in our experiment was found to be comparable to the results of Lonnerdal *et al.*[6] on a Cu-Sepharose column. However, the

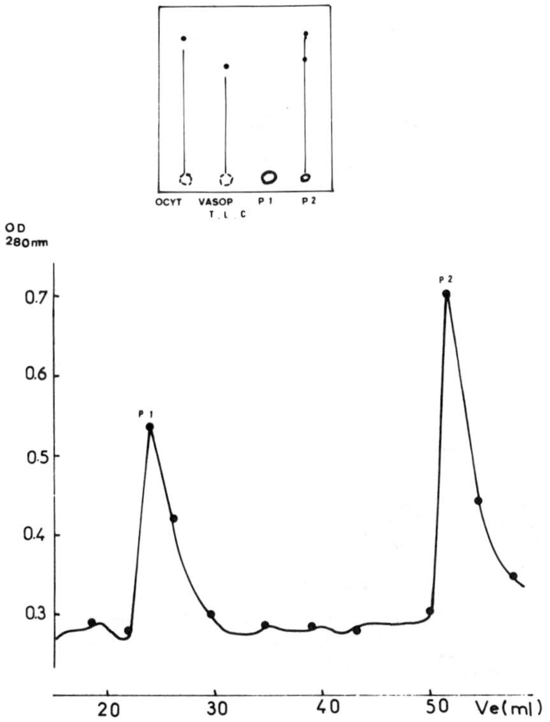

FIGURE 3. Separation of hormone peptides oxytocin and vasopressin on Cu column.

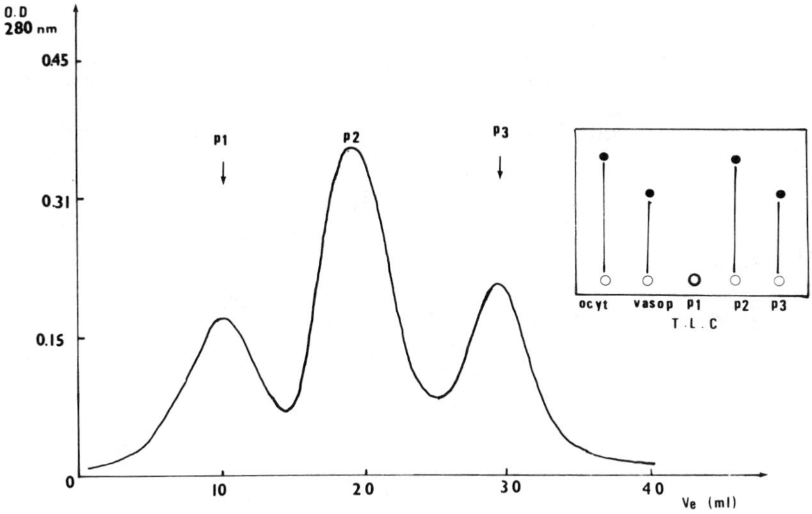

FIGURE 4. Dissociation and separation of neurophysin, oxytocin, and vasopressin on a Sephadex G 25.

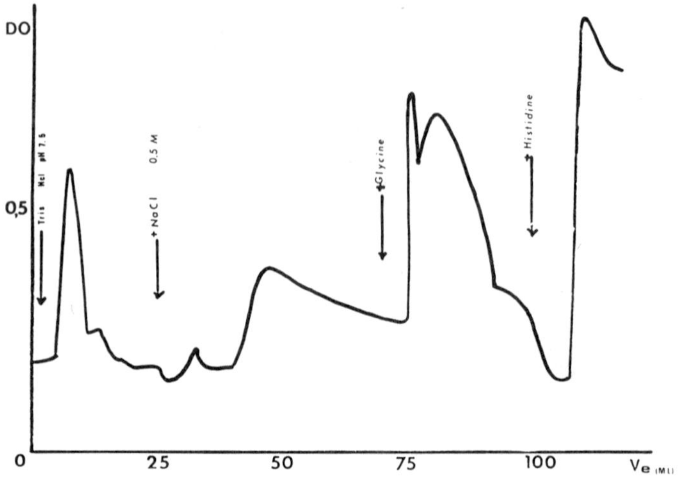

FIGURE 5. Separation of lactoferrin from total milk whey proteins. See text for elution conditions.

sequential elution with the buffer containing the amino acids Gly and His has the advantage of improving the resolution of proteins. The lactoferrin and the SIgA are found concentrated in the Gly and His peaks, respectively, as judged by immunoelectrophoresis.

This gel has an adsorption capacity of 2 mg of lactoferrin/ml and the reproducibility is very good as the same lot of adsorbent can be used for more than 2 years with successive adsorption, desorption, and regeneration.

CONCLUSIONS

These examples of peptides and protein purification demonstrate the usefulness of using silica-based metal chelate adsorbents as a separation tool for separation of biomolecules on a preparative scale.

ACKNOWLEDGMENT

The technical assistance of Mrs. L. E. Moullec is greatfully acknowledged and the authors thank Mrs. Domurado for her useful discussions on the immunoelectrophoresis characterization of proteins.

REFERENCES

1. Porath, J., J. Carlsson, I. Olsson & G. Belfrage. 1975. Nature **258**(5536): 598–599.
2. Vijayalakshmi, M. 1980. Doctoral Thesis. University of Compièugne, France. pp. 82–87.
3. Rajgopal, S. & M. Vijayalakshmi. 1982. J. Chromatogr. **243:** 164–167.
4. French Pharmacopia, VIII Ed. 1965. p. 1071.
5. Lowry, O. H., N. J. Rosebrough, A. L. Fara & R. J. Randall. 1951. J. Biol Chem. **193:** 265–275.
6. Lonnerdal, B., J. Carlsson & J. Porath. 1977. FEBS Lett. **75**(1): 89–92.

Ultrafiltration Affinity Purification

BO MATTIASSON AND MATTS RAMSTORP

Pure and Applied Biochemistry
Chemical Center, University of Lund
S-220 07 Lund, Sweden

Purification techniques based on biospecific affinity interactions are characterized by their high resolving power, but also by being laborious, time consuming, and having low capacity. Ever since it first was reported that these interactions were possible, affinity purification columns with biospecific sorbents have been used. These are operated in a cyclical fashion; first a loading step, then a washing step followed by elution and finally, a reconditioning of the matrix. One way to improve the output has been to use a set of columns controlled by a microcomputer.

So far, however, no serious attempts have been made to utilize affinity purification

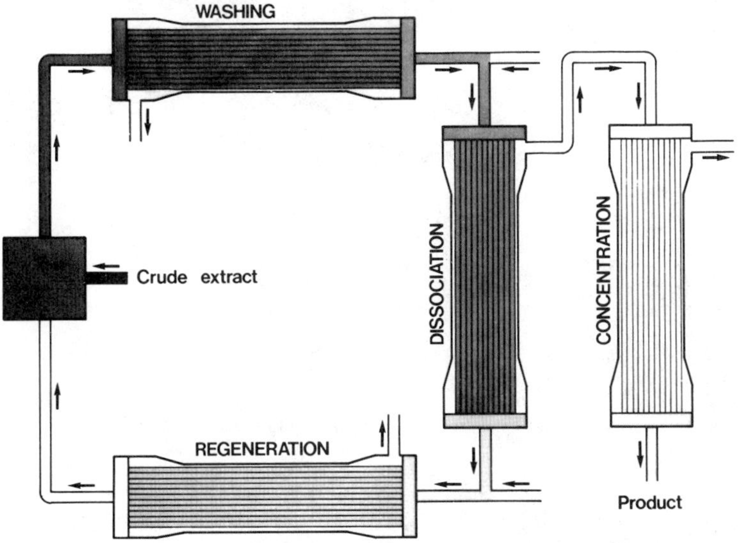

FIGURE 1. Presentation of ultrafiltration affinity purification. The cutoff of the various membranes is selected in such a manner that the ligands are continuously recycled on the inner side of the hollow fibers, thereby creating possibilities for continuously operated systems.

Incubation: A crude extract, containing the protein of interest is incubated with the ligand solution.

Washing: The ligand-containing solution is then transported to a hollow-fiber unit where nonbinding material is washed out.

Dissociation: The ligand-protein complex is treated with a buffer that dissociates the bound materials from the ligands. The dissociated protein leaves the ligand stream in a separate membrane unit and is then concentrated and dialyzed in another hollow-fiber unit.

Regeneration: The ligand stream after the dissociation stream is reconditioned before entering the binding unit.

in a continuous fashion. This report concerns itself with ultrafiltration affinity purification, a technique combining the specificity of affinity techniques with the advantages of membrane technology. The basic principle behind this technique is that the substance to be purified, when free in solution, passes unimpeded through the pores of the membrane, whereas when bound to a macromolecular ligand, it is restricted to the side of the membrane where the ligand is placed. The purification process is then a sequence of unit operations, binding takes place in a separate unit (see FIG. 1), and unbound material is eliminated as the ligand-binder complex is transported through a washing unit (see FIG. 1). After the ligand-binder complex has been rinsed of all contaminants, a dissociating medium is introduced and the freed binder can pass

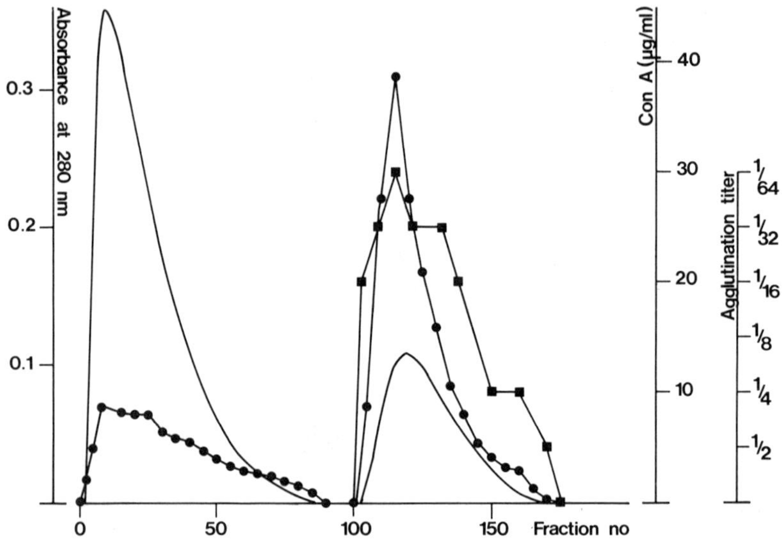

FIGURE 2. A typical elution profile obtained during a purification of concanavalin A from *Canavalia ensiformis*.

A crude homogenate of *Canavalia ensiformis* was incubated with heat-killed yeast cells. Washing was performed to eliminate material not binding to the sugar residues on the cell surface (Peak I).

When nonbinding material has been washed out, the bound material was eluted with 0.5 *M* glucose (Peak II).

Fractions were collected, dialyzed and analyzed for concanavalin A content (●) (immunological binding assay, ELISA) as well as biological activity (■) (ability to agglutinate red blood cells).

through the pores of a new membrane. The ligand is regenerated before entering the binding unit again whereas purified binder is treated in an ultrafiltration unit to remove dissociation medium and also to concentrate it. The model system reported here is the purification of concanavalin A from a crude extract of *Canavalia ensiformis* using heat-killed yeast cells (*Saccharomyces cerevisiae*) as macromolecular ligand.

As yet, the units have not been optimized in relation to each other with respect to surface area and so forth. This is why a constant, smooth flow through the whole system was not possible. Instead, reservoirs were used between the different unit operations. The membranes used for binding, washing, and separation were declared to have a molecular weight cutoff of 10^6 and those used for reconditioning and concentration had cutoffs of 10^3. All membranes were generously supplied by Gambro AB, Lund, Sweden.

FIGURE 2 is an example of a typical elution profile. Peak 1 is registered during the washing procedure directly after binding and peak 2 is the elution profile of concanavalin A from yeast cells. In order to obtain high yields in this purification process, it is important to eliminate all low-molecular-weight ligands before binding. With the yeast cells, this was achieved by a prewashing step in the membrane system.

The eluted material in peak 2 was tested for biological affinity as well as immunological reactivity (indicated in FIG. 2). Analysis of purity using SDS-polyacrylamide gel electrophoresis revealed that one single band was obtained. Yield in the total process was 70%. This level of yield has also been found in other systems.

CONCLUSION

Ultrafiltration affinity purification is a gentle and efficient method that can be operated continuously on a large scale. This is achieved by constantly recirculating the ligand and by keeping an ordered sequence for the different unit operations in the purification cycle.

ACKNOWLEDGMENTS

This project was supported by the National Swedish Board for Technical Development.

Recovery of Strategic Elements by Biosorption

B. VOLESKY, M. SEARS, R. J. NEUFELD, AND M. TSEZOS

Biochemical Engineering Unit
McGill University
Montreal, Quebec

INTRODUCTION

Biosorption is a property of certain types of microbial biomass in which heavy metallic and actinide elements are sequestered. This particular behavior has been traced to the cell wall of some filamentous fungi that exhibit high levels of metal uptake whether they are living or dead.[1,2] The chitin (or chitosan) component of *Rhizopus arrhizus* cell wall structure has been linked with the squestering activity for uranium and thorium,[3] which are deposited through a combined mechanism of coordination, adsorption, and microprecipitation. Chemical bonding evidence led to proposing the uranium and thorium deposition hypotheses supported by uptake kinetic data.[4,5] Raw biomass of the entire genus *Rhizopus* exhibits proven uranium or thorium uptakes on the order of 20% of dry weight making it an attractive biosorbent material.[6]

EXPERIMENTAL AND RESULTS

Seawater Extraction

A continuous-downflow, packed-bed column (i.d. 5 mm, 110 mm long) containing a mixture of 2 g of glass powder (200 mesh, to prevent compacting) and 60 mg of powered *Rhizopus arrhizus* biomass was in contact with 82 liters of seawater during 262 hours. A regulated 82-lb/in^2 line pressure was required to maintain a steady flow rate through the column. The exposed column packing was divided into upper and lower halves, separately digested by concentrated HNO_3. Glass powder removed by a preconditioned filter was washed with concentrated reagent-grade HNO_3.

Acid (digesting) solutions, seawater, and "blank" controls, were analyzed for the metals by an Inductively Coupled Plasma Atomic Spectrometer (Jarrell-Ash, simultaneous, multielement) and respective results are shown in TABLE 1, where the last column is indicative of the "partition coefficient" (k) order of magnitude. The coefficient is expressed as a ratio of the concentration in the upper half of the sorption bed biomass to that found in the seawater for each individual element. This preliminary experiment indicates the highest affinity of *R. arrhizus* raw biomass for uranium ($k \geq 10^6$).

Sorption-Desorption of Uranium

Compaction and high pressure drops in the flow-through, packed-bed contactor system using finely powdered *R. arrhizus* biomass led to application of the same biomass but propagated and applied in the form of 20–30 mesh pellets that, dried, filled the column (i.d. 8 mm, 6 ml active volume). Solutions of either 90 mg/l uranium or acid were alternatively passed through the contactor at low pressures and 0.83 ml/min. FIGURE 1 illustrates the elution of the pellet-packed column with 0.1 *M* H_2SO_4 and 1 *M* HNO_3. The sorption-desorption cycle was repeated 8 times, outlet uranium concentration monitored continuously by a modified Arsenazo III method.[7] After the first cycle, the new uranium uptake was ~50% lower but 85% reversible in all 8 subsequent cycles. Under the elution conditions applied, some uranium remained irreversibly bound to the biomass. In these preliminary experiments the eluting

TABLE 1. Analytical Results of the Seawater Extraction Biosorption Experiments

	Concentration in the Sample			Partition
Element	Sea Water (mg/l) or (mg/g) × 10^3	Lower Bed (mg/g)	Upper Bed (mg/g)	Coefficient (max.) k
Al	<1	1.0807	1.6244	>10^3
Be	<0.005	same	*	
Ca	360	10.27	12.67	
Cd	<0.4	same	*	
Cr	0.07	3.14	4.09	5.8 × 10^4
Co	<0.1	same	*	
Cu	<0.04	3.76	6.77	>1.7 × 10^5
Fe	<1	55.12	50.35	>5 × 10^4
Pb	<0.3	1.837	1.787	>6 × 10^3
Mg	1130	14.59	19.33	
Mn	<0.07	0.983	1.007	<1.4 × 10^4
Mo	<1	same	*	
Ni	<0.07	1.037	1.128	>1.6 × 10^4
P	<3	4.32	8.12	
K	410	<3	<3	
Ba	<0.05	same	*	
St	2.0	3.46	34.92	1.7 × 10^4
Ag	<0.3	same	*	
Na	9880	52.96	58.48	
Sr	6.70	0.129	0.178	26.5
B	3.9	0.075	0.195	50
Sn	<1	same	*	
Ti	<0.07	same	*	
Th	<0.3	<0.3	9.42	>3.1 × 10^4
Zr	<0.2	same	*	
V	<0.02	same	*	
Zn	<0.1	1.0807	1.462	>1.5 × 10^4
As	<0.005	0.0324	0.0325	>6.5 × 10^3
U	0.002	0.0184	2.274	1.14 × 10^6

solution concentrated uranium almost 6 times. Also tested was bimass of *R. oligosporus,* which exhibited nearly total reversibility of the uranium uptake.

Clearly not only the uptake capacity, but the degree of reversibility must be factors in evaluating biosorbent materials. While a metal-concentrating biosorbent may have numerous applications in metal recovery operations, the irreversible binding may be of interest in the production of nonleachable solid concentrated wastes for environmentally undesirable elements or isotopes.

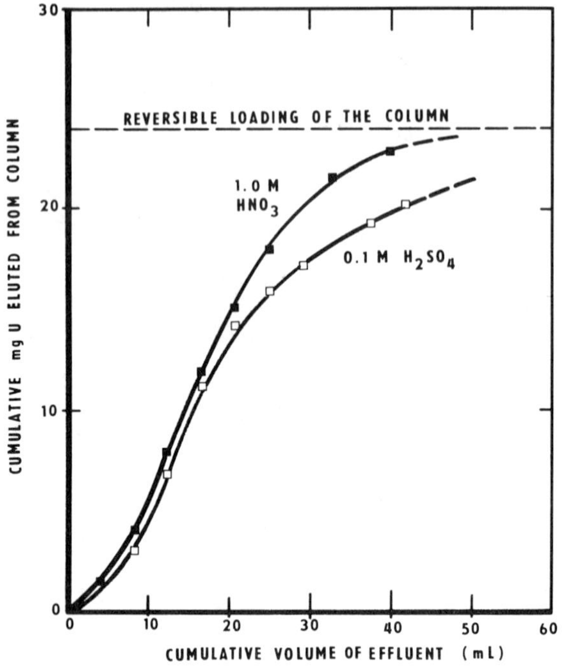

FIGURE 1. Elution of uranyl-loaded column with 1.0 and 0.1 *M* acid. Sorbent: *Rhizopus arrhizus* biomass, 20–30 mesh pellets; Volume = 6 ml; Acid Flow rate = 83 ml/min.

REFERENCES

1. STAMBERG, K., J. STAMBERG, J. KATZER, H. PROCHAZKA, *et al.* 1977. Canadian Patent. 1009600.
2. ZAJIC, J. E. & Y. S. CHIU. 1972. Dev. Ind. Microbiol. **13:** 91–100.
3. TSEZOS, M. & B. VOLESKY. 1981. Biotechnol. Bioeng. **23:** 583–604.
4. TSEZOS, M. & B. VOLESKY. 1982. Biotechnol. Bioeng. **24:** 385–401.
5. TSEZOS, M. & B. VOLESKY. 1982. Biotechnol. Bioeng. **24:** 955–69.
6. VOLESKY, B. & M. TSEZOS. 1982. U.S. Pat. 4,320,093.
7. TREEN, M. 1981. M. Eng. Thesis. McGill University. Montreal, Canada.

Integrating Biochemical Separation and Purification Steps in Fermentation Processes

HENRY Y. WANG

Department of Chemical Engineering
The University of Michigan
Ann Arbor, Michigan 48109

INTRODUCTION

Owing to recent developments in molecular biology and biotechnology, we have seen rapid growth in the development of industrial processes to produce valuable substances by microbial means. The idea is not new to the fermentation technologist because they are the means for producing beverages, antibiotics, and many other complex organic materials for commercial use. There are some inherent engineering disadvantages of fermentation processes as compared to conventional chemical processes that need to be solved in order to make these processes economically attractive. The traditional fermentation process involves growing a pure culture of a microorganism that produces the product of interest in a fermenter after sterilization and inoculation. The fermentation can be carried out in batch, semicontinuous, and continuous modes. Depending on whether the desired product is intracellular or extracellular, the harvested broth is then subjected to various separation and purification steps until the specific product of interest is isolated from the fermentation broth. The common steps involved are cell removal, volume reduction, isolation, and purification until the specified product purity has been achieved.[1] The fermentation and the subsequent recovery steps have traditionally been carried out in separate departments and interaction among these groups has been minimal. Kaufman and Paul[2] pointed out the interactive nature of these steps, especially between the fermentation and its subsequent product recovery scheme. In this paper, this interactive nature can be examined and utilized to optimize the entire biochemical process. A hybrid extractive fermentation system developed at the University of Michigan will be used as a model system to illustrate this point.

PROBLEMS ASSOCIATED WITH FERMENTATION AND RECOVERY

The microbial fermentation process has become an integral part of the recent advances in genetic engineering and biotechnology. Most of the fermentation processes are complex processes that involve multiple substrate, gas-liquid mass transfer, and sometimes non-Newtonian broth rheology. The final desired product concentration in the aqueous medium can vary over a wide range. Some of the organic acid fermentations such as citric acid and glutamic acid, a concentration of 100 grams per liter or above can be achieved. On the other hand, some more complex macromolecular product fermentations, such as vitamin B_{12} and human insulin, a concentration of 1 gram per liter or less is satisfactory because these products command premium prices. As shown in FIGURE 1, the fermentation product cost is more or less dictated by the final product concentration in the fermentation broth. It is always desirable to increase the final product concentration in the fermentation broth. The primary causes of the dilute nature of the fermentation end product are that the product itself can stop its own synthesis (product regulation) and product degradation in the fermenting broth. Even though these effects that occur at high product concentration are known, approaches for major improvement have been primarily limited to genetic manipulation. By means of random mutation and selection, strains tolerant to a high concentration of product may be selected to avoid such problems.

Isolation of microbial products produced by fermentation is a specialized field because of the unique character of the fermentation broth from which the products are synthesized. The complexity of the heterogenous mixture of soluble and insoluble materials is required to maintain or improve the fermentation, but it also causes problems for the separation scientists and engineers. Although not always, the product usually is present in a relatively low concentration compared to the concentration of other materials present. The desired product is also prone to degradation by extremes of temperature and pH. Most important of all, the product must be protected from both enzymatic/microbial and chemical degradations. Depending on whether the fermentation product is intracellular or extracellular, the sequence of recovery and purification steps is shown in FIGURE 2. The extraction efficiency of a product recovery and purification scheme is dictated not only by the final fermentation product concentration but also by the number of steps required to achieve the specified product purity that, in turn, is dictated by the market demand. Increasing the number of steps in the isolation of microbial product from the fermentation broth will increase the

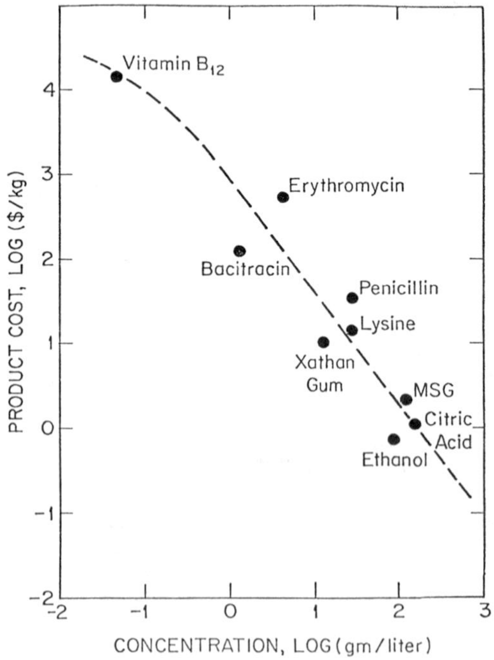

FIGURE 1. Sales price of various fermentation products versus the final fermentation product concentrations.

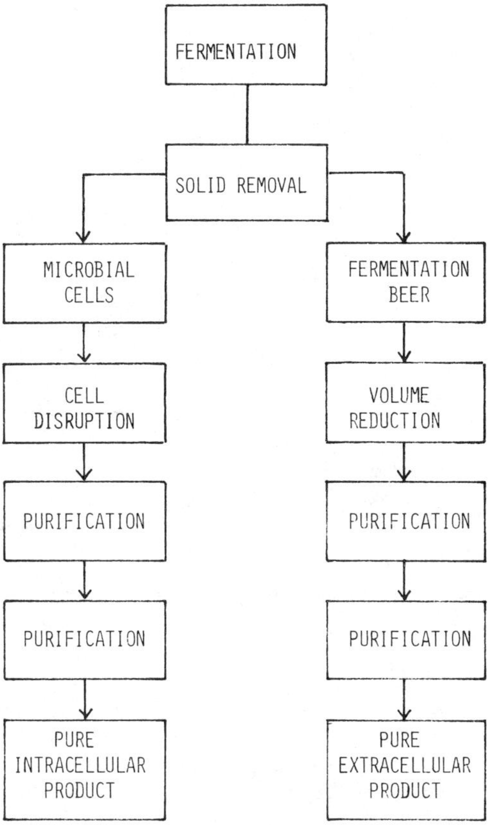

FIGURE 2. Various steps involved in fermentation processes and subsequent recovery processes.

product purity, but it also decreases the overall extraction efficiency (FIG. 3). The initial isolation steps such as cell removal and volume reduction generally cause a more drastic drop in the overall extraction efficiency as depicted in FIGURE 3.

HYBRID EXTRACTION-FERMENTATION SYSTEM

The need to achieve high product concentration in the fermentation broth is essential for both the fermentation (product yield) and recovery (extraction efficiency). One engineering approach is to couple the extraction with the fermentation step so that the inhibitory metabolic products can be continuously removed from the aqueous broth so that the inhibitory effects and/or degradation phenomena can be avoided or minimized throughout the fermentation. In this manner, the product formed may also be concentrated, which will facilitate the eventual purification steps as well as avoiding the metabolic inhibition and degradation phenomena. Various attempts have been made to improve fermentation yield and alleviate product inhibition using ion-exchange resins, solvents, and membrane processes (TABLE 1). Certain criteria of the specific extractant must be met in order to carry out this type of

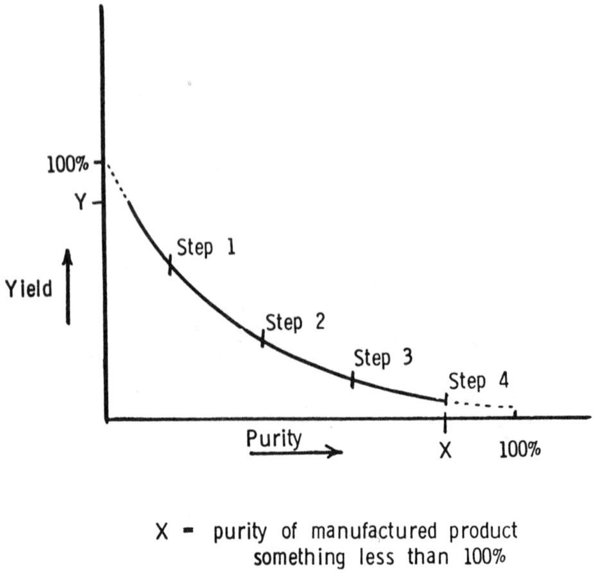

FIGURE 3. Yield-Purity relationship. The extraction efficiency as a function of product purity.

extraction-fermentation process (TABLE 2). After reviewing various separating agents, such as nonaqueous solvents, ion-exchange resins, and activated carbon, we found that the nonionic polymeric resins such as XAD-2 and XAD-4 (Rohm and Haas, Philadelphia, PA) that do not contain any specific function groups are most suitable for this purpose.

TABLE 1. Fermentation with On-line Extraction

Fermentation	Separation Techniques	Advantages	Disadvantages	Reference
Antibiotics	Ion-exchange resins	Eliminate filtration	Excess ions	3
Salicyclic acid	Dialysis	Alleviate product inhibition	Increased volume	4
Salicyclic acid	Ion-exchange resins	Alleviate product inhibition	Excess ions	5
Cycloheximide	Dialysis solvent	Alleviate product inhibition	Solvent toxicity	6
Carotenoids	Kerosene grease	Increase product yield	Limited application	8
Ethanol	Vacuum distillation	Increase productivity	Limited application	8
Ethanol	Activated carbon	Increase productivity	Toxicity	9
Anaerobic digestion	Activited carbon	Process stability	Recovery	10
Anaerobic digestion	Membrane	Increase productivity	Membrane fouling	11

TABLE 2. Criteria for Extractants Used in On-line Extraction-Fermentations

1. The extraction must be selective.
2. The extractant must be nontoxic to the growing culture.
3. The extraction system must be capable of aseptic operations.
4. The extractant must have a high loading capacity for the product.
5. The extractant must be reusable if it is expensive.

IMMOBILIZED PRODUCT FERMENTATION (IPF)

Cycloheximide fermentation, antifungal and antibiotic, has been used in our investigation. Biochemical and fermentation studies of the cycloheximide fermentation have been extensively investigated.[6,12] The time course for the conventional cycloheximide fermentation has been observed to be quite reproducible. The cycloheximide concentration in the fermentation broth reached a peak at log 168 (FIG. 4). This

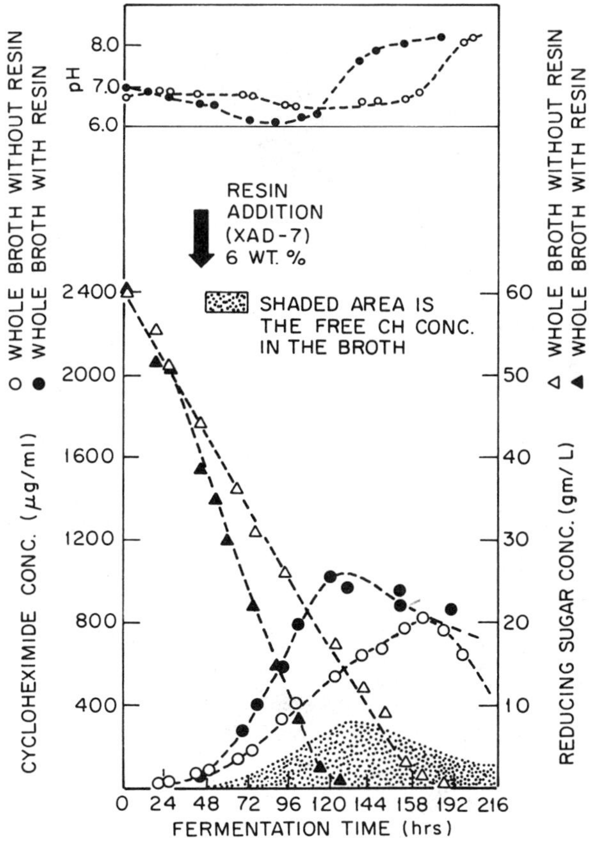

FIGURE 4. Comparison between a normal cycloheximide fermentation using 6 wt % XAD-7 resins. The adsorbents were added at log 48.

corresponds to the depletion of glucose in the medium. The medium pH rose above 7.0 and product degradation occurred (cycloheximide is less stable under alkaline conditions). The maximum cycloheximide titer that we can achieve in the laboratory fermenter is about 850 μg/ml. Kominek[12] has shown this fermentation to be feedback regulated.

By adding sterilized polymeric resins such as XAD-7 and XAD-4 directly into the fermentation broth, the profiles for both glucose utilization and cycloheximide production were changed. When 6 wt % of XAD-7 were added at the 48th hour of a cycloheximide fermentation (FIG. 4), the glucose utilization rate was increased from 0.36 g/1-hr to 0.5 g/1-hr indicating that the presence of these solid adsorbents somehow change the physiological activities of the microbial culture. Since the addition was made around the 48th hour when the cell growth was completed, this enhanced glucose utilization was not accompanied with additional cell growth. The peak product yield was higher with a value of 27.8 mg CH/g of glucose instead of the normal fermentation yield of 18.6 mg/g of glucose. Obviously, the adsorption rate by the XAD-7 resins was not fast enough, thus accumulation of free cycloheximide in the broth was observed (FIG. 4).

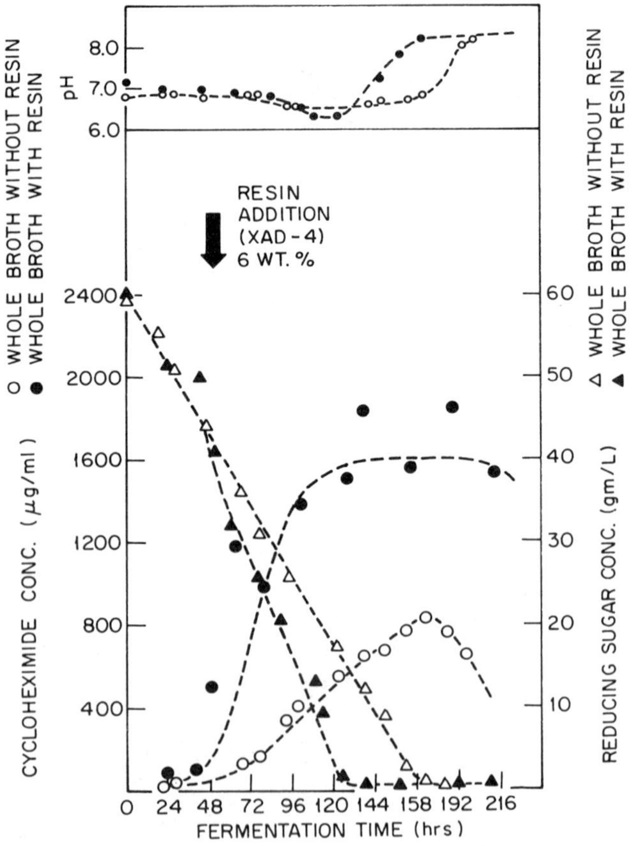

FIGURE 5. Comparison between a normal cycloheximide fermentation and resin addition cycloheximide fermentation using 6 wt % XAD-4 resin. The adsorbents were added at log 48.

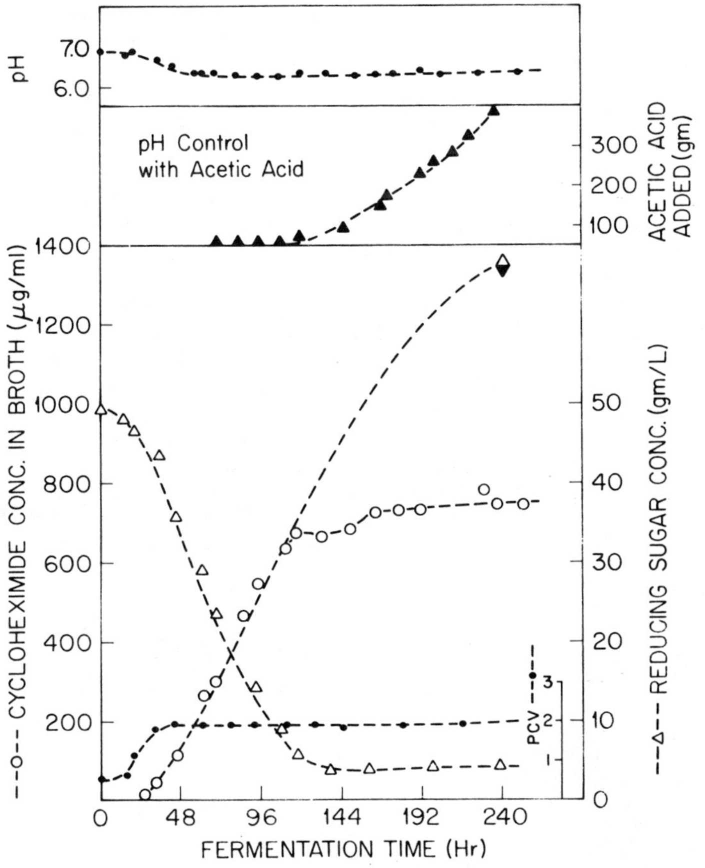

FIGURE 6. Cycloheximide accumulation in broth of the immobilized-resin fermentation.

When we used the same amount of XAD-4 resin instead of XAD-7 resin (FIG. 5), a similar phenomenon was observed. Since XAD-4 polystyrene- based polymeric resin is more hydrophobic than XAD-7 and contains more surface area, all cycloheximide produced was adsorbed onto the resins instantly and the free broth activity of cycloheximide was low throughout the fermentation period. The maximum cycloheximide production rate was 17.5 mg/1-hr with a product yield of 36.5 mg CH/g of glucose. These were substantially higher than the control fermentation run. Over 98% of the produced cycloheximide adsorbed onto the resin particles. We could conceivably harvest the antibiotic-loaded resins and drastically reduce the volume of materials that need to be processed for further separation and purification.

RECOVERY OF THE POLYMERIC RESINS

Dispersing the resins in the culture broth has several engineering problems such as clogging the valves and attrition of the resin particles which are also abrasive to the tank wall and the shaft seals. We have been experimenting with keeping the resins in dialysis membrane bags (Spectrapor M.W. cut off 8,000) that were mounted on the baffles inside the fermenter. The resin and the culture medium were sterilized together. Dense resin cakes were formed in these bags during the fermentation and diffusion into the cakes appeared to significantly limit the removal rate, and as a result cycloheximide accumulated in the culture broth (FIG. 6). Higher final cycloheximide titer was achieved, but a significant amount remained in the fermentation broth. Visual inspection of the resins in the membrane bag indicates very limited penetration by the fermenting broth. An alternative to this approach is to physically separate the resin from the culture medium in an external extractor. Broth can then be pumped between the fermenter and extractor during fermentation.

The antibiotic-loaded polymeric resins can easily be separated from the harvested culture broth using a vibrating screen or similar solid-liquid separation devices. For our studies, butyl acetate was used to elute the product from the resin. The purity of the butyl-acetate-eluted cycloheximide was then compared (TABLE 3). The cycloheximide eluted from the resins was comparable to those extracted directly from the culture broth. On the other hand, the product eluted from the resins that were enclosed in a dialysis membrane bag was considerably more pure (TABLE 3). These polymeric adsorbents can be reused after cleaning with caustic and methanol. Ideally, we would like to develop a system that would selectively adsorb the desired product from the fermentation broth. This can reduce subsequent purification steps. Currently, we are studying this possibility.

TABLE 3. Purity of Extracted Cycloheximide from the XAD-4 Resin

Fermentation	Solvent	Purity (%)
Shake flask (control)	Butyl acetate	39
Fermenter (control)	Butyl acetate	24
XAD-4 Resin (dispersed)	Butyl acetate	54
XAD-4 Resin in dialysis bag	Butyl acetate	77
XAD-4 Resin in dialysis bag	Methylene chloride	74
XAD-4 Resin in dialysis bag	Methanol	42

CONCLUSION

The need to increase the final product concentration in fermentation processes leads us to develop this immobilized product fermentation (IPF). By directly adding a separating agent such as XAD-4 resin that is nontoxic to the microbial cells into the fermentation broth, the antibiotic synthesis rate has been shown to increase drastically and the final product yield increases by a factor of 3 or more. The antibiotic-concentrated resins can easily be separated from the fermentation broth and the antibiotic can be extracted using solvents. Volume reduction and partial purification were being demonstrated in the case of cycloheximide fermentation. Currently, we are looking into the possibility of extending this antibiotic synthesis and optimizing this IPF process through computer control.

ACKNOWLEDGMENTS

The author would like to acknowledge the technical assistance of F. Robinson and K. Sobnosky. The financial support from NSF CPE-8010868 is also acknowledged.

REFERENCES

1. EDWARDS, V. 1969. Adv. Appl. Microbiol. **11:** 157–164.
2. PAUL, E. L., A. KAUFMAN & W. A. SKLARG. 1981. Ann. N.Y. Acad. Sci. **369:** 181–186.
3. DENKEWALTER, R. G. & J. GILLIN. 1959. Auslegeschrift 1062891.
4. ABBOTT, B. S. & P. GERHARDT. 1970. Auslegeschrift 1062891.
5. TONE, H., A. KITAI & A. OZAKI. 1968. Biotechnol. Bioeng. **10:** 689–692.
6. KOMINEK, L. A. 1975. Antimicrob. Agents Chemothera. **7:** 861–866.
7. CIEGLER, A., G. E. N. NELSON & H. H. HALL. 1962.
8. CYSEWSKI, G. R. & C. R. WILKE. 1977. Biotechnol. Bioeng. **19:** 1125–1130.
9. WANG, H. Y., F. ROBINSON & S. S. LEE. 1981. Biotechnol. Bioeng. Symp. **11:** 555–560.
10. SPENCER, R. R. 1978. Biotechnol. Bioeng. Symp. **8:** 257–268.
11. OLMSTEAD, D. R., T. W. JEFFRIES, R. NAUGHTON & H. P. GREGOR. 1980. Biotechnol. Bioeng. Symp. **10:** 247–258.
12. KOMINEK, L. A. 1975. Antimicrob. Agents Chemother. **7:** 856–860.

The Integration of Unit Operations for Bulk Product Manufacture by Continuous-Flow Fermentation Processes—Lessons from SCP Process Development

GEOFFREY HAMER

Swiss Federal Institute for Water Resources and Water Pollution Control
Swiss Federal Institutes of Technology
CH-8600 Dübendorf, Switzerland

INTRODUCTION

The concept of producing Single-Cell Protein (SCP) as a protein-rich component for inclusion in animal feeds and/or human foods dominated developments in biotechnology between the mid-1960's and the mid-1970's. More recently, SCP projects have remained important vehicles for further technological developments, and, contrary to popular opinion, SCP production will probably become a strategically important industrial microbiological process in some regions of the world. SCP production process development demonstrated, essentially for the first time, that both imaginative process-oriented microbiological research and appropriate and imaginative process engineering development could result in fully integrated, continuous-flow, biotechnological production processes without resort to either the traditional technology or the folklore of any established sector of the microbiological industries. In fact, SCP process development introduced an important and much-needed breath of fresh air into biotechnology by demonstrating that effective process design depends on the successful application of well-tried engineering principles that have been used to build the existing bulk chemicals industry.[1]

As the basis of this discussion on lessons that can be learned from the integration of unit operations for SCP manufacture and their application to other fermentation processes, three SCP process routes will be considered; those based on waxy n-alkanes, on methanol, and on methane as feedstocks. All three of these routes were researched and developed for application in technologically sophisticated industrial environments and resultant commercial manufacturing ventures either were or would have been subject to the conventional economic requirement of generating profit from product sales in return for risking capital investment in a production venture.

ECONOMIC EVALUATIONS AND PROCESS INTEGRATION

The economic evaluation of industrial processes and process routes is an essential activity that is fraught with difficulty. Evaluation procedures tend, for the most part, to be conservative with respect to the claims of proponents of either the process or process route undergoing evaluation. However, excessive stringency with respect to the evaluation of prospective manufacturing ventures can result in the loss of profitable manufacturing ventures to competitors.

Estimates of the total investment for any proposed manufacturing venture can be divided into three parts according to the degree of financial risk. The three parts are:

(1) the fixed investment in the immediate processing area, that is, investment inside battery limits;
(2) the investment in auxilliary services, that is, investment outside battery limits; and
(3) the investment in working capital.

Together, they comprise the money tied up and risked in the venture. The fixed investment carries the highest degree of risk, the working capital the least. Manufacturing costs are the costs incurred in operating a process plant and can also be divided into three parts, which are:

(1) costs proportional to the fixed investment, that is, factors such as labor and material aspects of maintenance, insurance, and administration that are independent of the rate of production and can be expressed as a percentage of the fixed investment;
(2) costs of raw materials, utilities, chemicals, other materials, quality control, maintenance costs related to operation, royalties and license fees that depend on the rate of production; and
(3) direct costs of maintaining the operating labor force including, in addition to salaries, overheads and the costs of supervision.

The gross profit obtained as a result of operating a particular process plant is the difference between the net income from the annual sales, after distribution, promotional, and sales costs have been deducted, and the annual manufacturing costs. The net profit is the expected annual return on the investment after the deduction of depreciation and taxes. Depreciation schedules are frequently complex, government-stipulated and location-dependent, and, frequently, tax incentives are such as to demand optimization of permitted depreciation schedules.

When evaluating a process with a view to improving its profitability by integration and optimization, it is essential to understand the technological factors that significantly affect the overall economics of manufacture. In the case of microbiological processes, the technological factors can be of either a microbiological or a process engineering nature. However, before discussing these factors, it is first appropriate to clarify what is understood by process integration.

Traditional process design practice considers the optimization of individual unit operations with little effort devoted to their integration as an overall process. For a decade, much of the world has been subjected to a series of step increases in the price of crude oil, the primary energy source for the process industries, and of feedstocks derived from oil. Before these politically mediated price increases, the bulk product process industries have always based their strategy on the premise that the cost of products will decrease with increasing scale of production. Escalation in both fuel and feedstock prices and the effects of associated inflation on plant construction costs has come close to destroying this principle and the need for savings in both capital investment and expenditure on energy and feedstocks by improved levels of process integration has become increasingly evident. Process integration can involve both the unit operations that comprise those parts of the process within battery limits and interactions between operations undertaken outside battery limits and those conducted within battery limits. Here, emphasis will be placed on the former interactions, but before embarking on further discussion of these, there are a few comments, concerning the latter, that are particularly appropriate to industrial microbiological processes and how they find difficulty in fitting into production sites developed primarily for bulk chemicals production by conventional technology.

Industrial microbiological processes are essentially low-temperature processes, with reactor operating temperatures usually less than 40°C, that require large quantities of both low-temperature cooling water and high-quality process water, significant quantities of clean steam, large quantities of clean air, feedstock and product storage facilities that minimize contamination of the stored materials, hygenic surroundings, specialized wastewater treatment facilities, and sophisticated quality-control laboratories. Typical production sites for bulk chemicals manufacture, particularly petrochemicals manufacture, are unlikely to be able to easily satisfy such requirements, thereby forcing an unusually large part of the essential service facilities to be constructed essentially inside battery limits with attendant increases in both the capital investment and the investment risk. The overall economics of microbiological process operation are obviously enhanced at locations where several such processes can receive appropriate services from optimized central facilities located outside the battery limits of the individual processes.

In commercial microbiological processes, the microbiological factors with the greatest impact on the process are:

(1) yield coefficients for biomass and/or product;
(2) growth rates and/or product production rates;
(3) affinity for carbon energy substrates; and
(4) stability and fastidiousness.

The process engineering factors with the greatest impact on the process are:

(1) feedstock conversions and conversion coefficients;
(2) productivities; and
(3) product concentrations.

Both the microbiological and the process engineering factors are intimately related in technologically and commercially successful processes.

MICROBIOLOGICAL PROCESSES FOR BULK PRODUCTS

Many of the established microbiological process industries are strictly conservative in nature, having successfully served mankind for centuries. Amongst such industries are those for the production of alcoholic beverages, of fermented dairy products, and of the fermented soy bean and fish products of the Orient. In such industries, product quality and market price are frequently assessed by subtleties of either taste or odor rather than by some more exactly definable criterion. In the future, the value of new products, particularly bulk products, is most unlikely to be established on a similar basis. As has been the case for SCP, the assessment of a product's market potential and value will be based exclusively on overall production costs, on predictable, but uncontrollable, variations in feedstock and energy costs and on product functionality.

Microbiological manufacturing ventures are usually initiated by ill-defined statements of need rather than by the discovery that some novel compound is produced by some newly isolated microbial culture. Examples of problem statements that have resulted in major microbiological process research and development programs are:

(1) that from the late 1960s onwards, both Europe and Japan would experience a critical shortage of protein-rich ingredients for animal feeds unless alternative sources of supply could be developed;
(2) that either resource depletion or production restrictions would, from the mid-1970s onwards, result in fluid fossil fuel price increases such that the production of fluid fuels from alternative indigenous resources would become essential in most developed and developing countries; and
(3) that the future recovery of a significant part of the proven reserves of crude oil, in strategically nonvulnerable regions, will depend on the development of enhanced oil recovery technology.

The first example cited was clearly responsible for the extensive research and development programs directed towards SCP production, the second was responsible for the establisment of extensive programs aimed at the conversion of renewable, indigenous, biomass resources into ethanol and/or methane, and the third, after the evaluation and rejection of synthetic polymers, the establishment of product and process research and development programs directed towards the production of biopolymers with appropriate properties for use in enhanced oil recovery. The commercially successful outcome for all these programs has either depended on or depends on both the long-term validity of the particular statement of need and on the absence of political decisions that completely alter the basic assumptions supporting the original statement of need.

Three examples of microbiological products based on the concept of novel compound discovery, but divorced from any justifiable statements of need, are the proposals for the production of poly-β-hydoxybutyric acid using *Alcaligenes eutrophus,*[1] of emulsifying agents, using numerous hydrocarbon-oxidizing bacteria,[2] and of microbial insecticides, using *Bacillus thuringiensis.*[3]

The future potential for the production of bulk products by microbiological processes is strictly limited. Recently, Humphrey,[4] when addressing this particular subject from the viewpoint of the potential of applied genetics, reported a consensus view with respect to only eight chemicals. At the present time, the only microbiological processes for bulk product production that are being researched and developed, and result from statements of need that can be substantiated, are those for biopolymers, ethanol, and methane, and it is these three processes that will be considered with respect to increases in understanding in process integration that have resulted from SCP process development.

OVERALL PROCESS SYNTHESIS

Irrespective of the feedstocks used, the products produced and the technology employed, all microbiological processes for bulk product production will be operated in the continuous flow mode and will comprise an ordered series of unit processes, all with a significant degree of dependency on each other. Such processes will be required to be sufficiently reliable so that they produce product that meets specification for between 300 and 330 days per annum. The major objective of process integration is to maximize resource utilization without incurring undue penalties with respect to either capital investment or manufacturing costs. Unfortunately, process integration usually reduces operability and can cause some problems in maintaining high on-stream factors.

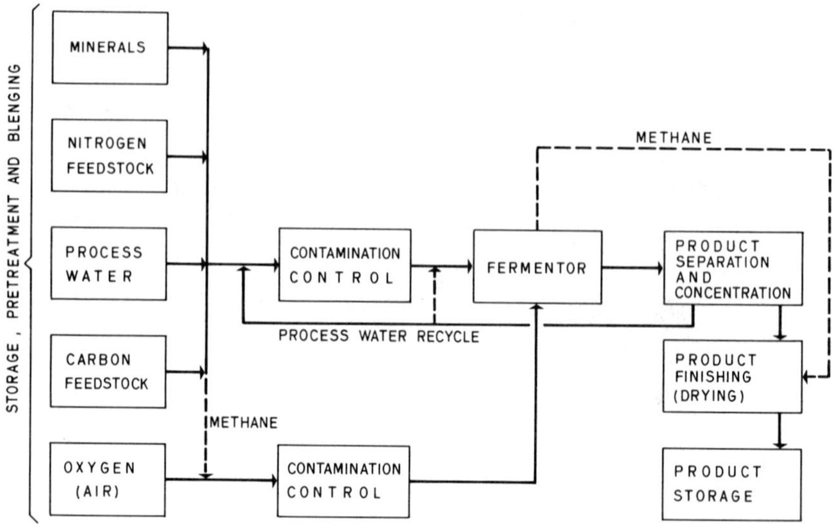

FIGURE 1. Block diagram for SCP production.

A block diagram of the several process steps comprising a typical SCP production process is shown in FIGURE 1. The several steps that comprise the overall process are:

(1) feedstock storage, pretreatment, and blending;
(2) contamination control;
(3) fermentation;
(4) product concentration and separation;
(5) product finishing; and
(6) product storage.

Several of the steps are multiple operations involving more than one unit process. Many minor, but essential, systems are omitted in FIGURE 1. In SCP manufacturing processes, virtually all major process innovations have occurred either in the fermentation step or in the product separation and concentration steps of the processes.

Bulk product production requires single-plant capacities in excess of 50,000 tons per annum and process engineering philosophy requires that maximum benefits be derived from economy of scale. Hence, the traditional approach of multiple process units, as practiced by the fermentation industry for fine chemical and intermediate volume products, was unacceptable, and, hence, very large single units for each process operation became an overall, but not necessarily attainable, objective. Perhaps the best example of this philosophy is the single fermenter used in the ICI Pruteen process, where production is within the range 54,000–70,000 tons dry product per annum.[5]

Relatively few realistic economic evaluations of SCP product processes have been published, but it has been estimated that for a methanol-based route, the fermentation step accounts for 41 percent of capital investment and utilities costs,[6] clearly indicating the significant role of the other process steps. The primary problems in the engineering design of SCP production plants have been the availability, the reliability, and the efficiency of appropriately sized equipment that could operate continuously. As a result of such restrictions, a marked similarity with respect to the unit processes employed in the several production processes developed has occurred. Perhaps the most important factor has been the absolute requirement that no excessive wet product holdup occurs at any stage, a feature that is also critical in biopolymer production.

In SCP production processes, the best examples of process integration are process water recycle and heat economy in the contamination prevention stages, which are features of all major processes, and the utilization of fermenter off gas containing residual methane, after enrichment, as fuel in the product-finishing (drying) step, specifically in methane-based SCP processes.[7]

The importance of process water recycle has been discussed elsewhere,[8] and the problems pertinent to process water recycle may be summarized as:

(1) recycle of inhibitory products resulting from either metabolic activities or cell lysis in the product concentration and separation stage;
(2) recycle of inhibitory products resulting from either metabolic activities or cell lysis in the fermentation stage;
(3) recycle of unutilized impurities present in the feedstock;
(4) recycle of unutilized feedstocks such that physical and/or chemical inbalances and dynamic instability occurs in the fermenter; and
(5) recycle of additives used to enhance product concentration and separation that adversely affect fermenter performance.

All these factors can adversely affect biopolymer and ethanol production processes in similar ways to the way in which they adversely affect SCP production.

Simple block diagrams of typical biopolymer, ethanol and methane production processes are shown in FIGURES 2, 3 and 4. As far as the number and overall types of process steps involved in continuous biopolymer production processes are concerned, marked similarities with respect to SCP production processes are evident. However, it must be stressed that the actual unit processes that comprise the individual process

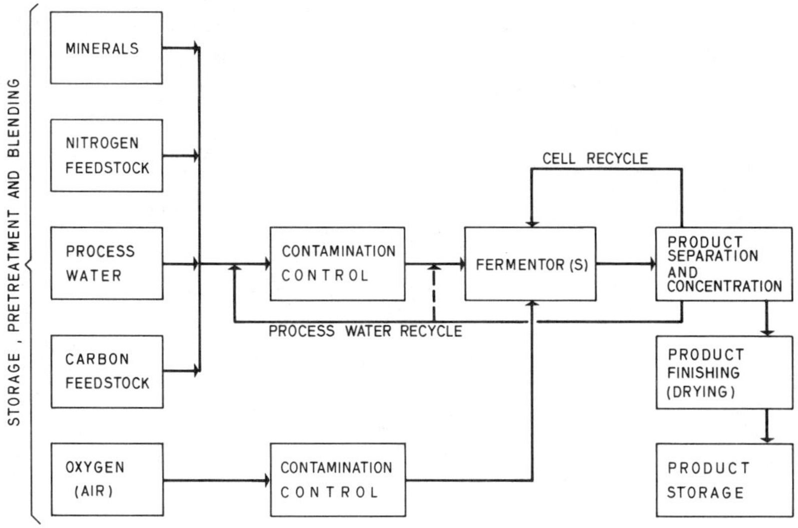

FIGURE 2. Block diagram for biopolymer production.

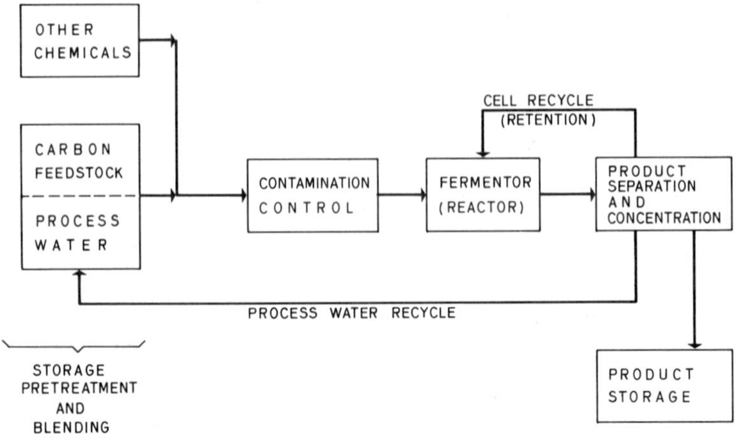

FIGURE 3. Block diagram for ethanol production.

steps are often distinctly different, particularly in the product separation and concentration step. In an SCP production process, the objectives of the product separation and concentration step are to separate either the protein-rich microbial biomass or the protein and other cellular components as a concentrated stream, suitable for product finishing by drying, from a dilute, essentially biomass and/or protein-free process water stream suitable for recycle. In a biopolymer production process, the objectives of the product separation and concentration step are to separate the water-soluble product in a concentrated stream, free from either microorganisms or microbial cell debris, from both a dilute-process water stream suitable for recycle and a concentrated active biomass stream that could be recycled to the fermenter to give enhanced process intensification.[9] Unit processes for the separation and concentration of biopolymers have been discussed by Smith and Pace.[10] Efficient production processes for the manufacture of bulk biopolymers can be expected to have similar levels of process integration to those developed for SCP production.

Bulk ethanol production, for use either as a fuel or as a feedstock for the chemicals

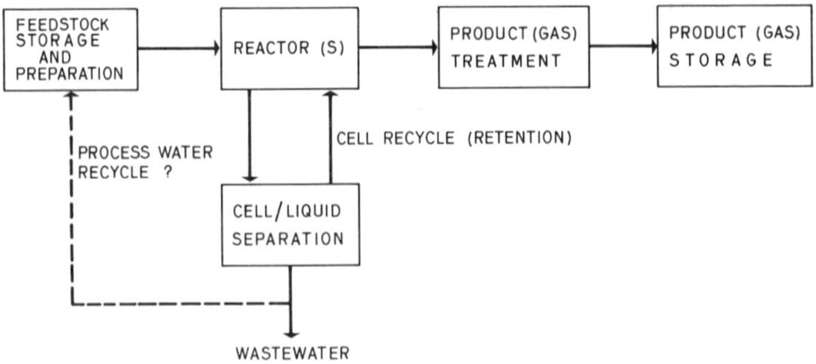

FIGURE 4. Block diagram for methane production.

industry, by any of the several proposed production technologies, offers extensive opportunities for process integration. All processes for fuel ethanol production should be energy efficient, although they need not necessarily exhibit an entirely favorable net energy balance, provided a favorable liquid fuel balance is achieved.[4] In order to achieve this latter objective, it is essential that the fermentation step is conducted under conditions that give both very high productivities and very high product concentrations with respect to ethanol, thereby allowing a high degree of optimization in the product separation and concentration step. To achieve the necessary intensity of operation in the fermenter (reactor), either microbial biomass recycle or microbial biomass retention, which is directly equivalent to recycle, will be an essential process feature. In addition, process water recycle and energy-efficient contamination prevention technology will also be of considerable importance. However, in commercial-scale processes for fuel ethanol production, the most important process step where integration to achieve high efficiencies with respect to both energy and resource utilization is the feedstock preparation (pretreatment) step. This step involves an extensive series of unit processes depending on the feedstock employed, and for economic process operation, both internal integration within the process step and integration with other process steps are required to achieve good process economics.

When compared with the other processes that have been discussed, the production of methane by anaerobic digestion, although a more biologically complex process, is usually carried out in relatively unsophisticated process plants. The feedstock for such processes can either be a concentrated biodegradable waste stream or biomass produced especially as feedstock for the process. In cases where biomass is employed as the feedstock, the most economically interesting source of the biomass is from the mass culture of photosynthetic algae on carbon dioxide. However, if one considers that only a decade ago, the production of protein-rich biomass, that is SCP, by the mass culture of bacteria on methane and of photosynthetic algae on carbon dioxide were considered potentially attractive commercial ventures, current proposals for methane production from algal biomass demonstrate the impact of politically mediated changes in energy and energy-associated feedstock prices during the decade. Stuckey[11] has reviewed the technological status of anaerobic digestion processes for methane production, dealing, almost exclusively, with the reactor (fermenter). As far as the overall process is concerned, the several process steps are shown in FIGURE 4. Obviously a considerable degree of integration is required to achieve appropriate levels of process stability, intensification, and productivity if such processes are to become suitable for operation in sophisticated industrial environments. Significant effort directed towards such objectives is presently being lavished on anaerobic digestion processes and apparent advances are resulting from improved active microbial biomass retention and/or recycle and from thermophilic operation, although, in the latter case, energy balance questions arise. If methane is to be produced for widespread distribution by anaerobic digestion processes, the gas mixture produced in the reactor will require treatment to remove carbon dioxide and hydrogen sulfide so that the gas supplied for distribution meets appropriate specifications for fuel gases. Some degree of process integration in anaerobic digestion processes might be possible by introducing process water recycle, but the inclusion of this will require markedly better digestor performance, particularly with respect to residual carboxylic acids in the digestor effluent stream.

As far as all bulk-product, continuous-flow fermentation processes are concerned, significant cost reductions can be achieved by the integration of several processes on a production site so that feedstocks and products can be exchanged on an "over the fence" basis, thereby reducing investment in storage facilities by either sharing or eliminating such facilities.

CONCLUDING REMARKS

The logic of processing requires that added value be conferred on the products produced with respect to the feedstocks and energy utilized so that the operation of processes gives satisfactory returns on the capital invested and risked. Process technology is developed as a specific solution to an original statement of need. However, if the bases of the original statement of need and the specific solution proposed are either incorrect or invalidated by subsequent changes, the resultant process, irrespective of technological innovation, may become unsuitable for commercialization. Processes developed for bulk fermentation product manufacture have suffered such fates. Frequently, the elapsed time required for research, development and commercialization exceeds planning horizons. However, a greater use of techniques based on the selection, construction, and quantification of scenarios[12] and the application of such scenarios to an examination of future possibilities for energy, feedstock, and product supply and demand would seem appropriate when developing proposals for novel biotechnological processes.

SUMMARY

The commercial development of processes for the manufacture of Single-Cell Protein (SCP) facilitated a massive influx of process-oriented technology into the fermentation industry and notable improvements with respect to fermenter design and operation on the one hand, and in the integration and optimization of the several unit operations that comprise the overall process on the other, occurred. Even though the production of SCP, particularly from either hydrocarbon or hydrocarbon-derived feedstocks, has not been an outstanding commercial success, this state of affairs results from politically mediated changes in feedstock prices rather than from any failure of the technology that was developed. Several process routes for bulk fermentation products for the energy and industrial chemicals market sectors are presently being researched and developed. Many such projects are marked by a total emphasis of effort on the fermentation stage and little regard for the several other unit operations that will inevitably comprise the overall process. This paper seeks to examine how lessons learned with respect to the interactions between and the integration of multiple process stages in the optimization of SCP production processes can be applied to the development of commercially interesting biotechnological process routes for fuels and industrial chemicals manufacture.

ACKNOWLEDGMENTS

The author thanks the Engineering Foundation and the Swiss Federal Institute for Water Research and Water Pollution Control for financial support to attend the Biochemical Engineering III Conference.

REFERENCES

1. King, P. P. 1982. J. Chem. Tech. Biotechnol. **32:** 2–8.
2. Kretschmer, A., S. Lang, G. Marwede, E. Ristau & F. Wagner. 1981. Formation of surface-active glycolipids by n-alkane-utilizing microorganisms. *In* Advances in Biotechnology. M. Moo-Young, Ed. Vol. **3:** 475–479. Pergamon Press. Toronto, Canada.

3. Huang, H. T. 1969. Microbial insecticides as fermentation products. *In* Fermentation Advances. D. Perlman, Ed. Academic Press. New York. pp. 591–610.
4. Humphrey, A. E. 1982. J. Chem. Tech. Biotechnol. **32:** 25–33.
5. Taylor, I. J. & P. J. Senior. 1978. Endeavour. **2**(1): 31–34.
6. Mogren, H. 1979. Process Biochem. **14**(3): 2–7.
7. Hamer, G. & D. E. F. Harrison. 1980. Single-cell protein: The technology, economics, and future potential. *In* Hydrocarbons in Biotechnology. D. E. F. Harrison, I. J. Higgins & R. Watkinson, Eds. Heyden & Son. London, England. pp. 59–73.
8. Hamer, G. 1982. Biotechnol. Bioeng. **24:** 511–531.
9. Atkinson, B. & N. W. F. Kossen. 1978. DECHEMA Monogr. **82:** 37–54.
10. Smith, I. H. & G. W. Pace. 1982. J. Chem. Tech. Biotechnol. **32:** 119–129.
11. Stuckey, D. C. 1982. Anaerobic treatment of industrial wastewater in the developing nations. *In* Management of Industrial Wastewater in Developing Nations. D. Stuckey & A. Hamza, Eds. Pergamon Press. Oxford, England. pp. 200–222.
12. Dumoulin, H. & J. Eyre. 1979. Energy Econ. **1:** 76–86.

Scale-up of Chick Cell Growth on Microcarriers in Fermenters for Vaccine Production

E. M. SCATTERGOOD, A. J. SCHLABACH,
W. J. MCALEER, AND M. R. HILLEMAN

Division of Virus & Cell Biology Research
Merck Institute for Therapeutic Research
Merck Sharp & Dohme Research Laboratories
West Point, Pennsylvania 19486

INTRODUCTION

One of the major problems in the traditional method of manufacturing a killed human virus vaccine is obtaining the large amount of viral antigen needed on a commercial scale. Large numbers of viable anchorage-dependent cells are needed for infection and replication of the virus.[1] One of the cell types used for production of human vaccines is the chick cell. A mincing and trypsinization procedure is carried out on chick embryos to obtain a primary chick cell suspension. These primary cells can then be used to plant a fermenter containing suspended microcarriers.[2,3] However, it will be projected for the case of a 1000-liter fermenter with suspended microcarriers that the primary cells from 1100 chick embryos would be required for planting and growth before viral infection. This would be a large number of embryos to handle without introducing contamination. To avoid this problem, secondary cells or further passaged cells could be used. "Secondary cells" is the terminology given to primary cells that have been grown, enzymatically removed from the surface, and replanted.

The objective of this experimental work was to demonstrate in fermenters that primary chick cells could be removed from the microcarriers in a small 9-liter fermenter and the secondary chick cells passed on to a larger 140-liter fermenter. The secondary chick cells could then be infected or passed further and many fewer chick embryos would be required. In the case of the 1000-liter fermenter, from the experimental results, it will be projected that only 70 chick embryos would be needed if secondary cells were used in the 1000-liter fermenter.

MATERIALS AND METHODS

Primary Cell Suspension

Embryos from thirty 11-day-old SPAFAS COFAL, Marek-free embryonated chicken eggs obtained from SPAFAS, Inc. were used for the primary cell suspension.

Microcarriers

Cytodex 3 microcarriers[4] obtained from Pharmacia, Inc. were used at concentrations of 3 g/l. Cytodex 3 microcarriers consist of a surface layer of denatured collagen covalently bound to a matrix of crosslinked dextran. This concentration provides 12,000 microcarriers/ml and 14 cm^2/ml of surface area.

Tissue Culture Medium

Both a 199 type of medium[5] with 2% v/v fetal bovine serum and Dulbecco's medium[6] with 5% v/v fetal bovine serum, 1% v/v chick serum, and 10% v/v tryptose phosphate broth[4] were used.

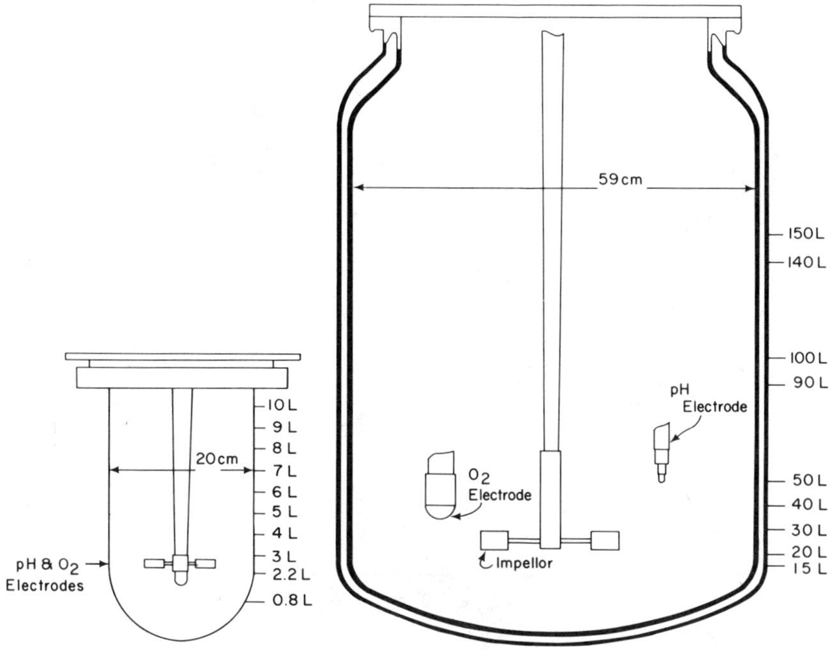

FIGURE 1. Nine-liter and 140-liter fermenters.

Vessels for Cell Culture

Tissue-culture flasks (75 cm^2) and 1-liter spinner vessels were utilized for growth tests of the media and cells and as controls. Furthermore, they were used to establish initial conditions for the fermenters in regard to planting concentration, pH set points, agitator speeds, and cell passage methodology. They also predicted cell attachment, growth rates, and medium utilization in the fermenters.

FIGURE 1 shows the two fermenters that were used.[7,8] The primary cell growth was done in the 9-liter fermenter and secondary cell growth in the 140-liter stainless steel jacketed fermenter. The location of the pH and O_2 electrodes and variable speed impellors are noted in the drawing. The surface to volume ratios are 0.036 cm^{-1} and 0.020 cm^{-1} for the 9-liter and 140-liter fermenters, respectively. Control of dissolved oxygen at 30% of air equilibrium and control of pH depended on surface aeration. The dissolved oxygen was controlled by solenoid valves so that air or oxygen gas was flowing, and the pH was controlled by introducing CO_2 gas to obtain a 15% mixture of CO_2 in the air or oxygen when the pH was above the set point. The 9-liter and 140-liter fermenters were refed by allowing the microcarriers to settle and removing the spent medium by a tube positioned at the level of 0.8 liters and 21 liters, respectively.

Assays

Samples were taken from the fermenters to determine medium glucose level (Beckman glucose analyzer), pH, unattached cells, and debris in the medium, concentration of cells attached to the microcarriers, and to monitor the microcarriers and cells visually.

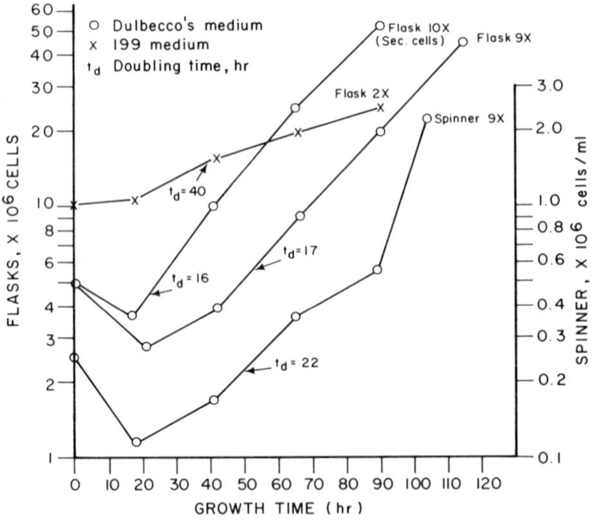

FIGURE 2. Cells versus growth time: flasks and spinner.

The concentration of cells per volume of microcarrier medium mixture was measured by a procedure involving trypsinization of the cells from the microcarriers and counting the cells using two independent counting methods. Staining the cells with trypan blue and microscopic counting gave a visual determination of the number of viable and nonviable cells, cell size, and quality. Electronic counting using a Coulter counter yielded total cell number and gave a good quantitative check of the microscopic method.

Cell Passage

Cell passage of the grown cells from the 9-liter fermenter to the 140-liter fermenter was carried out using a 3.5-liter trypsinization apparatus[7,8] equipped with a 320 cm^2,

60-micron mesh, stainless steel screen and a small vibromix impeller above the screen. The microcarriers from the 9-liter fermenter were collected on the screen (about 350 ml of bed volume with a bed thickness of 1.1 cm), washed with agitation, trypsin was added and agitated just enough to expose all the microcarriers, trypsin was then removed, the trypsin-exposed microcarriers were held at room temperature for one-half hour, and then the cells were removed from the microcarriers by several washes of complete medium with agitation and counted microscopically. These cells were then planted in the 140-liter fermenter at the desired level.

RESULTS

Preliminary Experiments

Preliminary experiments were done in 75 cm^2 tissue culture flasks and 1-liter spinners with typical growth curves shown in FIGURE 2. The three flask curves along with the doubling times and cell increase numbers (final cell yield divided by cell input) show the advantage of the Dulbecco's medium compared with the 199 medium. Dulbecco's medium was used for all subsequent work. The flask plots also show the improved planting efficiency and less lag time of secondary cells compared with primary cells, and the better planting efficiency with a larger cell input. The spinner experiment using 3 g/l of Cytodex 3 microcarriers indicated similar performance when compared to the flasks. In all cases, the flasks and spinners were gassed with an appropriate mixture of CO_2 in air to control the initial medium pH. It was found that the initial cell attachment on the spinner microcarriers was slightly enhanced at a pH of 7.5 versus 7.3. The cells from the spinner microcarriers were recovered as described previously with a 75% recovery.

All of this preliminary information—planting concentration, planting efficiency, effect of pH, growth kinetics, medium utilization, agitator speed, and efficiency of cell recovery—was used in the planning of the scaled-up equipment: growth of primary cells in a 9-liter fermenter, trypsinization of cells from the microcarriers with passage to a 140-liter fermenter, and growth of the secondary cells in the 140-liter fermenter.

Growth of Primary Cells in 9-liter Fermenter

FIGURE 3 summarizes the experimental results of primary cell growth in the 9-liter fermenter and calculated glucose utilization (mg glu/10^6 cells per day). TABLE 1 lists the operating volumes, impeller RPMs, pH set points, and glucose concentrations. The growth curve is based on the concentration at 9-liter volume. The percent dissolved oxygen was set at 100% initially when the culture medium without growing cells was under an air overlay at equilibrium.

The 9-liter fermenter was planted at 0.25×10^6 cells/ml, which amounted to 10 embryos from the 30-embryo primary cell suspension. The growth curve that shows cell attachment and growth was similar to the 1-liter spinner growth curve. The dissolved oxygen was controlled at 30%, but the demand exceeded the supply at 90, 105, and 126 hours and decreased to about 10%. The loss of control correlated with the depletion of glucose in the medium. Glucose depletion is indicated also by the rise in pH to the pH set point at 105 and 126 hours after the pH had been decreasing out of control. The pH set point was decreased from the high of 7.50 for attachment to 7.20 at 91 hours. Three refeeds of the 800 ml of "spent" medium and microcarriers were carried out at 91, 115, and 139 hours. The last refeed was done just before passage.

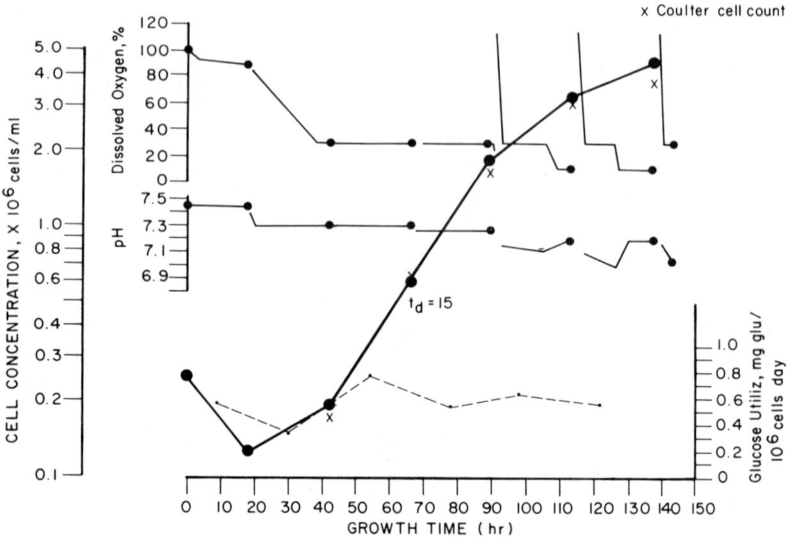

FIGURE 3. Cell concentration, dissolved oxygen, pH, glucose utilization versus growth time: 9-liter fermenter.

More than enough cells were available at the time of passage for planting the 140-liter fermenter at the 0.15 $\times$ 10^6 cells/ml level even assuming only a 75% recovery of cells from the 9-liter fermenter.

The impeller RPM was set initially at 38 in order just to keep the microcarriers suspended, but later, at 66 and 91 hours, increased to 52 and 79 RPM respectively to keep the microcarriers with cells suspended and to control microcarrier clumping. The decrease in growth rate after 89 hours was probably due to contact inhibition and/or some loss of pH and oxygen control. Sinskey *et al.*[9,10] have discussed the problems of

TABLE 1. Experimental Data

9-liter Fermenter

Growth time, hr	0	18	42	66	91	115	139
Volume, liters	3.1	to 6.0	—	to 9.0	Refeed	Refeed	Refeed to 8.5
RPM set point	38	—	—	52	79	—	—
pH set point	7.50	7.30	—	7.25	7.20	—	—
Glucose conc., mg/100 ml	110	90 to 110	92	47 to 70	8	8	7

140-liter Fermenter

Growth time, hr	0	17	24	40	43	46	49	65	89	100	113
Volume, liters	70	—	to 104	—	to 140	—	—	Refeed	Refeed	—	—
RPM set point	27	—	—	—	—	50	—	—	—	—	—
pH set point	7.40	7.25	—	—	—	—	—	—	—	—	—
Glucose conc., mg/100 ml	—	78	—	46	—	—	42	8	7	12	6

low dissolved oxygen concentration and high agitation rates on secondary chick cells and FS-4 cells grown on microcarriers.

Trypsinization of Cells

After the last refeed, the microcarriers with attached cells were removed from the 9-liter fermenter, washed, exposed to trypsin, and the cells removed into complete medium with agitation as described previously. The stripped microcarriers were left behind on the screen of the trypsinization apparatus. Of 40 × 10^9 potential cells, 5 × 10^9 were recovered in the trypsin and 26 × 10^9 in the medium for a 78% overall recovery.

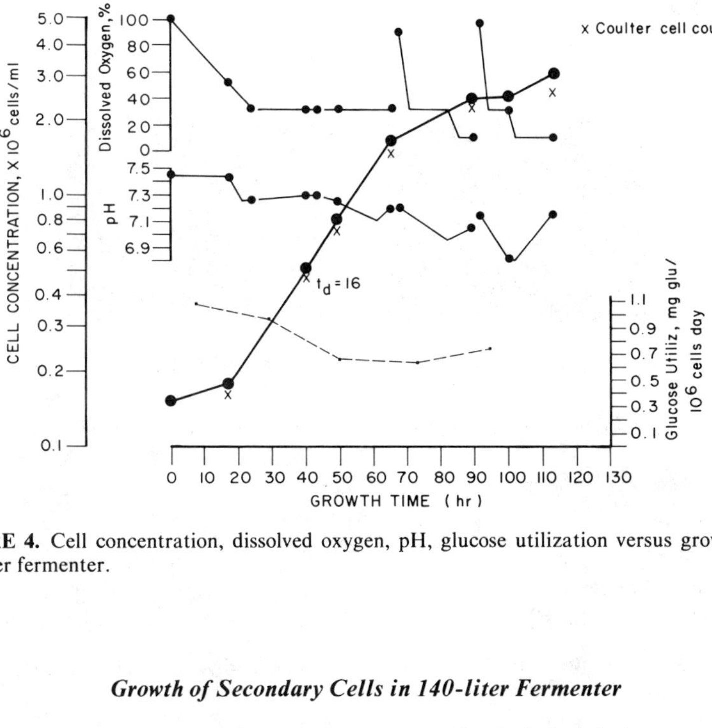

FIGURE 4. Cell concentration, dissolved oxygen, pH, glucose utilization versus growth time: 140-liter fermenter.

Growth of Secondary Cells in 140-liter Fermenter

The 140-liter fermenter was planted at 0.15 × 10^6 cells/ml with the secondary cells as shown in FIGURE 4. The plots in FIGURE 4 are similar to the plots in FIGURE 3 for the 9-liter fermenter except that pH control was lost sooner and decreased to lower levels (6.8 vs. 7.0). Again, control of oxygen was lost when glucose was depleted except for the initial glucose depletion at 61 hours. The growth curve is similar to the growth curve for secondary cells in flasks as shown in FIGURE 2; the attachment was excellent and there was little lag time. It is speculated that the decrease in growth rate after 65 hours is due to the low pH.

In FIGURE 5, photographs of the stained cells on microcarriers are pictured with calculated cells attached per microcarrier and cells per cm^2. There was very little microcarrier clumping at 17 and 40 hours, but some doublets and triplets are observed

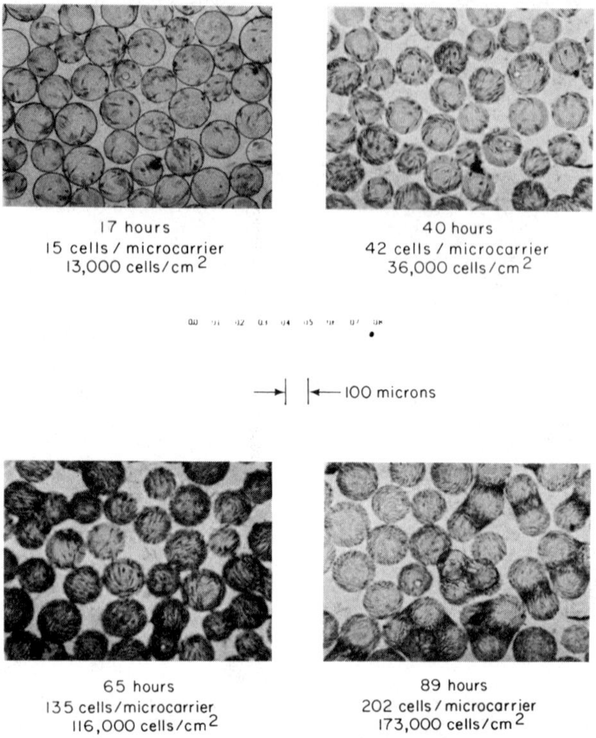

FIGURE 5. One-hundred and forty-liter fermenter microcarriers.

TABLE 2. Summary of Demonstration Experiment

9-liter Fermenter (Primary Cells)

0.25×10^6 → 4.5×10^6

2.2×10^9 —(6 days growth)→ 40×10^9

Trypsinize Cells → 31×10^9

140-liter Fermenter (Secondary Cells)

0.15×10^6 → 3.0×10^6 cells/ml

21×10^9 —(5 days growth)→ 420×10^9 viable cells

Fermenter Size, Liters	Embryos Required	
	Use Primary Cells	Use Secondary Cells
9	10	0.6
140	160	10
1000	1100	70

at 65 hours and larger clumps at 89 hours. There are very few empty microcarriers and the number of cells per microcarrier was fairly uniform.

DISCUSSION

It is feasible, with the described methodology, to reduce greatly the number of chick embryos required to manufacture a killed vaccine. TABLE 2 summarizes the demonstration experiment and gives an estimate of the approximate number of embryos needed for three sizes of fermenters based on the experimental data summarized in the table. The number of primary chick cells increased 18-fold in the 9-liter fermenter. It was possible to trypsinize the chick cells from the microcarriers with good recovery and pass them on to a 140-liter fermenter where a 21-fold increase was realized. It can then be projected, if this performance is maintained, that only 70 embryos in a 60-liter fermenter would be required to provide enough secondary cells for a 1000-liter fermenter. Preparing a sterile 70-chick-embryo cell suspension is a routine procedure compared with an 1100-embryo cell suspension.

ACKNOWLEDGMENT

We gratefully acknowledge the technical assistance of Ms. D. Williams.

REFERENCES

1. PERKINS, F. T. 1970. Virus vaccines: Introduction. Lab. Pract. **19**(1): 39–67.
2. VAN WEZEL, A. L., C. A. M. VAN DER VELDEN-DE GROOT & J. A. M. VAN HERWAARDEN. 1980. The production of inactivated polio vaccine on serially cultivated kidney cells from captive-bred monkeys. *In* Developments in Biological Standardization. Proc. Third Gen. Meet. Eur. Soc. Anim. Cell Technol. B. Griffiths, W. Hennessen, F. Horodniceanu & R. Spier, Eds. Vol. **46:** 151–158. S. Karger, Basel.
3. VAN WEZEL, A. L. 1981. Present state and developments in the production of inactivated poliomyelitis vaccine. *In* Developments in Biological Standardization. Int. Symp. Reassessment of Inactivated Poliomyelitis Vaccine. W. Hennessen & A. L. Van Wezel, Eds. Vol. **47:** 7–13. S. Karger, Basel.
4. Microcarrier Cell Culture. Principles and Methods. 1981. Pharmacia Fine Chemicals. Uppsala, Sweden.
5. MORGAN, J. F., H. J. MORTON & R. C. PARKER. 1950. Nutrition of animal cells in tissue culture. I. Initial studies on a synthetic medium. Proc. Soc. Exp. Biol. Med. **73:** 1–8.
6. DULBECCO, R. & G. FREEMAN. 1959. Plaque production by the polyoma virus. Virology **8:** 396–397.
7. (Brochure) Contact-Holland. Bilthoven-unit system. pp. 1–10. Ridderkerk, Holland.
8. VAN HEMERT, P. 1974. Vaccine production as a unit process. *In* Progress in Industrial Microbiology. D. J. D. Hockenhull, Ed. Vol. **13:** 151–271. Churchill Livingston, Edinburgh.
9. SINSKEY, A. J., R. J. FLEISCHAKER, M. A. TYO & D. I. C. WANG. 1981. Production of cell-derived products: Virus and interferon. Ann. NY Acad. Sci. **369:** 47–64.
10. FLEISCHAKER, R. J. & A. J. SINSKEY. 1981. Oxygen demand and supply in cell culture. Eur. J. Appl. Microbiol. Biotechnol. **12:** 193–197.

Kinetics of Isomaltose Formation by Amyloglucosidase and Purification of the Disaccharide by Fermentation of Undesired By-Products

A. HARDER, B. NOORDAM, AND A. M. BREKELMANS

Gist-Brocades N.V.
2600 MA Delft, the Netherlands

INTRODUCTION

Isomaltose is a disaccharide consisting of two glucose units, α, 1–6 linked together (FIG. 1). This compound has a bitter taste and therefore it is a favored flavoring for some drinks like Japanese sake. Isomaltose can be hydrogenated by Raney Nickel-H_2 to yield isomaltitol, which is useful as a low-calorie sweetening agent. In addition, pure isomaltose is a useful model compound for studying the kinetics of the starch-hydrolyzing enzyme, amyloglucosidase (exo-1,4-α-D-glucosidase, E.C. 3.2.1.3).

For these reasons, an exploratory lab-scale process has been developed for the production and purification of isomaltose. FIGURE 2 shows the overall scheme of the process. The first step entails the saccharification of maltodextrins into a mixture of glucose, isomaltose and maltose by incubation with amyloglucosidase. This saccharification process will be discussed by formulating a kinetic model in order to explain the phenomena observed and to optimize the production of isomaltose. The second step deals with the fermentative removal of glucose and maltose in the saccharified syrup. For this purpose, a special yeast strain was selected. The third and fourth steps enhance the purity of the isomaltose solution.

MATERIALS AND METHODS

Amyloglucosidase (Maxydrase®, Gist-Brocades) from *Aspergillus niger* was used in our experiments. The activity of the preparations used was 26700 GAU/g where 1 GAU (Glucose Amylase Unit) is defined as the amount of enzyme that, under assay conditions, degrades 800 mg of soluble starch for 1% to glucose.

The DE-15 maltodextrin substrate was received from Glucoseries Réunies, Amylum. The experimental conditions for the saccharification of the DE-15 substrate with Maxydrase® were 60°C, pH 4.5 (0.02 *M* HAc/Ac buffer), enzyme dosage 5 mg per g of substrate and substrate concentrations of respectively 10, 20, 30 and 40% w/w. Samples were taken after 1, 2, 3, 4, 5, 6, 7, 8, 24 and 144 hr. For measurements of initial kinetics of the reverse reactions, incubations were carried out with 40% w/w glucose, 38% w/w maltose, and a mixture of 32% w/w glucose and 7.6% w/w maltose with amyloglucosidase (3000 GAU/100 g incubation mixture) at 45°C at pH 4.5.

In our experiments concerning the production of isomaltose, immobilized amyloglucosidase was used because of the known preference of this system to make isomaltose. Moreover, reusability and improved stability of the enzyme are obtained

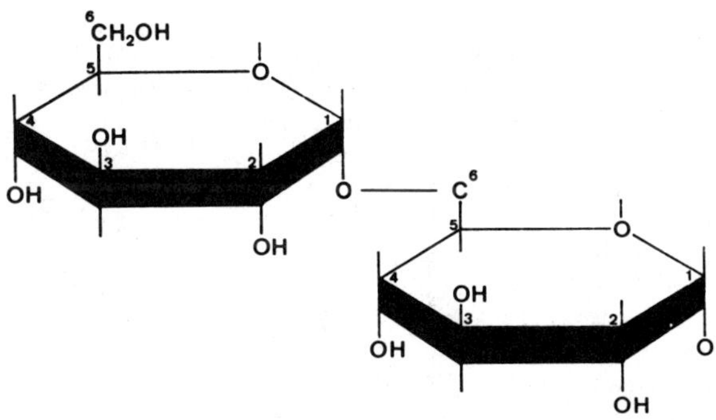

FIGURE 1. Isomaltose, a disaccharide of two glucose units α, 1–6 linked together.

besides continuous operation and easy separation of the product. The enzyme was entrapped into a support material of gelatine crosslinked with the bifunctional reagent glutardialdehyde.[1] The beads formed were packed into a column (3 × 30 cm) and the thinned starch pumped down slowly at a temperature of 60°C and pH 4.5.

For elimination of the by-products, that is glucose and maltose, a special yeast

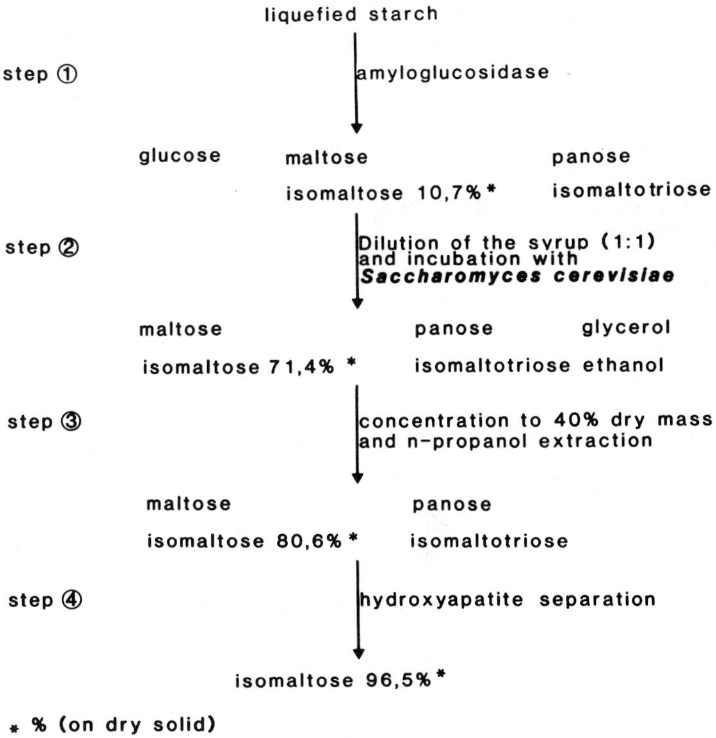

FIGURE 2. Overall scheme of the isomaltose process.

strain (*Saccharomyces cerevisiae* var.) was selected that was able to consume only glucose and maltose. Yeast cells were entrapped into alginate according to the method of Kierstan and Buck.[2] The cells (1330 g packed cells) were suspended in 2% alginate (4770 g solution). The suspension (6100 g of yeast-alginate slurry) was next added dropwise to 2% (w/w) $CaCl_2 \cdot 2\,H_2O$ (5 liters). The particles (5.1 liters) were packed into a column (10 × 100 cm) and 20% w/w saccharified substrate pumped up the column with an average flow rate of 120 ml/hr. The temperature of the column was 28°C, the pH of the column was not adjusted.

Discontinuous reactivation of the column material was achieved by incubating the particles for a period of 30 hr with 8 liters of a growth medium of 10% w/w biomalt, 15 g/l $CaCl_2 \cdot 2\,H_2O$, 2.7 g/l KH_2PO_4 and 3.75 g/l yeast extract (Difco), pH 4.5.

The carbohydrate composition of the samples as well as ethanol and glycerol has been analyzed by means of HPLC (column BioRad HPX87C).

KINETIC MODEL OF THE SACCHARIFICATION OF LIQUEFIED STARCH

Isomaltose is formed during the process of enzymatic saccharification of liquefied starch. This liquefied starch or maltodextrin solution is obtained by random hydrolysis of the molecular chains of starch by cleaving the endo α, 1–4 bonds of the polymers by the enzyme α-amylase (EC 3.2.1.1). Mixtures of oligosaccharides of definite lengths can be obtained under controlled conditions.[3] The oligosaccharides can be distinguished in two types of polymers of α-D-glucopyranose; first, amylose-type molecules, which are flexible linear chain molecules containing α, 1–4 linked glucose units and, secondly, amylopectin-type molecules which are branched chain molecules having α, 1–4 as well as α, 1–6 linked glucose units (FIG. 3). The average length of these molecules can be calculated from the concept of dextrose equivalent, *DE*. The dextrose equivalent is defined as the amount of D-glucose (g/kg) calculated from the reducing power of the dextrin solution divided by the amount of dry substance of the dextrins (g dry mass/kg) multiplied by 100.

Mathematically,

$$\mathrm{DE} = \frac{\sum_{ij}^{\infty} C_{ij} * M_{\text{glucose}}}{\sum_{ij}^{\infty} C_{ij} * (i\,M_{\text{glucose}} - (j-1)\,M_{H_2O})} \tag{1}$$

Here, j is the number of glucose units, C_{ij} is the amount of i th (oligo) saccharide with j glucose units in moles/kg.

Combining the definition of dextrose equivalent (DE) and Equation 1, the average length of the DE-15 maltodextrin solution employed in our experiments is calculated to be 7–8 glucose units per molecule.

Amyloglucosidase is an exo-splitting enzyme that removes glucose units consecutively from the nonreducing ends of the oligosaccharide chains.[4,5] The enzyme acts predominantly by a multichain mechanism, that is hydrolysis of only one bond per effective encounter.[4] This process proceeds more rapidly if the terminal unit is linked by an α, 1–4 bond rather than by an α, 1–6 linkage. The overall content of α, 1–6 linkages is generally less than 5%.[6] So we have to run the reaction in a dextrin solution with amylose and amylopectin in a ratio of 2 to 1.

The site of the α, 1–6 linkage in the amylopectin molecule (FIG. 3) is a hypothetical one and assumed to be an average amylopectin molecule in a DE-15 maltodextrin solution. Because of the exo-splitting action of amyloglucosidase, the subchains A and

B of the amylopectin molecule have to be depolymerized before the enzyme is able to attack the α, 1–6 linkage (FIG. 3). After the breakdown of the α, 1–6 bond, the subchain C is hydrolyzed.

During the hydrolysis of oligosaccharides two rate-limiting steps exist:[3,5] first, the α, 1–6 attack to the amylopectin molecules, and secondly, the splitting of maltose into two glucose molecules.

At higher glucose concentrations, reverse reactions proceed (FIG. 3). The main reverse reactions are the conversion of glucose into maltose and isomaltose, and also into panose and isomaltotriose but in much smaller amounts. Results of rate measurements of the hydrolysis of maltose and isomaltose have shown that even the slow

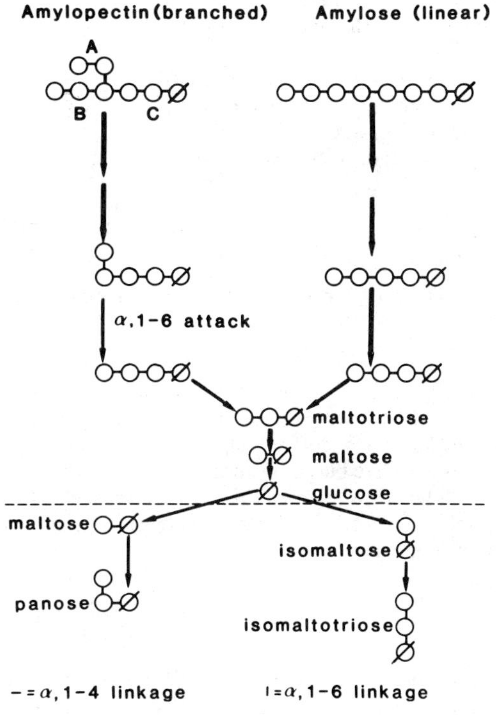

FIGURE 3. Enzymatic hydrolysis of a DE-15 maltodextrin substrate incubated with amyloglucosidase.

hydrolysis of maltose proceeds 30 times faster than the hydrolysis of isomaltose for the enzyme preparation used.[3]

In the case of the hydrolysis of the DE-15 maltodextrin substrate the general type of reaction is

$$DP_j + H_2O \underset{k_{\alpha,1\text{-}4;-1}}{\overset{k_{\alpha,1\text{-}4;+1}}{\rightleftharpoons}} DP_{j-1} + DP_1 \tag{2}$$

in which DP_j is an (oligo) saccharide with j glucose units, DP_1 is glucose and $k_{\alpha,1\text{-}4;+1}$ is the rate constant of the degradation reaction of α, 1–4 bonds and $k_{\alpha,1\text{-}4;-1}$ is the rate constant of the polymerization reaction of α, 1–4 bonds.

The next equations reflect both rate-limiting steps in the breakdown of the maltodextrin substrate.

$$DP_5 + H_2O \underset{k_{\alpha,1\text{-}6;-1}}{\overset{k_{\alpha,1\text{-}6;+1}}{\rightleftharpoons}} DP_4 + DP_1 \tag{3}$$

$$DP_{2,\mathrm{mal}} + H_2O \underset{k_{\mathrm{mal};-1}}{\overset{k_{\mathrm{mal};+1}}{\rightleftharpoons}} 2DP_1 \tag{4}$$

in which DP_5 is the oligosaccharide with five glucose units before the α, 1–6 attack, $DP_{2,\mathrm{mal}}$ is maltose, $k_{\alpha,1\text{-}6;+1}$ is the rate constant of the α, 1–6 degradation reaction, $k_{\alpha,1\text{-}6;-1}$ is the rate constant of the α, 1–6 polymerization reaction, $k_{\mathrm{mal},+1}$ is the rate constant of the hydrolysis of maltose and $k_{\mathrm{mal},-1}$ is the rate constant of the polymerization reaction.

In the present treatise, we will assume the rate constant $k_{\alpha,1\text{-}4}$ in Equation 2 to be independent of the degree of polymerization of the maltodextrin molecules except DP_5 and both disaccharides maltose ($DP_{2,\mathrm{mal}}$) and isomaltose ($DP_{2,\mathrm{iso}}$).

The hydrolysis reaction of isomaltose can be expressed by

$$DP_{2,\mathrm{iso}} + H_2O \underset{k_{\mathrm{iso};-1}}{\overset{k_{\mathrm{iso};+1}}{\rightleftharpoons}} 2DP_1 \tag{5}$$

The rate constant of the depolymerization reaction of α, 1–4 bonds is defined by

$$r_{\alpha,1\text{-}4;+1} = k_{\alpha,1\text{-}4;+1} \cdot C_{\mathrm{DP_j}} \cdot C_{H_2O} \cdot C_x \tag{6}$$

here $C_{\mathrm{DP_j}}$ is the concentration of maltodextrins with j glucose units, C_{H_2O} is the concentration of water and C_x is the concentration of amyloglucosidase. The other rate equations are defined in the same way. Equation 6 emphasizes that the kinetic constants are based on simple, mass-action law relationships.[7]

The kinetic model for the formation of glucose, maltose, and isomaltose on incubation of DE-15 maltodextrin solutions with amyloglucosidase is given in vector notation as

$$\frac{d\,\mathbf{C}_{ij}}{d\,t} = \boldsymbol{\alpha} \cdot \mathbf{r} \tag{7}$$

in which $\boldsymbol{\alpha}$ is the stoichiometric matrix of the defined conversion reaction of order m (reactions) $\times$ i (components) and $\mathbf{r}$ is the matrix of reaction rates of order $i \times m$. The number of reactions and components can be read from FIGURE 3. In general, the stoichiometry of the hydrolysis reactions is $\alpha = 1$ but for maltose or isomaltose, $\alpha = 2$. The reaction rates are defined by constitutive equations like Equation 6.

The model presented has been used to simulate some of our experiments and to optimize the production of isomaltose. The most important feature of this model in contrast with the model of Roels and Van Tilburg[8] is the use of initial kinetics.

RESULTS AND DISCUSSION

From our initial kinetic experiments, the kinetic rate constants of the model could be calculated. These values and the listing of the computer program are given in the

APPENDIX. Using this program and the rate constants obtained the degradation of maltodextrin solutions into glucose incubated with amyloglucosidase at four different dry mass concentrations could be simulated (FIG. 4). It is obvious that at high concentrations of dry mass, the concentration of glucose goes through a transient peak value and subsequently decreases until equilibrium is reached.

FIGURE 5 explains these observations. The course of the concentrations of maltose and isomaltose at different concentrations of substrate are shown. Any rise in the dry mass content of maltodextrin solution leads to higher concentrations of the two saccharides and therefore to lower concentrations of glucose, because the rates of both reverse reactions are proportional to the concentration of glucose. The higher the concentration of glucose, the higher the rates of maltose and isomaltose formation. On incubation with immobilized enzyme, mass transfer limitation of the polysaccharides as compared to glucose increases the rate of resynthesis reactions.[8] So the rate of

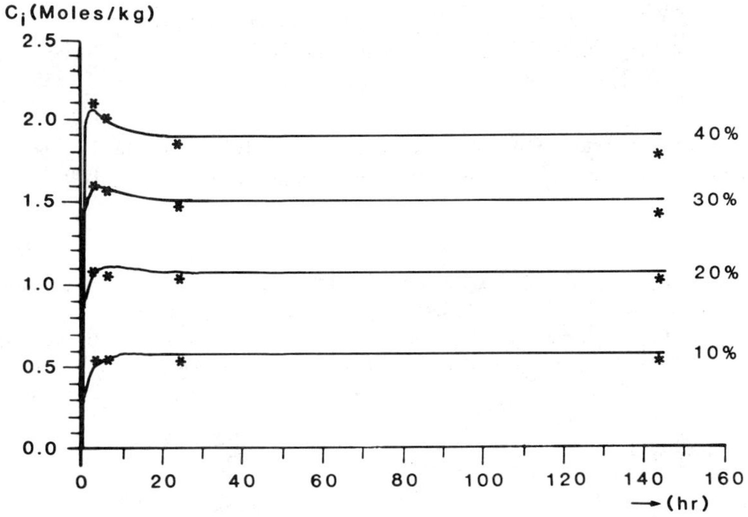

FIGURE 4. Experimental and simulated results of glucose vs. time for 15% DE maltodextrin solutions incubated with amyloglucosidase at four different dry mass concentrations. T = 60°C.

isomaltose production up to corresponding equilibrium concentration (FIG. 5) will be enhanced. FIGURE 6 shows experimental results and simulations of equilibrium concentrations of maltose and isomaltose versus different concentrations of DE-15 maltodextrin solutions. In conclusion, a process for optimal production of isomaltose from a liquefied starch should be carried out at high dry substance and with immobilized enzyme.

The results of the maltodextrin solution (400 g/kg) incubated with immobilized amyloglucosidase are summarized in TABLE 1. The concentration of isomaltose in the syrup is 10% on dry solid.

The following steps emphasize the elimination of the by-products. It was hardly possible to use chromatographic techniques like preparative HPLC because of the high concentration of glucose. Therefore the second step in the process (FIG. 2) deals with the elimination of glucose and maltose using the selected yeast strain. From the

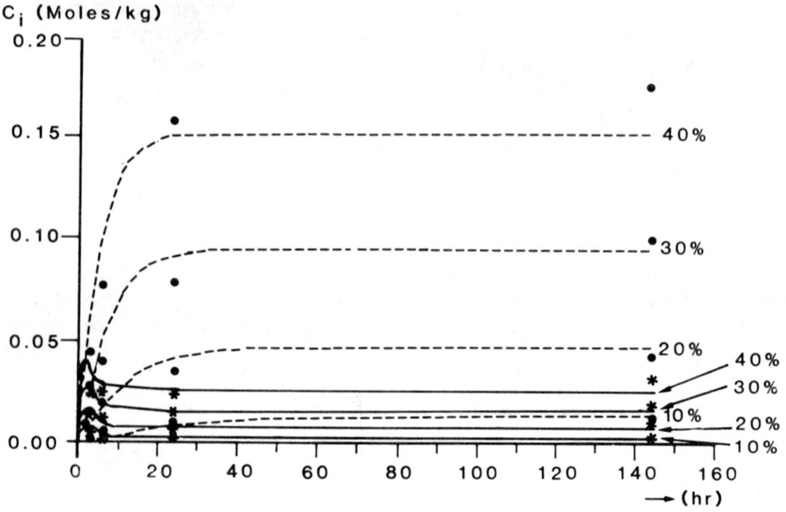

FIGURE 5. Experimental and simulated results of maltose (*, —) and isomaltose (●, -----) vs. time for 15% DE maltodextrin solutions incubated with amyloglucosidase at four different dry mass concentrations. T = 60°C.

literature, it is known that panose and isomaltotriose are nonfermentable sugars that will not be metabolized by yeast. Often isomaltose is also called a nonfermentable sugar, but in general it will be consumed too. Surprisingly, in a screenings procedure, a species of *Saccharomyces cerevisiae* has been found that converts, under anaerobic

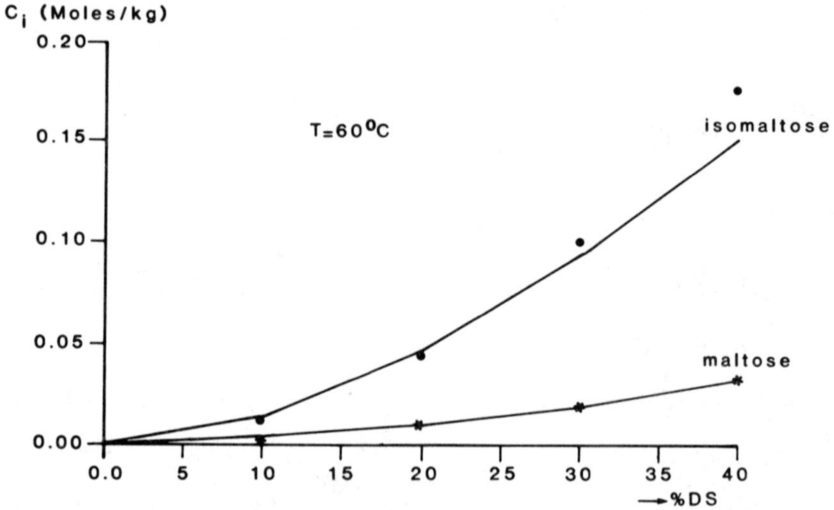

FIGURE 6. Equilibrium concentrations of maltose (*) and isomaltose (●). Solid lines are simulation curves.

TABLE 1. Composition of a Liquefied Starch after Incubation with Immobilized Amyloglucosidase

	% (On Dry Solid)
Glucose	85.7
Maltose	1.9
Isomaltose	10.7
Panose	0.4
Isomaltotriose	1.3

conditions, glucose and a part of the maltose except the isomaltose and the nonfermentable sugars. The mono- and disaccharide are converted into ethanol, glycerol, and CO_2.

From reasons mentioned before in favor of immobilized complexes, the yeast cells were immobilized into a matrix of Ca-alginate with a final concentration of 200 g (wet weight biomass) per liter of particles. In view of the kinetic advantages of a plug-flow reactor, experiments were carried out in a packed-bed reactor. Solutions of saccharified syrup were pumped up the packed-bed column. TABLE 2 summarizes the experimental results. Using the selected yeast strain, it was possible to enhance the concentration of isomaltose calculated on dry substance base from approximately 100 g/kg to 700 g/kg by selective consumption of glucose and maltose.

An interesting aspect of the packed-bed reactor was the concentration of ethanol in the effluent. The ethanol production of the column reached levels between 10 and 12 vol. %, that is, between 80 and 100 g of ethanol per liter. If it is assumed that one mole of glucose will theoretically be converted into two moles of ethanol, then the biochemical efficiency of the ethanol production was 93%. The total conversion of fermentable sugars in the column was 99%. In the packed-bed reactor, some aspects were investigated that pertain to the use of an immobilized, living biocatalyst.

First attention was paid to reactivation after use of the yeast cells and cell growth in the particles. For that purpose, the beads in the packed-bed reactor were incubated overnight with a growth medium. The medium used for reactivation was a mixture of biomalt, yeast extract, phosphate at a pH of 4.5, and $CaCl_2$, added for stabilizing the alginate beads. After the reactivation procedure, the column was washed with tap water. Then saccharified syrup was pumped up again. FIGURE 7 shows the consecutive reactivation steps, the consumption of glucose and maltose in time. It can be concluded that it is possible to reactivate the beads many times.

The average amount of yeast cells in one particle has been counted too during the

TABLE 2. Composition of the Saccharified Syrup after Treatment with *Saccharomyces cerevisiae* var.

	Packed-bed Reactor
Glucose	0.1[a]
Maltose	7.8
Isomaltose	71.4
Panose	1.1
Isomaltotriose	8.2
Glycerol	11.5
Ethanol	9.7 %(v/v)

[a] % (on dry solid)

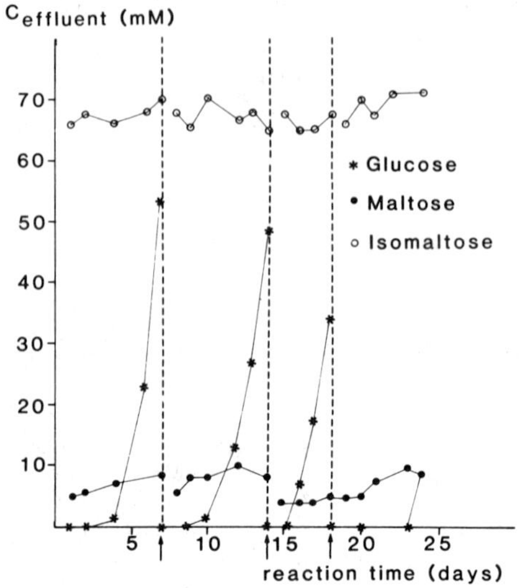

FIGURE 7. Effluent concentrations of sugars before and after discontinuous reactivation (↑) of the yeast-immobilized, packed bed.

reactivation experiments by plate counting technique. FIGURE 8 shows that after each reactivation procedure the average amount of yeast cells present in one particle has increased up to 150 million cells per particle. In this system, no leakage of cells out of the beads was observed. The packed-bed reactor fermented all the glucose and part of the maltose resulting in an increased purity of isomaltose by a factor of 7.

The third step of the isomaltose process (FIG. 2) deals with the extraction of the glycerol out of the syrup. It is known that glycerol mixes well with *n*-propanol, which therefore was used to extract glycerol from the mixture of maltose, isomaltose, and both trisaccharides. For treating large amounts of carbohydrate solution, a continuous

FIGURE 8. Discontinuous reactivation (↑) of yeast cells embedded in Ca-alginate particles.

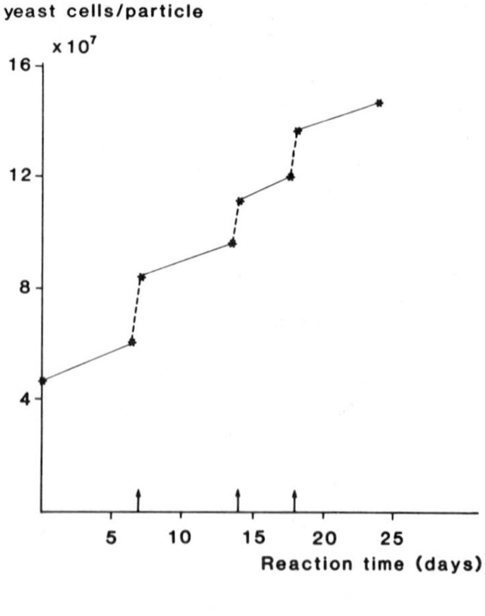

percolation system was used. Due to the extraction of the glycerol, the purity of the isomaltose was enhanced from 71.4% to 80.6% (w/w) on dry solid.

For further purification of the isomaltose in the syrup, hydroxyapatite (tricalcium phosphate) could be used (FIG. 2, step 4). An ethanol-water mixture is used as the liquid phase in this column chromatographic procedure. With 90% (v/v) ethanol, maltose, and isomaltose could be eluted selectively. The isomaltose fractions were combined and the solution concentrated.

The HPLC analysis of the end product and also ^{13}C-NMR analysis showed that the main component was isomaltose with a final purity of 96.5% (w/w) on dry solid.

CONCLUSIONS

The development of the kinetic model based on initial kinetics proved to be a useful tool in the optimization of the production of isomaltose.

This process of isomaltose production and purification serves as an example of the general possibility of applying biocatalytic systems to the purification of organic compounds. In addition to physicochemical methods, the biocatalytic approach of product purification may be of great value in the fine-chemical industry.

REFERENCES

1. VAN TILBURG, R. 1983. Engineering aspects of biocatalysts in industrial starch conversion technology. Ph.D. Thesis. Delft University Press. Delft, The Netherlands.
2. KIERSTAN, M. & C. BUCKE. 1977. Biotechnol. Bioeng. **19:** 387.
3. TAKAHASHI, T., N. INOKUCHI & M. IRIE. 1981. J. Biochem. **89:** 125.
4. PAZUR, J. H. & T. ANDO. 1959. J. Biol. Chem. **234:** 1966.
5. PAZUR, J. H. & K. KLEPPE. 1962. J. Biol. Chem. **237:** 1002.
6. SOLOMON, B. 1978. Adv. Biochem. Eng. Vol. **10:** 131. Springer-Verlag. New York.
7. HARDER, A. & J. A. ROELS. 1982. Adv. Biochem. Eng. Vol. **21:** 55. Springer-Verlag. New York.
8. ROELS, J. A. & R. VAN TILBURG. 1977. Starch/Stärke. **31:** 338.

APPENDIX

Symbols of CSMP

$k_{\alpha,1\text{-}4;+1}$ = K1 rate constant of degradation of α,1–4 bonds
$k_{\text{mal};+1}$ = K2 rate constant of degradation of maltose
$k_{\text{mal};-1}$ = K3 rate constant of polymerization of maltose
$k_{\text{iso};+1}$ = K4 rate constant of degradation of isomaltose
$k_{\text{iso};-1}$ = K5 rate constant of polymerization of isomaltose
$k_{\alpha,1\text{-}6;+1}$ = K6 rate constant of degradation of α,1–6 bonds
$k'_{\alpha,1\text{-}6;+1}$ = K7 rate constant of degradation of panose and isomaltotriose
$k'_{\alpha,1\text{-}6;-1}$ = K8 rate constant of polymerization of panose and isomaltotriose
$k'_{\alpha,1\text{-}4;+1}$ = K9 rate constant of degradation of isopanose and maltotriose
r_{sub} = RL_{sub}; RB_{sub}; R_{sub} specific activity (moles/kg/hr)

DP_j = concentration of linear polymer with j glucose units (moles/kg)
DPB_j = concentration of branched polymer with j glucose units (moles/kg)
$DP_j\mathrm{O}$ = initial concentration of linear polymer (moles/kg)
$DPB_j\mathrm{O}$ = initial concentration of branched polymer (moles/kg)
$DP2$ = maltose (moles/kg)
$DPB2$ = isomaltose (moles/kg)
$DPB3\mathrm{A}$ = panose (moles/kg)
$DPB3\mathrm{B}$ = isopanose (moles/kg)
$DPB3\mathrm{C}$ = isomaltotriose (moles/kg)
H_2O = concentration of water (moles/kg)
C_x = concentration of enzyme (g(ds)/kg)
DS = dry solids (g/kg)
TDS = total dry solids (g/kg)

```
*       CSMP PROGRAM OF THE KINETIC MODEL OF THE HYDROLYSIS
*       OF A DE-15 MALTODEXTRIN SUBSTRATE BY INCUBATION
*       WITH AMYLOGLUCOSIDASE
INITIAL
PARAM K2=0.2284, K3=0.06281, K5=0.003927, H20=40., J=4,
*       INITIAL CONCENTRATION OF (OLIGO) SACCHARIDES
INCON DP10=0., DP20=0., DPB20=0., DP30=0., DPB3AO=0., ...
        DPB3BO=0., DPB3CO=0., DP40=0., DP50=0., DPB50=0., ...
        DP60=0., DPB60=0., DP70=0., DPB70=0.
        DS     = J*100.0
        CX     = J*0.5
        DPB80  = DS/(1314.*3.)
        DP80   = 2.*DPB80
*       RATE CONSTANTS
        K1     = 6.*K2
        K4     = 0.59*K5
        K6     = Ki/30.
        K7     = 6.*K4
        K8     = 6.*K5
        K9     = 6.*K3
DYNAMIC
*       INTEGRATORS
        DP2    = INTGRL(DP20,DP2DOT)
        DPB2   = INTGRL(DPB20,DPB2DOT)
        DP1    = INTGRL(DP10,DP1DOT)
        DP4    = INTGRL(DP40,DP4DOT)
        DP3    = INTGRL(DP30,DP3DOT)
        DPB3A  = INTGRL(DPB3AO,DPB3ADOT)
        DPB3C  = INTGRL(DPB3CO,DPB3CDOT)
        DPB3B  = INTGRL(DPB3BO,DPB3BDOT)
        DPB8   = INTGRL(DPB80,DPB8DOT)
        DPB7   = INTGRL(DPB70,DPB7DOT)
        DPB6   = INTGRL(DPB60,DPB6DOT)
        DPB5   = INTGRL(DPB50,DPB5DOT)
        DP8    = INTGRL(DP80,DP8DOT)
        DP7    = INTGRL(DP70,DP7DOT)
        DP6    = INTGRL(DP60,DP6DOT)
        DP5    = INTGRL(DP50,DP5DOT)
*       CONSTITUTIVE RATE EQUATIONS
        RB8    = K1 * DPB8 * H20 * CX
        RB7    = K1 * DPB7 * H20 * CX
        RB6    = K1 * DPB6 * H20 * CX
        RB5    = K6 * DPB5 * H20 * CX
        RL8    = K1 * DP8 * H20 * CX
        RL7    = K1 * DP7 * H20 * CX
        RL6    = K1 * DP6 * H20 * CX
        RL5    = K1 * DP5 * H20 * CX
        R2     = K1 * DP4 * H20 * CX
        R3     = K1 * DP3 * H20 * CX
        R4     = K2 * DP2 * H20 * CX
        R5     = K3 * DP1 * DP1 * CX
        R6     = K5 * DP1 * DP1 * CX
        R7     = K9 * DP1 * DP2 * CX
        R8     = K8 * DP1 * DP2 * CX
        R9     = K8 * DP1 * DPB2 * CX
```

```
            R10   = K9 * DP1 * DPB2 * CX
            R11   = K7 * DPB3A * H2O * CX
            R12   = K7 * DPB3C * H2O * CX
            R13   = K1 * DPB3B * H2O * CX
            R14   = K4 * DPB2 * H2O * CX
      *     MASS BALANCE EQUATIONS
            DPB8DOT  = -RB8
            DPB7DOT  = RB8-RB7
            DPB6DOT  = RB7-RB6
            DPB5DOT  = RB6-RB5
            DP8DOT   = -RL8
            DP7DOT   = RL8-RL7
            DP6DOT   = RL7-RL6
            DP5DOT   = RL6-RL5
            DP2DOT   = R3-R4+R5-R7-R8+R11
            DPB2DOT  = R6-R9-R10+R12+R13-R14
            DP1DOT   = R2+R3+2.*R4-2.*R5-2.*R6-R7-R8-R9-R10+R11+R12+R13+ ...
                         2.*R14+RB8+RB7+RB6+RB5+RL8+RL7+RL6+RL5
            DP4DOT   = RB5+RL5-R2
            DP3DOT   = R2-R3+R7
            DPB3ADOT = R8-R11
            DPB3CDOT = R9-R12
            DPB3BDOT = R10-R13
      *
            TDS = ((DPB8+DP8)*1314.0+(DPB7+DP7)*1152.0+(DPB6+DP6)*990.0 ...
                  +(DPB5+DP5)*828.0+DP4*666.0+(DP3+DPB3C+DPB3A+DPB3B)*504.0...
                  +(DP2+DPB2)*342.0+DP1*180.0)
            DE  = ((DPB8+DPB7+DPB6+DPB5+DP8+DP7+DP6+DP5+DP4+ ...
                  DP3+DP2+DP1+DPB2+DPB3C+DPB3A)*180.0/TDS)*100.0
      *
      TIMER FINTIM=145.0,OUTDEL=1.0
      METHOD RKS
      PREPAR DP1,DP2,DPB2,DE,DPB3A,DPB3C
      ENDJOB
```

A Novel Process for High-Maltose Syrup Production from Barley Starch

Y.-Y. LINKO, H. MÄKELÄ, AND P. LINKO

Laboratory of Biochemistry and Food Technology
Department of Chemistry
Helsinki University of Technology,
SF-02150 Espoo 15, Finland

INTRODUCTION

The interest in high-maltose syrup production is increasing because of its special properties such as low, pleasant sweetness, low hygroscopicity, low tendency to crystallize, good heat stability, low solution viscosity, high fermentability, antiseptic properties, and, in particular, its physiological properties.[1–3] High-maltose syrups may be applied in the food industry for the manufacturing of confectionery and frozen desserts, baking and brewing, as well as in the pharmaceutical industry for replacing glucose in intravenous feeding.

Maltose syrup is produced from acid and/or enzyme liquefied starches by saccharification with β-amylase. The maximum yield of maltose is about 60%. For high-maltose syrup production, a debranching enzyme such as pullulanase is required in the addition to β-amylase to saccharify the thinned starch, which should have as low a DE-value as possible to avoid excessive maltotriose formation.[1] It is difficult to control the economics of this process by either acid or by enzyme hydrolysis. We had previously applied HTST-extrusion cooking techniques in glucose syrup production[4,5] and observed that this technique may also be applied to obtain low DE thinned starches.[6,7] A twin-screw extrusion cooker with closely intermeshing self-wiping screws operates as a positive displacement pump, and can be regarded as a continuous high-temperature, high-pressure, high-shear, short-time reactor.[7–9] The applicability of this new technique in high-maltose syrup production has now been investigated.

PROCESS

A Creusot-Loire BC-45 twin-screw extruder was used for thinning of commercial barley starch at various moisture levels and temperatures with or without the addition of thermostable α-amylase Termamyl. Barley β-amylase was then used either alone or with pullulanase for the saccharification.

The extrudate characteristics are determined by the temperature, pressure and shear profiles, water activity, and residence time distribution during extrusion, while feed rate, screw speed, and die diameter largely determine the residence time and, thus, the overall severity of the treatment. One of the dominating factors in extrusion cooking of biopolymer materials is moisture content (water activity).[7]

The decrease in starch crystallinity with a decrease in feed moisture from 60% to 25% and an increase in mass temperature from 125°C to 150°C could be clearly shown by X-ray diffraction techniques, while the effect of various processing conditions on starch gelatinization could be followed by scanning electron microscopy and enzymatic susceptibility.

HIGH-MALTOSE SYRUPS

Barley starch was extruded at 19% to 65% moisture, 105° to 160°C at a screw rate of 150 rpm and varying feed rate from 175 to 225 g (d.m.) min^{-1}. The DE-values varying from 0.76 to 4.0 were obtained, when starch was extrusion cooked without the addition of thermostable α-amylase, suggesting that α-amylase treatment could be completely eliminated by employing extrusion cooking technology.

TABLE 1 shows the composition, as determined by HPLC-techniques, of high-maltose syrups produced from variously thinned barley starch substrate at 10% dry solids, saccharified at 55°C, pH 6.0 with barley β-amylase (ABM 1500 l, 1638 units/100 g d.m.) and pullulanase (ABM pulluzyme S 2000, 33 units/100 g d.m.).

TABLE 1. The Composition of Various High-Maltose Syrups Obtained

Liquefaction	DE	Saccharification hr	DE	Glucose %	Maltose %	Maltotriose %	Higher Oligomers %
α-Amylase, 95°C, pH 6.0	3.0	22	52.1	—	84.0	10.9	5.0
Extrusion at 40% H_2O, 135°C without α-amylase	3.2	26	46.2	—	86.1	7.0	6.9
α-Amylase, 95°C, pH 6.0	9.2	26	45.0	0.6	69.6	15.7	14.1
Extrusion at 40% H_2O, 135°C with α-amylase 2.5 ml/100g d.m.	10.1	26	50.6	0.7	71.2	22.5	5.7
α-Amylase, 95°C, pH 6.0	14.5	26	45.5	0.7	60.8	19.1	19.5
Extrusion at 55% H_2O, 135°C with α-amylase 2.5 ml/100g d.m.	12.0	23	57.5	0.8	72.3	21.9	5.0
Extrusion at 19% H_2O, 125°C without α-amylase	1.3	25	45.5	—	87.5	8.4	4.0

One can see that by selecting optimal extrusion-processing parameters, enzymatic thinning can be replaced by thermomechanical treatment to obtain high maltose level and lower quantities of high oligomers.

REFERENCES

1. TAKASAKI, Y. & T. YAMANOBE. 1981. Production of maltose by pullulanase and β-amylase. *In* Enzymes and Food Processing. G. G. Birch, N. Blakebrough & K. J. Parker, Eds. Applied Science Publ. London. pp. 73–88.
2. NORMAN, B. E. 1981. New developments in starch syrup technology. *In* Enzymes and Food Processing. G. G. Birch, N. Blakebrough & K. J. Parker, Eds. Applied Science Publ. London. pp. 15–50.
3. FULLBROOK, P. D. 1982. Malt and maltose syrups. *In* Nutritive Sweeteners. G. G. Birch & K. J. Parker, Eds. Applied Science Publ. London. pp. 49–81.

4. LINKO, Y.-Y., A. LINDROOS & P. LINKO. 1979. Soluble and immobilized enzyme technology in bioconversions of barley starch. Enzyme Microb. Technol. **1:** 273–278.
5. LINKO, Y.-Y., H. VUORINEN & P. LINKO. 1980. The effect of HTST-extrusion on retention of cereal α-amylase and on enzymatic hydrolysis of barley starch. *In* Food Process Engineering, Vol. 2, Enzyme Engineering in Food Processing. P. Linko & J. Larinkari, Eds. Applied Science Publ. London. pp. 210–223.
6. LINKO, P. 1982, HTST (High Temperature-Short-Time)-Extruder als biochemischer Reaktor. Getreide, Mehl u. Brot. **36:** 326–332.
7. LINKO, P., Y.-Y. LINKO & J. OLKKU. 1983. Extrusion cooking and bioconversions. J. Food Proc. Eng. **2:**. In press.
8. LINKO, P., P. COLONNA & C. MERCIER. 1981. High-temperature short-time extrusion cooking. *In* Advances in Cereal Science and Technology. Y. Pömeranz, Ed. Vol. **4:** 145–235.
9. OLKKU, J., A. HAGQVIST & P. LINKO. 1983. Steady-state modelling of HTST-cooking extrusion employing response surface methodology. J. Food Proc. Eng. **2:** 105–128.

Large-scale Production of Mammalian Cells and Their Products: Engineering Principles and Barriers to Scale-up

M. W. GLACKEN, R. J. FLEISCHAKER, AND
A. J. SINSKEY

Department of Nutrition and Food Science
Biotechnology Laboratories and Cell Culture Center
Massachusetts Institute of Technology
Cambridge, Massachusetts 02139

INTRODUCTION

Biologicals manufactured from mammalian cells in culture have importance in the medical and clinical fields. Some of the mammalian-cell-derived products that have potential commercial importance are listed in TABLE 1. Interferon, a biological hailed as a potential cure for the common cold and cancer, has been produced at high titers and on a large scale via cell culture technology.[1]

Vaccines constitute the largest market outlet of any biological produced by mammalian cells, with world-wide sales of over 90 million dollars.[2] Animal vaccines account for the largest share of this market, with foot-and-mouth disease vaccine being the principal product.[3] Human vaccines produced from mammalian cells include polio, rubella, rabies, measles, and encephalitis.[4]

Monoclonal antibodies have the potential to be the most economically significant product of mammalian cells. In addition to diagnostic applications, monoclonal antibodies have already been used on a laboratory scale for the single-step purification of many proteins.[5] If this usage of monoclonal antibodies is translated to a larger scale, the present market for monoclonal antibodies will increase tremendously. Additionally, monoclonal antibodies have possible therapeutic value in the treatment of tumors and of various immunological diseases.[6,7]

A model of a mammalian cell is presented in FIGURE 1. Large-scale mammalian cell culture techniques have to provide suitable environmental conditions of oxygen, temperature, and pH. Nutritional demands of the cells must be satisfied and strategies developed for the control of waste metabolic products. Understanding how to control these extrinsic parameters is necessary in developing strategies for optimization of mammalian cell cultures in order to economically produce a variety of biologicals. As the field of hybridoma technology emerges and new recombinant DNA techniques are developed for mammalian cells, the need for large-scale mammalian cell culture will increase. The engineering problems that need to be solved in large-scale mammalian cell culture are summarized in FIGURE 2. The following discussion will address: (1) the various techniques used to cultivate mammalian cells; (2) the technical barriers to scale-up; and (3) comparison of various cell production methods with respect to their potential in overcoming these barriers.

TABLE 1. List of Potential Products from Mammalian Cell Cultures

Group I: Cells as an End Product	Group II: Cell-derived Products
Artificial skin	Growth factors
Artificial organs	Nerve growth factor
Beta-islet cells (pancreas)	Epidermal growth factor
Hepatocytes (liver)	Blood factors
Bone marrow	Monoclonal antibodies
Lymphocytes	Interferons
	Proteases (Urokinase, etc.)
	Vaccines
	Hormones
	Human growth hormone
	Insulin
	Calcitonin
	Parathyroid hormone

SYSTEMS FOR PRODUCING MAMMALIAN CELLS ON A LARGE SCALE

There are two types of mammalian cells, anchorage-dependent cells (cells that require a surface for growth), and suspension cells (nonanchorage-dependent). In general, mammalian cells capable of multiplying in suspension can be cultivated similar to microorganisms used by the fermentation industry. On the other hand,

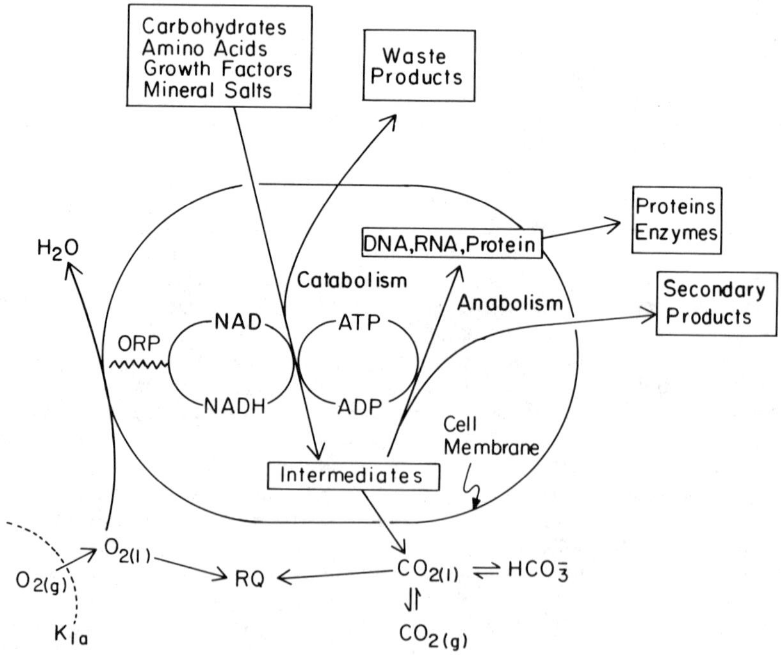

FIGURE 1. Generalized cell model.

anchorage-dependent cells, because of their surface attachment requirements, present unique problems for producing these cells on a large scale.

To reduce the costs of producing mammalian cell products on a large scale, it is desirable to maximize the cell density obtained in culture. The higher the cell density, the more "biological catalyst" is present, and the higher the product titer will be. As the product titer is increased, the costs of purification of the biological product decreases. Since serum components are thought to catalyze cellular reactions and are not depleted to a great extent in the medium, no additional serum components are needed at higher cell densities. Therefore, at higher cell densities, the unit yield of product per unit of serum used in the medium increases. Since serum is the predominant cost of culture media, high cell densites greatly reduce the cost of producing mammalian biologicals. Cultures of suspension cells using conventional techniques usually obtain cell densities of approximately 1×10^6 cells/ml. The following novel techniques for cultivating mammalian cells have been developed in an attempt to achieve higher cell densities and to subsequently reduce the cost of production.

Barriers to scale up that still need to be addressed:
- Oxygen transfer limitations
- Accumulation of toxic waste products
- Lack of regulatory hardware and software for process control
- High cost of serum

Barriers to scale-up that have been addressed to some extent:
- Low surface-to-volume ratios of systems that cultivate anchorage-dependent cells
- Shear sensitivity of mammalian cells

FIGURE 2. Some barriers to the scale-up of mammalian cell systems.

Anchorage-dependent Cells

The traditional method used for the large-scale production of anchorage-dependent cells employs the insides of rotating bottles as the surface area for growth. The rolling action ensures that the cells are alternately exposed to growth medium and oxygen in the air space. This method is very cumbersome and expensive for production of a large quantity of cells for many reasons. Since the surface area for growth is only a small percentage of the total bottle volume, many roller bottles are needed to produce even a small number of cells. For example, one bottle with 500 cm^2 of surface area only supports 3×10^7 cells. Thus, roller bottles require extensive handling, labor, and medium to produce a high quantity of cells. In addition, cells produced from a series of roller bottles are highly variable in the separate cultures. The variation in the cell cultures makes the monitoring of cellular kinetics and the ability to change the growth environment to obtain optimal production rates practically impossible.

To increase the ratio of the surface area available for cell growth to the total culture volume, VanWezel, in 1967, grew cells on charged dextran microcarrier beads suspended in liquid culture media.[8] The attempt was successful, except that the cells could not grow at very high bead densities due to the inhibitory effect that the microcarriers exhibited on the cells at concentrations greater than 1 g beads/l. By modifying the surface charge of the beads, Levine *et al.* eliminated the inhibitory effect, thereby allowing the cells to proliferate to densities of 5×10^6 cells/ml at microcarrier densities up to 5 g/l.[9] Thus, it was then feasible to grow anchorage-dependent cells to high cell densities in traditional suspension systems on a large scale.

Other schemes for growing mammalian cells are shown in FIGURE 3. In all of these methods, the surface-to-volume ratio has been increased over that of roller bottles.

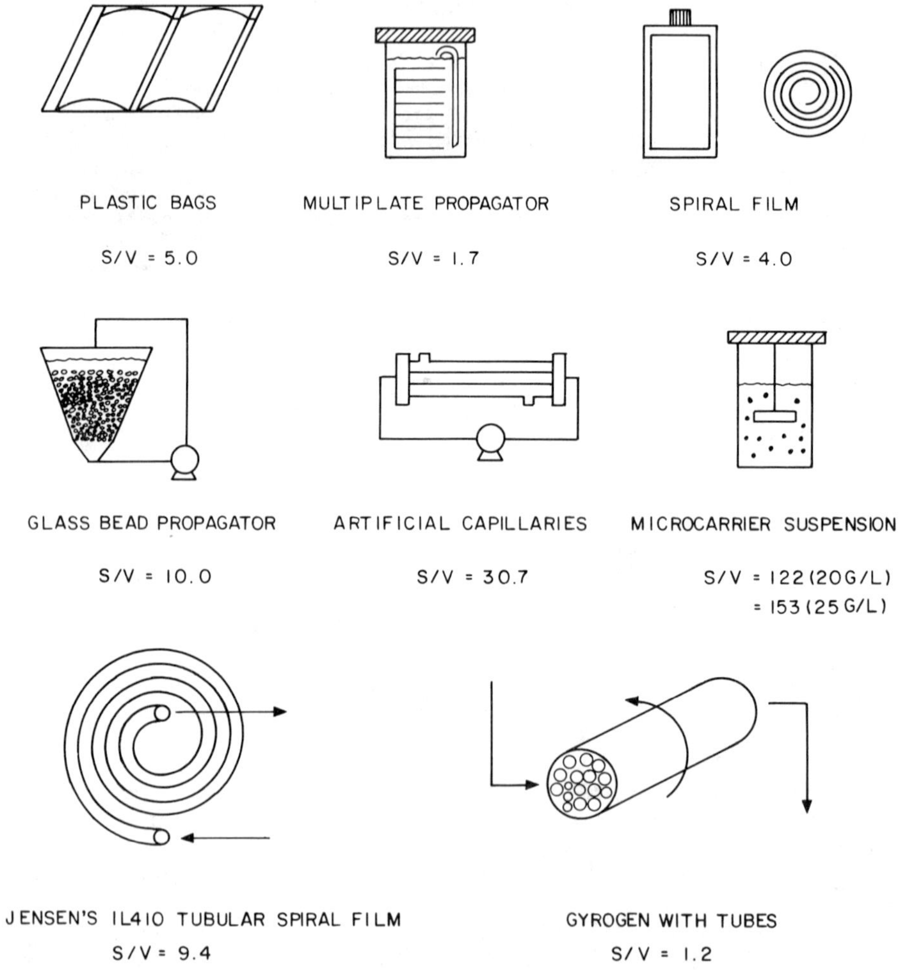

FIGURE 3. Surface-to-volume ratios (cm^{-1}) of various methods to culture anchorage-dependent cells.

However, the multiplate propagator,[10] the spiral film,[11] plastic bags,[12] and the Gyrogen with tubes[13] have limited utility for growing cells on a large scale, since their surface-to-volume ratios are much too low to be practical. The glass bed propagator[14] is not practical on a large scale, since the flow rate of medium that would be needed to supply nutrients and remove waste products would be severe enough at large scales to cause shifting in the glass bead bed, thereby shear damaging the cells on the beads. The most practical methods presently available for growing mammalian cells on a large scale are artificial capillaries,[15] Jensen's IL410 tubular spiral film,[16] and the microcarrier suspension procedure, previously described.

Jensen's IL410 tubular spiral film consists of a tube made of gas-permeable film through which medium flows. The cells grow on the inside of the tube and oxygen is fed to the cells from the outside of the tube through the gas-permeable film, which is wrapped into a coil to decrease the total volume occupied by the culture system.

An artificial capillary system consists of a bundle of dialysis tubes packed inside of a hollow cylinder. Culture medium flows through the insides of the "capillaries," while the cells grow on the outside surface. Nutrients reach the cells by diffusing through the capillary wall, while oxygen is supplied to the cells via air flow through the shell side of the culture vessel.

Suspension Cell System

Tumor cell lines such as HeLa cells and lymphoblastoids do not require a surface on which to grow. Essentially, these tumor cells can be grown in equipment similar to that utilized for bacterial fermentations. However, since mammalian cells do not have a protective cell wall, they are much more sensitive to shear than microbial cells. Thus, while increased surface-to-volume ratio was a goal that catalyzed the development of new techniques for producing anchorage-dependent cells, shear reduction was the goal that catalyzed the development of novel techniques for cultivating suspension cells.

In an attempt to reduce the shear to which the cells are exposed, two different types of agitators were developed. The first, the vibromixer, is composed of a horizontal disk with conical apertures attached to a vertical shaft that rapidly oscillates up and down.[3] This vibromixer circulates the medium in a vertical, rather than a horizontal direction, resulting in less shear force to obtain an adequate dispersion of cells. The other agitator design consists of four slowly turning flexible sheets that span the depth of the culture fluid.[17] Again, this type of agitator provides adequate dispersion of the cells at low shear rates. Feder and Tobert have demonstrated that these agitator designs are essential if high cell densities are to be achieved.[17]

Another strategy to reduce the shear exposure of the cells is to create an artificial cell wall, analogous to the cell wall that protects microbial cell membranes from shear damage. This is done by encapsulating cell suspensions inside a polymer-reinforced calcium alginate membrane.[18] The capsules produced can be as small as 500 microns and can contain as many as 10,000 cells.[19] High cell densities, increased cell viability, and increased product yields can be obtained using encapsulated cells. This technology is now commercially available for the large-scale production of suspension cells, including hybridoma cells.

Mode of Operation: Batch, Fed-batch, or Perfusion

Microbial or mammalian cells can be produced using three types of cultures, batch, fed-batch, or perfusion (FIGURE 4). In a batch system, no additional nutrients are replenished except oxygen. The only parameters that can be controlled are the temperature, aeration, and the pH. Thus, the cells in batch culture are subject to a constantly changing environment where nutrients are being depleted while waste products accumulate. Cell growth and/or product formation can be prematurely inhibited due to nutrient limitation or to the toxic build-up of waste products. This problem can be partially alleviated by replacing the "spent" medium with "fresh" medium. However, the cells are still subjected to a constantly changing environment unless the medium is replaced frequently, which can be very expensive if the medium consists of relatively expensive components, for example serum. In addition, when replacing the medium, some components that are not exhausted at the time of the medium change are being discarded needlessly.

A better mode of operation would be a fed-batch system that feeds vital components only as needed to the culture, thus obtaining a constant nutrient concentration. The fed-batch process requires that methods exist to monitor and

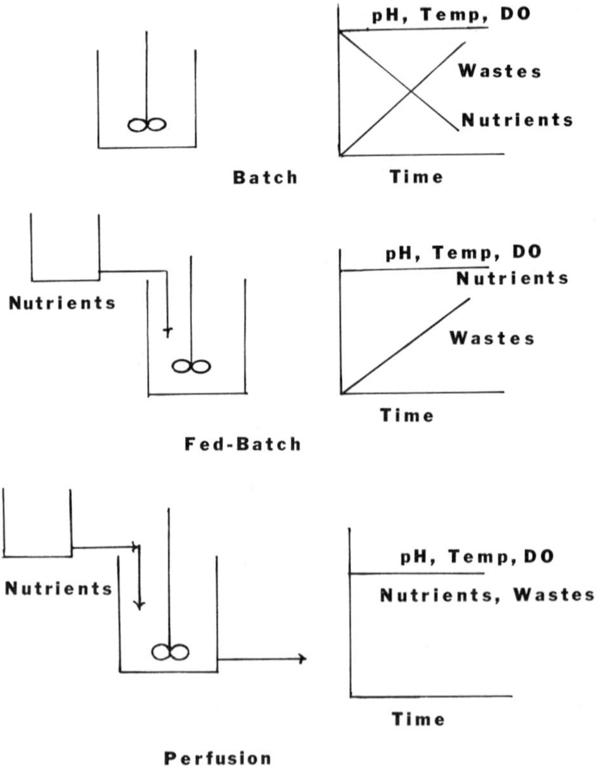

FIGURE 4. Environmental characteristics of batch, fed-batch, and perfusion cultures.

control the nutrients in question. More environmental parameters are controllable in this sytem, and thus, the cell culture technologist can better optimize cell growth and product generation. Although cellular waste products are still allowed to accumulate in this system, it may be possible to limit the accumulation of these waste products by adjusting the nutrient flow. Examples illustrating this point will be given later in the presentation.

Perfusion is the third method for growing mammalian cells in culture. Both nutrient and waste product concentrations can be controlled by varying the dilution rate of the system. Increased dilution rates result in increased nutrient and decreased waste-product concentrations. Thus, a high degree of control can be exhibited over the environment of this system. However, perfusion suffers from some of the same disadvantages as the batch method with periodic replenishment of the medium; that is, nutrients are being discarded in the effluent of the culture vessel. Again, this is not very cost effective. Thus, perfusion could only be economical in comparison to the fed-batch mode if, because of the decreased waste product concentrations, the cell densities obtainable are much higher than that obtainable in a fed-batch culture.

TECHNICAL AND ENGINEERING BARRIERS TO SCALE-UP

The three main barriers to the scale-up of mammalian cell production processes are: (1) oxygen transfer limitations; (2) the accumulation of toxic compounds; and (3) present lack of regulatory hardware and software to control the process for production of cells and/or products. All of these barriers become more pronounced as the culture volume increases.

Oxygen Supply

The ability to provide an adequate oxygen supply to mammalian cells at increasing culture volumes is the most critical barrier to the scale-up of mammalian cell cultures, especially suspension systems. Unless oxygen is constantly supplied to the medium, oxygen limitation will result. Oxygen is only sparingly soluble in cell culture medium (0.2 mmoles O_2/l).[20] Oxygen demand varies depending upon the cell type, ranging between 0.0531 mmole O_2/l-hr-10^6 cells/ml[21] to 0.59 mmole O_2/l-hr-10^6 cells/ml.[22]

For small-scale suspension systems, oxygen can be supplied from the head space of the culture vessel. For example, HeLa cells at 1×10^7 cells/ml respire at a rate of 5.0 mmoles O_2/l-hr.[24] (Cell densities of this magnitude are not usually found in typical suspension cultures, but have been achieved under certain conditions.[23]) The oxygen transfer rate of the culture vessel must be capable of supplying oxygen at this rate to the cells in order to avoid oxygen limitation. The oxygen transfer rate of any vessel can be expressed as:[25]

$$N_O = k_L a(C^* - C_L)$$

where N_O = oxygen transfer rate, mmole O_2/l-hr
$k_L a$ = mass transfer coefficient, mmole O_2/atm-l-hr
C^* = concentration of oxygen in the gas phase in equilibrium with the saturated concentration of oxygen in the liquid phase, atm.
C_L = concentration of oxygen in the gas phase in equilibrium with the actual concentration of oxygen in the liquid phase, atm.

It can be assumed as an approximation that the "a" term in the oxygen transfer coefficient for unaerated vessels is a function of the ratio of the surface area available for aeration to the total volume of the culture. Using this assumption, and some values of k_L calculated from unaerated vessels of mammalian cells,[20] it can be calculated that oxygen limitation for 10^7 HeLa cells/ml will occur at a volume of less than 1.0 liters, when only air is in the headspace of the vessel ($C^* = 0.21$ atm.). If pure oxygen instead of air is used in the headspace, the maximum allowable volume before oxygen limitation only increases to 3.5 liters. Thus, for any culture volumes over 3.5 liters, additional oxygen must be added to the medium.

Unlike microbial fermentations, air sparging can not easily be used to provide oxygen to mammalian cells. Air sparging causes lysis of the cell membrane and foaming of the medium.[26]

An alternative to sparging, developed by Sinskey *et al.* was to use silicone tubing through which air diffuses into the liquid medium.[27,35] Fleischaker used this method to provide oxygen to a 7.5-l culture of FS-4 cells at a density of 10^6 cells/ml.[20] We may now ask to what scale can this procedure be easily translated. Suppose we wish to oxygenate 1000 liters of our hypothetical culture of HeLa cells. Since it is known that the $k_L a$ for oxygen transfer through the silicone tubing is 0.35 mmole O_2/atm-cm^2-hr, it would take 100 feet of 1.0-inch silicone tubing to provide enough oxygen to a culture of that size. This solution may be quite feasible, but may be somewhat expensive and quite unwieldy.

Another strategy for oxygenating mammalian cell cultures, demonstrated by Clark and Hirtenstein,[28] is to circulate the medium through the culture vessel, while simultaneously oxygenating it in an external loop. Clark externally oxygenated the medium by allowing it to equilibrate with the oxygen in the headspace of another vessel, but there is no reason why the medium could not be oxygenated externally via silicone tubing. The advantage of this method is that the circulation rate of the medium would dictate the amount of oxygen that is supplied to the cells. This can be described mathematically by:

$$F(C_i - C_o) = OUR$$

where
F = Flow rate of medium, l/hr
C_i = Dissolved oxygen concentration, inlet, mmole/l
C_o = Dissolved oxygen concentration, outlet, mmole/l
OUR = Oxygen uptake rate of the cells, mmole/hr

Clark and Hirtenstein used small culture volumes to illustrate external oxygenation, so it would be proper to inquire how feasible this strategy would be on a larger scale. For a 1000-liter volume of our HeLa cell culture, a flow rate of 30,000 l/hr would be needed to ensure a dissolved oxygen concentration of 10% of saturation is maintained in the vessel medium. Such a high flow rate would surely cause cell damage and excessive foaming. Thus, this method does not seem practical on a large scale.

Another possible method of oxygenating mammalian cell cultures would be to increase the surface area of the medium that is exposed to air. Essentially, this is the principle of using silicone tubing. The entire surface area of the mammalian culture vessel could be composed of an oxygen-permeable material, thus increasing the "a" term in the oxygen transfer coefficient($k_L a$). For example, the artificial capillary and Jensen's tubular spiral film culture techniques utilize an oxygen-permeable membrane as the structural material that confines the culture medium. For suspension cultures, a concentrical lining of the inside walls of the culture vessel with a silicone membrane would facilitate oxygen diffusion to the medium from a stream of air flowing between the walls of the vessel and the silicone membrane (FIG. 5). For our 1000-l suspension culture of HeLa cells, 68,000 cm^2 of surface area would be required to prevent oxygen limitation. This method of oxygenation would require culture vessels with unorthodox dimensions (a vessel with a diameter of 60 cm and a height of 360 cm), and multiple impellers to obtain sufficient agitation. In addition, the cost of the silicone membrane would have to be taken into account in order to analyze the validity of this method.

Waste-Product Accumulation

The products of mammalian cell metabolism include lactic acid, ammonia, and carbon dioxide. If lactic acid and ammonia are left to accumulate to high enough concentrations, cell metabolism or even cell growth can be inhibited. Lactic acid, if produced in excess, can overcome the buffering capacity of the medium, and result in a pH lower than optimal. In addition, it has been recently shown that prolonged exposure to ammonia concentrations in excess of 4 mM can also inhibit growth.[29] Thus, ammonia generation can be a serious problem, since it is estimated that a culture of HeLa cells growing from a cell density of 3.4×10^5 cells/ml to a cell density of 1×10^7 cells/ml could generate approximately 30 mM of ammonia. On a small scale, frequent replacement with fresh medium would keep the concentration of these toxic compounds from reaching inhibitory levels. On a large scale, this is not desirable, since serum-supplemented medium is very expensive, and thus, it would be very wasteful to discard serum components (which account for the great majority of the medium cost) to remove lactate and ammonia.

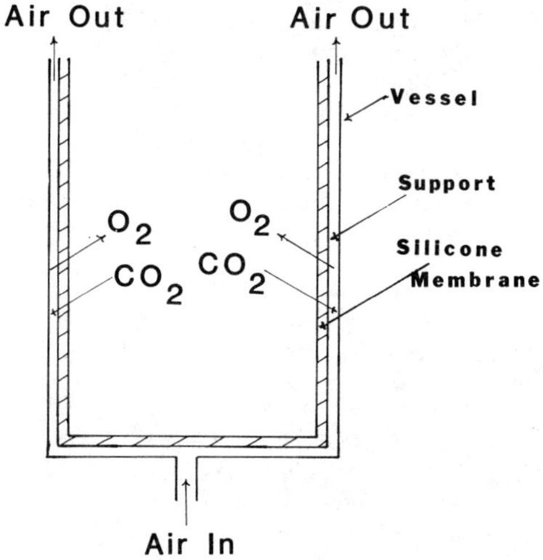

FIGURE 5. Strategy for increasing surface area available for oxygen transfer.

Two strategies have been developed to overcome the problem of ammonia and lactate generation. The first strategy is to modify the culture medium so that the cell's metabolism is altered to generate less waste products. Eagle *et al.* in 1958, substituted galactose for glucose as the carbohydrate source in the medium, resulting in a 67-fold decrease in lactic generation.[30] In 1982, Fleischaker reported the same effect by feeding glucose to FS-4 cells in a fed-batch type fashion, such that the glucose concentration in the medium was relatively low at all times (0.5 m*M* vs. 20 m*M*).[20] Glacken and Sinskey in 1983, demonstrated that ammonia generation can be reduced by over 60% by continually feeding glutamine to the culture so as to keep the glutamine concentration at a consistently low level (0.2 m*M* vs. 4 m*M*).[29]

Perfusion culture is the second strategy used to limit the accumulation of metabolic end products. As stated previously, this strategy discards medium needlessly, and if it is to be economical, this technique must support higher cell densities than is achieved by other techniques, or must employ dialysis to recover expensive macromolecular serum components. Thus, whether a fed-batch system with a controlled environment designed to limit end-product accumulation, or a perfusion system should be the method of choice, depends on: (1) the optimal medium turnover rate of the perfusion system that balances high final cell densitities with low medium usage; (2) the extent of reduction in waste-product accumulation that can be achieved in a controlled, fed-batch system; (3) the number of medium changes required to keep the concentration of waste products below the maximum level in the fed-batch system, if the generation of waste products cannot be sufficiently reduced through adjustments to environmental parameters only; (4) the cost of medium dialysis; and (5) the cost of equipment peculiar to perfusion systems, that is, pumps, spin filters, and so forth.

Lack of Regulatory Hardware and Software

When cultivating cells on a large scale, it is highly desirable to produce cells and cell products consistently at a maximal rate and level with every production run. One 1000-l culture that produces less than optimal cell and product titers may greatly increase production costs. To ensure consistent performance at optimal conditions, computer monitoring and control should be implemented. To date, most mammalian cell production units described in the literature have not utilized process control. One practical reason for this is that many production facilities are located in Third World countries, which do not have adequate resources to maintain complex computer hardware.[3] Even in developed nations, process control is usually only applied to monitoring the culture temperature and pH.

The lack of sophisticated process-control strategies for mammalian cell cultures is not a problem of scale. The problem is that, except for a few limited cases, process-control strategies designed to optimize cellular metabolism have not been developed, and that this deficiency is more serious as the scale of production increases.

In 1982, Fleischaker utilized computer monitoring and control to optimize the production of interferon by FS-4 cells in a 7.5-l culture.[20] Oxygen uptake rate, lactic acid production, carbon dioxide formation, and glucose consumption were monitored by means of a microcomputer to determine the ATP flux with time. From the calculated ATP flux, the cell concentration in the culture could be continuously estimated. The rate of glucose fed to the vessel was varied by the computer in response to the glucose concentration of the culture medium (measured using an on-line spectrophotometer after automatic derivitization of the glucose in the sample stream), compared with the glucose concentration set point, which was adjusted according to the lactate production rate. This strategy accomplished two goals: (1) a reduction in the lactic acid production rate; and (2) determination of the optimal culture time to induce the cells for interferon production. Thus, through the use of computer monitoring and control, maximum interferon titers could be consistently obtained. This same strategy could be used to determine the optimum time to induce cells for virus production or other products so as to reduce the run-to-run variability of virus titers that so often affects biological manufacture.

A paradigm elucidating process control strategies involving biological products is shown in FIGURE 6. Variables pertinent to the mammalian cell production process must be monitored, and the values compared with predetermined set points, which are estimated to be optimal as a first approximation. If the values of the variables differ from the set points, changes based on information provided by mathematical models relating the system process parameters to one another adjust the variables back to the set-point values. The variables measured are then used to calculate parameters that characterize the state of the cellular metabolism, such as nutrient uptake rates and waste-product excretion rates. These values are used to estimate the growth rate of the cells, which in turn are utilized to establish an elemental mass balance for the cell population. The mass balance, in turn, can be used to estimate the rate of product formation in the culture. The productivity thus calculated is compared to the desired productivity. If this productivity is too low, the set points of the process variables are

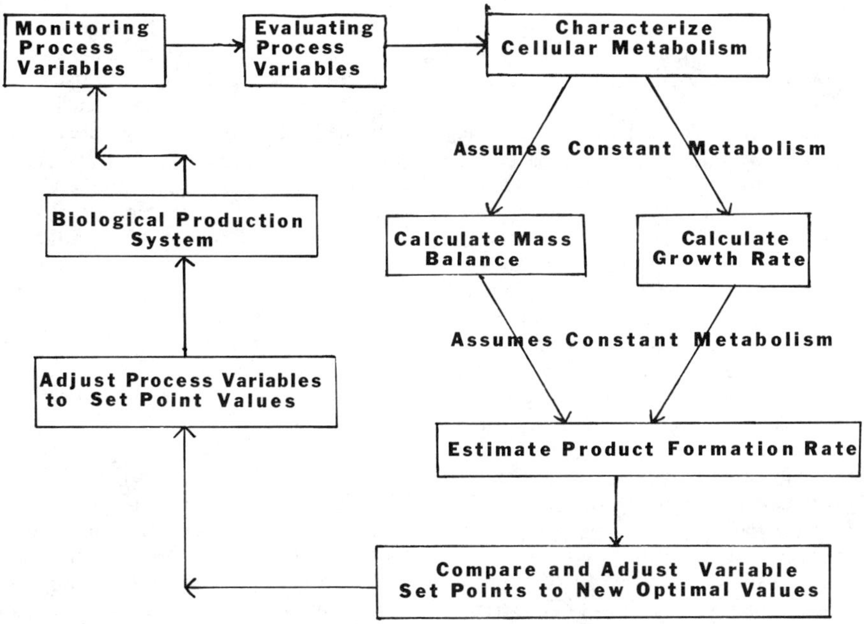

FIGURE 6. Process control paradigm.

altered, in a structured fashion according to techniques developed from chemical engineering process optimization theory. This feedback-feedforward type of algorithm can zero in on the optimal culture conditions.

In practice, it may be quite difficult to use an elemental mass balance to estimate mammalian cell-product formation, since most cell products of interest are produced in very small quantities. The magnitude of the error inherent in the mass balance calculation may be greater than the absolute amount of the product produced. Nevertheless, this paradigm outlines the basic research strategy needed to determine the optimum process parameters that result in maximum product generation. Off-line measurements of the product generation rate would be needed as feedback information to implement the use of a mass balance to maximize product titers.

COMPARISON OF SOME PROMISING MAMMALIAN CELL CULTURE TECHNIQUES

Anchorage-dependent Cells

As mentioned previously, three promising techniques for the large-scale production of anchorage-dependent mammalian cells are microcarrier suspension culture, artificial capillary, and Jensen's spiral film (TABLE 2).

There are numerous advantages to using a microcarrier suspension system to produce anchorage-dependent cells on a large scale. First, a suspension system of microcarriers has the highest surface-to-volume ratio of any technique developed to date (FIG. 3). Thus, all other factors being equal, less labor, materials and medium are required to produce a given quantity of cells in microcarrier cultures as compared to other systems. For example, Fleischaker has demonstrated that one 7.5-liter microcarrier culture using a microcarrier density of 5 g beads/l can produce as many cells as 500 roller bottles,[20] using only 15% of the medium that would have been used in a comparable roller-bottle system. Microcarrier cultures can support cell densities of 5–6 $\times$ 10^6 cells/ml in batch cultures,[9] and 4 $\times$ 10^7 cells/ml in perfusion cultures.[31] The cell densities obtainable from batch cultures could probably equal that of perfusion systems if adequate oxygen transfer rates are maintained in the batch cultures.

Another advantage of microcarrier suspension systems is that no discrete cell populations exist and the environment is uniform. This implies that this technique is very amenable to system monitoring and process control. Adjustments to the culture environment can be easily made, so as to continually maximize cell growth and product generation, while at the same time keeping waste-product formation to a minimum. In addition, since the environmental parameters can be controlled, greater consistency from culture to culture can be expected. Fleischaker demonstrated that the average interferon titers produced per production run of FS-4 cells in computer-controlled microcarrier cultures were higher than those produced in cultures without computer control.[20]

Microcarrier suspension cultures have a few unattractive characteristics. First, not every cell line can be easily removed from the microcarriers while still maintaining high cell viability. Since small microcarrier cultures cannot be used to inoculate larger microcarrier cultures, a large-scale production facility utilizing microcarriers still must depend on cells produced from roller bottles for inocula. For example, approximately 200 roller bottles would be needed to seed a 10-l microcarrier culture at 4 $\times$ 10^5 cell/ml. This limitation may restrict the scale of the microcarrier culture vessel that may be used. Researchers at M.I.T. have made some progress in solving this problem, but to date, no literature has been published that adequately addresses this topic. Additionally, the microcarriers add to the material cost of producing cells, which is not present in other systems. For example, a 5 g beads/l culture would cost \$50/l if the beads are not reprocessed and reused. A reprocessing step, although less expensive than simply discarding the used beads, would still add to the final cost of producing the cells.

Microcarrier suspension systems suffer from the same poor oxygen transfer characteristics as observed for any suspension culture of mammalian cells. Methods described previously can improve the oxygen transfer characteristics of suspension cultures, but to each of these methods is ascribed a certain inherent cost. The main advantage of the two cell production methods yet to be described is that they have excellent oxygen transfer potential. Both artificial capillary and the tubular spiral film propagators incorporate an oxygen-permeable membrane as the support for cell

TABLE 2. A Comparison of Three Promising Systems for Producing Mammalian Cells on a Large Scale

	Microcarrier Suspension	Artificial Capillaries	Jensen's IL410 Tubular Spiral Film
S/V Ratio	31—153	30.7	9.35
O_2 Transfer, large volumes, high cell densities	Difficult, requires much silicone tubing or another design	Good, but some cells may be O_2 limited due to O_2 gradient through cells	Excellent
Uniform environment	Yes	No—gradients will exist.	
Ability to control environment	Excellent, fed-batch or perfusion	Moderate—perfusion only—gradients	
Optimization via computer control	Excellent—has been achieved for many parameters	Difficult—amenable for only a few parameters	
Ability to monitor cell growth	Excellent, by sampling or instrumentation	Difficult for both	
Maximum reported cell densities	$5–6 \times 10^6$ cells/ml (batch) $4–5 \times 10^7$ cells/ml (perfusion)	1×10^8 cells/ml	6×10^6 cells/ml
Ability to reduce medium (serum) consumption	Excellent, via fed-batch system with computer control	Excellent, via high cell densities	Moderate
Scale-up possibilities	Good, limited by oxygen transfer and bead cost and handling	Good—limited by difficulty in monitoring and control, nutrient + O_2 gradients can limit scale, pressure drop problems	Good—limited by low S/V and nutrient gradients

attachment. Thus, oxygen can diffuse through the membrance and come directly in contact with the cell mass. Because of the high oxygen transfer characteristics of these systems, high cell densities can be achieved (TABLE 2). The low surface-to-volume ratio of Jensen's spiral film limits the maximum cell density achievable with this system.

The very high cell densities achieved with the capillary system (10^8 cells/ml) implies that cells are not only growing on the surface of the capillaries, but they are also growing in the interstitial space between the capillaries, in a tissue-like cell clump. One can speculate that the cells in the center of such a clump may be oxygen limited, due to the transfer resistance of the cell mass. Thus, the metabolism of these oxygen-limited cells may be entirely different from the cells on the outer layers of the "tissue," if they

TABLE 3. Microencapsulated Cell Systems versus Traditional Cell Systems

Characteristic	Free Suspension	Microencapsulated Suspension
Cell density	2×10^6 cells/ml	1×10^7 cells/ml
Product titer per cell	+	++
Cell separation	Centrifugation—expensive	Gravity settling
Cell recovery	No extra step	Capsule lysis
Product separation very high MW	Standard	Easier, trapped in capsule. Performs separation *in situ*.
Lower MW	Standard	Easier, separated from cells *in situ*—important if using cancer cells.
Media (Serum) usage	Standard	Lower, because of higher cell density.
Computer control and optimization	Yes	Yes
Protection of product from shear denaturation	No	Yes
Oxygen transfer	Limited by K_La	Limited by O_2 diffusivity in cell mass.
Scale-up potential	In use in large-scale vaccine production	O_2 transfer problem must be solved, must weigh cost of encapsulation vs. reduced costs of serum usage and product separation.

are indeed still viable. The extent of the oxygen limitation of an individual cell would depend on the thickness of the cell mass that separates it from the capillary wall. Therefore, even though very high cell densities may be reached, it may be advisable to limit the final cell density to a lower value, since cells in anoxic conditions may produce large quantities of acid, which may then inhibit the product formation of cells still under aerobic conditions. However, if the product is the cell mass, high cell densities are, of course, very desirable, since the cost of medium utilization is much less at the higher densities.

The cellular environment in both the artificial capillary and the tubular spiral film is not uniform. Gradients in nutrient and waste-product concentration will exist. If these gradients are large, cell growth and metabolism may be adversely affected in some sections of the system. In addition, these gradients complicate the successful implementation of process control. Since cellular metabolism is a function of nutrient and waste-product concentrations, many different populations of cells will exist in these systems. Thus, monitoring the gradients would only provide an average indication of cellular activity. Cell populations that are operating less than optimally would be masked by this averaging. Therefore, actions taken by the control system based on the values of the gradients may serve only to optimize certain cell populations. Of course, gradients may be kept to a minimum by increasing the flow rate, with a concomitant increase in pressure drop across the propagator, which would eventually become prohibitive at increasing flow rates. In addition, the gradients cannot be kept too small, since the accuracy of subsequent calculations estimating cellular kinetics (e.g., oxygen uptake rate) would suffer.

Monsanto has reported the development of a flat-bed, artificial capillary system that minimizes the magnitude of the gradients developed.[33] The medium flows axially on the outside of the capillaries, instead of flowing longitudinally on the inside of the capillaries. However, gradients would still exist, and process control would still be somewhat complicated.

In addition, it is quite difficult to take representative cell samples with which to estimate cell growth in these two systems. Cells cannot be removed during standard operation conveniently in these systems. In contrast, with the microcarrier suspension system, all that is required is to withdraw a given volume of the suspension medium. Cells on the beads can be viewed under a microscope, or can be removed from the microcarriers and counted.[32] Cells can be viewed under a microscope in the IL410, but not in the artificial capillary propagator.

In summary, microcarrier suspension culture has the best scale-up potential of the systems mentioned, provided the problem of oxygen transfer limitation is solved, since it is very amenable to process control. If the oxygen transfer problem cannot be easily solved, then the artificial capillary and the tubular spiral film systems will become attractive alternatives.

Suspension Cells

The only drastically novel procedure to date for cultivating suspension cells has been developed by Lim[18] and applied commercially by Damon Corporation, Needham Heights, MA.[19] This procedure utilizes cells encapsulated in a semipermeable membrane, which are cultivated in standard suspension culture vessels. The advantages of this system are that: (1) higher cell densities are obtained compared to standard suspension systems; (2) increased product generation per cell can be obtained; (3) cells can be separated from the medium simply by means of gravity settling, as opposed to centrifugation; and (4) product can be partitioned into either the microcapsule or the medium, separate from the cells, whichever is desired (TABLE 3).

A disadvantage of this system is, of course, the cost of encapsulation and removal of cells and products from the capsule. A more serious limitation to the use of this technique on a large scale is that of oxygen transfer. In addition to the low oxygen transfer potential of typical suspension systems, the cell mass in the capsule may limit oxygen to cells in the center of the capsule. As stated earlier, this could have adverse effects on cell metabolism and product formation, as well as complicating the implementation of any process control system. The magnitude of the oxygen transfer resistance of the cell mass in the capsule is a function of the capsule radius according to the following relation:[34]

$$\Phi = r_s r^2 / 9 D_s S_O$$

where Φ = modified form of Thiele modulus, dimensionless
r_s = respiration rate of cells, mmole O_2/l-s
r = radius of cell floc, cm
D_s = diffusivity of oxygen through cell mass, cm^2/hr
S_O = dissolved oxygen concentration in the liquid medium, m*M*

The Thiele modulus is functionally related to the effectiveness factor, η, which is the ratio of the observed oxygen uptake rate to the oxygen uptake rate that could be obtained in the absence of any oxygen transfer resistance in the cell floc.[34] For example, an effectiveness factor equal to one implies that no transfer resistance exists, while an $\eta = 0.5$ implies the oxygen transfer resistance of the cell floc decreases the respiration rate of the cells by one-half. In a 500-micron microcapsule, it is known that 10,000 cells can exist.[19] If one assumes that the respiration rate of the cells is a zero-order function of the oxygen concentration, equal to 0.59 mmole O_2/l-hr and also assumes that the diffusivity of oxygen through the cell floc equals 2.1×10^{-5} cm^2/s,[34] and that the dissolved oxygen concentration can be kept at saturation, one can calculate that the Thiele modulus is 1.84, which corresponds to an effectiveness factor of approximately 0.6. This implies that 40% of the cells in the capsule would be under oxygen limitation. To avoid any oxygen limitation, the Thiele modulus must be less than 0.3. This would occur at a capsule diameter of 170 microns. Thus, not only must techniques be developed to transfer oxygen from the gas phase to the liquid phase at a sufficiently rapid rate, strategies must also be developed to produce smaller diameter microcapsules. This could possibly be achieved by varying the surface tension of the alginate membrane mixture.

SUMMARY

Mammalian cell products have great medical and clinical importance, but to date, production methods employed to manufacture these products on a large scale are not as cost efficient as they could be. The implementation of process control would greatly improve the productivity of these products. Recently developed methods to produce cells on a large scale, such as microcarriers, artificial capillaries, tubular spiral film, and microencapsulation must be optimized, and the problem of oxygen transfer limitation must be solved. The accumulation of potentially toxic waste products can inhibit growth and reduce productivity. This effect can be reduced by either adjusting the environmental parameters of a fed-batch culture, so that the cell's metabolism is shifted away from producing these compounds, or by continually perfusing medium through the culture. If these technical barriers can be overcome, the cost of producing products derived from mammalian cells can be greatly reduced.

ACKNOWLEDGMENTS

We greatfully acknowledge Dr. Carl Batt for editing the text of this manuscript. Support for M. W. Glacken and R. J. Fleischaker was provided by NCI Training Grant 5-T32-CA09258-05.

REFERENCES

1. JOHNSTON, M. D., G. CHRISTOFINIS, G. D. BALL, K. H. FANTES & N. B. FINTER. 1979. A culture system for producing large amounts of human lymphoblastoid interferon. *In* Developments in Biological Standardization. R. H. Regamey, R. Speir & F. Horodniceanu, Eds. Vol. **42**: 189–192. S. Karger. Basel.
2. AHARONOWITZ, Y. & G. COHEN. 1981. The microbiological production of pharmaceuticals. Sci. Am. **245**: 141–152.
3. GIRARD, H. C. 1977. Problems encountered in large-scale production plants. *In* Cell Culture and Its Applications. R. T. Acton & J. D. Lynn, Eds. Academic Press. New York. pp. 111–127.
4. JACOBS, J. B. 1979. Serially propagated human diploid cells: A synopsis of the present position concerning their use for producing viral vaccines and interferon. *In* Developments in Biological Standardization. R. H. Regamey, R. Spier & F. Horodniceanu, Eds. Vol. **42**: 13–18. S. Karger. Basel.
5. STAEHEHELIN, T., D. S. HOBBS, H. F. KUNG & S. PESTKA. 1981. Purification of reccombinant human leukocyte interferon (IFLrA) with monoclonal antibodies. *In* Methods of Enzymology. S. Pestka, Ed. Vol. **78**: 505–512. Academic Press. New York.
6. OLSSON, L. & G. MATHE. 1982. Emerging immunological approaches to treatment of neoplastic diseases. *In* Recent Results in Cancer Research. G. Mathe, G. Bonadonna & S. Salmon, Eds. Vol. **80**: 334–337. Springer-Verlag. Berlin.
7. RITZ, J. & S. F. SCHLOSSMAN. 1982. Utilization of monoclonal antibodies in the treatment of leukemia and lymphoma. Blood **59**: 1–11.
8. VAN WEZEL, A. L. 1967. Growth of cell strains and primary cells on microcarriers in homogeneous culture. Nature (London) **216**: 64–65.
9. LEVINE, D. L. 1977. Ph.D. thesis. M.I.T.
10. WEISS, R. E. & J. B. SCHLEICHER. 1968. A multisurface tissue propagator for the mass-scale growth of cell monolayers. Biotechnol. Bioeng. **10**: 601–615.
11. HOUSE, W., M. SHEARER & N. G. MAROUDAS. 1972. Method for bulk culture of animal cells on plastic film. Exp. Cell Res. **71**: 293–296.
12. MUNDER, P. G., M. MODOLELL & D. F. H. WALLACH. 1971. Cell propagation on films of polymeric flurocarbons as a means to regulate pericellular pH and pCO_2 in cultured monolayers. FEBS Lett. **15**: 191–196.
13. GIRARD, H. C., M. SUTCU, H. ERDEM & I. GURHAN. 1980. Monolayer cultures of animal cells with the gyrogen equipped with tubes. Biotechnol. Bioeng. **22**: 477–493.
14. WOHLER, W., H. W. RUDIGER & E. PASSARGE. 1972. Large-scale culturing of normal diploid cells on glass beads using a novel type of culture vessel. Exp. Cell Res. **74**: 571–573.
15. KNAZEK, R. A., P. M. GULLINO, P. O. KOHLER & R. L. DEDRICK. 1972. Cell culture on artificial capillaries. Science **178**: 65–67.
16. JENSEN, M. D. 1981. Production of anchorage-dependent cells—problems and their possible solutions. Biotechnol. Bioeng. **23**: 2703–2716.
17. FEDER, J. & W. R. TOLBERT. 1983. The large-scale cultivation of mammalian cells. Sci. Am. **248**: 36–43.
18. LIM, F. & R. D. MOSS. 1981. Microencapsulation of living cells and tissues. *In* Microencapsulation: Selected Papers from the 47th International Symposium on Microencapsulation. M. H. Furguson, Ed. American Pharmaceutical Society. Washington, D.C. pp. 1–4.
19. DAMON CORPORATION. 1981. Tissue Encapsulation. Press release.
20. FLEISCHAKER, R. J. 1982. An experimental study in the use of instrumentation to analyze metabolism and product formation in cell culture. Ph.D. Thesis, M.I.T.
21. KATINGER, H. W., W. SCHEIRER & E. KROEMER. 1978. Der blasensaulenfermenter fur die massensuspension Kultur tierlscher zellen. Chem. Ing. Tech. **50**: 472–473.
22. GREEN, M., G. HENLE & F. DEINHARDT. 1958. Respiration and glycolysis of human cells grown in tissue culture. Virology **5**: 206–219.
23. KEAY, L. & C. W. BRUTON. 1979. Recent advances in the technology of animal cell production. Proc. Biochem. pp. 17–21.

24. DANES, S. B., M. M. BROADFOOT & J. PAUL. 1963. A comparative study of respiratory metabolism in cultured mammalian cell strains. Exp. Cell. Res. **30**: 369–378.
25. WANG, D. I. C., C. L. COONEY, A. L. DEMAIN, P. DUNNILL, A. E. HUMPHERY & M. D. LILLEY. 1979. Chapter 9. *In* Fermentation and Enzyme Technology. John Wiley and Sons. New York.
26. KILBURN, D. G. & F. C. WEBB. 1968. The cultivation of animal cells at controlled dissolved oxygen partial pressure. Biotechnol. Bioeng. **10**: 801–814.
27. SINSKEY, A. J., R. J. FLEISCHAKER, M. A. TYO, D. J. GIARD & D. I. C. WANG. 1981. Production of cell-derived products: Virus and Interferon. Ann. N.Y. Acad. Sci. **369**: 47–60.
28. CLARK, J. M. & M. D. HIRTENSTEIN. 1981. Optimizing culture conditions for the production of animal cells in microcarrier culture. Ann. N.Y. Acad. Sci. **369**: 33–46.
29. GLACKEN, M. W. & A. J. SINSKEY. Ammonia production and glutamine metabolism of cultured mammalian cells. In press.
30. EAGLE, H., S. BARBAN, M. LEVY & H. O. SCHULZE. 1958. The utilization of carbohydrates by human cell cultures. J. Biol. Chem. **233**: 551–558.
31. THOMAS, J. T. & W. G. THILLY. Personal communication.
32. VAN WEZEL, A. L. 1973. Microcarrier culture of animal cells. *In* Tissue Culture: Methods and Applications. P. F. Kruse, Jr. & M. K. Patterson, Eds. Academic Press. New York. pp. 372–377.
33. KU, K., M. J. KUO, J. DELENTE, B. J. WILDI & J. FEDER. 1981. Development of a hollow-fiber system for large-scale culture of mammalian cells. Biotechnol. Bioeng. **23**: 79–95.
34. BAILEY, J. E. & D. F. OLLIS. 1977. Biochemical Engineering Fundamentals. McGraw-Hill, New York. pp. 393–401.
35. FLEISCHAKER, R. J. & A. J. SINSKEY. 1981. Oxygen demand and supply in cell culture. Eur. J. Appl. Microbiol. Biotechnol. **12**: 193–197.

Entrapped Plant Cell Tissue Cultures

M. L. SHULER, O. P. SAHAI, AND G. A. HALLSBY

School of Chemical Engineering
Cornell University
Ithaca, New York 14853

INTRODUCTION

The potential of using plant cell tissue culture for the commercial production of speciality chemicals has been well recognized.[1-3] Extraction of medicinals, flavoring agents, essential oils, pesticides, or herbicides from whole plants presents numerous problems.[4] Conceptually, the alternatives to whole-plant extraction are organic synthesis, plant cell tissue culture, or growth of simpler organisms with the appropriate plant genes. The compounds of commercial interest are too complex to be economically synthesized. These compounds are generally the result of the coordinated expression of many genes. With our current knowledge of plant cell genetics, the movement of such genes into bacteria or yeast is impracticable. Consequently, plant cell tissue culture is the only viable alternative to extraction from whole plants.

The "factory production" of such compounds from tissue culture would insure a more continuous supply of product—the raw materials would be more homogenous in nature, and their supply would be independent of weather, political, or disease considerations. The yields could be more easily and rapidly manipulated and improved with a "factory" scheme rather than with wild or domesticated plants. It is also possible to use tissue cultures to modify unnatural substrates (e.g., conversion of 6-ethyl analogue of thebaine to codeinone).[5]

Because of difficulties in the culturing of plant cells and in obtaining high product yields, the commercial adoption of plant cell culture has not yet been accomplished.[1,2,4] Essentially, those conditions favoring rapid growth suppress secondary metabolism and product formation; those factors that might increase cellular differentiation and product formation are detrimental to cellular growth and replication.

More recently, high yields of secondary products have been obtained by selecting naturally occurring variants in tissue culture.[6] The genetic stability of such variants can be a significant problem, but some variant lines have proved to be relatively stable. Even with these high-yielding variants, the engineer needs to build a reactor allowing the most complete expression possible of the cell's genetic potential. To do this, we believe it will be necessary to build reactors separating growth and product formation. Then the environmental conditions for each function can be decoupled and optimized separately. With suspended-cell culture systems, the separation can be with respect to time (e.g. fed-batch) or with respect to space (e.g., multistaged continuous culture). An alternative would be an immobilized-cell process where cells are first grown optimally and then immobilized, and environmental conditions optimized for product formation. What are the advantages and characteristics of these alternatives?

Cells in suspension culture grow slowly (doubling times of about 20 to 60 hr are common). They tend to form large aggregates in suspension cultures. The cells in the center of the aggregate are morphologically distinct from those on the periphery. The center cells are partially differentiated and may act as feeder cells supplying the cells on the periphery with growth-promoting compounds.[7] Other observations have indicated that the formation of embryoids in response to alterations of exogenous hormonal concentrations occur only from peripheral cells on an aggregate—not from single cells.[8] Cell aggregation in the fairly simple media used in tissue culture may be essential to rapid growth and cellular differentiation[4,9] and consequently may be necessary for high product formation.

But aggregation presents practical problems. The degree of aggregation depends on the fluid dynamics within the culture vessel and, to a certain extent, on the characteristics of the inoculum used. Clearly, scale-up of such processes would be difficult. Currently, even more important is that the unambiguous interpretation of basic experiments is complicated by the diversity of cell types present and the difficulty in reproducing the same distribution of cell types from one investigator to another and from one time to another. In essence, the chemical environment that is manipulated by the investigator need not result in a unique response.

Thus suspended-cell cultures are difficult to control and to scale-up, and such cells grow slowly. Slow growth would necessitate large reactors and these would be quite susceptible to contamination.

ADVANTAGES AND DISADVANTAGES OF ENTRAPPED PLANT CELL CULTURES

The alternative to suspended-cell cultures is immobilized-cell cultures. Immobilization with plant cells is usually by physical entrapment. Plant cell immobilization would be preferred to the immobilization of enzymes except in the cases where the conversion involved a single enzyme (or possibly two or three under very special circumstances) and cofactor regeneration was not a problem. For most of the plant products of commercial interest, whole-cell immobilization will be required.

One of the main disadvantages to the use of suspended cultures is the low growth rate that can be obtained. Not only does such slow growth impair the rate at which an investigator may determine the response of cells to manipulated parameters, it presents a significant economic barrier to the production of secondary metabolites by suspended cultures.[3,10] With immobilized cells, the growth rate is immaterial. Cell stability is far more imporant. High cell densities can be used as well as a continuous flow process. Flow rates well in excess of washout can be used. High volumetric productivities are possible—perhaps on the order of immobilized bacterial processes, if the cellular concentrations of key enzymes and reactivities per enzyme are similar. A significant increase in volumetric productivity will greatly increase the potential for the development of commercial plant tissue culture processes.

The commercial development of tissue culture will also require the use of high-yielding variants. The maintenance of such variants can be difficult,[6] and their reversion to less productive strains in commercial scale, suspended-culture processes such as fed-batch or continuous culture would be a problem. Since growth in immobilized cultures is discouraged, there is far less probability of reversion of high-yielding variants to less productive strains.

The large-scale culture of plant cells in stirred fermenters presents technical challenges in the maintenance of sufficient oxygen transfer capability without excessive liquid shear.[11] Plant cells are shear sensitive. Cell immobilization removes such complications since the cell is "protected" from shear effects.

Immobilized cultures present fewer operational and scale-up problems than suspended cell cultures. We have used both continuous-culture and continuous-flow, immobilized-cell reactors. With suspended cultures, wall attachment of cells, plugging of outflow lines, and contamination presented significant obstacles to be overcome.[12,13] With continuous flow, immobilized-cell reactors, we have encountered none of these problems. An immobilized plant cell reactor should have no greater susceptibility to contamination than one with immobilized bacterial cells.

The problems of wall attachment compounded with the heterogeneous nature of suspended cultures interferes with the rational scale-up of suspended-cell plant processes. Plant cells form aggregates, and the size of such aggregates may greatly influence the observed biochemical capabilities of the culture.[4] The degree of aggregation and the range of aggregate sizes is a function of physical parameters such as the fluid hydrodynamics in the fermenter. Such factors might lead to much different responses of cells in large vessels versus that in smaller ones. A possible example comes from the study by Zenk *et al.*[14] of serpentine production from cells of *Catharantheus roseus*. In shake flasks with 25 ml of medium, serpentine production closely paralleled growth; while in airlift fermenters with 22 liters of medium, serpentine formation was strongly decoupled from cell growth. With immobilized-cell reactors, the level of aggregation is essentially a design parameter (e.g., spacing between fibers in a hollow-fiber membrane unit). There will probably be an optimal thickness or "aggregate size." The nature of the catalyst in a large unit will be the same as in a small unit.

Another potential advantage is the possible "unmasking" of metabolic pathways. Davidson and Yeoman[15] have indicted that the build-up of certain metabolic products can inhibit the expression of other pathways. Those suppressed pathways may be the ones responsible for the formation of a desired product. With entrapped-cell cultures, the metabolic by-products can be removed (basically like dialysis) and potentially allow the expression of the suppressed pathway. Consequently, the yield of the products of the suppressed pathway might be significantly higher than in a comparable suspended-cell process.

The major disadvantage of immobilized cells is that most plant products are stored intracellularly in vacuoles. Clearly, efficient product release is essential to the performance of an immobilized-cell reactor. Product release may be obtainable through variations in pH, the use of permeabilizing agents,[17–19] and perhaps due to the natural saturation of the vacuoles.

Renaudin[16] has studied alkaloid accumulation in suspension cultures of *Catharanthus roseus* and *Acer pseudoplatanus*. The results indicated that the accumulation of charged species can be easily controlled by manipulating the external (or medium) pH values. Cyclic changes in pH could possibly be used to release alkaloids stored in vacuoles.

Recently Felix, Brodeluis, and Mosbach[17] and Felix and Mosbach[18] have demonstrated the feasibility of using permeabilized cells for biochemical transformations. Felix[19] has recently reviewed techniques and solvents used for permeabilization.

Alferman *et al.*[20] have noted that purpureaglycoside was stored within the cell at the beginning of the cycle, but after six days and increasing amounts of substrate, the purpureaglycoside formed was excreted into the medium. It could be that vacuoles of resting cells can become saturated, and further increase in intracellular metabolic products is prevented by excretion.[7]

Thus, cell immobilization appears to offer distinct advantages over suspended-cell operations. Release of product into the medium can probably be obtained in most cases and is not a significant barrier to the commercial development of immobilized plant cell cultures.

TABLE 1. Summary of Previous Work on Immobilized Plant Cells

Author	Species	Matrix	Reaction	Reactor	Length of Experiment
Brodelius *et al.*[21,22]	*Morinda citrifolia*	Alginate beads	*De novo* anthraquinones	Batch, shake flasks	22 days
	Catharanthus roseus	Alginate beads	Indole alkaloids from tryptamine & secologanin	Recirculating packed column	90 hours
	Digitalis lanata	Alginate beads	Digitoxin to digoxin	Batch, shake flasks	33 days
Alfermann *et al.*[20]	*Digitalis lanata*	Alginate beads	Digitoxin & β-methyldigitoxin to purpureaglycoside, digoxin & β-methyldigoxin	Batch, shake flask	61 days
Brodelius & Nilsson[23]	*Catharanthus roseus*	Alginate	Ajmalicine from precursors & *de novo*	Batch, shake flasks	14 days
		Agarose	″	″	″
		Agar	″	″	″
		Carrageenan	″	″	″
		Gelatin	″	″	No respiration or growth
		Polyacrylamide	″	″	″
Shuler[4]	*Glycine max*	C-DAK hollow fiber unit	*De novo* phenolics	Continuous flow	30 days
Jones & Veliky[24]	*Daucus carota* Ca68	Alginate beads	Digitoxigenin to periplogenin	Batch, shake flask	24 days
	Ipomoea sp.	″	Test for Viability	″	8 days
	Cannabis sativa a5c	″	″	″	8 days

Jones & Veliky[25]	*Daucus carota* Ca68	"	Digitoxigenin to periplogenin	" & semi-continuous	6 sequential batch runs of 2-day duration
Veliky & Jones[26]	*Daucus carota* Ca68	Alginate beads	Gitoxigenin to 5β-hydroxygitoxigenin	Column bioreactor, semi-continuous	>30 days with periodic recharging of growth medium
Jirku *et al.*[27]	*Solanum aviculane*	Polyphenyleneoxide beads	*De novo* steroid glyco-alkaloids	Recirculating packed-column, semi-continuous feed	11 days
Felix *et al.*[17]	*Catharanthus roseus*	Agarose	Isocitrate dehydrogenase	Semicontinuous column, permeabilized cells	Half-life of 4 days
Felix & Mosbach[18]	*Catharanthus roseus*	Agarose with & without hardening agents	Isocitrate dehydrogenase & cathenamine reductase	Semicontinuous column, permeabilized cells	Half-life iso. dehyd. = 72 days agarose/HMDA/GA
		Polyurethane (Hypol 3000) with & without hardening agents	"	"	Half-life with hypol/HMDA/GA = 35 days. Use of HMDA/GA destroyed cath. red. activity in all cases.

REVIEW OF PUBLISHED REPORTS ON PLANT CELL IMMOBILIZATION

TABLE 1 summarizes literature reports on immobilized cell systems. The papers by Brodelius[20] and by Pederson,[29] which are included in this volume, provide two additional examples. Brodelius and Mosbach[30] have recently reviewed much of the literature.

The first reports by Brodelius *et al.*[21,22] established the feasibility of using immobilized cells for product formation from either precursors or from *de novo* synthesis. When using *Morinda citrifolia* cells for *de novo* synthesis of anthraquinones, growth hormone was removed to reduce cellular expansion and growth that in initial experiments had caused the beads containing the cells to burst. In medium without growth hormone, immobilized cells produced about ten times more anthraquinone per cell than freely suspended cells did. The reason for the enhanced productivity is not known, but quite possibly the microenvironment about the immobilized cells was sufficiently different to allow greater expression of the cells' genetic potential to make anthraquinones. In examining the production of ajmalicine from *C. roseus,* they found that 85% of this normally intracellular compound was released by the cell. The authors speculated that the presence of chloroform in the recirculating medium were responsible for the product release. Chloroform was used to extract the lipophilic product as rapidly as possible.

Others have observed problems with the mechanical integrity of alginate beads. Alfterman *et al.*[20] have reported the longest run with immobilized cells—60 days. Their experiment was terminated due to the disintegration of the alginate beads. Veliky and Jones[26] also had to modify their system to use a buffer to reduce cell growth and bead breakage; the cells were periodically recharged with growth medium.

Other possible matrixes for immobilization have been considered.[23] The plant cells are sensitive to polymerizing agents. Crosslinking with gluteraldehyde or polyacrylamide gels are not feasible. "Soft" matrixes such as alginate, agarose, agar, and carrageenan are compatible with plant cell immobilization. One report, by Jirku *et al.*[27] made use of covalent linkage with gluteraldehyde to polyphenylene oxide beads. It is not clear whether the cells retain their viability or not. Products were released in an oscillatory manner for eleven days.

MEMBRANE-ENTRAPPED CELLS

All of the reports listed in TABLE 1 made use of polymeric beads for all entrappment or attachment except for that of Shuler[4] who used membrane entrapment (see also Pederson,[29] this volume, for another report on membrane-entrapped cells). The potential advantages of membrane-entrapped cultures include a greater degree of control over the biological system. The thickness of the cell layer is uniform no matter what scale of operation is used. Each cell layer experiences a constant fluid environment. Bead size cannot be perfectly controlled with alginate beads. Problems such as channeling and large pressure drops that might be experienced in a packed bed are avoided. Mechanical abrasion of particles could be a problem with the types of polymers suitable for plant cell entrapment. The membrane system should have a greater degree of mechanical integrity and stability. The membrane system is more flexible in that a cell layer that is exhausted can be flushed out and replaced with fresh cells; the immobilizing matrix remains intact. Scale-up is potentially easier with membrane systems because of the greater control over fluid dynamics and cell-layer size.

Possible disadvantages of membrane systems may reside in high initial membrane cost and with the availability of commercial-scale membrane units with appropriate membrane spacing. Custom large-scale membrane devices are potentially available if vender interest can be made sufficiently high. There is not any rigorous economic analysis available of membrane-entrapped cell units versus systems with cells entrapped in polymeric matrixes. As more basic reaction data become available, it should be possible to complete such an analysis.

Our interest in membrane reactors is related to their high level of controllability and ease of characterization. Both characteristics are important to basic studies to elucidate the interaction of alterations in the chemical environment with the degree of aggregation (i.e., depth of the cell layer) and with the level of differentiation and product formation. We propose to use a flat plate geometry in specially constructed reaction chambers. Essentially, the system would consist of a removable plate, spacers, gaskets, membrane and membrane support, and a nutrient flow chamber. The depth of

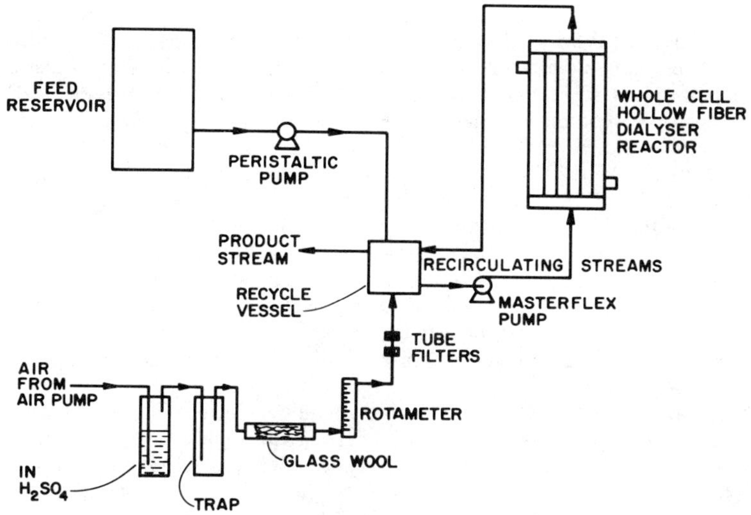

FIGURE 1. The flow diagram for the continuous-flow, hollow-fiber entrapped plant cell system is shown.

the cell layer would be determined by the number and size of spacers. The cells would be entrapped between the removable plate and the membrane. The removable plate could be removed periodically to allow direct sampling of the cell layer. A magnetic stirring bar in the nutrient flow chamber is proposed to improve mass transfer. Such a flat-plate device is currently being tested.

The membrane-entrapped cell reactor is used as part of an overall system as shown in FIGURE 1. In FIGURE 1, cells are entrapped in a hollow-fiber cartridge (C-DAK-5 hollow-fiber dialyzer unit, Cordis-Dow Corp.) instead of the flat-plate unit currently being developed. The design of a recycle chamber and system operation with bacteria has been described elsewhere.[31–33] The C-DAK unit is less useful for basic studies because the close-fiber spacings makes it impossible to completely fill the available volume outside of the fibers. Also, sampling of the cell layer and uniformity in cell

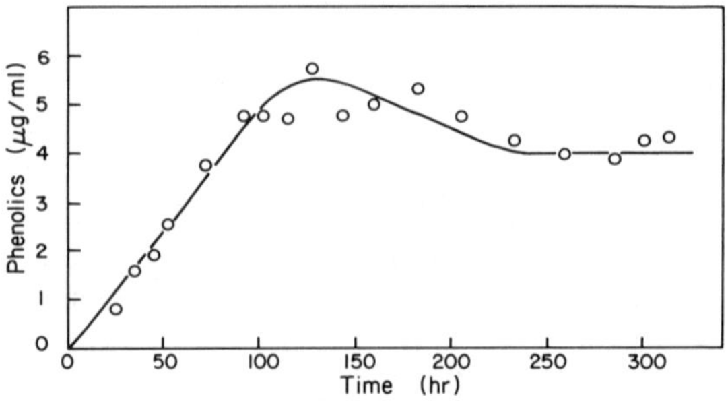

FIGURE 2. The variation of phenolics production with time for an immobilized tobacco cell system is shown. The dry weight of entrapped cells was 860 ± 20 mg. The temperature was maintained at 25 ± 1°C. The flow rate through the system was 3.6 ml/hr, giving a liquid residence time of 92 hr.

thickness are problems. Nonetheless, exploratory experiments using the C-DAK unit have been accomplished.

Using tobacco cells and the production of extracellular phenolics (as a model secondary product), we have tested the system with a C-DAK-5 as the reaction chamber. The inoculum used was from a stationary-phase culture of tobacco cells grown in MS medium free of 2,4-D and kinetin. The removal of hormones alters the pattern of product formation (i.e., makes it appear less growth associated); cells transferred for two or more generations in such medium typically fail to grow and make little phenolics. The dry weight of these cells pumped into the hollow-fiber unit was 860 ± 20 mg. About 35% of the available volume was filled with cells. The reaction mixture pumped into the system was MS medium free of hormones but supplemented with chloroform (25% of saturation) as a permeabilizing agent.

Results are shown in FIGURES 2 and 3. A steady-state production of soluble phenolics was established and maintained until 312 hr after start-up when the system was intentionally perturbed by adding 35 mg phenylalanine to the recycle chamber (about 0.01 weight %). The response to this pulse was a decline in phenolics production that was irreversible as evidenced by the failure of the system to recover as the phenylalanine washed out.

The productivity of the cells before phenylalanine addition (0.017 μg extracellular phenolics/mg cells-hr) was significantly higher than in single or two-stage, continuous-flow systems with LS medium containing hormones and no chloroform (0.012 μg extracellular phenolics/mg cells-hr) or in batch cultures where sequential transfer in hormone-free medium resulted in cell death.

The response to phenylalanine was unexpected since a pulse of phenylalanine in the second stage of a two-stage, continuous system resulted in a slight enhancement of phenolics formation (accompanied by a decrease in cell mass). With the entrapped

FIGURE 3. This plot indicates the effect of a pulse input of phenylalanine on phenolics production. At 312 hours, 35 mg of phenylalanine was added to the recycle chamber.

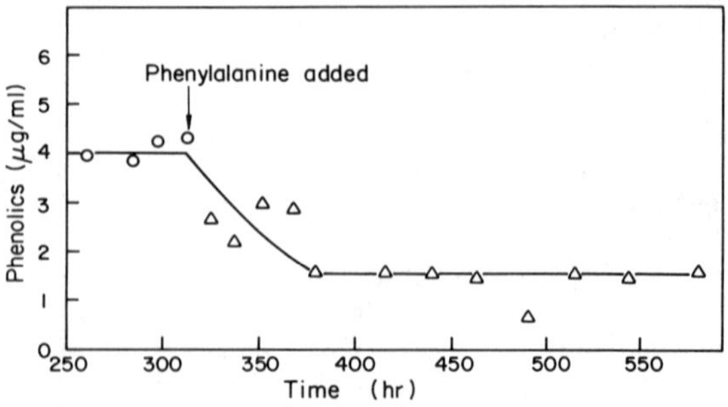

cells in an environment without hormones and with chloroform present, the cells may have not had sufficient metabolic vigor to respond positively to the challenge with phenylalanine.

SUMMARY

Cell entrapment appears to be an attractive method of cultivating plant cells for secondary product formation. The use of membrane-entrapped cultures offers greater controllability and ease of scale-up as compared to polymer-entrapped cells. The membrane-entrapped cultures are well adapted to basic studies on the interaction of the chemical environment with cellular physiology.

ACKNOWLEDGMENT

This work was supported in part by NSF Grant CPE-8114995.

REFERENCES

1. LEE, S. L. & A. I. SCOTT. 1979. Dev. Ind. Microbiol. **20:** 381–391.
2. DOUGALL, D. K. 1979. *In* Cell Substrates. J. C. Petricciani, H. E. Hopps & P. J. Chapple, Eds. Plenum Publishing Corp. pp. 135–152.
3. ZENK, M. H. 1978. *In* Frontiers of Plant Tissue Culture. T. A. Thorpe, Ed. University of Calgary. Canada. pp. 1–13.
4. SHULER, M. L. 1981. Ann. N.Y. Acad. Sci. **369:** 65–79.
5. BROCHMANN-HANSSEN, E. & C. Y. CHENG. 1982. J. Nat. Prod. (Lloydia) **45:** 437–439.
6. DEUS, B. & M. H. ZENK. 1982. Biotechnol. Bioeng. **24:** 1965–1974.
7. STREET, H. E., G. G. HENSHAW & M. C. BUIATTI. 1965. Chem. Ind. **1:** 27–33.
8. MCWILLIAM, A., S. M. SMITH & H. E. STREET. 1974. Ann. Bot. **38:** 243–250.
9. KUBEK, D. J. & M. L. SHULER. 1978. Can. J. Bot. **56:** 2521–2527.
10. GOLDSTEIN, W. 1983. Ann. N.Y. Acad. Sci. **413:** 394–408. This volume.
11. TANAKA, H. 1981. Biotechnol. Bioeng. **23:** 1203–1218.
12. SAHAI, O. P. & M. L. SHULER. 1982. Can. J. Bot. **60:** 692–700.
13. SAHAI, O. P. 1982. Ph.D. Thesis. Cornell University. Ithaca, N.Y.
14. ZENK, M. H., H. EL SHAGI, H. ARENS, J. STOCKIGT, E. W. WEILER & B. DEUS. 1977. *In* Plant Tissue Culture and Its Biotechnological Application. W. Barz, E. Reinhard & M. H. Zenk, Eds. Springer-Verlag. New York. p. 27.
15. DAVIDSON, A. W. & W. M. YEOMAN. 1974. Ann. Bot. **38:** 545–554.
16. RENANDIN, J. P. 1981. Plant Sci. Lett. **22:** 59–69.
17. FELIX, H., P. BRODELIUS & K. MOSBACH. 1981. Anal. Biochem. **116:** 462–470.
18. FELIX, H. R. & K. MOSBACH. 1982. Biotechnol. Lett. **4:** 181–186.
19. FELIX, H. 1982. Anal. Biochem. **120:** 211–234.
20. ALFERMANN, A. W., I. SCHULLER & E. REINHARD. 1980. Planta Med. **40:** 218–223.
21. BRODELIUS, P., B. DEUS, K. MOSBACH & M. H. ZENK. 1979. FEBS Lett. **103:** 93–97.
22. BRODELIUS, P., B. DEUS, K. MOSBACH & M. H. ZENK. 1979. Enzyme Eng. **5:** 373–381.
23. BRODELIUS, P. & K. NILSSON. 1980. FEBS Lett. **122:** 312–316.
24. JONES, A. & I. A. VELIKY. 1981. Can. J. Bot. **59:** 2095–2101.
25. JONES, A. & I. A. VELIKY. 1981. Eur. J. Appl. Microbiol. Biotechnol. **13:** 84–89.
26. VELIKY, I. A. & A. JONES. 1981. Biotechnol. Lett. **3:** 551–554.
27. JIRKU, V., T. MACEK, T. VANEK, V. KRUMPHANZL & V. KUBANEK. 1981. Biotechnol. Lett. **3:** 447–450.
28. BRODELIUS, P. 1983. Ann. N.Y. Acad. Sci. **413:** 383–393. This volume.

29. JOSE, W., H. PEDERSEN & C-K CHIN. 1983. Ann. N.Y. Acad. Sci. **413:** 409–412. This volume.
30. BRODELIUS, P. & K. MOSBACH. 1982. Adv. Appl. Microbiol. **26:** 1–26.
31. KAN, J. K. & M. L. SHULER. 1978. Biotechnol. Bioeng. **20:** 217–230.
32. WEBSTER, I. A., M. L. SHULER & P. RONY. 1979. Biotechnol. Bioeng. **21:** 1725–1748.
33. WEBSTER, I. A. & M. L. SHULER. 1979. Chem. Eng. Sci. **34:** 1273–1282.

Production of Biochemicals with Immobilized Plant Cells: Possibilities and Problems

PETER BRODELIUS

Institute of Biotechnology
Swiss Federal Institute of Technology
Honggerberg
CH-8093 Zurich, Switzerland

PLANT TISSUE CULTURES

Plant tissue cultures can be established from most higher plants as schematically shown in FIGURE 1. A piece of tissue (explant) is surface sterilized and placed on a solid medium. The composition of this medium may vary and the optimal growth conditions have to be empirically established for each plant species. The cells within the explant grow and divide; after a period of time, a clump of undifferentiated cells forms, which then can be transferred to a liquid medium in order to obtain a suspension culture. Cells of *Daucus carota* grown in a suspension culture are shown in FIGURE 2.

Plant tissue cultures have been employed for various purposes during the last decades, in particular by plant physiologists and plant breeders. Some of these applications of plant tissue cultures are summarized in FIGURE 3. Of particular interest from the biotechnological point of view is the use of plant tissue cultures for the possible production of biochemicals.[1,2] Many compounds with complex structures, that is, pharmaceuticals and fragrant compounds, have been isolated from higher plants.[3] The production of these compounds by fermentation of plant cells is a potential alternative that has some distinct advantages.[1,2]

The biosynthesis of valuable compounds in plant tissue cultures may be achieved in various ways (FIG. 3). In one procedure, the entire metabolism of the cell can be used for *de novo* synthesis. A relatively inexpensive carbon source (e.g. sucrose) is metabolized by the cultured cells and converted into very complex molecules. A wide range of products may be obtained in this way, but the yield of a particular substance is often low. During recent years, methods have been developed for the selection of high-producing cell lines to overcome this problem.[4–6] When the biosynthetic pathway is known and when appropriate precursors are available, a more selective and efficient synthesis may be achieved by feeding the precursors to the cultured cells. In yet another procedure, plant tissue cultures can be employed for stereospecific biotransformations, such as hydroxylations or methylations.[7] These latter two approaches allow for the biosynthesis of completely new compounds by using analogues to the natural substrates. It may thus be possible to modify existing plant pharmaceuticals by feeding substrate analogues to the cell culture, thereby obtaining compounds with new and/or improved pharmaceutical properties.

IMMOBILIZED PLANT CELLS

The advantages of immobilized biocatalysts are well established,[8,9] and during the last years, we have been investigating whether these advantages also apply for

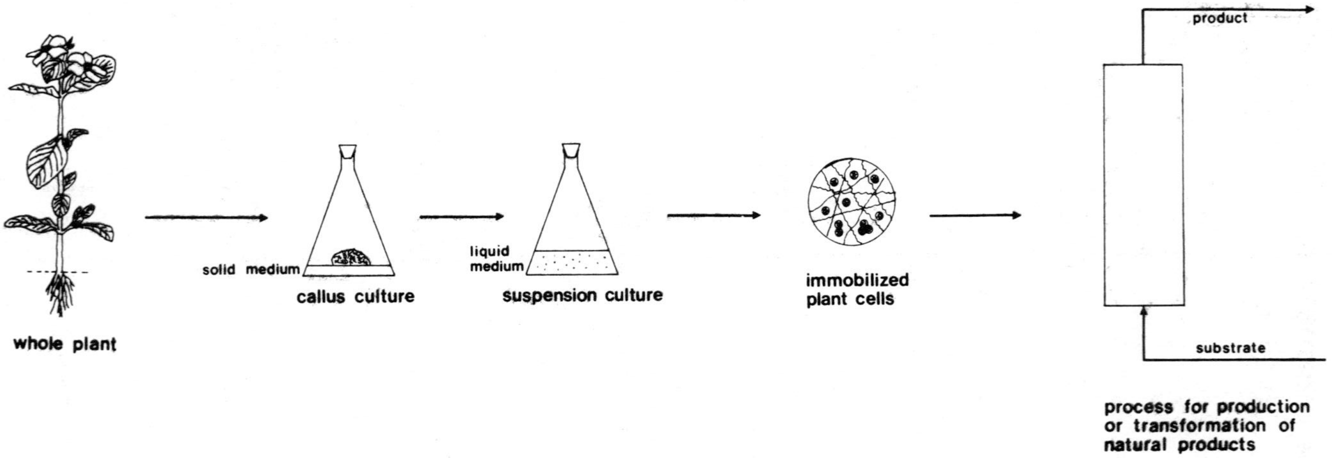

FIGURE 1. Schematic diagram showing the procedure for immobilizing cells in a bioreactor from an intact plant.

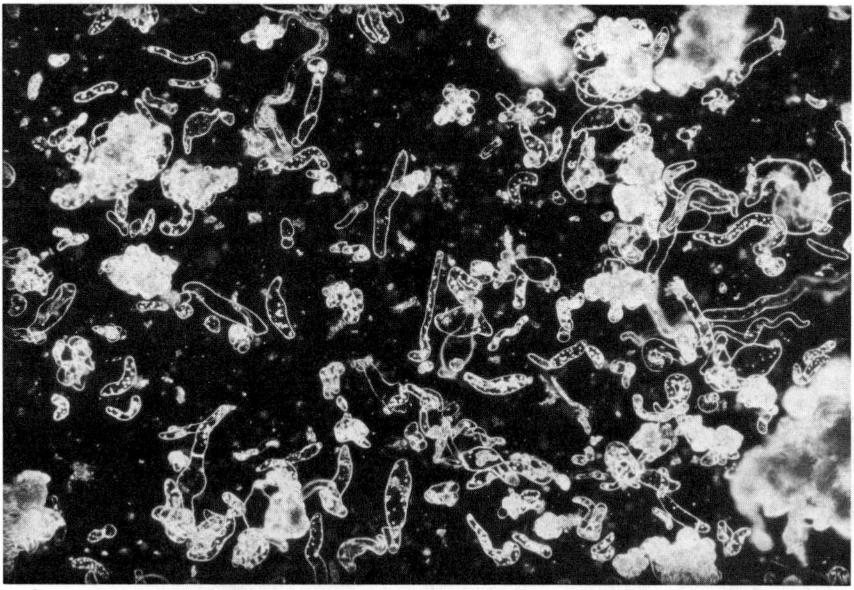

FIGURE 2. Suspension culture of *Daucus carota*.

immobilized whole plant cells.[10,11] This has been done by studying the biosynthesis of various compounds in model systems that were selected to cover various aspects of biosynthesis, as mentioned above. For a preserved biosynthetic capacity, a viable preparation of immobilized cells is of importance. We have, therefore, investigated various immobilization procedures and studied the viability and biosynthetic capacity of these preparations.

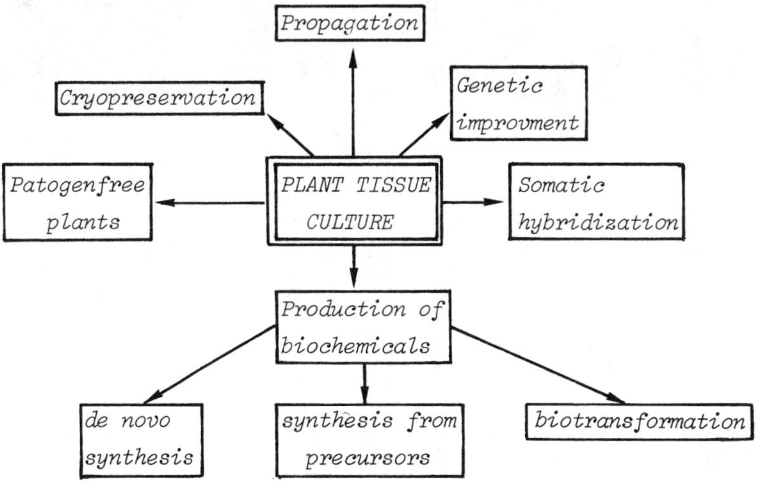

FIGURE 3. Various applications of plant tissue cultures.

Immobilization of Plant Cells

Immobilization work has concentrated on the entrapment of cells in various matrices, since this method appears to be the most convenient for immobilizing sensitive, large plant cells. The polymers tested can be classified according to three principal groups:

(a) gel formation by ionic cross-linking of a charged polymer;
(b) gel formation by cooling of a heated polymer; and
(c) gel formation by chemical reactions.

A classic example from the first group is alginate, and examples from the second group are agar and agarose. Gelatin (cross-linked with glutaraldehyde) and polyacrylamide belong to the third group.

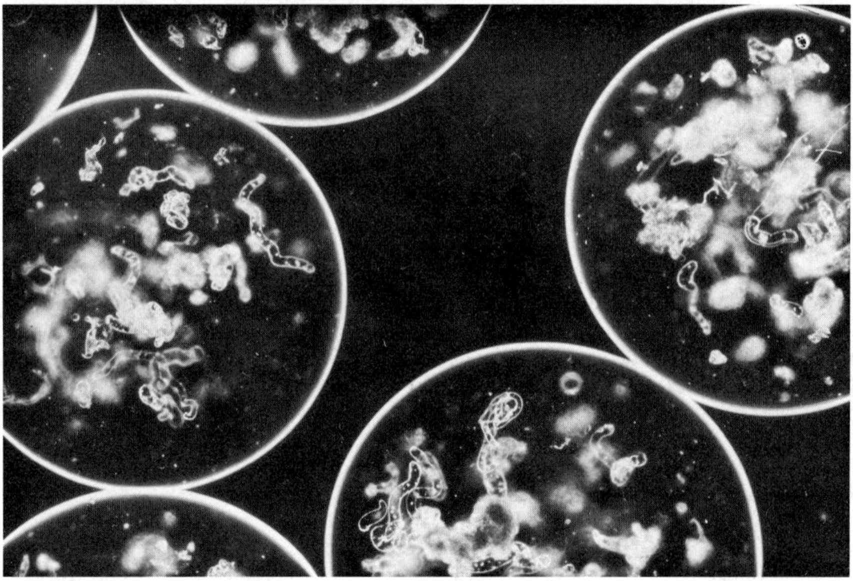

FIGURE 4. Agarose-entrapped cells of *Daucus carota.*

Beads of alginate-entrapped cells are readily prepared by mixing the cells and the alginate solution and subsequently dripping this suspension into a calcium-containing medium. A new method has been developed for the immobilization of viable cells in other matrices.[12] This method is based on a two-phase system in which the hydrophobic phase is relatively compatible with the cells. In a typical immobilization, a low-melting agarose is mixed with plant cells at 40°C, and the suspension is poured into the hydrophobic phase, which is stirred and also at 40°C. Droplets are formed and upon cooling, the agarose solidifies and the cells are entrapped within the resulting beads. The size of the beads can be readily controlled by varying the stirring speed. Various hydrophobic phases have been tested, vegetable oils and paraffin oil have been found to be suitable for the immobilization of plant cells. Cells of *D. carota* entrapped in agarose

TABLE 1. Comparison of Various Preparations of Immobilized *C. roseus* Cells

Preparation	Respiration	Cell Growth	Cell Division
Alginate	+	+	+
Agarose	+	+	+
Agar	+	+	n.d.[a]
Carrageenan	+	+	n.d.
Gelatin	−	−	n.d.
Polyacrylamide	−	−	n.d.

[a]n.d. = not determined.

are shown in FIGURE 4. The method described is applicable to a wide range of polymers and can be generally employed for the immobilization of viable cells.[12]

Viability of Immobilized Plant Cells

The viability of immobilized cells can be tested in various ways. Respiration is monitored in an oxygen electrode, cell growth is determined by the increase in dry weight, and cell division is indicated by the mitotic index. The results from qualitative studies on cells of *Catharanthus roseus* entrapped in various matrices are summarized in TABLE 1. While the cells within chemically prepared gels (gelatin and polyacrylamide) appear to be dead, the cells entrapped in various polysaccharides are fully viable.

Quantitative studies have shown that agarose- or agar-entrapped cells of *C. roseus*[13] or *Glycine max*[14] respire (FIG. 5) and grow (FIG. 6) at approximately the same rate as freely suspended cells. Alginate-entrapped cells are, however, somewhat restricted in their rates of respiration (FIG. 5) and growth (FIG. 6). This does not appear to be due to reduced viability of the cells as a result of immobilization, but rather is caused by diffusion barriers within this gel type. The limited growth can be verified by inspecting the cell distribution within the beads. In freshly prepared beads of agarose or alginate, the plant cells are evenly distributed throughout the gel. However, after incubation in a growth medium for one week, the cell distribution in the

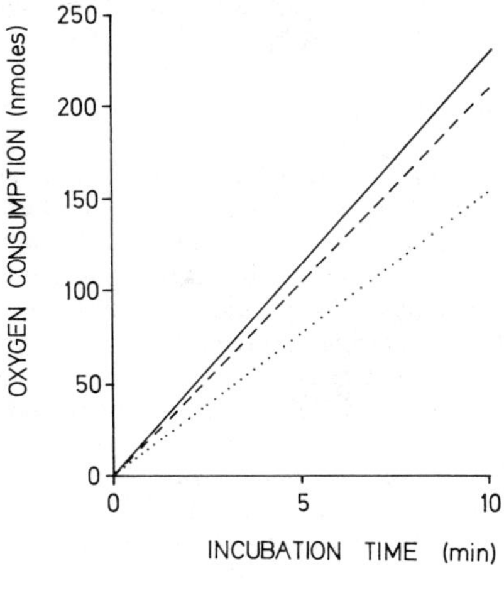

FIGURE 5. Respiration of various preparations of *Catharanthus roseus:* (——) freely suspended cells, (---) agarose-entrapped cells, and (· · · · ·) alginate-entrapped cells.

alginate beads has changed, while the cells within agarose are still growing throughout. The cells entrapped in alginate are not found in the core, only as a layer in the periphery of the bead. In both cases, the cell density has increased considerably. Entrapped cells grow to such an extent that the beads start to burst after incubation in a complete medium for a period of 8–12 days. Therefore, growth has to be restricted in long-term experiments. This can readily be achieved by modification of the medium.

Biosynthetic Capacity of Immobilized Plant Cells

The biosynthetic capacity of the viable preparations of entrapped cells (cf. TABLE 1) has been studied in various model systems as outlined below.

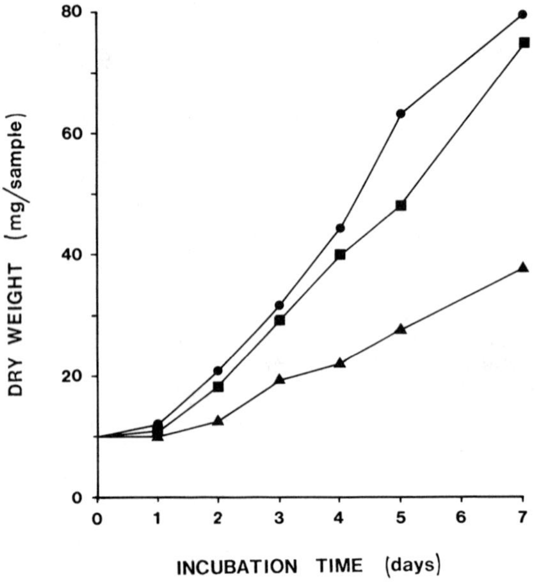

FIGURE 6. Dry weight of various preparations of *Catharanthus roseus* cells as a function of incubation time: (—●—) freely suspended cells, (—■—) agarose-entrapped cells, and (—▲—) alginate-entrapped cells.

Biotransformation

Recent studies have shown that cells of *D. carota* Ca68 (This cell line was kindly supplied by Dr. I. Veliky, NRCC, Ottawa) entrapped in alginate, agar, agarose, or carrageenan retain the capacity to 5-β-hydroxylate digitoxigenin to periplogenin.[15]

In our initial studies on immobilized plant cells, it was shown that alginate-entrapped cells of *Digitalis lanata* can efficiently hydroxylate digitoxin to digoxin.[16] More extensive studies on the conversion of digitoxin to digoxin by immobilized *Digitalis* cells with high 12-β-hydroxylase activity have been reported by Alfermann *et al.*[17] A major finding in this study is that immobilized cells can be utilized for this biotransformation over an extended period of time.

Finally, agarose-entrapped cells of *C. roseus* reduce cathenamine to isomers of ajmalicine (reductions of double bond) at approximately the same rate as freely suspended cells as shown in FIGURE 7.[18]

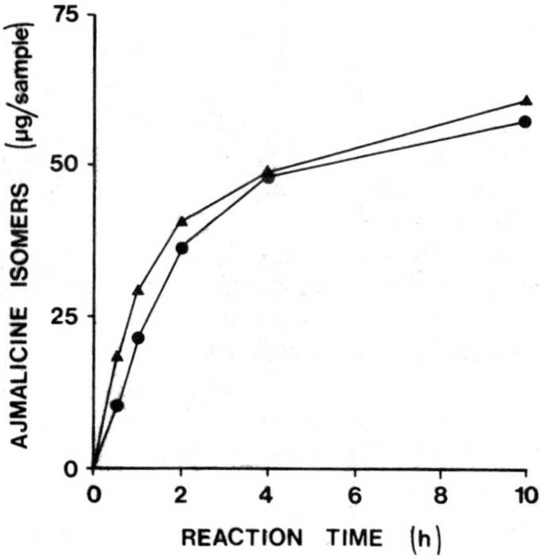

FIGURE 7. Biotransformation of cathenamine to ajmalicine-isomers by *Catharanthus roseus:* (—▲—) freely suspended cells, (—●—) agarose-entrapped cells.

Synthesis from Precursors

The biosynthesis of complex compounds from simpler precursors added to immobilized cells may be classified as a multistep biotransformation. An example of this kind of reaction, the synthesis of ajmalicine isomers from the relatively distant precursors tryptamine and secologanin by various preparations of immobilized *C. roseus* cells has been investigated,[19] with radiolabeled trypamine used to follow the synthesis. The results have been summarized in TABLE 2. As can be seen, the activity of the immobilized cells varied between 82 and 176% of that observed for freely suspended cells. Obviously, the immobilized cells can condense the two precursors to strictoside, which in a number of steps is transformed to ajmalicine isomers. Strictosidine is a common precursor for indole alkaloids and therefore a higher incorporation rate may be expected if the total alkaloid content were determined.

De Novo *Synthesis*

The *de novo* synthesis of complex secondary products from simple carbon sources is probably the most important feature of plant tissue cultures from the biotechnological point of view. The synthesis of ajmalicine isomers from sucrose as sole carbon source by various preparations of immobilized *C. roseus* has been determined under growth-limiting conditions (a hormone-free medium was used). The results are summarized in

TABLE 2. Synthesis of Ajmalicine Isomers from Precursors by Free and Immobilized Cells of *C. roseus*

Cell Preparation	Relative Incorporation (%)	Tryptamine Incorporation (%)
Free cells	100	7.9
Alginate	176	13.9
Agarose	114	9.0
Agar	95	7.5
Carrageenan	82	6.5

TABLE 3. A similar pattern to that observed with the addition of precursors is obtained when the cells produce the alkaloids by *de novo* synthesis. In this case, the activity of the immobilized cells varies between 62 and 140% of that of freely suspended cells. Also, the alginate-entrapped cells show considerably increased productivity here. In other model systems in which various methods to restrict growth were investigated, increased synthesis by alginate-entrapped cells has been observed.[20,21] For instance, cells of *Morinda citrifolia* entrapped in alginate produce up to 10 times more anthraquinones than freely suspended cells under the same conditions (hormone-free medium).[20] We ascribe this increased synthesis by alginate-entrapped cells as resulting from the previously discussed limitation of growth, so that the metabolism is shifted towards secondary product formation.

In conclusion, a number of model studies have shown that various preparations of immobilized plant cells can carry out biotransformations and synthesis from precursors, as well as synthesis *de novo* at approximately the same rate as freely suspended cells.

IMMOBILIZED PERMEABILIZED PLANT CELLS

In most cases, the products formed by cultured plant cells are stored within the cells. One of the inherent advantages of an immobilized biocatalyst is its possible utilization in a continuous process. This distinct advantage cannot be realized if the product or products are stored within the immobilized cells. We have, therefore, initiated studies on the permeabilization of the cell membranes for the release of synthesized product(s) into the surrounding medium.[18,22]

TABLE 3. Synthesis of Ajmalicine Isomers *de Novo* by Free and Immobilized Cells of *C. roseus*

Cell Preparation	Relative Production (%)	μg/Sample
Free cells	100	4.2
Alginate	140	5.9
Agarose	100	4.2
Agar	88	3.7
Carrageenan	62	2.6

The permeability of the plasma membrane can be monitored by measuring the activity of various coenzyme-requiring enzymes within the cells (e.g., glucose-6-phosphate dehydrogenase, isocitrate dehydrogenase, citrate synthase).[18] As an example, the time course of the hexokinase/glucose-6-phosphate dehydrogenase reaction for intact and permeabilized cells of *C. roseus* is shown in FIGURE 8. Various permeabilization agents have been tested[18] and DMSO has been investigated further since the cells appear to survive treatment by this agent.[22] Permeability as a function of DMSO-concentration is shown in FIGURE 9 for agarose-entrapped cells of *C. roseus*. The lowest concentration of DMSO leading to full permeabilization is in this case around 5% (v/v) under the conditions used (treatment for 30 min). The DMSO concentration required for permeabilization may vary depending on the plant species. *Catharanthus* cells treated with 5% DMSO recover relatively fast and grow almost at the same rate as untreated cells, as shown in FIGURE 10. Therefore, it appears possible

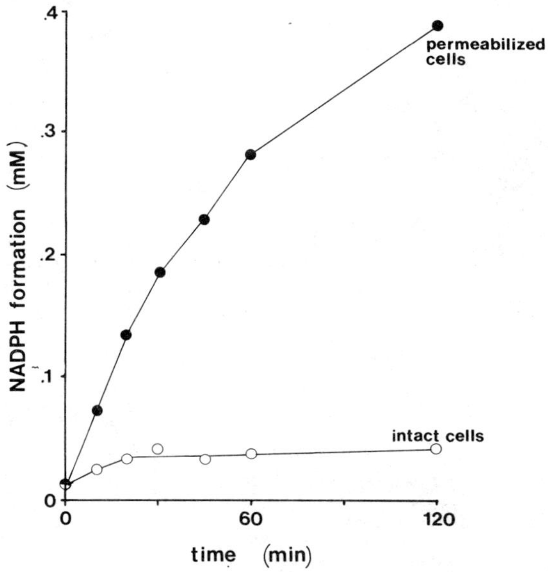

FIGURE 8. Time course of hexokinase/glucose-6-phosphate dehydrogenase reaction in intact and permeabilized cells of *Catharanthus roseus.*

to fully permeabilize immobilized plant cells and subsequently let them recover and again grow if necessary. In principle, it may be possible to operate a reactor of immobilized plant cells continuously with intermittant release of product(s), as schematically illustrated in FIGURE 11. In preliminary studies, we have observed that the product yield actually increases for each cycle, according to FIGURE 11. This increase is most likely due to an increase in the biomass within the beads because of the growth phase included in the cycle between permeabilization and the production phase. Treatment with 5% DMSO results in the release of 85–90% of alkaloid into the medium. For a steady-state production, further optimization of the cycle is required.

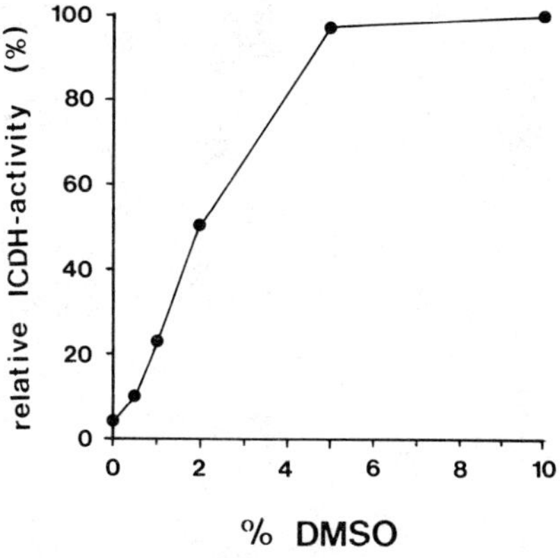

FIGURE 9. Relative isocitrate dehydrogenase (ICDH) activity within agarose-entrapped cells of *Catharanthus roseus* as a function of DMSO concentration.

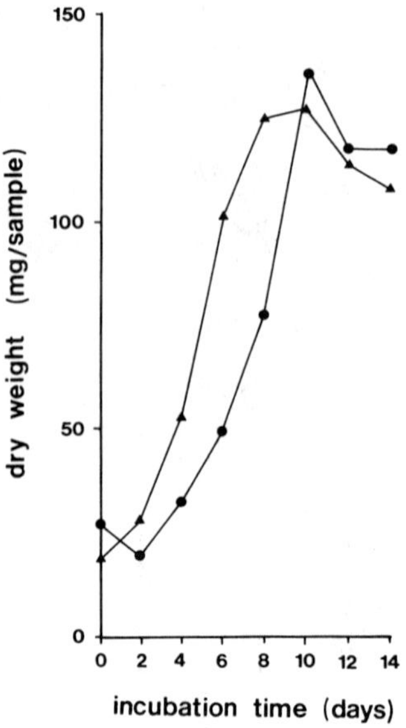

FIGURE 10. Growth curves of *Catharanthus roseus* cells: (—▲—) untreated cells, (—●—) cells treated with 5% DMSO for 30 min.

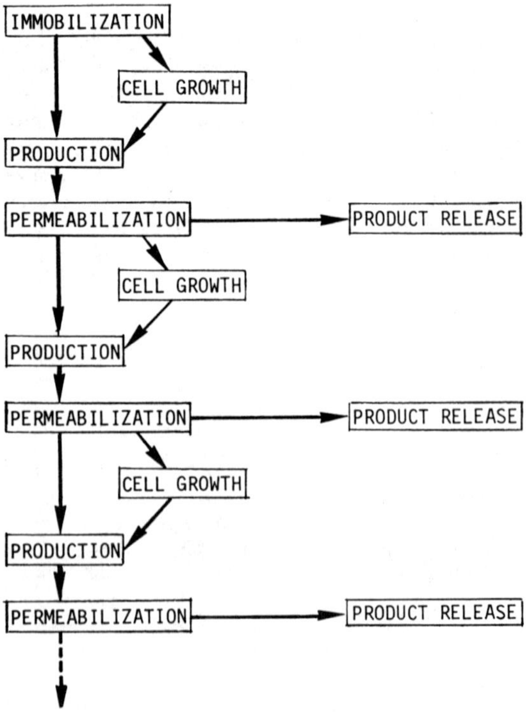

FIGURE 11. Schematic diagram of intermittent product release from immobilized plant cells by permeabilization.

CONCLUSIONS

The model systems discussed in this paper clearly demonstrate that plant cells remain viable and biosynthetically active after immobilization. Some advantages that have been observed with the immobilized cells are:

(a) increased product yield,
(b) prolonged stationary (production) phase,
(c) protection of sensitive cells, and
(d) intermittant release of product through permeabilization.

Point d may be of fundamental importance for the development of a biotechnological process based on plant cell culture since it has been reported that the biomass has to be reused in such a process.[23] With intermittant permeabilization, the formed product(s) may be recovered without loss of biomass. Some recent reviews on immobilized plant cells may be consulted for further reading.[10,11,24]

REFERENCES

1. ZENK, M. H. 1978. *In* Frontiers of Plant Tissue Culture. T. A. Thorpe, Ed. International Association for Plant Tissue Culture. pp. 1–13.
2. KURZ, W. G. W. & F. CONSTABEL. 1979. *In* Advances in Applied Microbiology. D. Perlmann, Ed. Vol. **25:** 209–240. Academic Press. New York.
3. BELL, E. A. & B. V. CHARLWOOD, Eds. 1980. Secondary Plant Products. Springer-Verlag. Berlin, Heidelberg, New York.
4. ZENK, M. H., H. EL-SHAGAI & U. SCHULTE. 1975. Planta Med. Suppl. pp. 79–101.
5. DEUS, B. & M. H. ZENK. 1982. Biotechnol. Bioeng. **24:** 1965–1974.
6. YAMAMOTO, Y., R. MIZUGUCHI & Y. YAMADA. 1982. Theor. Appl. Genet. **61:** 113–116.
7. REINHARD, E. & A. W. ALFERMANN. 1980. *In* Advances in Biochemical Engineering. A. Fiechter, Ed. Vol. **16:** 49–83. Springer-Verlag. Berlin, Heidelberg, New York.
8. BRODELIUS, P. 1978. *In* Advances in Biochemical Engineering. T. K. Ghose, A. Fiechter & N. Blakebrough, Eds. Vol. **10:** 75–129. Springer-Verlag. Berlin, Heidelberg, New York.
9. MOSBACH, K., Ed. 1976. Methods in Enzymology, Vol. **44.** Academic Press. New York.
10. BRODELIUS, P. & K. MOSBACH. 1982. *In* Advances in Applied Microbiology. A. Laskin, Ed. Vol. **28:** 1–26. Academic Press. New York.
11. BRODELIUS, P. & K. MOSBACH. 1982. J. Chem. Tech. Biotechnol. **32:** 330–337.
12. NILSSON, K., S. BIRNBAUM, S. FLYGARE, L. LINSE, U. SCHRODER, U. JEPPSSON, P.-O. LARSSON, K. MOSBACH & P. BRODELIUS. 1983. Eur. J. Appl. Microbiol. Biotechnol. In press.
13. BRODELIUS, P., F. CONSTABEL & W. G. W. KURZ. 1982. *In* Enzyme Engineering. L. B. Wingard, Ed. Vol. **6:** 203–204. Plenum Press. New York.
14. BRODELIUS, P., L. LINSE & K. NILSSON. 1982. *In* Proceedings of the Vth International Congress of Plant Tissue and Cell Cultures. Tokyo, Japan. pp. 371–372.
15. LINSE, L. & P. BRODELIUS. 1982. Unpublished results.
16. BRODELIUS, P., B. DEUS, K. MOSBACH & M. H. ZENK. 1979. FEBS Lett. **103:** 93–97.
17. ALFERMANN, A. W., I. SCHULLER & E. REINHARD. 1980. Planta Med. **40:** 218–223.
18. FELIX, H., P. BRODELIUS & K. MOSBACH. 1981. Anal. Biochem. **116:** 462–470.
19. BRODELIUS, P. & K. NILSSON. 1980. FEBS Lett. **122:** 312–316.
20. BRODELIUS, P., B. DEUS, K. MOSBACH & M. H. ZENK. 1980. *In* Enzyme Engineering. H. H. Weetall & G. P. Royer, Eds. Vol. **5:** 373–378. Plenum Press. New York.
21. BRODELIUS, P., B. DEUS, K. MOSBACH & M. H. ZENK. 1979. Swedish Patent Application 7905613-6.
22. BRODELIUS, P. & K. NILSSON. 1983. Eur. J. Appl. Microbiol. Biotechnol. In press.
23. GOLDSTEIN, W. E. 1983. Ann. N.Y. Acad. Sci. **413:** This volume.
24. BRODELIUS, P., 1983. *In* Immobilized Cells and Organelles. B. Mattiasson, Ed. CRC Press. Boca Raton, FL. In press.

Large-scale Processing of Plant Cell Culture

WALTER E. GOLDSTEIN

Research and Development
Biotechnology Group
Miles Laboratories, Inc.
Elkhart, Indiana 46515

INTRODUCTION

The extensive effort in development of plant cell culture is evident from the literature on the subject. However, except for example cases, such as, tobacco cells,[1] this area has been relegated to laboratory investigation; the basis for broad translation to large-scale processing is not yet evident. The theme of this paper is to establish a proposed chain of events or circumstances in technology necessary to allow realization of the promise of plant cell culture. The speculations shared herein are evaluated through means of economics, and the vehicle is the assumption of analogies with microbial fermentation processes. This assumption can be tenuous in certain cases, but still valid providing suitable design considerations and qualifications are invoked.

RESEARCH AND SCALE-UP

The component steps of plant cell culture at laboratory scale may be represented by the block diagram (FIG. 1). Nutrients are supplied for culturing of the biomass, assumed to be separated from the stage of product synthesis by the plant cell culture. Biomass and product are then conveniently separated, before purification of the product in an appropriate manner.

During this phase of research, the study of the plant cell (or microbial) culturing process will involve a search for particular attributes conducive to product formation; or in turn, researchers may seek to modify the host through recombinant genetics.[2] Studies in solid-phase (surface) media will proceed to liquid phase and use of measurements with cell suspension devices; this then indicating that aspects of bioengineering as well as plant genetics and microbiology should be considered.

The matrix of essential nutrients for suspension culture of plant (or microbial) cells may include the study of provision of complex entities (polypeptide or polycarbohydrate) and their influence on control of growth relative to product formation. This is particularly important to plant cells, which are subject to noticeable differentiation as compared to microbial cells.[3] The introduction of factors such as auxins must be added to the list of required nutrients (carbon, nitrogen, other elements, vitamins). Environmental conditions (pH and temperature) are also to be defined as well as needs for mixing and oxygen transfer.

Appropriate studies then provide information needed in large-scale processing TABLE 1. Rates of cell growth and product formation and associated factors should be developed along with assayable means of detection. Programing of nutrients and conditions may be relevant in terms of repression or stimulation, along with needs for

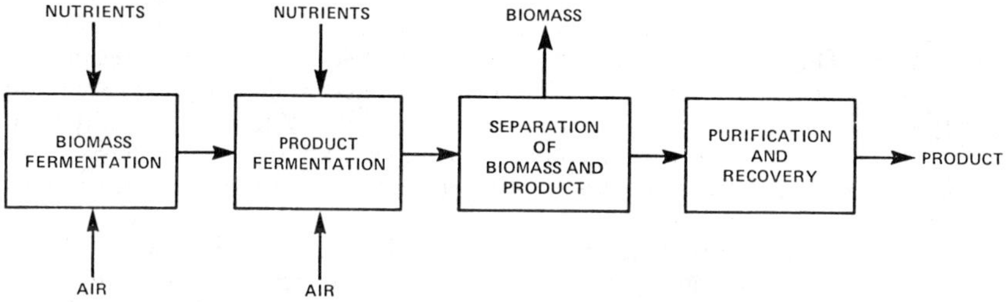

FIGURE 1. Block of diagram plant cell culture.

stringency in control and treatments of nutrients or products. Asepsis needs are particularly important since, for example, low rates of formation invite contamination; therefore, practical means to control contamination would have to be identified. As output from this phase of development, the cost of the cell culture media and value of the product could be estimated—a first basis to determine if further research and scale-up is justified.

Quantitation of entities in TABLE 2 is effected through bioengineering studies at laboratory scale; for example, stoichiometry (nutrients supplied and used), and heat generation effects expected as rates and equipment sizes increase. Oxygen requirements are defined in terms of rheological constraints (e.g., cell fragility). Aspects of recovery are considered; for example, suitability of the biomass as a product and whether the product is extracellular or intracellular. These factors thus affect definition of the process concept. Finally, retention of product stability is evaluated and conditions of protection defined; this is done much in the manner employed for a microbial product such as an enzyme.

EVALUATION OF A SYMBOLIC LARGE-SCALE PROCESS

A positive net result of the above is a large-scale process concept. A representation of such a concept is provided in FIGURE 2. As indicated, cell and product formation

TABLE 1. System Information for Scale-up

- Organism Growth Rate
- Product Formation Rate
- Means of Detection
- Programing Needs for Nutrients and Conditions
  - Repression–Nutrient or Product Limitation of an Entity
    - Constant
    - Variable
- Control in Processing
  - pH
  - Temperature
  - Nutrients
    - Treatments
  - Product
  - By-products
  - Asepsis
- Media Cost Projection and Final Product Value

stages are distinct; biomass is separated from product and treated for product residing in the biomass. The biomass can be removed from the system, or in theory, recycled for benefit to increased rates in suspension culture. Treatment of the biomass for product recovery could involve an aqueous or organic solvent cycle. Concentration of product fluid is assumed possible through coupling of standard water removal techniques (evaporation) and membrane processes. Fluids from biomass separation are similarly concentrated and may be recycled for use or withdrawn from the system. Process specifics would include prevention of contamination (proper design in recycling of carbohydrate-laden solutions) and provision of suitable drag streams to control accumulation of entities potentially deleterious to long-term system performance. This speculative view of a large-scale plant cell manufacturing process is thus useful in assessing research needs in augmentative ways, and is also useful for examination of economic criteria.

TABLE 2. Information from Laboratory-Scale Bioengineering

- Projected Stoichiometry
- Projected Heat Generation from Stoichiometry
- Oxygen Requirements
    - Rheology
    - Dissolved Oxygen
    - Respiration
    - Off-gas
- Fermentation
- Sterilization Needs & Conditions
- Recovery
    - Biomass as Product
    - Intracellular Metabolite
    - Extracellular Metabolite
    - Process Definition
    - Product Stability

Suspension Culture

Culture residence time is analyzed in three modes for the economic evaluation; that is, CSTR (Continuous Stirred Tank Reactor), plug flow, and batch (semi-batch or feeding being a special case). The expressions for these (TABLE 3) follow from classical catalysis, autocatalytic growth assumed. The critical parameter indicated is the specific growth rate of the cell mass, μ_B—the implicit dependence of μ_B on other parameters is assumed to be analogous to findings from microbial/plant cell culture experience. In theory, the CSTR mode is preferred for autocatalysis. In practice, for low roles of growth and complexities in design for plant cells, the economic comparison can indicate preference for another mode. TABLE 3 also presents similar expressions for product formation; in this case, it is presumed that product forms in proportion to biomass present. Critical parameters are the amount of product contained in the biomass (γ_p) and the specific rate of formation of the product (μ_p). Design of the plant cell system to facilitate extracellular expression of the product is also key to the process. Similarly, the product formation stage is dependent on the previous biomass synthesis stage since maximal formation of cell quantity per unit volume of fluid is desired.

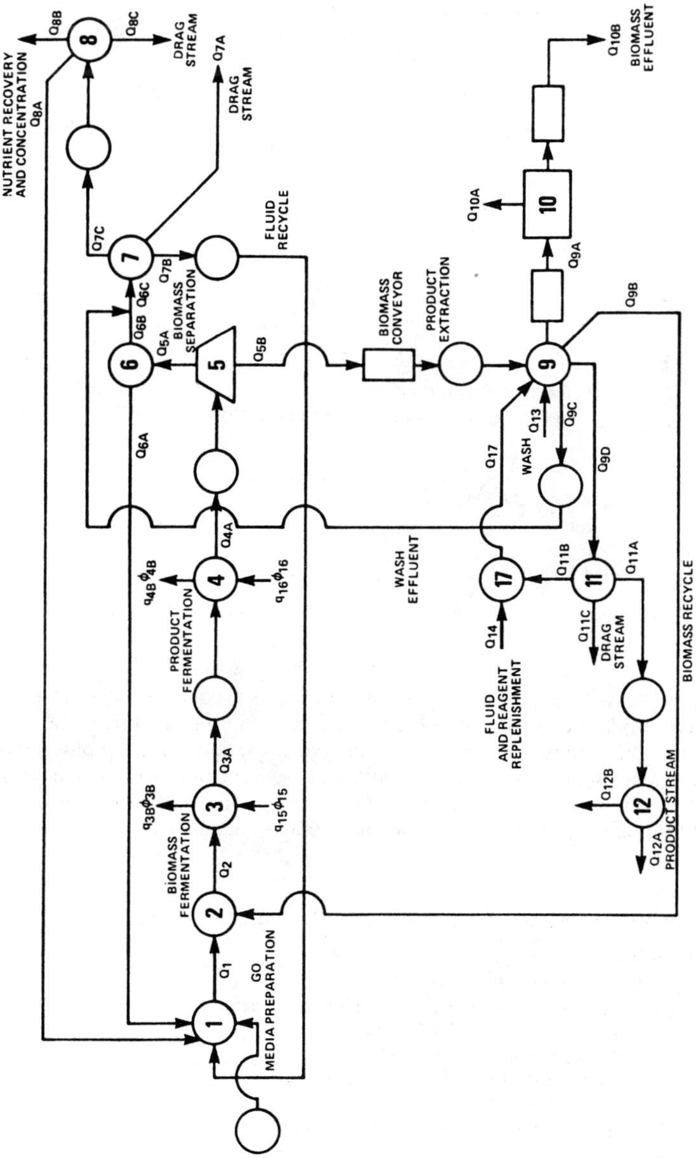

FIGURE 2. A representation of a plant cell culture/product manufacturing process.

TABLE 3. Residence Times for Suspension Cell Culture

Process Parameters for Fermenters

Residence Time

Biomass Formation

Idealized Autocatalytic Growth

$$\frac{V}{Q} = \frac{1 - \frac{C_{BI}}{C_{BO}}}{\mu_B} \quad \text{CSTR}$$

$$\frac{V}{Q} = \frac{\mathrm{1N}\frac{C_{BO}}{C_{BI}}}{\mu_B} \quad \text{Plug Flow}$$

$$\tau = (1 + \Psi)\frac{\mathrm{1N}\frac{C_{BO}}{C_{BI}}}{\mu_B} \quad \text{Batch}$$

Product Formation

In Proportion to Biomass Present (Zero Order)

$$\frac{V}{Q} = \frac{C_P}{\gamma_P C_{BO} \mu_P} \quad \text{CSTR Plug Flow}$$

$$\tau = \frac{C_P}{\gamma_P C_{BO} \mu_P} \quad \text{Batch}$$

Factors relevant to achieving desired biomass and product in scale-up (microbial fermentations or plant cell culture) include aspects of continuous sterilization, biomass propagation, broth circulation and air bubble patterns, and energy input and output. Heat input at large scale can become significant due to the inherent chemical reactions and heat of mixing. Oxygen utilization will vary throughout the fermentation where it is essential that oxygen be sufficient to meet system demands.[4]

TABLE 4. Process Parameters for Scale-up of Plant Cell Culture

- Rate of product effluent from process
- Yield of product/unit of biomass
- Maximal biomass concentration in the fermenter
- Fraction of biomass that can be reused
- Consumption of nutrients to form biomass
- Consumption of nutrients to form product
- Inlet concentration of nutrients to the fermentations
- Liquid removal ratio at the biomass separator
- Extractant needs per quantity of biomass
- Washing needs of biomass
- Disposition of fluids for recycle or waste after product extraction
- Fluid removal needs for biomass and product
- Specification of drag streams, constituents, and fluid

Investigation of Parameters

Solution of mass balances[5] for the process depicted in FIGURE 2 yields the set of parameters relevant for system description and for research (TABLE 4). As indicated in TABLE 4, the rate of product effluent from the process is the specification required to project economics, as developed further in this paper. In turn, the yield of product/unit of biomass and maximal biomass concentration possible in suspension culture or an equivalent mode must be specified from research. Extracellular expression and/or nondestructive treatment of biomass, and the need to relieve the process of impurities will dictate how much biomass reuse is possible, immobilization being an extreme of biomass reuse. Nutrient supply and consumption are required for system specification. Finally, processing requirements for separation and disposition of fluids complete the information necessary for system specification.

TABLE 5. Factors of Importance: Energy Requirements for Large-scale suspension Cell Culture

- Vessel Height to Diameter
  - Cost
  - Operation
- Pressure Losses in Delivery of Air
  - Piping, Filters
- Mechanical Impeller Sizing
  - Power Input
  - Shear
- Number of Fermenters
  - Practical Size
  - Balance of Risk
  - Cost as a Function of Geometry
- Compressor Power Consumption Dependence on Air Flow, Vessel Dimensions, and Broth Properties
- Desired Gas/Liquid Holdup
  - Oxygen Mass Transfer Relationship
- Relationship of Mechanically Induced Power for Gassed and Ungassed Conditions
- Depreciation (Capital Cost) Dependencies for Mechanical Mixing and Aeration for Mixing or Oxygenation
- Economic Optimum
  - Mechanical Mixing
  - Aeration

The required analysis includes determination of energy requirements for fermenters (suspension culture)—factors of importance noted in TABLE 5. These include vessel dimensions since geometry bears on establishing proper gas/liquid distribution patterns. Energy input is determined by pressure loss incurred in supply of air and needs for mechanical power input. Fermenter quantities are based on practicality of size, the risk of downtime in using too few units, and the cost incurred as a result of engineering complexity required for cell culture. Oxygen transfer requires knowledge of gas/liquid holdup patterns and mechanical power input reduction effected in the presence of gas.[4] The cost of aeration/agitation would also include the effect of depreciation due to capital requirements. In general, for convenience, calculations noted herein assume Newtonian behavior and an economic optimum is derived for the combination of mechanical agitation and aeration yielding the minimum cost. The expression so derived[5] for purposes of the economic evaluation is exhibited in TABLE 6.

TABLE 6. Expression for Minimum Energy Cost in Suspension Cell Culture

$$\text{Annual Cost}^{a} = \frac{3514.8\, N_F\, V^{1.026}}{Q^{.7}} + 136\,(N_FQ)^{.488} + 1108.2\, C_EN_FQ\,[(.4984\, V^{1/3} + 1)^{.274} - 1] + \frac{1{,}291{,}812\, C_C\, N_F V^{1.833}}{Q^{1.25}}$$

$$\frac{d\,[\text{Annual Cost}]}{dQ} = 0 = \frac{-2460.36\, N_F V^{1.026}}{Q^{1.7}} + \frac{66.368\, N_F^{.488}}{Q^{.512}} + 1108.2\, C_EN_F\,[(.4984\, V^{1/3} + 1)^{.274} - 1] - \frac{1614765\, C_EN_F\, V^{1.833}}{Q^{2.25}}$$

[a] Nomenclature:
N_F = Number of fermenters
V = Fermenter volume (m^3)
Q = Airflow (m^3/hr)
C_E = Cost of electricity ($/kWhr)

TABLE 7. Calculation Cases for Suspension Cell Culture

Case	Fermenter	Aeration Quantity m^3/hr	Power Input Aeration	Power Input to Overcome Liquid Head	Mechanical Horsepower Input, hp	Agitator Diameter m	Rotational Speed RPM	Superficial Velocity m/hr	Volume Air/ Vol. Liquid Min. VVM	Tank Diam. m	Liquid Volume m^3
10,000 kg/yr D.S. product											
A	Biomass (batch)	853	45.1	18.8	32.5	1.91	37	133	.38	2.86	37
	Product (batch)	1062	60.4	25.5	43.0	2.11	34	135	.35	3.17	50
B	Biomass (CSTR)	183	5.5	2.1	4.2	0.98	57	108	.72	1.47	4.2
	Product (CSTR)	176	5.2	2.0	4.0	0.95	59	111	.73	1.42	4
1,000,000 kg/yr D.S. product											
B	Biomass (CSTR)	2950	233	105	167.5	3.41	24	143	.23	5.12	211
	Product (CSTR)	2950	233	105	167.5	3.41	24	143	.23	5.12	211

The components are agitator and compressor depreciation, and the cost of electricity, expressed in terms of adiabatic loss in supply of air and in terms of power input from mechanical mixing. As indicated, a derivative of the expression relative to air flow determines the minimum cost.

Examples of cases considered in calculations are presented in TABLE 7. The variables indicated are type of fermentation (batch or CSTR) and quantity of product. The calculations provide power input values and implicitly allow specification of agitator diameter and rotational speed. Agitator rotation is predicted to be relatively low to impart adequate mixing and broth turnover without excessive shear—this is then a necessity for calculation validity. This is assumed achieved by use of special impellers with a diameter/tank diameter ratio larger than normally employed for turbine impellers.[6]

Recovery and Purification

Particular areas of recovery are common to scale-up in both microbiological and anticipated plant cell culture processes. In either case, the product must be protected against destabilization in processing and thus necessitating treatments. Pecularities of biomass separation suggest a need for testing different solid/liquid separation devices, and that polishing, clarification, economical water removal, and final purification may be needed. The processing thereby resembles that common to large-scale bioproduct recovery (e.g., enzymes or antibiotics).

The costing of the facility requires consideration of control of contamination influences, heating of facilities, and cooling to remove sizeable heat generation at larger scale for processes with hopefully higher rates of reaction.

COST ANALYSIS AND TECHNOLOGICAL ADVANCE

Four cases are chosen (TABLE 8) for the cost analysis; each representing a different state of technological advance. All cases are evaluated at different assumed levels of annual production in order to introduce economies of scale. Biomass and product formation rates increase through the four cases in the manner shown; that is, 140 to 350 g/1/day for biomass (hydrated, 80% water) and .07 to 17.5 g/1/day for product.

The latter figures are influenced by assumed product content levels/unit of biomass from .25% (case A) up to 2.5% (case D); case D may represent a significant change in technology[7]. Achievable biomass content/unit of fermentation broth was

TABLE 8. Cases for Cost Analysis and Technological Advance

	Case			
	A	B	C	D
Biomass Formation g/l/day	140	140	350	350
Product Formation g/l/day	.07	.7	1.75	17.5
Biomass Reuse	50%	90%	90%	90%
Nutrient Reuse	—	X	X	X
Product (g)/Biomass (g), DB	.0025	.0025	.0025	.025
Biomass (g), DB/liter	40	40	100	100
μ_B (days^{-1})	.693	.693	.693	.693
μ_P (days^{-1})	.693	6.93	6.93	6.93

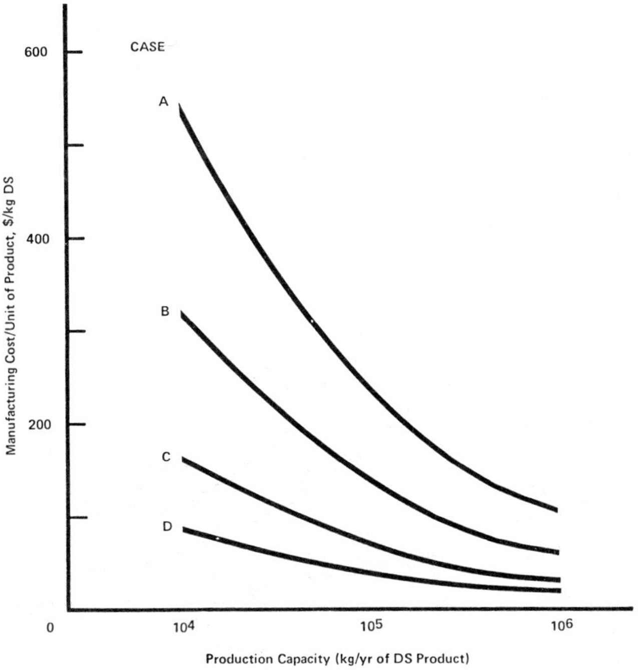

FIGURE 3. Manufacturing cost as a function of product capacity—plant cell culture.

assumed to be at 40 g/l (dry basis) up to 100 g/l. The latter figure may represent a significant change in technology (e.g., concentration of cells through immobilization on a surface).[7] Biomass reuse presumes nondestructive release of product; a limiting extension of cell reuse, immobilization, could allow the effective percentage reuse of cells to increase beyond 90%. Specific rates of biomass formation (.693 $days^{-1}$) represent one-day doubling time; increasing plant-cell-specific rates of propagation beyond this value may be difficult. Product formation rates were assumed to range from once/day (.693 $days^{-1}$) to ten times/day (6.93 $days^{-1}$). The latter figure represents a substantial advance in technology. Nutrient reuse is also programed in as indicated since this has a substantial bearing on manufacturing cost.

Cost analyses were completed[5] resulting in FIGURES 3 and 4, which show manufacturing and capital cost dependencies on scale of operation and state of technology. From this analysis, the effect of production scale on costs per unit of product is substantial. The effects of changes in technology (TABLE 8) impact significantly, particularly at lower levels of production. The results of this analysis indicate that costs associated with large-scale plant cell culture can be lowered to acceptable levels. At such lower cost levels, the cost of product from plant cell culture may be in proximity of the costs to produce certain amino acids by microbial culture. This is illustrated in FIGURE 5, a graph of capital and manufacturing cost trends for varied products produced by microbial cells.[8] FIGURE 6 (factors contributing to manufacturing cost), in turn, illustrates that at reduced production levels, labor- and capital-related costs increase in importance compared to raw material cost.[8] This may in turn suggest an avenue of process automation for cost control in these latter cases.[9]

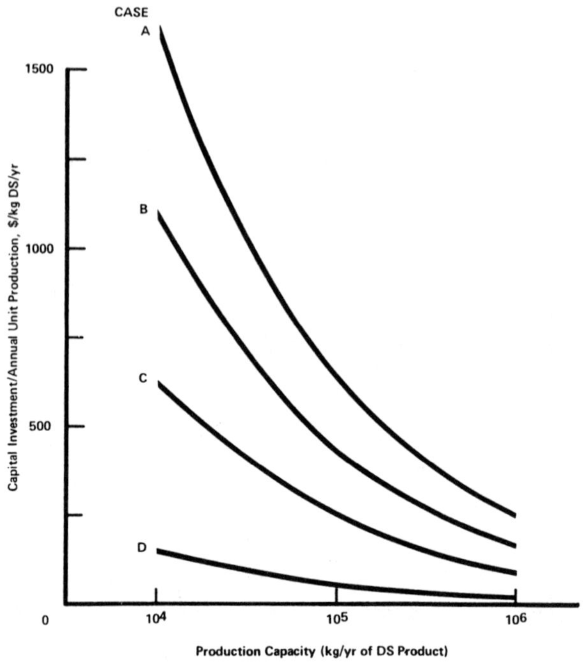

FIGURE 4. Capital cost as a function of product capacity—plant cell culture.

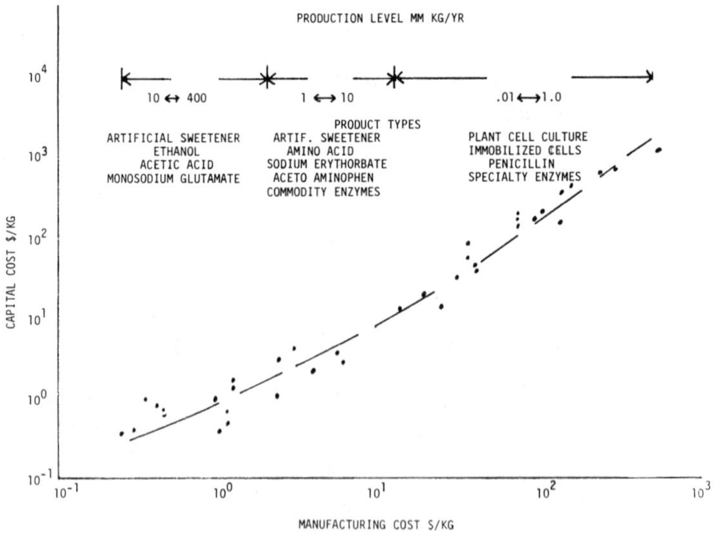

FIGURE 5. Relationships between unit capital and unit manufacturing costs.

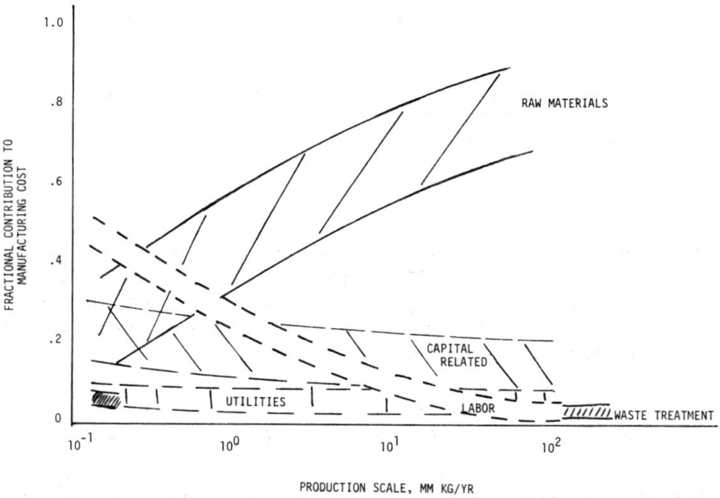

FIGURE 6. Fractional contribution to manufacturing cost.

CONSIDERATIONS IN DESIGN AND TECHNOLOGY TRANSLATION

Aspects of contrast between plant and microbial cell culture include extreme fragility, optimal aggregation, tendency for differentiation, and programing of suspension culture conditions for expression of products.[3] However, metabolic effects of digestion of proteins within the core aggregates may relate to similar occurrences with pelleted fungi.[10,11] It is possible that cell culture can be accomplished by provision of a suitable carbon source, but fermenter design may be affected substantially if photosynthesis is required and provision for light absorption must be included.[3]

MASS TRANSFER

Oxygen transfer to plant cell aggregates involves transport effects similar to those encountered with mycelial pellet aggregates, or aggregated flocs in waste treatment.[12] FIGURES 7 and 8 illustrate such common analogies. Thus, the cell aggregate is effectively a spherical catalyst of changing size; the changing size then alters the inherent transport characteristics of the catalyst. It is plausible that such transport

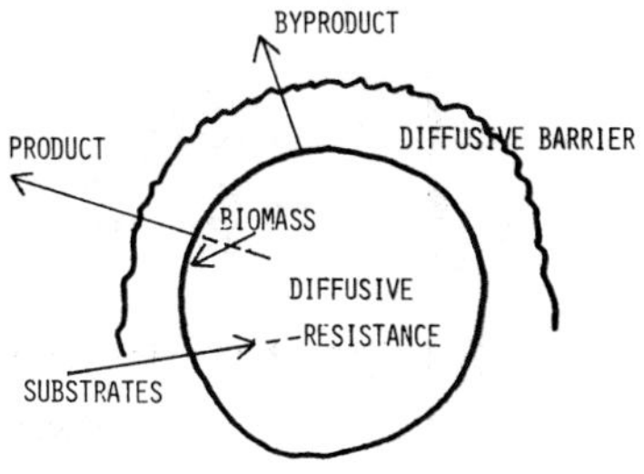

FIGURE 7. The plant cell aggregate as a catalytic entity.

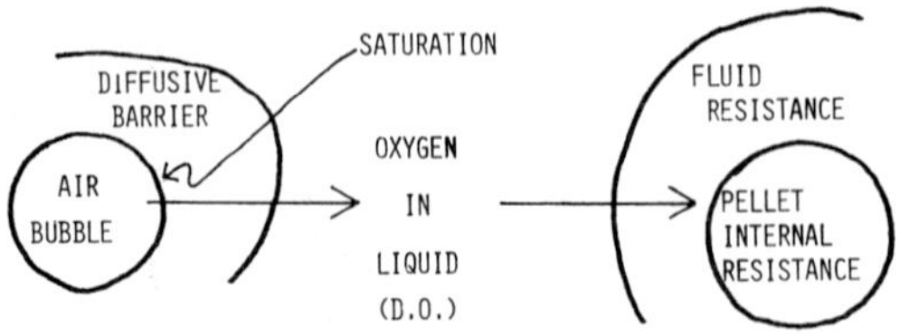

FIGURE 8. The oxygen transfer train.

effects could result in differentiation dependent on a best aggregate size or profile nutrient composition. FIGURE 8 portrays the more complex role of oxygen transfer from bubble to aggregate interior, suggesting that at higher cellular densities, difficulties of oxygen supply in microbial systems and effects of scale-up may also bear on phenomena inherent to plant cell culture.[10,11]

REACTOR DESIGN AND IMMOBILIZED CELLS

As other workers have noted,[3] and as a direction suggested in the previously described economic analysis, "anchoring" of cells to beads, particles, or surfaces may permit findings from immobilized microbial cell systems to be translated to plant cell culture. Fixed-bed reactor design could be employed, assuming the design insures acceptable resistance of the packed bed to compression.[13] Expanded or near-fluidized-bed design may be preferred for reasons of supply of oxygen from the liquid phase. In this latter case, certain expanded-bed reactors investigated elsewhere may be usefully applied to plant cell culture.[14] A further advantage of immobilization is that distinct separation of biomass and product formation stages would then be possible.

Commercial experience from the application of immobilized cells with glucose isomerase (GI) activity to production of high fructose corn syrup (HFCS) could aid research and development efforts in plant cell cultures (a typical GI reactor is displayed in FIG. 9); parameters important to operation of such reactors include substrate refining, means for temperature and pH control, incidence and control of microbial contamination, and design modifications to accommodate changes in the fixed bed, effects of occluded matter, and channeling considerations.[15] The manner of

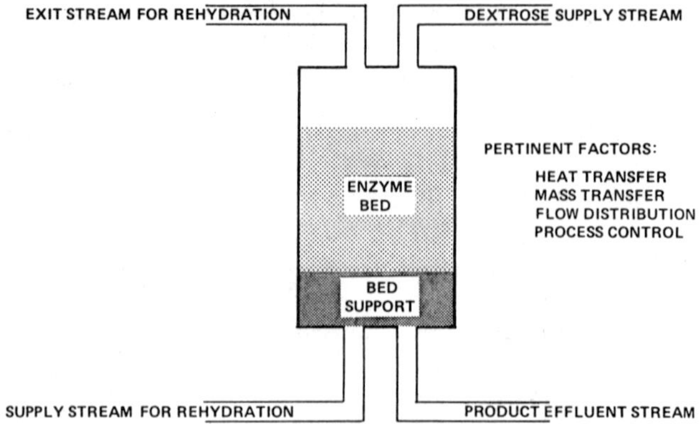

FIGURE 9. Glucose isomerase reactor.

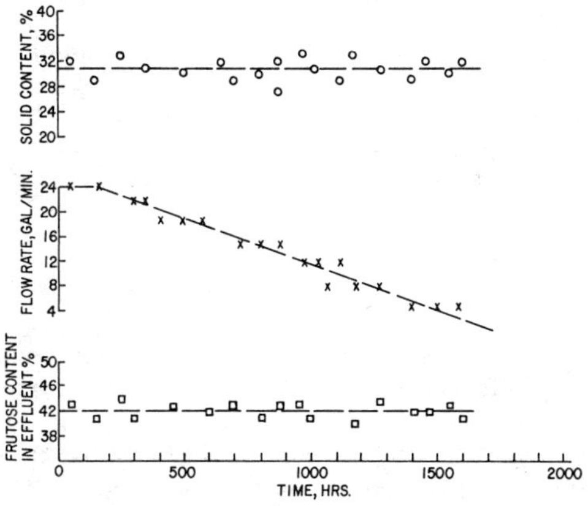

FIGURE 10. Activity depletion with time.

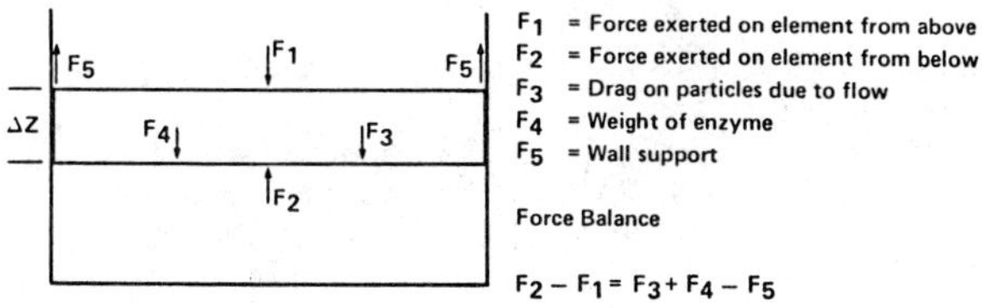

FIGURE 11. Force balance across reactor bed.

$$\frac{dPs}{dZ} = (\rho s - \rho L)\,(1-\varepsilon)g \qquad \text{STATIC WEIGHT}$$

$$+\ \frac{k\mu U\,(1-\varepsilon)^2}{2Rp\ \ 3} \qquad \text{DRAG}$$

$$-\ \frac{2\mu' K'}{R}\,Ps \qquad \text{WALL SUPPORT}$$

$\varepsilon = \varepsilon\,(Ps)$ MEASUREMENT OF COMPRESSIBILITY

μ' = WALL FRICTION

K' = FORCE TRANSMITTED TO WALL DUE TO AXIAL SOLIDS PRESSURE

FIGURE 12. Solution of the force balance.

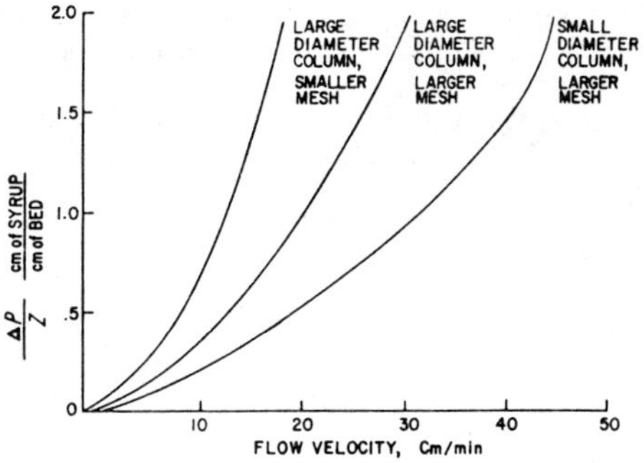

FIGURE 13. Pressure drop–flow characteristics through compressible packed beds.

compensation for activity loss with time (FIG. 10) and ways employed to maintain constant product output in a multicolumn format would also apply to beds of immobilized plant cells. Similarly, the technology developed in evaluating bed compression could apply, as indicated in the force balance portrayed in FIGURE 11, and then expressed in analytical form in FIGURE 12.[13] The graphs representing the expression in FIGURE 12 are portrayed in FIGURE 13.

SUMMARY

This paper has presented a view on the relationship of technological advance and cost in future processing of plant cells at large scale. Analogies with microbial culture and products thereof are used to develop the concept. Past and current work in immobilized cell technology can be addressed in order to establish further guidelines for development of this area.

REFERENCES

1. KATO, A., *et al.* 1978. Viscosity of the broth of tobacco cells in suspension culture. J. Ferment. Technol. **56**(3): 224.
2. AUSUBEL, F. M. 1982. Molecular genetics of symbiotic nitrogen fixation. Cell **29:** 1.
3. SCHULER, M. L., *et al.* 1981. Production of secondary metabolites from plant tissue culture—problems and prospects. Ann. N.Y. Acad. Sci. **369:** 65.
4. BAILEY, J. C. & OLLIS, D. F. 1977. Biochemical Engineering Fundamentals. McGraw-Hill. New York.
5. GOLDSTEIN, W. E., *et al.* 1980. Product cost analysis. *In* Plant Tissue Culture as a Source of Biochemicals. E. John Staba, Ed. CRC Press. Boca Raton, FL.
6. RUSHTON, J. H., *et al.* 1950. Chem. Eng. Prog. **46:** 395, 467.
7. BARZ, W. & B. E. ELLIS. 1981. Plant cell cultures and their biotechnological potential. Ber. Dtsch. Bot. Ges. Bd. **94:** 1.
8. GOLDSTEIN, W. E. Private information.
9. GREINER, B., 1974. Mess-und regel technik zur automatisierung von fermentations betrieben. Chem. Ing. Tech. **46**(16): 680.
10. WHITAKER, A. & P. A. LONG. 1973. Fungal pelleting. Proc. Biochem. p. 27.
11. MUKHOPADHYAY, S. N. & T. K. GHOSE. 1976. Oxygen participation in fermentation part 1: Oxygen microorganism interactions. Proc. Biochem. p. 19.
12. JENKINS, D. & M. G. RICHARD. 1983. Factors influencing the selection of filamentous organisms in activated sludge. Proc. conf. on Impact of Applied Genetics in Pollution Control. Univ. Notre Dame. In press.
13. VERHOFF, F. H. & J. J. FURJANIC. 1983. Compressible packed-bed fluid dynamics with application to a glucose isomerase reactor. Ind. Eng. Chem. Proc. Des. Dev. In press.
14. SCOTT, C. D., *et al.* 1975. A tapered fluidized-bed bioreactor for treatment of aqueous effluents from coal conversion processes. Oak Ridge National Laboratory. Oak Ridge, TN.
15. VERHOFF, F. H. & GOLDSTEIN, W. E. 1982. Diffusion resistance and enzyme activity decay in a pellet. Biotechnol. Bioeng. **24:** 703.

Immobilization of Plant Cells in a Hollow-Fiber Reactor

WILFREDO JOSE, HENRIK PEDERSEN,
AND CHEE-KOK CHIN[a]

Department of Chemical and Biochemical Engineering
[a]Department of Botany and Plant Physiology
Rutgers University
New Brunswick, New Jersey 08903

INTRODUCTION

Plants are a major source of medicinals, flavors, essential oils (fragrances), and other fine chemicals.[1-3] The production of these materials is carried out primarily with intact plants that are susceptible to a variety of climatological as well as political variation. In recent years, the suspension culturing of isolated plant cells is being increasingly considered to avoid the problems associated with whole plants and to develop an industry producing biochemicals by fermentation processes similar to those employed with microorganisms. Therefore, our work deals with the adaptation of current fermentation techniques and biochemical reactors to plant cell cultures in order to enhance their potential for the synthesis and transformation of natural compounds.

FREE AND IMMOBILIZED PLANT CELLS

Plant cell and tissue cultures are usually maintained in small-batch shake flasks. For large-scale cultivation, fermenters similar to those used for other microorganisms would be employed. Since plant cells are much bigger (30 to 100 microns) than the bacteria familiar to the fermentation industry and, furthermore, since they exist almost entirely as aggregates, grow slowly, and have rigid walls that make them susceptible to shear forces, current fermentation systems have to be modified. The above factors, and particularly the slow growth, makes immobilization or entrapment an attractive policy for plant cell utilization.

The first immobilization of plant cells was carried out by entrapping plant cells in alginate beads.[4] Brodelius *et al.*[4] demonstrated the biosynthetic capabilities of immobilized plant cells with the following examples:. (1) *Morinda citrofolia* for the *de novo* synthesis of anthraquinones, (2) *Catharanthus roseus* for the formation of indole alkaloids from a distant precursor, and (3) *Digitalis lanata* for a specific biotransformation reaction. Other matrix carriers that have been used since that time are carrageenan, agarose, agar, gelatin, polyacrylamide, and polyphenyleneoxide.[5-7] Alginate, and the first three carriers above, maintain fully viable cells.[5] The entrapped cells usually grow, however, when the medium composition is complete with the necessary nutrients, and this growth could result in the disintegration of the matrix.[8]

A more gentle entrapment technique is the use of hollow-fiber membrane reactors originally designed for the culturing of mammalian cells. For example, Shuler[9] employed commercially available, bundled hollow-fiber cartridges in a continuous-flow system for the production of phenolics by *Glycine max*. For all these immobilized plant cells, the desired product should be able to permeate out through the cell wall and be recovered from the medium.

HOLLOW-FIBER MEMBRANE REACTORS

The use of hollow-fiber reactors is particularly suited to fragile plant cells. The cells are simply entrapped outside the fibers and no special carrier matrix is needed. The medium flows inside the fibers and provides the necessary nutrients and the substrate. The reactants and the products with molecular weights less than a few thousand Daltons are freely exchanged. Growth can be maintained if the essential nutrients are provided, or it can be suppressed by altering the composition of the medium. Unlike other entrapment procedures, the diffusional limitation imposed by the carrier is minimal and the pressure drop across the reactor is small. Plant cells do not have a tendency to stick to the fibers so that the cells could easily be washed away after an experiment and, with care against puncturing the walls, the tube bundle reused.

EXPERIMENTAL

Vitafiber^{TM} hollow-fiber units (Amicon Corporation, Lexington, MA: Type 3P10) were used in this work. *Daucus carota* (carrot) cells grown in suspension with

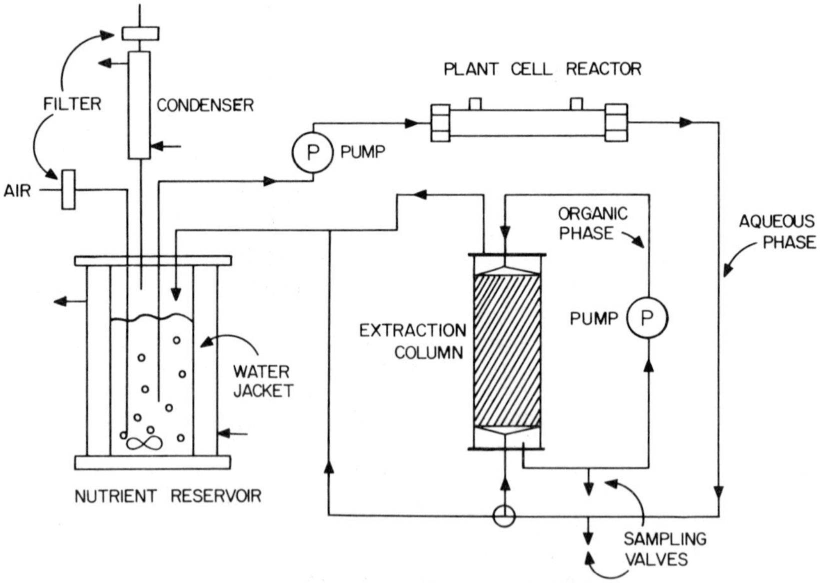

FIGURE 1. Flow system used to study the biotransformation reactions with immobilized plant cells. The reservoir contains about 100 ml of nutrient maintained at 30°C, whereas the reactor volume is about 2 ml. In all experiments reported here, the extraction column was by-passed.

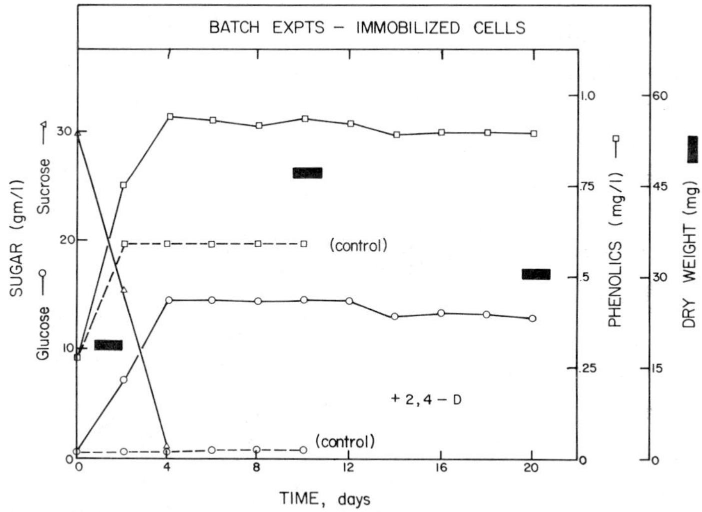

FIGURE 2. Experimental results obtained with *Daucus carota* cells. A control experiment was also run as shown by the dashed lines. The medium contained sucrose and essential nutrients provided by the medium of Murashige and Skoog as well as the growth factor 2,4-D.

Murashige and Skoog medium in rotary shake flasks (280 rpm) were subcultured every seven days. Approximately 1.4 ml of the seven-day-old suspension was loaded in the shell side of the hollow-fiber cartridge with a sterile syringe and the shell side was sealed. Aerated medium (100 ml) was recirculated through the reactor using the flow system shown in FIGURE 1. The medium is aerated in the reservoir and recycled through the reactor, which could be regarded as being in a batch configuration. Products can be removed using the extraction column. Samples were taken every two days and analyzed for phenolics, sucrose, and glucose concentration in the medium. The cell growth was also noted where possible. The system was run for 20 days while a control without plant cells in the reactor was run for 10 days. The flow rate was kept at a value of 10 ml/min, the maximum recommended for the hollow-fiber bundle.

RESULTS AND DISCUSSION

Sixteen mg (dry weight) of cells were loaded into a reactor and followed for 10 days. At the same time, 13.9 mg of cells were loaded into a reactor and followed for 20 days with the sampling starting on day 12, in order to conserve medium. The cell weight in the first reactor increased by 36 mg, whereas the increase measured in the second experiment was only 16 mg. The results are shown in FIGURE 2. The phenolic concentration (with gallic acid as a standard) increased from an initial value of 0.31 mg/l to 0.90 mg/l after 4 days and stayed at that level for the rest of the experiment. Half of the increase, however, comes from the nutrient as the control shows a value of 0.51 mg/l of phenolics after two days. Although not in very large amounts, the carrot cells are excreting intermediates as measured by the phenolics concentration in the medium and maintaining the level reached after four days. The invertase activity of the cells is indicated by the conversion of 3 g of sucrose to glucose and fructose within four days. The invertase activity of the carrot cells can therefore be used to evaluate the performance of the hollow-fiber membrane reactor with immobilized plant cells using a relatively simple reaction and substrate. Our present work is proceeding in this direction with results to be applied to studies on the production of more exotic biochemicals.

REFERENCES

1. STREET, H. E., G. G. HENSHAW & M. C. BINATTI. 1965. Chem. Ind. Jan. 2: 27–12.
2. FOWLER, M. W. 1981. Chem. Ind. April 4: 229–233.
3. LEE, S. L. & A. I. SCOTT. 1979. Dev. Ind. Microbiol. **20:** 381–391.
4. BRODELIUS, P., B. DEUS, K. MOSBACH & M. H. ZENK. 1979. FEBS Lett. **103:** 83–97.
5. BRODELIUS, P. & K. NILSSON. 1980. FEBS Lett. **122:** 312–316.
6. JIRKU, V., T. MACEK, T. VANEK, V. KRUMPHANZL & V. KUBANEK. 1981. Biotechnol. Lett. **3:** 447–450.
7. BRODELIUS, P. & K. MOSBACH 1982. Adv. App. Microbiol. **28:** 1-26.
8. ALFERMANN, A. W., I. SCHULER & E. REINHARD. 1980. Planta Med. **40:** 218–223.
9. SHULER, M. L. 1981. Ann. N. Y. Acad. Sci. **369:** 65–79.
10. SUGANO, N., R. IWATA & A. NISHI. 1965. Phytochem. **14:** 1205–1207.
11. JONES, A. & I. A. VELIKY. 1981. Eur. J. Appl. Microbiol. Biotechnol. **13:** 84–89.
12. VELIKY, I. A. & A. JONES. 1981. Biotechnol. Lett. **3:** 551–554.

New Microcarriers for Culturing Mammalian Cells

S. REUVENY, A. MIZRAHI, M. KOTLER[a]
AND A. FREEMAN[b]

Department of Biotechnology
Israel Institute of Biological Research
Ness Ziona, 70450, Israel

[a]Department of Microbiological Chemistry
Hadassah Medical School
Hebrew University
Jerusalem 91000, Israel

[b]Center for Biotechnology
The George S. Wise Faculty for Life Sciences
Tel-Aviv University
Tel-Aviv 69978, Israel

A working system for studying the effects of factors involved in the chemical nature of a microcarrier, on cell attachment, spreading, and growth was established. The system is based on polyacrylamide beads, prepared by the emulsion polymerization technique. Sieved beads of desirable mean diameter were derivatized to generate controlled amounts of primary and tertiary amino groups:[1]

I Primary amine derivatives

$$\Big)\!\!-CONH_2 + NH_2(CH_2)_nNH_2 \xrightarrow[\Delta]{\text{ethylene glycol}} \Big)\!\!-CONH(CH_2)_nNH_2 + NH_3$$

$$n = 2, 4, 6, 8$$

II Tertiary amine derivative:

$$\Big)\!\!-CONH_2 + NH_2CH_2CH_2N(CH_2CH_3)_2 \xrightarrow[\Delta]{\text{ethylene glycol}}$$

$$\Big)\!\!-CONHCH_2CH_2N(CH_2CH_3)_2 + NH_3$$

Following derivatization and determination of amino groups content, the beads were sterilized and used for the propogation of five different cell strains. Cell attachment kinetics, cell spreadings, and cell growth (in stationary and submerged cultures) were studied, employing the various polyacrylamide microcarriers.

The effect of the type of the positively charged group (primary vs. tertiary amines) was investigated at first. We have found that primary amino groups allowed for faster cell attachment rates, at lower degree of charging, as compared to the tertiary amino

[b]Author to whom correspondence should be addressed.

derivatives. Cell spreading began 1–2 hr after inoculation with the primary-amino-derivatized microcarriers as compared to 5–6 hr observed for the tertiary-amino-derivatized beads. Cell growth on the primary-amino-derivatized beads could be achieved at low charging, exhibiting "threshold effect." It seems that cells with pronounced epithelial morphology grow better on the primary amino derivative while cells with pronounced fibroblast morphology achieved higher cell yields on the tertiary-amino-derivatized beads.

In the second phase of this research, we investigated the effect of the amount and location of hydrophobicity accompanying the primary-amino-charged groups on cell attachment, spreading, and growth. Hydrophobicity was introduced onto the side chains (via increase of n) and into the polyacrylamide backbone (by replacing 10% of the acrylamide monomer ($CH_2{=}CHCONH_2$), with methacrylamide ($CH_2{=}CCH_3CONH_2$). The effect of increased hydrophobicity at a fixed degree of charging (0.5 meq/g (A) and 1.0 meq/g (B)) on cell attachment kinetics is shown in FIGURE 1. The data show that a gradual increase of hydrophobicity in the side chain leads to a parallel gradual increase in cell attachment. Introduction of hydrophobicity onto the polymeric backbone had a negative effect on cell attachment.

Cell growth on derivatized polyacrylamide microcarriers carrying various degrees of hydrophobicity is demonstrated in FIGURE 2, for chicken embryo fibroblast cells (CEF). Similar results were obtained for BHK (baby hamster kidney cell line), MDCK (a canine kidney epithelial cell line), and FS (a human diploid foreskin cell strain). FIGURE 2 clearly shows that there is an optimum for cell growth for side chains with n = 4–6, at a limited range of charging. On the other hand, the introduction of hydrophobicity onto the bead backbone led to a decrease in cell yields.

Out of the series of derivatized polyacrylamide microcarriers that have been tested, the diaminohexane-derivatized beads (n = 6), seem to be an alternative to the

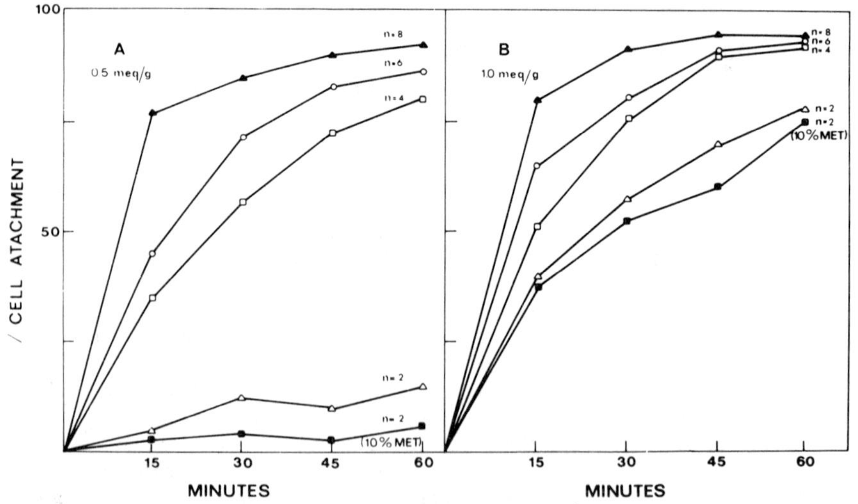

FIGURE 1. BHK (baby hamster kidney) cell attachment kinetics on primary-amino-derivatized polyacrylamide microcarriers, with different degrees of hydrophobicity, at fixed exchange capacity. Cell concentration: 2.5×10^5/ml. Microcarrier concentration: 5 mg dry polyacrylamide/ml.

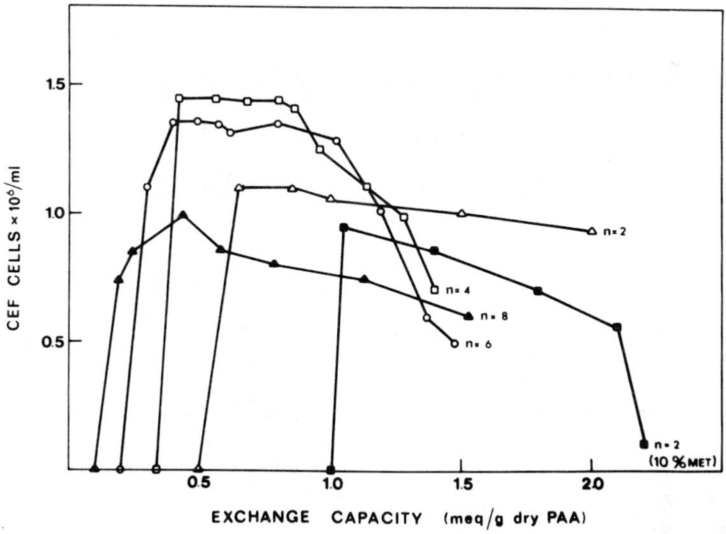

FIGURE 2. CEF (chicken embryo fibroblasts) cell growth on primary-amino-derivatized microcarriers, with different degrees and locations of hydrophobicity (stationary culture). Inoculum concentration: 1×10^5 cells/ml. Microcarriers concentration: 5 mg dry polyacrylamide/ml. Duration of growth: 3 days.

commonly employed tertiary-amino-derivatized microcarriers. We have found that this microcarrier allows for improved growth of cells with pronounced epithelial morphology. Moreover, yields of cells with pronounced fibroblast morphology were comparable to those achieved with the DEAE-derivatized beads. In addition, it is reasonable to assume that this microcarrier will be cheaper than commercially available microcarriers.

REFERENCES

1. Reuveny, S., A. Mizrahi, M. Kotler & A. Freeman. 1983. Factors affecting cell attachment spreading and growth on derivatized microcarriers: I establishment of a working system and effect of the type of the amino-charged groups. Biotechnol. Bioeng. **25**:469.

Ceramic-supported Hybridomas for Continuous Production of Monoclonal Antibodies

A. MARCIPAR, P. HENNO, E. LENTWOJT, A. ROSETO,[a]
AND G. BROUN

Institut de Technologie des Surfaces Actives
Université de Compiègne
BP. 233, 60206 Compiègne, France
[a]Institut de Recherches sur les Maladies du Sang—CNRS
Unité n⁰ 107 INSERM, L.O.I., Hôpital Saint Louis
75475 Paris Cédex 10, France

INTRODUCTION

The use of hybridomas for the production of monoclonal antibodies has deeply modified approaches in immunochemical research. It provided the researchers with biological research methods and is a promising tool for the fractionation and preparative purification of biological molecules.

The use of continuous-flow, fixed-bed reactors, containing whole immobilized cells, permits recycling of cells, high cell concentrations in the reactor, a reduction of feed-back effects, better substrate utilization, and easier control of prevailing conditions. Most of the applications of such systems have been applied to the treatment of waste waters and bacterial conversions, some of them being presently used at the industrial scale.[1-3]

Few applications have been described for processes applying animal cells, such as the production of exoantigens by protozoa and of monoclonal antibodies by hybridoma cells. Both applications of that type of reactor offer major advantages. The cells can be bound onto different supports very easily (sometimes, they require immobilization to keep their normal metabolism). Furthermore, high valorization products can be obtained, which can be operated at a relatively small scale with the use of well-controlled equipment.

MATERIAL AND METHODS

Strain

Murine hybridoma producing[4] IgG-type immunoglobulins were obtained by the fusion of BALB/c lymphocytes with the $SP_2/0$ myeloma strain, and then subcultured on RPMI 1640 culture medium, supplemented with 20% fetal calf serum in conventional Falcon flasks.

TABLE 1. Operating Conditions of the Reactor

Matrix mass	20 g
Number of immobilized cells	6.9×10^7
Medium	RPMI 1640 with 10% of calf fetal serum
Volume of recycled culture medium	50 ml
Temperature	37°C
Composition of gas O_2	20%
N_2	74%
CO_2	6%
Flow conditions	1.3 cm/min

Matrix

"Biogrog A" ceramic support was prepared as described in a previous paper. Twenty grams of this matrix were packed in a thermostated column (FIG. 1), and a suspension of hybridoma cells harvested during the growth phase was passed through it. As concerns the operating conditions of the reactor, flow conditions, final cell concentrations, and temperature during the immobilization process were the same as described in TABLE 1.

Reactor

The reactor consisted of the thermostated tower packed with Biogrog A, connected with a gas exchanger. Culture medium was continuously recirculated. Three-way

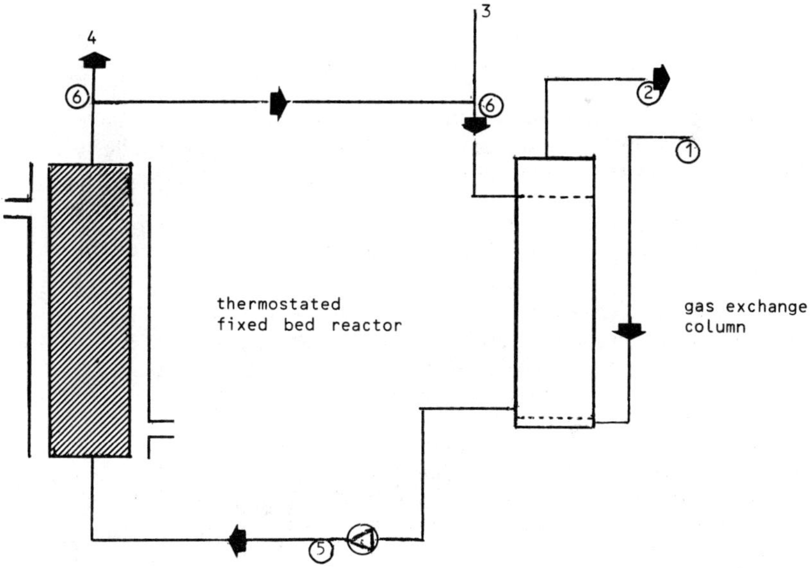

FIGURE 1. Schema of the reactor: (1) gas input (2) gas exit (3) fresh medium input (4) products output (5) peristaltic pump (6) three-way valves.

valves and sterilizing filters permitted the introduction of fresh sterile medium and the release of products (FIG. 1).

Analytical methods

Cell concentration was monitored in a hematimeter. ATP measurements have been performed by a bioluminescent probe in a "Pico ATP[b]," according to the method described by Hysert *et al.*[6] and Jakubczak.[7] This method of ATP measurement is based on the measurement of light emission during oxidation of luciferin by luciferase in the presence of oxygen, Ma^{2+}, and ATP. The peak of emitted light was proportional to the concentration of ATP in the medium. One ml of each cell suspension was diluted in nine volumes of DMSO (dimethyl sulfoxide). After one min of stirring in a vortex, 4 ml of MOPS buffer was added. 200 ml of this solution was mixed with 50 μl of the enzymatic mixture containing the luciferin luciferase complex and transferred to the instrument. The emitted light was read immediately. Secreted immunoglobulins have been monitored by a classical ELISA-quantitative immunosassay, using prederivatized nylon plaques sensitized by anti-mouse antibodies.[8]

TABLE 2. Control of Input and Output Cell Concentration Values during Immobilization onto "Biogrog A"

	Input Culture Medium	Output Culture Medium
Cells concentration 10^5/ml	4.6	0
ATP (pg/ml)	8.6×10^3	0

Electron microscopy

Immobilized cells were fixed with 2.5% glutaraldehyde in phosphate buffer overnight, then dehydrated with increasing concentrations of ethyl alcohol and coated with gold for an electron-scanning probe.

RESULTS AND DISCUSSION

A secretion of IgG by the immobilized cells was observed in this classical whole-immobilized-cell system. Cells are fully adsorbed on the support by one passage, as can be seen on TABLE 2: no cell or ATP is detectable in the effluent.

In FIGURE 2, a typical scanning electron micrograph of a hybridoma cell immobilized on the ceramic matrix can be seen.

Biosynthetic activity of these immobilized cells, followed by an ELISA test, shows that kinetics parameters of monoclonal antibody secretion are different from conventional cultures (TABLE 3): the yield of the reactor is increased 4–5 times in these experiments mainly due to higher cell concentrations per volume unit of reactor obtained by using this process.

The support used (Biogrog A) is a cheap, inert material, already applied to the immobilization of other prokaryotic and eukaryotic cells. It seems well adapted to the

[b]Instruments S.A., France.

FIGURE 2. Scanning electron micrograph of a hybridoma cell immobilized on the ceramic support (× 2000).

growing and immobilization of this type of cells. Viability of the reactors and life cycles of the secretory cells are now being followed, in order to determine conditions for continuous production of monoclonal antibodies. General conditions for the recycling of the matrix are also yet to be optimized.

ACKNOWLEDGMENTS

This investigation received financial support from the UNDP/World Bank/WHO special program for research and training in tropical diseases, DGRST (Délégation

TABLE 3. Secretion Parameters for IgG Secretion by Hybridoma Cells in a Conventional Culture and in an Open Tubular, Immobilized-Whole-Cell Reactor

	μg IgG/ml[a] · hr
A Immobilized whole cell reactor	0.099
B Conventional Falcon flask culture	0.022
A/B Report	4.5

[a]Volume of the reactor.

Générale de la Recherche Scientifique et Technique, France), and INSERM (Institut National de la Santé de la Recherche Médicale, France). We are indebted to Mrs. D. Mistro and Mrs. F. Masrour for their important technical cooperation.

REFERENCES

1. Kolot, F. B. 1981. Proc. Biochem. **16**(5): 2–9.
2. Kolot, F. B. 1981. Proc. Biochem. **16**(6): 30–33.
3. Marcipar, A., N. Cochet, L. Brackenridge & J. M. Lebeault. 1979. Biotech. Lett. **1**(2): 65.
4. Roseto, A., R. Scheerer, J. Cohen, M. C. Guillemin, A. Charpilienne, S. Seynerol & J. Peries. 1983. J. Gen. Virol. **64:** 237–240.
5. Henno, P., A. Marcipar, Geinaert & E. Segard. 1980. Microbial Adhesion to Surfaces. Ellis Horwood Publishers. London. p. 543.
6. Hysert, D. N., F. Dovecses & N. M. Morrison. 1976. J. Ann. Soc. Biochem. Chem. **34**(4): 145–150.
7. Jakubczak, E. & H. Leclerc. 1980. Ann. Biol. Clin. **38:** 297–304.
8. Marcipar, A., M. A. Vijayalakshmi, D. Afchain & E. Segard. 1979. IRCS Medical Science. **7:** 178.

A Theoretical Model for Insulin Secretory Dynamics in a Hybrid Artificial Pancreas

NAOMI L. WEINLESS AND CLARK K. COLTON

Department of Chemical Engineering
Massachusetts Institute of Technology
Cambridge, Massachusetts 02139

INTRODUCTION

Lack of normal insulin release by the pancreas may contribute to the long-term complications associated with diabetes. In an effort to restore physiological insulin delivery, an artificial hybrid pancreas is being developed for implantation in the circulatory system as an arteriovenous shunt. The device (FIG. 1) consists of insulin-secreting pancreatic cells from the islets of Langerhans cultured on the outside surface of a semipermeable tubular membrane and surrounded by an external shell for loading cells.[1] Glucose diffuses from the tubular lumen across the membrane to the islet cells, which secrete appropriate amounts of insulin that diffuse back into the lumen and into the surrounding shell. The membrane is permeable to insulin and glucose but excludes the larger antibodies and lymphocytes, thus circumventing the problem of immune rejection.

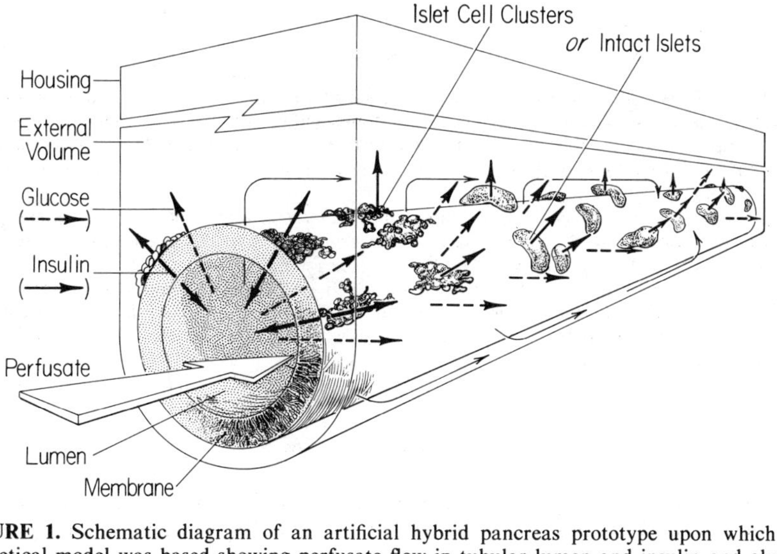

FIGURE 1. Schematic diagram of an artificial hybrid pancreas prototype upon which the theoretical model was based showing perfusate flow in tubular lumen and insulin and glucose diffusion (*heavy arrows*) through membrane and shell phases. Convection (*thin arrows*) may also occur but is not included in the model.

The use of a wide-bore tube to avoid clotting, a thick membrane wall to provide mechanical strength, and the presence of an external shell all introduce mass transfer time lags into the dynamics of insulin secretion in response to changes in lumenal glucose concentration. In order to more closely approach physiological control, we have investigated the effects of transport delays with a mathematical model that is physically more realistic than that previously described.[2]

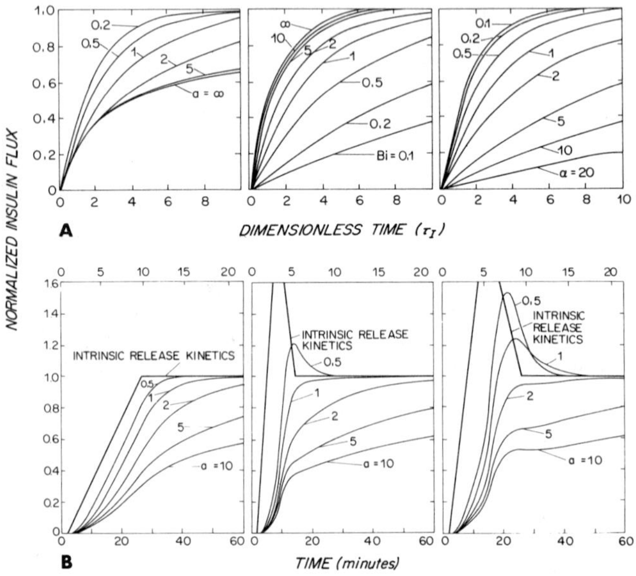

FIGURE 2. Effect of dimensionless transport parameters on predicted normalized insulin flux (relative to steady-state value) as a function of dimensionless and real time. The parameter a is proportional to the ratio of shell to membrane thickness; α is the ratio of the shell to membrane permeability; and Bi (Biot number) is the ratio of the mass transfer resistance in the membrane to that in the lumenal boundary layer.

(A) (*Upper Panels*) Effect of variations in a, Bi, and α on the response to a step change in glucose concentration at the membrane-shell interface with instantaneous insulin secretion kinetics. Parameter variations are about a base case of a = Bi = α = 1.

(B) (*Lower Panels*) Effect of variations in a, with Bi = ∞ and α = 2, on the response to a step change in lumenal glucose concentration with intrinsic insulin secretion kinetics represented by three different idealized patterns incorporating a time delay and either a ramp or an initial peak secretion rate of variable duration.

MODEL DEVELOPMENT

Solute transport is represented by transient one-dimensional diffusion in both the membrane and shell phases with quasi-steady transport in the lumenal concentration boundary layer. Diffusional resistance in the membrane is associated with the stagnant fluid in the pores, the skin providing negligible resistance. Both curvature and convection are neglected. The islet cells are represented by an infinitely thin layer at the membrane/shell interface, and idealized patterns, representative of those found in the literature, are used to describe their intrinsic secretion kinetics. Insulin secretion rate is assumed to be linearly dependent on glucose concentration. The time dependence of the insulin secretion rate is taken as independent of glucose concentration. An analytical solution for insulin lumenal flux is obtained by the method of Laplace transformations with the successive use of Duhammel's superposition integral to account for both the time-dependent secretion kinetics and the time-varying glucose stimulus at the interfacial cells.

RESULTS AND CONCLUSIONS

Theoretical simulations show how transport resistances associated with the membrane (α), the lumen (Bi), and the shell (a) affect observed insulin secretory dynamics. The influence of these three transport parameters is shown in FIGURE 2A where normalized insulin flux is plotted as a function of dimensionless insulin time for the case of instantaneous islet cell secretion kinetics and a glucose step change at the membrane-shell interface. The observed increase in device response time and decreased rate of approach to steady-state secretion rate is reflected in the separate contributions of decreasing membrane permeability, increasing shell thickness, and increasing lumenal boundary layer resistance. With finite secretion kinetic models, the magnitude of any rapid-phase peak secretion rate is diminished and can disappear entirely if these transport resistances are sufficiently large, as is shown for the case of increasing shell thickness in FIGURE 2B.

These results suggest that transport resistances can have substantial effects on observed insulin secretion kinetics. Further modeling studies are under way to help guide and interpret experiments and to provide a rational basis for device design.

REFERENCES

1. COLTON, C. K., B. A. SOLOMON, P. M. GALLETTI, P. D. RICHARDSON, C. TAKAHASHI, S. P. NABER & W. L. CHICK. 1980. Development of novel semipermeable tubular membranes for a hybrid artificial pancreas. *In* Ultrafiltration Membranes and Applications. A. R. Cooper, Ed. Plenum Publishing Corp. New York. p. 541.
2. COLTON, C. K. & L. H. UNGER. 1980. Transient diffusion in a two-phase composite with a glucose concentration-dependent insulin source at the interface. *In* Proceedings of the 8th Annual Northeast Bioengineering Conference. Cambridge, MA. p. 547.

Ethanol Production with Immobilized Cell Reactors

P. LINKO, M. SORVARI, AND Y.-Y. LINKO

Laboratory of Biochemistry and Food Technology
Department of Chemistry, Helsinki University of Technology
SF-02150 Espoo 15, Finland

INTRODUCTION

Recent advances in biotechnology have resulted in a renewed interest in renewable carbohydrate sources as an alternative feedstock for chemical synthesis as well as a source both for energy and liquid fuel. It is not surprising, therefore, that the interest in biotechnical ethanol production has significantly increased during the past few years.[1] Continuous fermentation, combined with ethanol stripping and cell recycling as in the vacuum fermentation process developed by Cysewski and Wilke,[2] allows a significant increase in ethanol productivity in comparison with conventional batch fermentation. Several techniques for continuous ethanol production have been developed and also applied both in industrial ethanol[3] and alcoholic beverage[4] manufacture. Additional advantages may be realized with flocculating yeasts together with special settling devices[5] or in a tower fermenter.[6] Research with ethanol-producing microorganisms other than yeasts, such as *Zymomonas mobilis* and *Clostridium thermocellum* bacteria, and *Rhizopus* species and other related molds may also open up new possibilities.[7] Nevertheless, the recent developments in immobilized biocatalyst engineering and, in particular, in continuous heterogeneous biocatalysis techniques involving immobilized living microbial cells is likely to bring about significant economic improvements in biotechnical conversions.[8,9] Total costs of biotechnical ethanol production are largely determined by the price of raw materials, which may account for as much as 60 or even 70% of total production costs.[10] Consequently, it is important to aim for a maximum possible conversion. On the other hand, in a batch process fermentation may account for more than 80% of the total capital costs.[11] Immobilized cell reactors allow high cell densities with little cell washout even at very short residence times (at high dilution rates) thus eliminating costly cell removal before distillation. Such reactors may be operated for long periods without biocatalyst replacement. High cell density results both in an increased productivity and decreased bioreactor volume for the same ethanol production rate, reducing both production and capital costs. Furthermore, no energy is required for agitation, and the reactor may be easily reactivated by intermittent addition of necessary nutrients, if needed. Production rate may be controlled by residence time and by optimizing the nutrient content and other process parameters for minimum cell growth and maximum ethanol production. The current state-of-the-art of ethanol fermentation has been recently reviewed by Kosaric *et al.*[1] and by Rhigelato.[12] Kolot[10] has reviewed work on immobilized yeasts, and Linko and Linko[13] have written a comprehensive review covering literature on immobilized cells for ethanol production until the second half of 1981. The present article is intended to update this review, illustrating some details with examples from our own most recent work.

IMMOBILIZATION OF LIVING YEASTS AND BACTERIA

Saccharomyces cerevisiae and *Kluyveromyces fragilis* yeasts were immobilized in calcium alginate gel beads essentially employing the method described by Linko *et al.*[14] except that the final crosslinking step with glutaraldehyde, originally intended for eliminating the ethanol fermentation without interfering with invertase activity, was omitted. Various quantities from 32 to 128 g wet yeast cell mass were typically suspended in 8% sodium alginate solution, and extruded under slight pressure through hollow needles of ϕ 0.6 mm into 0.5 M calcium chloride to form biocatalyst beads of about ϕ 2 mm. The beads were allowed to harden for about 30 min before collecting. The high viscosity of 8% alginate solution caused no technical difficulties, and was

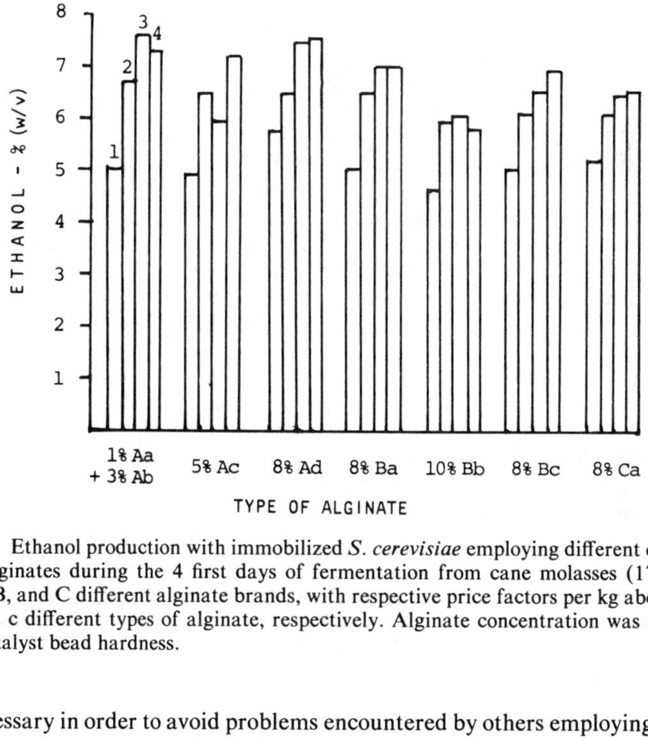

FIGURE 1. Ethanol production with immobilized *S. cerevisiae* employing different quantities of different alginates during the 4 first days of fermentation from cane molasses (17.5% wt/vol sugar). A, B, and C different alginate brands, with respective price factors per kg about 1, 2, and 3; a, b, and c different types of alginate, respectively. Alginate concentration was adjusted for same biocatalyst bead hardness.

found necessary in order to avoid problems encountered by others employing 2% or less alginate.

A large number of commercially available sodium alginates were tested for their suitability for immobilized biocatalyst carrier. Although the chemical composition of the alginate is of importance, it was found that alginates of widely varying price range could be employed. FIGURE 1 illustrates this with a number of alginates tested. The alginate concentration was chosen to obtain as close as possible equal bead hardness for similar reactor performance, and was found to vary from 4 to 10%. In most cases, nearly the same ethanol production was obtained during the 4-day fermentation trials. BDH sodium alginate (BDH Chemicals Ltd, Poole, UK) was employed throughout the rest of this work. Although most work was done using 8% sodium alginate, some experiments were also carried out employing 2% sodium alginate solution, with subsequent partial drying of the biocatalyst beads at room temperature for further hardening. The shrinking of beads during drying also resulted in a decreased bioreactor volume, another advantage, although some reswelling could be noticed during long-term operation. TABLE 1 shows that partial drying resulted, optimally, in almost doubling the volumetric productivity to approximately 40g l^{-1} hr^{-1}, based on biocatalyst bed volume (the liquid phase including bead porosity but excluding space taken up by carbon dioxide is maximally about 80% of the bed volume). We have previously shown[15] that with cane molasses containing adequate nutrients, initial yeast inoculum does not significantly affect bioreactor performance after the first 5 days, with a typical final yeast cell level in the biocatalyst of about 2 to 5 $\times$ 10^9 cells per gram. The maximum cell count in the effluent was of the order of 10^7 cells per ml.

Veliky and Williams[16] treated calcium alginate beads in aqueous solution of a polycation such as polyethyleneimine for increased resistance towards phosphate ions. Some recent experiments with other carriers than alginate may also be cited here. Sitton and Gaddy[17] employed gelatin-coated Rashing rings but the reactor liquid void volume decreased to one-fifth in 2 weeks due to excessive cell growth, severely interfering with the reactor performance. Wada *et al.*[18] have pioneered the entrapment of living cells in κ-carrageenan gel, claiming about 50g l^{-1} hr^{-1} productivity at nearly 100% conversion and 1-hr residence time from 10% glucose. Wang and Hettwer[19] further improved the κ-carrageenan carrier system by coentrapping tricalcium phosphate. Good results have also been obtained by adsorbing yeast cells on strongly basic anionic resin (XE-352, Rohm & Haas), with 55g l^{-1} hr^{-1} productivity (based on void volume) at a steady-state effluent ethanol level of 48g l^{-1} from 12% (wt/vol) glucose.[20] Ghose and Bandyopadhyay[21] claimed a maximum productivity of about 25g l^{-1} hr^{-1} at 2.9 hr residence time, 75% conversion from molasses of 19.7% reducing sugar content. Typical volumetric ethanol productivity of about 22g l^{-1} hr^{-1} was reported by Moo-Young *et al.*[22] with a beechwood-chip adsorbed yeast reactor, but the basis for calculation was not given.

Zymomonas mobilis cells were entrapped with similar techniques to those employed for yeasts in κ-carrageenan and locust-bean-gum mixed gel matrix. The carrageenan system was chosen in this case in order to avoid problems easily encountered with alginate gels by the necessary nutrients in long-term operations. Calcium alginate gel has also been employed in such a case, but continuous[23] or intermittent[24] addition of calcium chloride with substrate was necessary to prevent excessive softening of the gel. On the other hand, we found κ-carrageenan alone, as recommended by Wada *et al.*[18] too soft for continuous operation, whereas a mixture of carrageenan and locust-bean gum proved satisfactory. In a typical experiment, 4g of κ-carrageenan and 1g of locust-bean gum were dissolved in 90 ml of 0.9% sodium

TABLE 1. Continuous Ethanol Production with Partially Dried Immobilized Yeast

Biocatalyst Dry Matter %	Initial Yeast Quantity g/100 g	Residence time hr	Productivity[a] g l^{-1} hr^{-1}
9.0 (not dried)	25	4.3	19
18.4	25	2.7	19
27.8	25	1.8	39
77.5	25	1.8	33
50.6	15	1.1	45
20.3	5	0.96	52

[a]Based on biocatalyst bed volume.

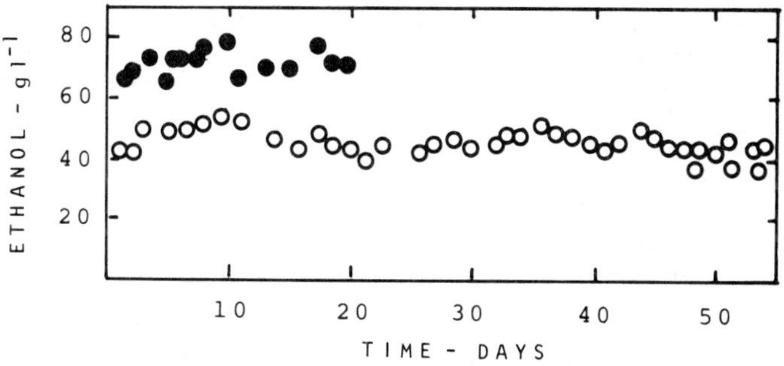

FIGURE 2. Continuous ethanol production from grape juice (●, 14.8% wt/vol sugar, $\tau \sim 5$ hr) and barley malt wort (○, $\tau \sim 2$ hr) with immobilized *S. cerevisiae.*

chloride, mixed with a suspension of 6.5g of wet *Z. mobilis* cells in 5 ml of 0.9% sodium chloride, and extruded through hollow needles at 45°C to 2% potassium chloride to obtain biocatalyst beads.

ALCOHOLIC BEVERAGES

The first applications of immobilized cell technology in ethanol fermentation were suggested in the field of alcoholic beverage manufacture, and as early as in 1969, a patent assigned to Intermag-Getränke Industri A.G. described a method of manufacturing alcoholic beverages with yeast cells attached to an insoluble carrier.[13,15] Similar techniques developed later by several others, particularly for beer production, have been reviewed earlier.[8,9] Problems encountered in beer flavor as a result of rapid continuous fermentation have until today prevented large-scale commercial applications of immobilized cell reactors in the brewing industry. Nevertheless, we observed that barley-malt wort is an excellent substrate for alginate-gel-immobilized living yeast, and that about 4.5% (wt/vol) product ethanol level could be maintained for more than 3 months at 2-hr residence time, and 3.8% (wt/vol) ethanol level at $\tau \sim 1$ hr[15] (FIG. 2). Hartmeier and Mücke[25] experimented with adsorbing and crosslinking *Rhizopus niveus* glucoamylase on yeast cell surface to obtain a living yeast/glucoamylase coimmobilized biocatalyst for continuous, low-dextrin beer production. It is also attractive to attempt to replace part of beer fermentation time by employing a continuous, packed-bed immobilized-yeast bioreactor.

Assuming an annual beer production of 50 million liters with a weekly production of 1 million liters, typically five, 200-m^3 batch fermenters would be needed for the primary fermentation step for 1 week at about 10°C. The secondary fermentation and storage for 9 weeks at 0°C would require an additional 45 fermentation tanks, making a total of 50 tanks with a total capacity of 10 million liters, not allowing necessary flexibility for washing, and so forth. In comparision, a continuous, immobilized-yeast bioreactor plant would require a total reactor bed volume of only 12 m^3 in order to produce 6000 l hr^{-1} at $\tau \sim 2$ hr on a 24-hr basis for 50 weeks. With such a system, it might be possible to cut the time required for the secondary fermentation and storage in half, with a considerable savings in capital costs.

Similar calculations may be made for cider and wine production. It has already been shown that continuous, industrial-scale wine production employing a tower fermenter with flocculating yeast is feasible.[6] With such a system, wine at up to 12% ethanol content may be obtained in 6 hr and be ready for bottling after a few days of storage. Similarly, cider of 5 to 6% ethanol requires a total fermentation time of only about 3 to 4 hr. We have also shown that the fermentation of grape juice with a continuous, immobilized-yeast packed-bed column fermenter is possible (FIG. 2). A mead-type beverage has also been successfully produced in such a system. In a conventional process involving 1 week primary fermentation and 4 weeks of storage after fermentation, producing 1 million liters in one operation would require, for example, five 200-m^3 fermenters. The same quantity could be theoretically produced in 10 days with a 2-m^3, immobilized-yeast packed-bed reactor capacity! A product fermented in laboratory scale was judged excellent with no observable off-flavor by a taste panel of 25 judges. In this case, 1.9% (wt/vol) product ethanol level could be maintained at $\tau \sim 0.9$ hr, and 1.2% (wt/vol) ethanol level at $\tau \sim 0.5$ hr. Although with some products flavor problems still remain to be solved, it is obvious that immobilized living yeast reactors should not be overlooked as an economic alternative.

POTABLE AND INDUSTRIAL ETHANOL

We have investigated a number of fermentable, carbohydrate-containing, by-product and waste materials as substrate for ethanol fermentation employing immobilized yeast bioreactors. Such continuous bioreactors are particularly suitable for processing dilute sugar streams, which by conventional batch fermentation techniques would require very large fermenter volume.[8] A steady-state optimal conversion may be easily maintained by continuous or intermittent nutrient addition, if necessary. FIGURE 3 illustrates ethanol production from sulfite spent liquor. The rate of ethanol production is lower than from pure glucose supplemented with the necessary nutrients, in which case 1% (wt/vol) product ethanol level could be maintained at $\tau \sim 0.5$ hr and nearly 100% conversion. Nevertheless, about 1% (wt/vol) ethanol level could be maintained at $\tau \sim 1$ hr from spruce sulfite spent liquor for extended periods. Inasmuch as pentoses are not fermented by *S. cerevisiae,* only 0.4% (wt/vol) product ethanol concentration was obtained from birch sulfite spent liquor, even at nearly 8-hr

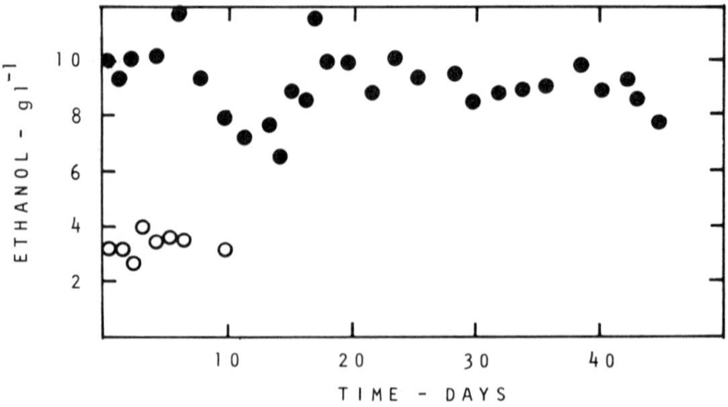

FIGURE 3. Ethanol production from spruce (●, $\tau \sim 1$ hr) and birch (O, $\tau \sim 8$ hr) sulfite spent liquor with immobilized *S. cerevisiae.*

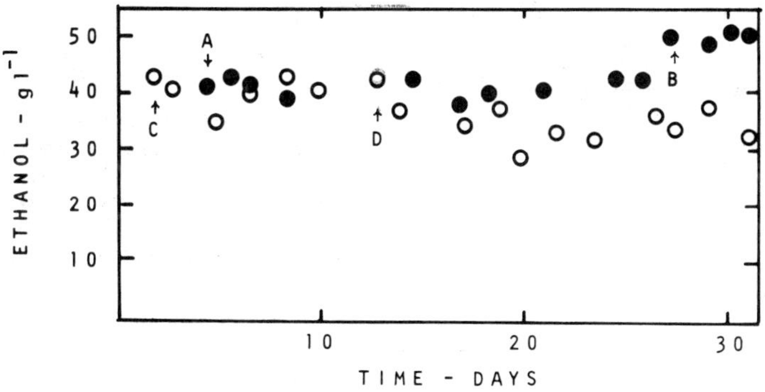

FIGURE 4. Ethanol production from commercial corn glucose syrup (●, A 17% wt/vol glucose, $\tau \sim 4.4$ hr; B 22% wt/vol glucose, $\tau \sim 8$ hr) and HTST-extrusion cooking pretreated barley starch hydrolysate (○, 17% wt/vol glucose; C $\tau \sim 4.2$ hr; D $\tau \sim 3.2$ hr) with immobilized *S. cerevisiae.*

residence time. In a commercial, conventional process, typically four 500m^3 fermenters at $\tau = 24$ hr are required to produce about 1.7 m^3 of ethanol per hour. For the same production rate, only about 130 m^3 total immobilized-yeast packed-bed bioreactor volume would be needed. Although the savings in total reactor volume are in this case not as striking as with some other substrates, they are, nevertheless, significant.

Satisfactory fermentation of pentose sugars is necessary also for complete economic bioconversion of lignocellulosic material hydrolysates. Maleszka *et al.*[26] compared the ability of free and calcium-alginate-gel-immobilized *Pachysolen tannophilus* yeast cells to ferment 2% xylose containing 0.67% yeast extract in batch experiments. With immobilized cells in small-scale experiments a little over 0.5% (wt/vol) ethanol concentration was reached in 9 days. Tsao and coworkers[27,28] employed a different approach. Having observed that calcium-alignate-gel-entrapped *Mucor* sp. or *Fusarium oxysporum* f. sp. lini fungus adsorbed on corn cob pieces fermented xylose poorly, they used *Actinoplanes missouriensis* whole-cell-immobilized glucose isomerase to convert xylose to xylulose, which could then be fermented to ethanol with immobilized *S. cerevisiae* yeast. In such a system, they were able to obtain 32 g l^{-1} of ethanol from 10% xylose, representing 64% yield. However, the half-life of the calcium-alginate-gel-immobilized yeast was only about 5 days, with the xylulose-fermenting ability totally lost in 8 days.

During the past few years, we have extensively investigated the possibilities of an HTST-extrusion cooker as a bioreactor for syrup manufacture from various starches and as an instrument for biopolymer pretreatment for subsequent enzymatic hydrolysis. For example, we have observed that such a thermomechanical treatment of cereal starch may be substituted for α-amylase liquefaction step in glucose syrup production.[31] Similar findings have been recently reported also by Korn and Harper.[32] FIGURE 4 shows that continuous ethanol production from hydrolysates obtained with glucoamylase from such extrudate is possible with an immobilized yeast bioreactor, with no inhibitory substances formed.

Cheese whey, a by-product of the dairy industry, appears to be a nearly ideal substrate for both technical and potable ethanol production, and we have previously reported that calcium-alginate-gel-entrapped *Kluyveromyces fragilis* cells can be employed in continuous conversion of whey lactose to ethanol.[33] A steady-state 2% (wt/vol) product ethanol level could be maintained at 5% (wt/vol) substrate lactose level for at least one month. Similar results were obtained with a reactor containing

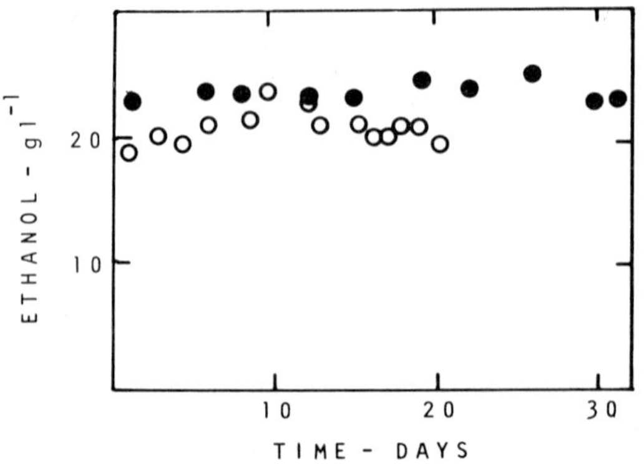

FIGURE 5. Ethanol production from commercial whey UF-permeate (5% wt/vol lactose) with immobilized *K. fragilis* (●) and with immobilized *S. cerevisiae* and *A. niger* β-galactosidase (○) (τ 4 to 5 hr).

calcium-alginate-gel-immobilized living *S. cerevisiae* yeast and phenol formaldehyde resin immobilized *Aspergillus niger* β-galactosidase enzyme (FIG. 5). Pilot plant tests with a 50-liter column reactor are currently being performed in cooperation with Valio Laboratory, employing immobilized *K. fragilis* yeast. In large-scale experiments with substrates such as whey UF-permeate contamination of the bioreactor may be a problem. Consequently, the effect of decreased pH on ethanol fermentation by *K. fragilis* was investigated. FIGURE 6 illustrates that lowering the pH from 4.5 to 3.0 did not significantly affect ethanol production.

It has already been mentioned that *Zymomonas mobilis* bacterium has recently gained a lot of attention because of its rapid ethanol production rate. We have recently reported that *Z. mobilis* cells entrapped in κ-carrageenan and locust-bean-gum mixed-gel matrix of good mechanical characteristics can be successfully used for ethanol fermentation in laboratory scale.[34] However, the very rapid carbon dioxide evolution, cell growth, and poor biocatalyst stability at high production rates present problems that need to be solved before scale-up. FIGURE 7 shows that when a conical column reactor was used, the reactor could be operated continuously for at least 40 days without significant technical difficulties, whereas an equivalent tubular column reactor could be operated for only 10 days with difficulty. All subsequent experiments

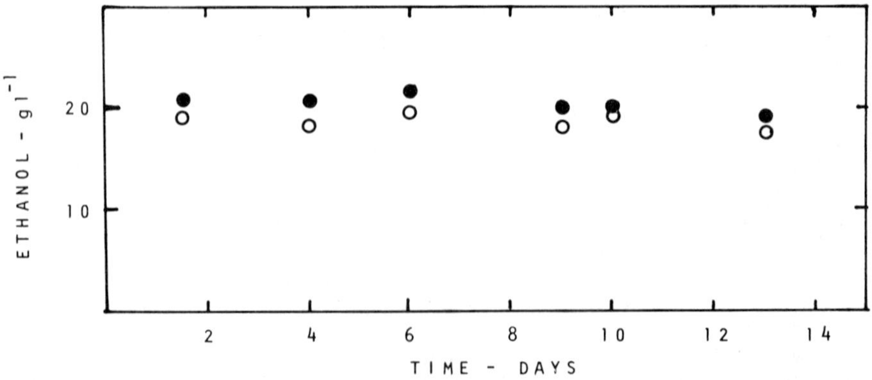

FIGURE 6. The effect of pH on ethanol production from whey UF-permeate (5% wt/vol lactose) with immobilized *K. fragilis* (●, pH 4.5; ○, pH 3.0).

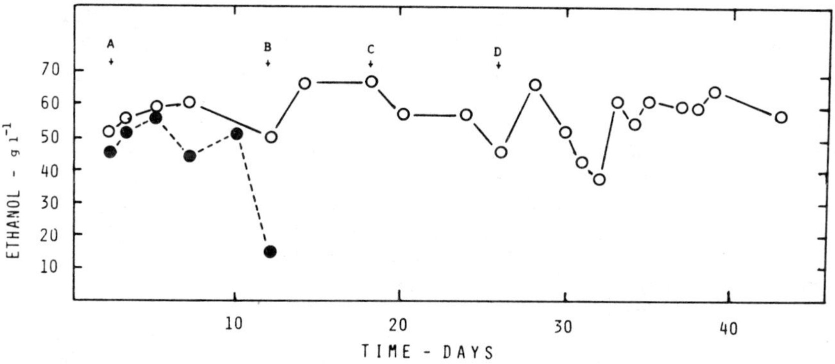

FIGURE 7. Continuous ethanol production from glucose (15% wt/vol; 1% yeast extract, 0.1% KH_2PO_4, 0.1% $(NH_4)_2PO_4$, 0.05% $MgSO_4 \cdot 7H_2O$) with immobilized *Z. mobilis* in a conical (○) and a tubular (●) column reactor 25°C, pH 5.0; τ A ~ 1 hr, B ~ 2 hr, C ~ 0.7 hr, D ~ 0.5 hr).

have been performed with a conical reactor. Productivities of at least 120 g l^{-1} hr^{-1} have been obtained. Margaritis *et al.*[24] employed a horizontal, slightly inclined, packed-bed reactor with a fine stainless steel grid along the length of the reactor to facilitate carbon dioxide removal. In such a reactor, the liquid-phase volume was reported as only about 41% of the total bioreactor volume, and was recommended as the basis of residence time and volumetric productivity calculations. Maximum productivity of 102 g l^{-1} hr^{-1} for an inlet glucose concentration of 10% (wt/vol) at 87% conversion was given. Volumetric productivity values from about 40 to 200 g l^{-1} hr^{-1} have been reported for immobilized *Z. mobilis* reactors but the basis of calculation has not always been given, making comparisions difficult. Arcuri *et al.*[35] reported a maximum volumetric productivity of 132 g l^{-1} hr^{-1} with a fluidized-bed type of reactor involving polystyrene bead bound cells, but the high flow rate required to keep the bed at expanded state also resulted in only 0.66% effluent ethanol concentration from 10% glucose. Grote *et al.*[23] on the other hand, either immobilized

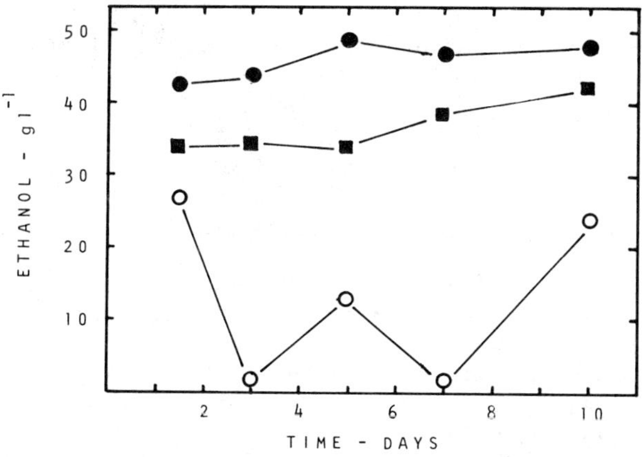

FIGURE 8. The effect of temperature (●, 25°C; ■, 30°C; ○, 37°C) on ethanol production from 10% (wt/vol) glucose with immobilized *Zymomonas mobilis* in a conical reactor.

Z. mobilis cells in calcium alginate gel that was precipitated directly into the reactor column as a fiber-like matrix or entrapped the cells in κ-carrageenan-gel-coated Rashing rings. Both biocatalysts behaved similarly, with maximum volumetric productivity of 53 g l^{-1} hr^{-1} obtained at about 1 hr residence time with the latter. However, it was also reported that a $\tau \sim 3$ hr was required for complete conversion of 15% (wt/vol) glucose. According to Amin and Verchatert,[36] the volumetric productivities obtained with carrageenan-gel-immobilized *Z. mobilis* were always higher than those obtained under similar conditions with *S. bayanus,* with the maximum of 58 g l^{-1} hr^{-1} from 17.5% (wt/vol) glucose at $\tau \sim 1.4$ hr (based on bed volume) when no sugar was found in the effluent. However, it was also obvious that the nutrient composition of the medium was not optimal.

We found maximum ethanol production with immobilized *Z. mobilis* at about 26°C (FIG. 8). An increase in temperature to 37°C was strongly inhibitory. In typical experiments, others have employed reactor temperatures of 24° to 30°C, but in all cases the temperature has not been given making comparisions difficult. The effect of

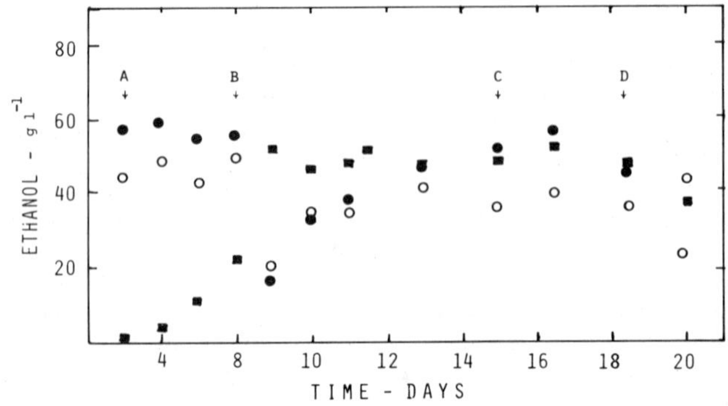

FIGURE 9. The effect of initial glucose concentration (●, 10%; ○, 15%; ■, 20%, wt/vol) on ethanol production with immobilized *Zymomonas mobilis* in a conical reactor (τ A ~ 2 hr, B ~ 1 hr, C ~ 0.7 hr, D ~ 0.5 hr).

initial glucose concentration is shown in FIGURE 9. One can see that the organism is slowly able to adapt to 20% (wt/vol) glucose.

CONCLUSIONS

It has been shown that heterogeneous biocatalysis principles involving immobilized living yeasts such as *Saccharomyces cerevisiae* or *Kluyveromyces fragilis* yeasts and bacteria such as *Zymomonas mobilis* can be applied in continuous ethanol fermentation employing packed-bed or fluidized-bed tubular or conical reactors. The process tested with yeast using a 50-liter pilot reactor can be modified to suit both alcoholic beverage and potable or industrial ethanol manufacture with considerable economic savings both in capital and processing costs. Large-scale applications involving *Z. mobilis* still require that a number of technical problems be solved. Depending on the organism, substrates may include very dilute waste streams such as sulfite spent liquor and cellulose hydrolysates, whey ultrafiltrate permeate, cane molasses, starchy materials pretreated with HTST-extrusion cooking both in the presence and absence of thermostable α-amylase, and others.

ACKNOWLEDGMENTS

The authors gratefully acknowledge the financial support of the Ministry of Trade and Industry (Finland) and the Neste Oy Foundation.

REFERENCES

1. Kosaric, N., D. C. M. Ng, I. Russell & G. S. Stewart. 1980. Ethanol production by fermentation: An alternative liquid fuel. *In* Advances in Applied Microbiology. D. Perlman, Ed. Vol. **26:** 147–227. Academic Press. New York.
2. Cysewski, G. R. & R. Wilke. 1978. Process design and economic studies of alternative fermentation methods for the production of ethanol. Biotechnol. Bioeng. **20:** 1421–1444.
3. Rosen, K. 1978. Continuous production of alcohol. Process Biochem. **13**(5): 25–26.
4. Hough, J. S. & A. H. Button. 1972. Continuous Brewing. *In* Progress in Industrial Microbiology. D. J. D. Hockenhull, Ed. Vol. **11:** 89–132. Churchill Livingstone. London.
5. Englebart, W. & H. Dellweg. 1976. Continuous alcoholic fermentation by circulating agglomerated yeast. *In* Abstracts of Papers. H. Dellweg, Ed. 5th Intl. Ferment. Symp. June 28–July 3, 1976. Berlin.
6. Wiesenberger, A. 1978. Wine making goes continuous. Food Eng. Intl. **3**(2): 26–28.
7. Anonymous. 1979. Researchers foment better ways to ferment. Chem. Week. **125**(6): 42–43.
8. Linko, P. 1980. Immobilized biological systems for continuous fermentation. *In* Food Process Engineering, Vol. 2, Enzyme Engineering in Food Processing. P. Linko & J. Larinkari, Eds.: 27–39. Appl. Science Publ. London.
9. Linko, P. 1981. Immobilized live cells. *In* Advances in Biotechnology. M. Moo-Young, C. W. Robinson & C. Vezzina, Eds. Vol. **1:** 711–716. Pergamon Press. New York.
10. Kolot, F. B. 1980. New trend in yeast technology—immobilized cells. Process Biochem. **15**(7): 2–8.
11. Bu'Lock, J. D. & D. M. Comberbach. 1981. A practical system for high-productivity ethanol fermentation. Paper presented at the 2nd Eur. Congr. of Biotechnol. April 5–10, 1981. Eastbourne, U.K. Abstracts of Papers p. 204.
12. Rhigelato, R. C. 1980. Anaerobic fermentation: Alcohol production. Philos. Trans. R. Soc. London Ser. B **290:** 303–312.
13. Linko, P. & Y.-Y. Linko. 1983. Applications of immobilized microbial cells. *In* Applied Biochem. Bioeng. I. Chibata & L. B. Wingard, Jr, Eds. Vol. 4. In press.
14. Linko, Y.-Y., L. Weckström & P. Linko. 1980. Sucrose inversion by immobilized *Saccharomyces cerevisiae* yeast cells. *In* Food Process Engineering, Vol. 2, Enzyme Engineering in Food Processing. P. Linko & J. Larinkari, Eds. Appl. Sci. Publ. London. pp. 81–91.
15. Linko, Y.-Y. & P. Linko. 1981. Continuous ethanol production by immobilized yeast reactor. Biotechnol. Lett. **3:** 21–26.
16. Veliky, I. A. & R. E. Williams. 1981. The production of ethanol by *Saccharomyces cerevisiae* immobilized in polycation-stabilized calcium alginate gels. Biotechnol. Lett. **3:** 275–280.
17. Sitton, O. C. & J. L. Gaddy. 1981. Ethanol production in an immobilized cell reactor. Biotechnol. Bioeng. **22:** 1735–1748.
18. Wada, M., J. Kato & I. Chibata. 1980. Continuous production of ethanol using immobilized growing yeast cells. Eur. J. Appl. Microbiol. Biotechnol. **10:** 275–287.
19. Wang, H. Y. & D. J. Hettwer. 1982. Cell immobilization in κ-carrageenan with tricalcium phosphate. Biotechnol. Bioeng. **24:** 1827–1838.

20. Daugulis, A. J., N. M. Brown, W. R. Cluett & D. B. Dunlop. 1981. Production of ethanol by adsorbed yeast cells. Biotechnol. Lett. **3:** 651–656.
21. Ghose, T. K. & K. K. Bandyopadhyay. 1980. Rapid ethanol fermentation in immobilized yeast cell reactor. Biotechnol. Bioeng. **22:** 1489–1496.
22. Moo-Young, M., J. Lamptey & C. W. Robinson. 1980. Immobilization of yeast cells on various supports for ethanol production. Biotechnol. Lett. **2:** 541–548.
23. Grote, W., K. J. Lee & P. L. Rogers. 1980. Continuous ethanol production by immobilized cells of *Zymomonas mobilis*. Biotechnol. Lett. **2:** 481–486.
24. Margaritis, A. P. K. Bajpai & J. B. Wallace. 1981. High ethanol productivities using small Ca-alginate beads of immobilized cells of *Zymomonas mobilis*. Biotechnol. Lett. **3:** 613–618.
25. Hartmeier, W. & I. Mücke. 1982. Basic trials to coimmobilize living yeast cells and glucoamylase for beer wort fermentation. Paper presented at Intl. Symp. on the Use of Enzymes in Food Technology. May 5–7, 1982. Versailles, France.
26. Maleszka, R., I. A. Veliky & H. Schneider. 1981. Enhanced rate of ethanol production from D-xylose using recycled or immobilized cells of *Pachysolen tannophilus*. Biotechnol. Lett. **3:** 415–420.
27. Gong, C. S., L. F. Chen, M. C. Flickinger & G. T. Tsao. 1981. Production of ethanol by yeast using xylulose. PCT Int'l. Appl. WO 81/03032. October 29, 1981.
28. Chiang, L. C., H. Y. Chiao, M. C. Flickinger, L. F. Chen & G. T. Tsao. 1982. Ethanol production from pentoses by immobilized microorganisms. Enzyme Microb. Technol. **4:** 93–98.
29. Linko, P. 1982. HTST-(High-Temperature-Short-Time) Extruder als biochemischer Reaktor. Getreide, Mehl Brot. **36:** 326–332.
30. Linko, P., Y.-Y. Linko & J. Olkku. 1983. Extrusion cooking and bioconversions. J. Food Eng. **2:** In press.
31. Linko, Y.-Y., A. Lindroos & P. Linko. 1979. Soluble and immobilized enzyme technology in bioconversion of barley starch. Enzyme Microb. Technol. **1:** 273–278.
32. Korn, S. R. & J. M. Harper. 1982. Extrusion of corn for ethanol fermentation. Biotechnol. Lett. **4:** 417–422.
33. Linko, Y.-Y., H. Jalanka & P. Linko. 1981. Ethanol production from whey with immobilized living yeast. Biotechnol. Lett. **3:** 263–268.
34. Linko, P. & Y.-Y. Linko. 1981. Immobilized microbial cells for ethanol and other applications. Paper No. 69a presented at 74th Annual AIChE Meeting. November 8–12, 1981. New Orleans, LA.
35. Arcuri, E. J., R. M. Worden & S. E. Shumate II. Ethanol production by immobilized cells of *Zymomonas mobilis*. Biotechnol. Lett. **2:** 499–504.
36. Amin, G. & H. Verachtert. 1982. Comparative study of ethanol production by immobilized-cell systems using *Zymomonas mobilis* or *Saccharomyces bayanus*. Eur. J. Appl. Microbiol. Biotechnol. **14:** 59–63.

Production of Ethanol from Biomass

E. C. CLAUSEN AND J. L. GADDY

Chemical Engineering Department
University of Arkansas
Fayetteville, Arkansas 72701

INTRODUCTION

Renewable alternatives to petroleum as a source of fuels and chemicals have been under intensive investigation for the past ten years. Biomass has been shown to have considerable promise as a raw material for gaseous fuels, liquid fuels, and certain petrochemical intermediates. It has been reported that biomass produced through photosynthesis, and available worldwide for processing, totals 2.5×10^{10} metric tons annually.[1] This resource converted into useful energy at a 65% efficiency, could supply all of the world's present energy needs.

The most readily available source of biomass is in the form of agricultural and municipal residues. The collectible agricultural residues have been estimated to total 300 million tons annually.[2] Corn stover is the single most abundant residue available, with 150 million tons produced each year.[3] This quantity of corn stover could furnish 10% of the U.S. liquid fuel requirements, or 50% of the nation's petrochemical needs. All of our petrochemical production could be supplied from the available agricultural residues.

The major components of cellulosic biomass are hemicellulose, cellulose, and lignin. The composition of the various lignocellulosic residues varies; however, most materials contain 15–25% hemicellulose, 30–45% cellulose, and 5–15% lignin. A recent analysis of corn stover showed: hemicellulose—27%; cellulose—43%; lignin—8%; and ash—2%.[4] Therefore, 60–70% of the biomass is available as carbohydrate and can be converted into fuels and chemicals by biochemical methods.

The hemicellulose of biomass contains varying quantities of glucan and xylan. Paper hemicellulose, for example, contains about 50% glucan and xylan. Corn stover hemicellulose contains 75–85% xylan. The cellulose fraction is almost entirely glucan. The constituents exist as polymers, which must be reduced to monomeric sugars before they can be converted biologically into useful chemicals. Therefore, the biological conversion of biomass is actually a two-stage process: hydrolysis followed by fermentation.

The carbohydrate hydrolysis can be carried out by contact with cellulase or xylanase enzymes, or by treatment with mineral acids. Enzymatic hydrolysis has the advantage of operating at mild conditions and producing a high-quality sugar product. However, the enzymatic reactions are quite slow, and the biomass must be pretreated with caustic or acid to improve the yields and kinetics. Acid hydrolysis is a much more rapid reaction, but requires higher temperatures or high acid concentrations to achieve good yields. Furthermore, under these conditions, xylose degrades to furfural and glucose degrades to 5-hydroxymethyl furfural (HMF), both of which are toxic to microorganisms. A two-stage acid hydrolysis process enables the hemicellulose and cellulose to be degraded separately under conditions appropriate for each reaction. High yields are obtained and mild conditions are required, with very little resultant degradation. Large quantities of acid are necessary, however, and acid recovery is an important part of this process.

The purpose of this paper is to describe a process for producing ethanol from corn stover. Data for the two-stage acid hydrolysis of corn stover are presented along with the fermentation results of the hydrolyzate to alcohol. The economics of this process are determined, and areas for potential improvements in the process are examined.

FEEDSTOCK PREPARATION

In order to speed up the hydrolysis reactions, the size of the biomass particles must be reduced to increase the accessibility to the polymeric structure. A high solids concentration is desirable since this concentration controls the sugar concentration and the size of the hydrolysis and fermentation equipment. The size of the particles also affects the fluidity of the solids/acid slurry. It is desirable to maintain fluidity of the slurry to promote mass transfer and to facilitate pumping and mixing. Therefore, the particle size is an important variable in the biomass conversion process.

TABLE 1 gives the maximum solids concentration to maintain fluidity of the slurry, as a function of particle size. A maximum concentration of about 10% is possible with particle sizes less than 40 mesh. Grinding to 20 mesh gives a particle size distribution in which 90% of the material is less than 40 mesh. Therefore, grinding the corn stover to pass 20 mesh gives the appropriate size and produces the maximum possible slurry concentration. Laboratory studies also show that grinding to smaller sizes does not improve the reaction rates.

TABLE 1. Maximum Solids Concentration for Fluid Slurry

Mesh Range (Sieve #'s)	Solids Conc. (wt. %)
0–1	4.6
12–20	4.6
20–30	8.4
30–40	8.2
40–45	10.2
45–70	10.4
70–100	10.6
100+	10.8

ACID HYDROLYSIS

The two-stage acid hydrolysis process involves a prehydrolysis to remove the hemicellulose fraction, followed by a hydrolysis reaction to break down the cellulose.

Prehydrolysis

The two major factors that control the prehydrolysis reaction are temperature and acid concentration. Data from the University of Arkansas laboratories showing the effect of these variables on the prehydrolysis of corn stover are presented in FIGURE 1. Ten percent slurry concentrations were maintained. Complete xylan conversion can be

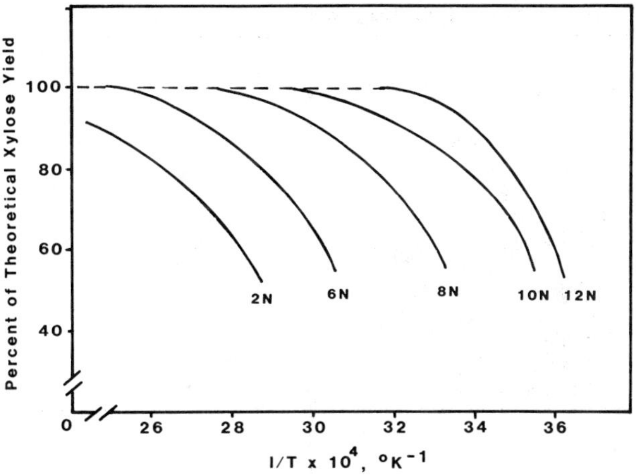

FIGURE 1. Effect of acid concentration and temperature on prehydrolysis.

achieved at a low acid concentration (2 N) at temperatures of about 100°C. The use of higher acid concentrations (10 N) results in complete conversion at about 60°C. For a given acid concentration, the conversion would be reduced at high temperature; therefore, the curves would all go to zero at the origin.

In deciding upon which conditions to use for acid hydrolysis, several factors must be considered. The acid cost would be prohibitively high, even at low concentrations (2 N), unless the acid was recovered for reuse. The degradation of xylose to furfural is more readily promoted by high temperature than by high acid concentration (in this range). Therefore, the preferred operating range is a moderate acid concentration and mild temperature.

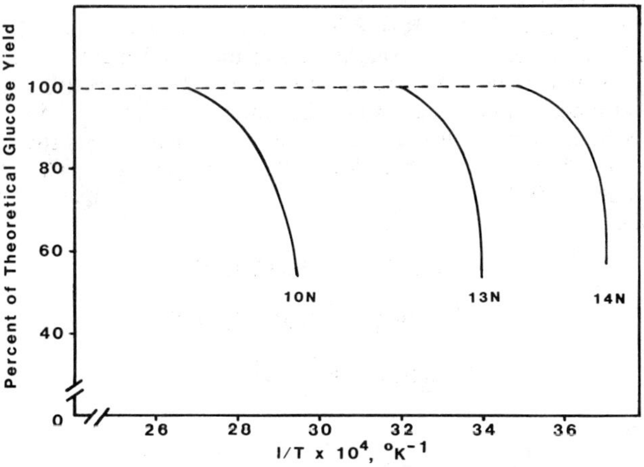

FIGURE 2. Effect of acid concentration and temperature on hydrolysis.

TABLE 2. Corn Stover Acid Hydrolyzates

	Corn Stover
Prehydrolyzate	
Xylose, g/l	22.0
g/100g	20.0
Glucose, g/l	6.0
g/100g	5.0
Hydrolyzate	
Xylose, g/l	0
g/100g	0
Glucose, g/l	40.0
g/100g	30.0
Combined	
Xylose, g/100g	20.0
Glucose, g/100g	35.0

The hydrolysis of cellulose to glucose is also highly dependent upon acid concentration and temperature. FIGURE 2 shows the effect of these variables on the hydrolysis of the corn stover residue from prehydrolysis. As noted, an acid concentration of 14 N gives total conversion at room temperature. High temperatures are required to give suitable conversions at low acid concentrations (3 N or less). These high temperatures also promote glucose degradation and repolymerization.

The sugar concentrations and yields from a typical prehydrolysis and hydrolysis stagewise process are given in TABLE 2. Very dilute (2–4%) sugar concentrations result from these reactions. The prehydrolysis step yields 20% of the initial stover as xylose. The combined yield of glucose is 36%. These yields represent nearly complete conversion of hemicellulose and cellulose to sugars.

Data for the reaction rate as a function of time have been compiled for both the prehydrolysis and hydrolysis of a number of substrates. Plots of these data show first-order kinetics with respect to carbohydrate concentration at constant acid concentration. Using these expressions, a relationship can be developed for the dependency of reaction rate on acid concentration.

FIGURE 3 is a plot of ln k, where k is the first-order rate constant, (min^{-1}) as a function of ln C_A, acid normality. Straight lines result, which dictate an equation of the form: $k = aC_A^b$, where a and b are the intercept and slope, respectively. Prehydrolysis data are given for peat, paper, and corn stover, and hydrolysis data are shown for corn stover. As expected, the prehydrolysis of corn stover is faster than paper and peat, and the hydrolysis reaction is somewhat slower than prehydrolysis. The following rate expressions result at 100°C:

Prehydrolysis

$$\text{Peat } k = .007\, C_A^{1.38}$$

$$\text{Paper } k = .013\, C_A^{.81}$$

$$\text{Corn Stover } k = .020\, C_A^{.67}$$

Hydrolysis

$$\text{Corn Stover } k = .001\, C_A^{.51}$$

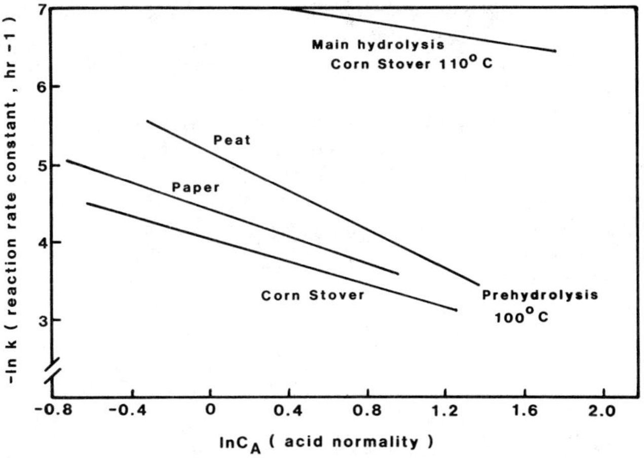

FIGURE 3. Kinetics of biomass hydrolysis.

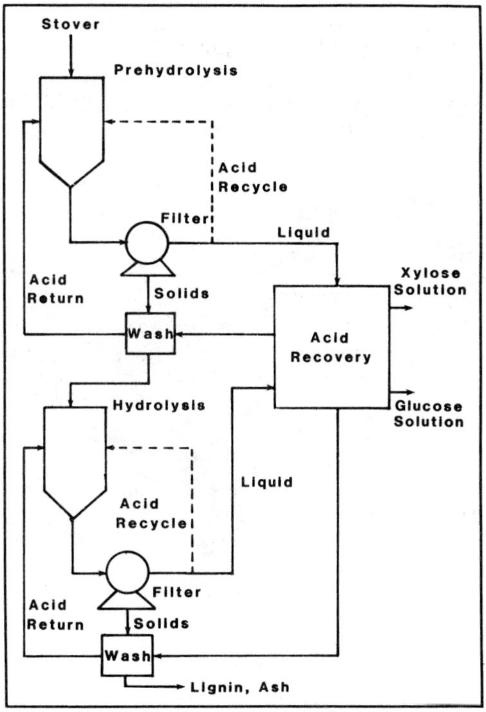

FIGURE 4. Schematic of acid hydrolysis process.

The rate coefficients can be used to design reactors to carry out these reactions. Reaction times of about thirty minutes are required to achieve 90% conversion in a series of two reactors each for prehydrolysis and hydrolysis at acid concentrations of 8–14 N with 10% solids feed.

The proposed process for stagewise acid hydrolysis of corn stover is shown schematically in FIGURE 4. Corn stover is fed continuously to the prehydrolysis reactor. Residual solids are separated by filtration, washed with fresh acid, and fed to the hydrolysis reactor. Solids from the hydrolysis reactor are filtered, washed, and discarded. Acid and sugars are separated and the acid is returned to the reactors. Two dilute sugar streams result: the prehydrolyzate, containing primarily xylose, and the hydrolyzate, containing only glucose. These sugars can then be fermented to ethanol or other chemicals.

Sugar Decomposition

The fermentability of the sugars is dependent upon the sugar decomposition that occurs during hydrolysis. Xylose decomposes to furfural and hexoses decompose to HMF, which are both toxic to yeast. Tolerance can often be developed, and toxicity is difficult to define. However, the toxic limit of furfural on alcohol yeast is reported to be .03–.046%.[5] HMF is reported to inhibit yeast growth at .5%, and alcohol production is inhibited at .2%.[6]

The rate of decomposition of xylose to furfural and hexoses to HMF were studied at varying sugar concentrations. Using the method of initial rates, these reactions were found to be first order. The ratio of rate constants for decomposition to formation are given in TABLE 3. These ratios appear to be small, and subsequent calculations and experiments show that the rate of HMF appearance is significant. However, the rate of furfural appearance can reach toxic limits, especially if acid recycle is utilized.

ETHANOL FERMENTATION

The acid hydrolysis of cellulosic residues, if performed at high temperature, can produce a sugar/degradation product mixture that is difficult, if not impossible, to ferment. However, if the hydrolysis conditions are mild, only small quantities of toxic substances are observed, and normal fermentation is expected. FIGURE 5 shows the comparison of a batch fermentation with *Saccharomyces cerevisiae* of both a synthetic

TABLE 3. Ratio of First-order Rate Constants for Sugar Decomposition to Formation under Prehydrolysis Conditions

Sugar	Acid Concentration	Rate of Decomposition/ Formation
Glucose	2 N	0.0053
	3 N	0.0090
	4 N	0.0074
Xylose	2 N	0.0257
	3 N	0.0402
	4 N	0.0374

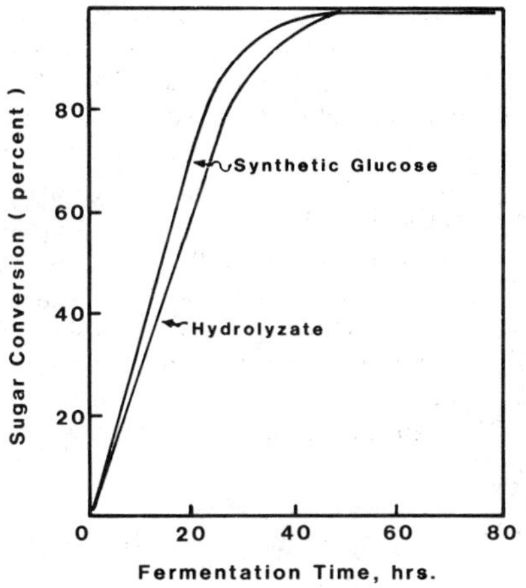

FIGURE 5. Comparison of fermentation of hydrolyzate and synthetic glucose.

glucose and corn stover hydrolyzate. Although these results are quite encouraging, some inhibition results. Fifty percent conversion of glucose is achieved after 18 hours for the synthetic medium and requires 20 hours for the hydrolyzate. The hydrolyzate is 90% converted after about 30 hours, whereas only 25 hours are required for 90% conversion of pure glucose. Acclimation is achieved rather quickly, and both substrates are 98% converted in 48 hours.

The concentration of furfural and HMF are both below .05% in the hydrolyzate. These low concentrations could be the cause of the slower fermentation of the hydrolyzate, or some other substance might be the problem. Studies are continuing to examine the effect of resins, ions, and so forth on the fermentation. Efforts have not been made to develop or utilize resistant strains at this point, as the problem is not felt to be serious.

ECONOMIC PROJECTIONS

To illustrate the economics of this process, a design has been performed for a facility to convert corn stover into 20 million gallons per year of ethanol, utilizing the acid hydrolysis procedures previously described. The capital and operating costs are summarized in TABLE 4.

Corn stover, in large round bales, would be stored in the field and delivered to the plant site as needed. Feedstock preparation consists of shredding, grinding, and conveying to the reactors. The hydrolysis section, as shown in FIGURE 4, consists of continuous reactors. Acid-resistant materials of construction are necessary for this equipment. Acid recovery is accomplished by electrodialysis; although other methods, such as evaporation, are equally applicable and cost about the same. The typical batch fermentation and ethanol distillation units are included. The total capital cost for this plant is $42.2 million, including all utilities, storage, and offsites.

The annual operating costs are also shown in TABLE 4. These costs are also given on the basis of unit production of alcohol. Corn stover is estimated to cost $20 per ton. The energy costs are substantial because of the dilute sugar and alcohol solutions resulting from the hydrolysis. A lignin boiler is used to reduce the energy requirements, and energy costs are \$.22 per gallon. Fixed charges are computed as a percentage of the capital investment and total 42% or \$.87 per gallon. The total cost, including profit, to produce alcohol by this process is \$1.85 per gallon.

Fuel-grade ethanol has sold for \$1.85 per gallon in the past, although the price has declined in recent months. It is highly unlikely that a pretax profit of 25% would be acceptable in today's economy, and that this process would be economically feasible. It should be noted that this process does not include utilization of the pentose stream. Acid recovery is included, but fermentation of the pentoses is not provided. Xylose could be fermented to alcohol, acids, or other valuable chemicals, which would improve the economics. However, since this technology is not perfected, such products have not been included.

In examining these economic figures to determine where to improve the process, it

TABLE 4. Economics of 20 Million Gallon per Year Ethanol Facility

A. Capital Cost

	Million \$
Feedstock preparation	2.1
Hydrolysis	8.0
Acid recovery	12.4
Fermentation & purification	8.1
Utilities/offsites	11.6
	\$42.2

B. Operating Cost

	Million \$/yr	\$/gal
Corn stover (\$20/ton)	6.7	.33
Utilities	4.3	.22
Chemicals	5.6	.28
Labor	3.0	.15
Fixed charges	17.4	.87
Maintenance (5%)	2.1	
Depreciation (10%)	4.2	
Taxes & insurance (2%)	.8	
Profit (25%)	10.3	
	\$37.0	\$1.85/gal

is evident that yields cannot be substantially improved, nor raw material costs reduced. It is significant to note that the raw material costs are quite important and that yields of only 50% (typical for dilute acid processes) would add significantly to the ethanol cost.

Other significant cost areas are utilities, chemicals (including acid and denaturants), and fixed charges. Therefore, if economic improvements in this process are possible, reductions in the energy and capital costs must be sought.

PROCESS IMPROVEMENTS

Perhaps the single most detrimental factor to the economics of this process is the very dilute solutions that result from acid hydrolysis. Dilute concentrations increase both the equipment size and the energy required for purification. Another pernicious economic factor in this, and other fermentation processes, is the use of large conventional batch fermenters. Methods for overcoming these problems will be considered next.

Solids Concentration

The ultimate sugar and alcohol concentrations are direct functions of the initial solids concentration in the hydrolysis. Since fluidity in a stirred reactor is a requirement, a 10% mixture has been considered maximum. Therefore, the resultant sugar concentrations have been only 2–4% and alcohol concentrations only half as much.

If the limiting factor is considered to be fluidity in the *reactor,* instead of the feed mixture, the feed concentration could be increased by roughly the reciprocal of one minus the solids conversion in the reactor. Of course, solids and liquid would have to be fed separately, which would also save equipment cost. For corn stover containing 30% hemicellulose and 35% cellulose, the reactor sizes could be reduced by 40% and 50% for prehydrolysis and hydrolysis, respectively. Attendant reductions would also result in the filtration and washing units.

Equally important are the resultant increases in sugar concentrations. The xylose concentration from the prehydrolysis reactor would be increased from 2.2% to about 3%. The glucose concentration in the hydrolyzate would be doubled to about 8%. Energy and equipment costs in the fermentation area would be reduced proportionately.

This simple alteration in the process has a profound impact on the economics. It is estimated that the capital cost would be reduced by 33% in the hydrolysis and acid recovery sections and 50% in the fermentation and utilities areas. Furthermore, the energy requirements for distillation are reduced by 50%.

Acid Recycle

Another method for increasing the sugar concentration is to recycle a portion of the filtrate (acid and sugar solution) in each hydrolysis step. The acid would catalyze further polysaccharide hydrolyses to increase the sugar concentration. Of course, recycle of the sugars increases the degradation to furfural and HMF.

Experiments have been conducted in our laboratories to determine the enhancement possible with acid recycle. Various amounts of the acid and sugar solution from the filtration were recycled to determine the resulting sugar and by-product concentrations. Acid and solids concentrations and temperatures were kept constant.

FIGURE 6 gives the sugar concentration as a function of the quantity of filtrate recycled for the prehydrolysis reactor. As noted, the xylose concentration is increased nearly fourfold and the glucose concentration is increased nearly sixfold at total recycle. Final concentrations of 8.7 and 2.8% are obtained for xylose and glucose, respectively. It should be noted that not all the filtrate can be recycled, since a portion adheres to the solids in filtration. Therefore, the yield of xylose begins to decrease due to losses with the filter cake and due to decomposition.

As expected, the concentration of by-product furfural and HMF increase as a function of the quantity of filtrate recycled. With total recycle, the composition of these by-products reaches concentrations of .2 and .17% for furfural and HMF, respectively. These values are above the toxic limit of furfural on most microorganisms;

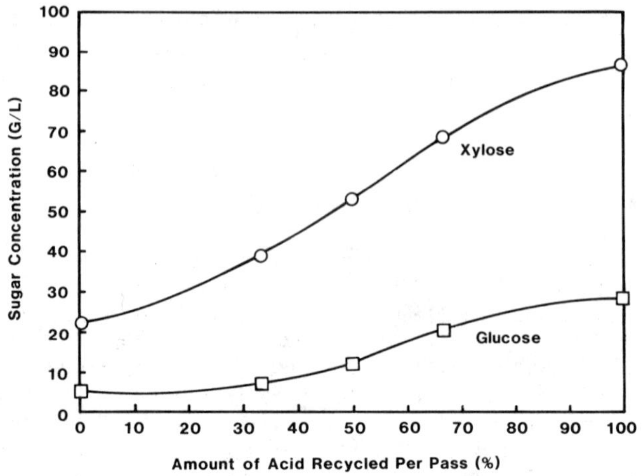

FIGURE 6. Sugar concentration with various amounts of recycle.

therefore, high recycle rates would not be possible. The xylose yield is maximized with a 50% recycle and the furfural level is not excessive in this range.

With 50% recycle, a steady-state concentration of 6% xylose is achieved, or about 2.7 times higher than with no recycle. Similar results were obtained for 50% acid recycle in the hydrolysis step. No furfural is present in this stream, since xylose is not present in the hydrolyzate.

The effect of acid recycle on the economics is significant. A recycle rate of 50% in both prehydrolysis and hydrolysis reactors, coupled with high solids concentrations, would result in a xylose concentration of 12% in the prehydrolyzate and a glucose concentration of nearly 25% in the hydrozylate. It should be noted that these concentrations have been exceeded in the laboratory, while maintaining furfural and HMF less than .05%. These high concentrations reduce the equipment size in the hydrolysis, acid recovery, and fermentation sections by another 60%. Energy consumption is also reduced another 40%.

Continuous Fermentation

One serious problem with the large-scale production of industrial chemicals by fermentation is the high cost of providing for batch fermenters and insuring sterile conditions. Most industrial fermentations, including ethanol for fuel usage, are conducted in batch equipment so that contamination and mutation can be controlled. Sterilization between batches and the use of a fresh inoculum insure efficient fermentation. However, most batch alcohol fermentations are designed for thirty hours (or more) reaction time, which results in very large and expensive reactors.

The reactor size can be reduced substantially by using continuous-flow fermenters. When fermenting acid hydrolyzates, the problems with maintaining sterile conditions are substantially reduced. Therefore, the use of continuous fermentation is a natural application for producing alcohol from corn stover hydrolyzate.

A number of continuous fermentation schemes are being studied, including the CSTR,[7] cell recycle reactor,[8] flash fermentation,[9] and immobilized cell reactors.[10]

Typical laboratory data for the fermentation of a synthetic corn stover hydrolyzate in a CSTR are shown in FIGURE 7. A sugar concentration of 3% is utilized, and a maximum alcohol concentration of 1.4% is achieved. The maximum productivity of 1.7 gm/1-hr occurs at a dilution rate of .2 hr^{-1}, or a 5-hour retention time.

FIGURE 8 shows the results of fermentation in the University of Arkansas laboratories of the same hydrolyzate in a 36-inch column with immobilized *Saccharomyces cerevisiae* (ATCC 25858). The column was packed with ¼-inch Raschig rings, coated with gelatin and cross-linked with glutaraldehyde. Dilution rates as high as 1.75 hr^{-1} were attained in the ICR, or 6 times the washout rate in the CSTR. Alcohol productivities of 15.9 gm/l-hr were achieved, or about 10 times the CSTR. Furthermore, alcohol inhibition and furfural toxicity were greatly reduced in the ICR.

Compared to a batch fermentation time of 30 hours, the ICR could perform the same fermentation in 2 hours with 93% less reactor volume. An estimated savings of 60% of the capital cost for batch fermenters should result when using an ICR.

ECONOMIC IMPACT OF PROCESS IMPROVEMENTS

It is instructive to determine the consequences on the economics of incorporating the above process modifications. The process, using concentrated feed, acid recycle, and continuous fermentation results in capital costs as given in TABLE 5. The plant investment is reduced by nearly 50%. Adjusted operating costs are also shown. On the basis of an ethanol price of $1.85 per gallon, a pretax profit of 70% results. Without interest or profit, the cost of producing alcohol is $1.08 per gallon. Sufficient margin should remain for commercialization.

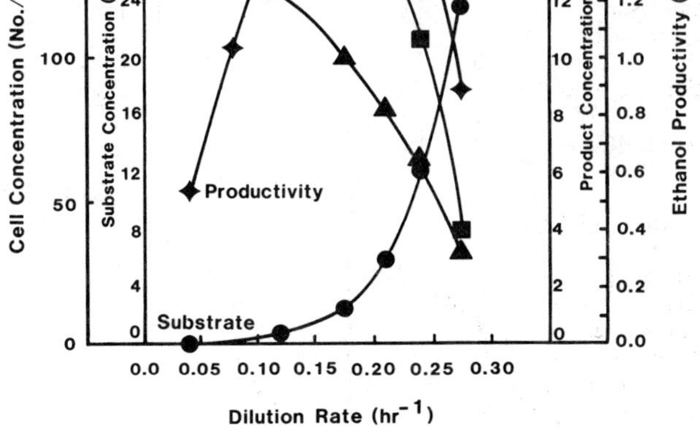

FIGURE 7. Hydrolyzate fermentation in a CSTR.

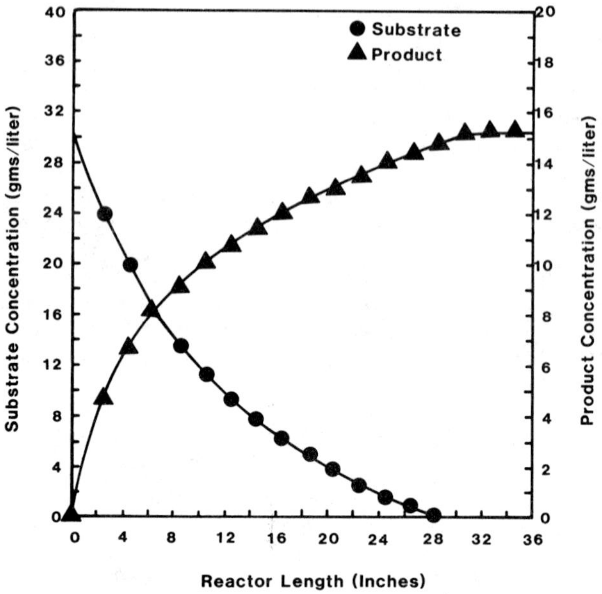

FIGURE 8. Hydrolyzate fermentation in an ICR.

TABLE 5. Improved Economics of 20 Million Gallon per Year Ethanol Facility

A. Capital Cost

	Million $
Feedstock preparation	2.1
Hydrolysis	4.0
Acid recovery	6.2
Fermentation & purification	4.1
Utilities/offsites	5.8
	22.2

B. Operating Cost

	Million $/yr	$/gal
Corn stover ($20/ton)	6.7	.33
Utilities	2.3	.12
Chemicals	5.1	.25
Labor	3.0	.15
Fixed charges	19.9	1.00
Maintenance (5%)	1.1	
Depreciation (10%)	2.2	
Taxes & insurance (2%)	.4	
Profit (25%)	16.2	
	$37.0	$1.85/gal

CONCLUSIONS

The two-stage acid hydrolysis of residues, such as corn stover, require mild temperatures but large quantities and high concentrations of acid. The resulting hydrolyzates, containing primarily xylose and glucose, can be fermented to ethanol or other chemicals. The economics of this process are marginal in the present economy.

The process economics can be improved by increasing the sugar concentration through high solids reactions and acid recycle, and by using continuous fermentation. These modifications result in reducing the capital cost by almost 50% and improving the profitability to about 35%, after taxes. It should be noted that these process modifications could be employed to improve the economics of similar processes to produce other products as well as ethanol.

SUMMARY

Lignocellulosic residues, such as corn stover, can be converted into ethanol and other chemicals by acid hydrolysis and fermentation. This paper presents laboratory data for the two-stage hydrolysis of corn stover, which results in nearly complete conversion of hemicellulose and cellulose to sugars with only moderate decomposition to furfural and HMF. Fermentation of these hydrolyzates can be accomplished with about the same efficiency as fermentation of synthetic glucose mixtures.

The economics of producing ethanol from corn stover with this process are marginal today. However, sugar concentrations can be increased sixfold by using high solids concentrations and acid recycle. Coupled with continuous fermentation, these process changes reduce the capital and energy costs by at least 50%. Under these conditions, the process appears to be commercially feasible.

REFERENCES

1. BLANCH, H. W. & C. R. WILKE. 1982. Rev. in Chem. Engr. **1:** 71.
2. FALKEHAG, I. Progress in Biomass Conversion. G. Saakamen & D. Tillman, Eds. Vol. **1**.
3. GREEN, F. L. 1975. Trans. Am. Nucl. Soc., **21:** 147.
4. Report prepared by Industrial Testing Laboratory, St. Louis, MO. 1982.
5. BANERJEE, N. & L. VISHWANATHAN. 1972. Proc. Annu. Sugar Conv. Tech. Conf. Calif. **61:** 1228f.
6. BANERJEE, N. & L. VISHWANATHAN. 1974. Proc. Annu. Sugar Conv. Tech. Conf. Calif. **82:** 123277W.
7. CYSEWSKI, G. R. & C. R. WILKE. 1978. Biotechnol. Bioeng. **20:** 1421.
8. ELIAS, S. 1979. Food Eng. Oct. p. 61.
9. CYSEWSKI, G. R. & C. R. WILKE. 1976. Lawrence Berkeley Laboratory Report 4480. March 1977. Biotechnol. Bioeng. **19:** 1125.
10. SITTON, O. C. & J. L. GADDY. 1980. Biotechnol. Bioeng. **22:** 1735.

Fluidized-Bed Bioreactors Using a Flocculating Strain of *Zymomonas mobilis* for Ethanol Production

CHARLES D. SCOTT

Chemical Technology Division
Oak Ridge National Laboratory[a]
Oak Ridge, Tennessee 37830

INTRODUCTION

Large-scale fermentation for ethanol production is conventionally carried out in batch, stirred-tank bioreactors containing active microorganisms (usually the yeast *Saccharomyces cerevisiae*). The microorganisms are maintained in suspension in an aqueous medium that is initially rich in the sugar (usually glucose) and finally contains primarily the product ethanol. More recently, such systems have been upgraded to allow continuous introduction of the feed material and withdrawal of the product. Such changes have resulted in ethanol production rates approaching 10 g/l · hr.[1] Additional modifications have been proposed to enhance productivity in stirred-tank bioreactors;[1,2] however, these systems have some inherent limitations that may be circumvented by using another class of bioreactors, namely, fluidized-bed columns composed of particulates containing immobilized cells.[3,4]

Recent research has also shown that microorganisms other than the *S. cerevisiae* may be more efficient for ethanol production, especially when the ethanol is to be used as a fuel. The bacterium *Zymomonas mobilis* appears to give a significantly higher production rate than the more common yeast.[5,6] A recently isolated strain of *Z. mobilis* is self-flocculating, yielding particles that have sufficient stability to be used as the active agent in a fluidized-bed bioreactor.[7] Preliminary results indicate that such a system may have extremely high productivity.

IMMOBILIZED MICROORGANISMS

Retained-cell bioreactors, such as fluidized beds, require that the active organism be immobilized into or onto a solid matrix that remains in the reactor. This allows very high levels of the microorganisms to be maintained even at flow rates that would ordinarily "wash out" organisms present in the form of a dispersed suspension. Although microorganisms have been immobilized by adsorption or chemical bonding on support surfaces[8] or by entrapment in a gel matrix,[9,10] some microbial strains also tend to self-aggregate, forming relatively stable floc particles. A recently isolated strain of the bacterium *Z. mobilis* (available as NRRL B-12526 from the parent strain ATCC 10988) has been shown to form stable floc particles in the size range of 0.5 to 1 mm (FIG. 1). These particles maintain their structural integrity even when subjected to

[a]Operated by Union Carbide Corporation under Contract W-7405-eng-26 with the U.S. Department of Energy.

the high shear forces generated in an active three-phase fluidized-bed bioreactor that is used to convert glucose to ethanol.

FLUIDIZED-BED BIOREACTOR

Description

Laboratory-scale, fluidized-bed bioreactors were utilized in this investigation. Most of the work was carried out in a tapered-colum bioreactor, in the shape of an

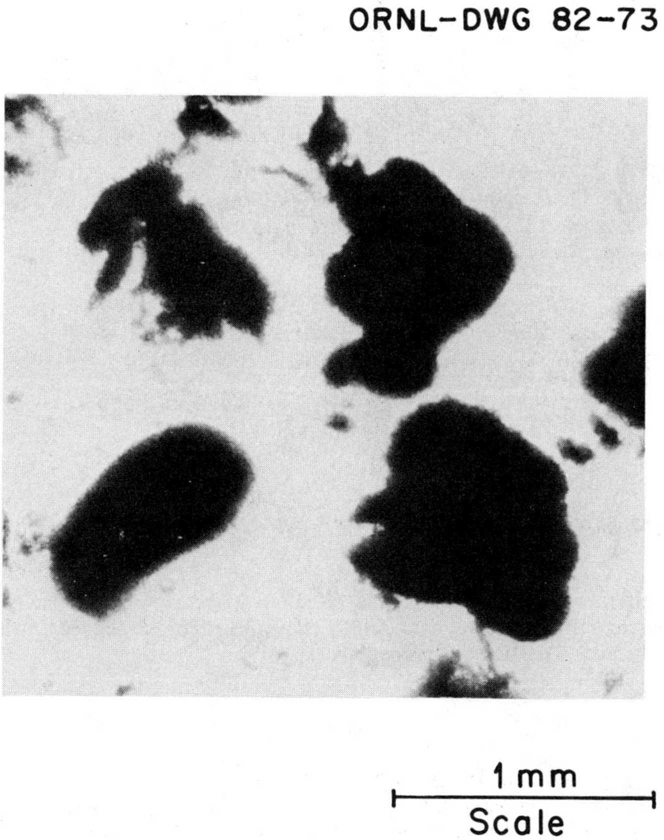

FIGURE 1. Stable *Zymomonas mobilis* floc particles formed in a fluidized-bed bioreactor.

inverted cone (FIG. 2), which was 100 cm long and had an inside diameter that ranged from 1.25 cm at the entrance to 3.81 cm at the widest part; its active volume was 690 ml. The temperature of the active portion of each bioreactor was controlled to ± 0.5°C via a heating jacket that contained a circulating fluid. The feed material could be introduced, or samples taken, at ports spaced along the column.

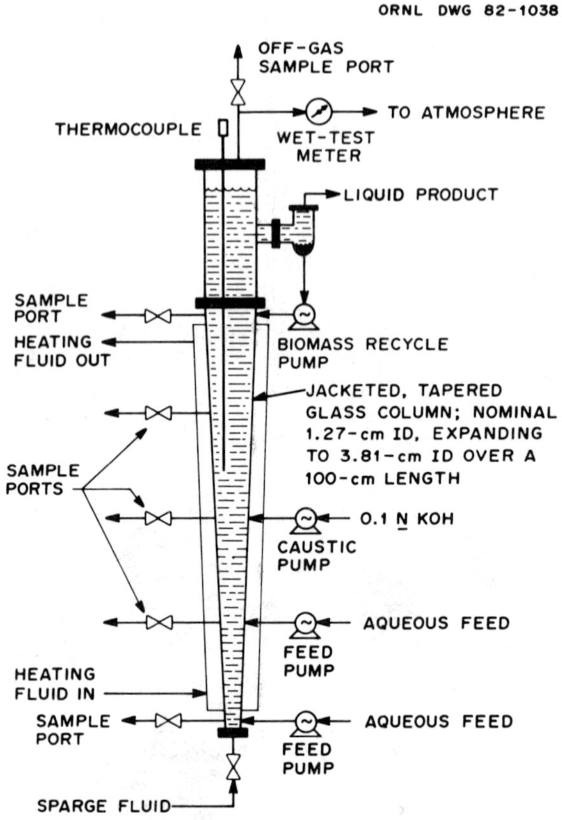

FIGURE 2. Typical tapered fluidized-bed bioreactor used during the laboratory investigation.

Although the tapered column was effective in retaining the floc particles, a biomassrecycle system was used to return those particles suspended in the product effluent to the active portion of the reactor.

Operation

The systems were operated with a synthetic feed stream containing glucose (laboratory-grade dextrose, Fisher Scientific Co.) in the nominal range of 100–150 g/l; 0–5 g/l of yeast extract (Difco Laboratories); and a constant level of 1 g/l $(NH4)_2SO_4$, 1 g/l KH_2PO_4, and 0.5 g/l $MgSO_4 \cdot 7H_2O$. The system was initially sterilized with a 70% ethanol solution, and the glucose feed solution was sterilized by autoclaving at 121°C for 45 min.

Each test was initiated by the introduction of 200–400 ml of a floc slurry that had been generated in a shaker flask with the feed solution over a period of 24–48 hr. The feed solution was then admitted to the column at progressively higher rates as additional biomass floc was developed. After 80–100 hr, a sufficient amount of biomass floc had been generated to enable controlled experiments to proceed.

Sampling and Analytical Methods

Liquid samples (0.1 to 0.5 ml) were withdrawn at each sample port by syringe with needle through a septum. The liquid was then immediately (within 10 sec) filtered through a filter with nominal 0.45-μ pores to remove the included active biomass. The glucose concentration was measured by an electrochemical method using a YS1 Model 27 Industrial Analyzer (Yellow Springs Instrument Co., Inc.). The aqueous ethanol concentration was determined by a gas chromatographic method suggested by Supelco, Inc.,[11] and organic acids were determined by a gas chromatographic procedure based on the method of Narkin and Henfeld-Furie.[12]

The off-gas constituents were measured by an on-line mass spectrometer or by a dedicated analytical gas chromatograph (Carle Instruments, Inc., Model III). Biomass content was determined by withdrawing ~10 ml of sample from each sample port, centrifuging at 1650 *g* for 10 min, washing with distilled water, recentrifuging and drying overnight at 105°C.

RESULTS AND DISCUSSION

A series of 300- to 400-hr runs was made in an attempt to determine optimum conditions. Typical results are presented in FIGURE 3, where ethanol and glucose concentrations measured at the first sampling port are shown as functions of the flow rate and glucose concentration of the feed. As shown in FIGURE 4, the biomass loading progressively increased during the run; ethanol productivity also increased as indicated in the first column section (from entrance to first sample port), reaching a maximum of

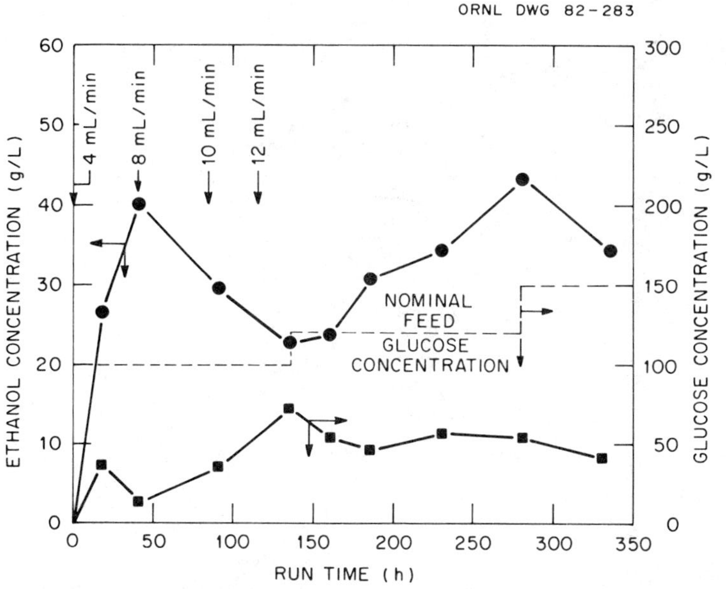

FIGURE 3. Ethanol and glucose concentrations for run F-2 at the first sample port in the fluidized-bed bioreactor.

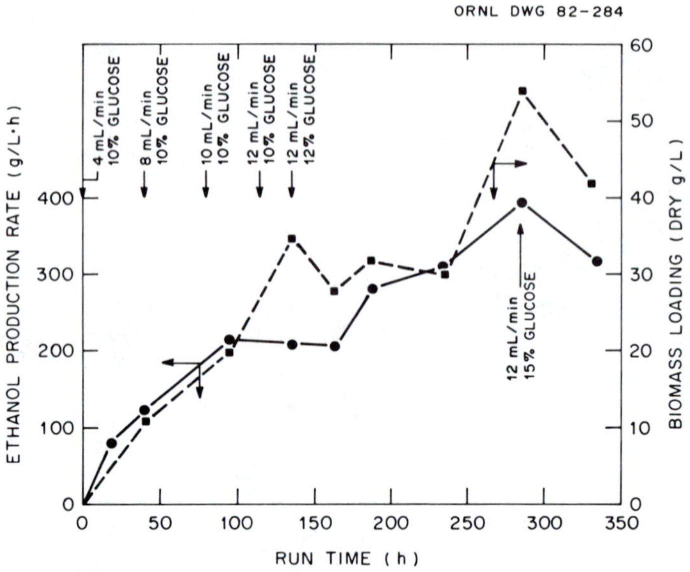

FIGURE 4. Ethanol production rate and biomass loading for run F-2 in the first section (up to the first sample port) of the fluidized-bed bioreactor.

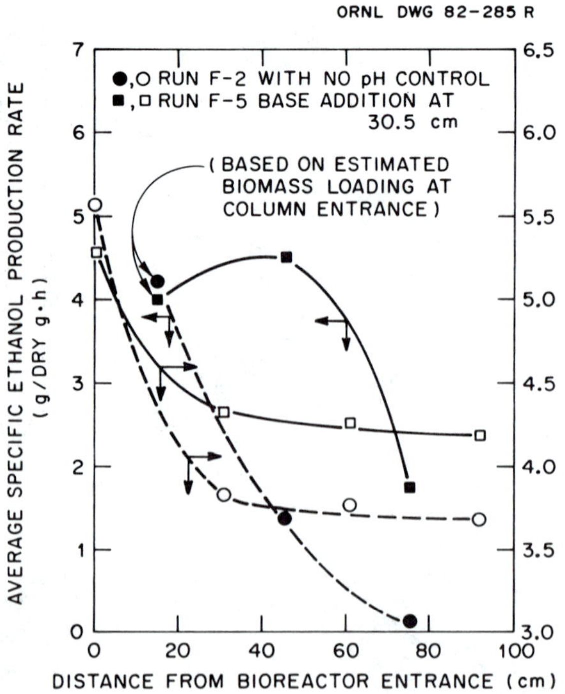

FIGURE 5. Effect of pH profile on the specific ethanol production rate.

395 g/l · hr (based on total volume). In this test, the higher glucose concentration (150 g/l) caused a decrease in biological activity; however, in later tests, the activity returned to original levels after 50–60 hr of operation at that concentration.

Effect of pH

The optimum pH for ethanol production for this organism was apparently in the range of 5.0 to 6.0, with appreciable activity at a pH greater than 4.0. When no pH control was exercised, a gradient was established up the column starting at about a pH of 5 and decreasing rapidly to a constant value of about 3.7 (FIG. 5). At the latter level,

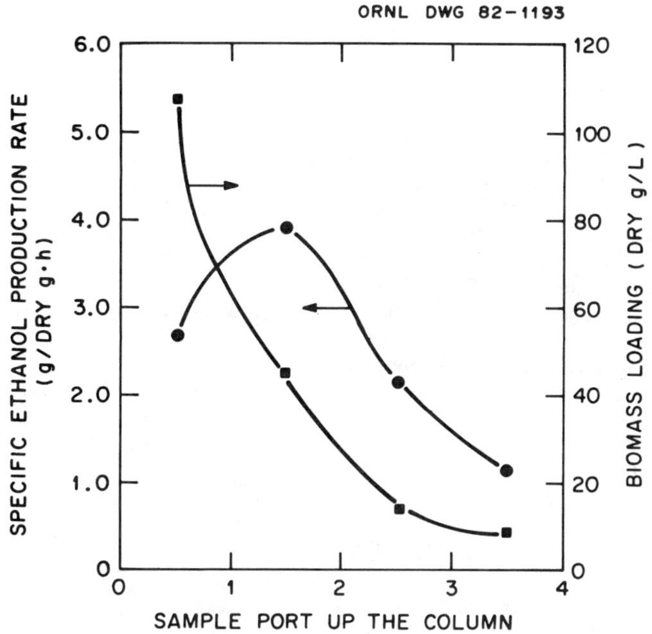

FIGURE 6. Biomass loading and specific ethanol production rate profiles in a tapered fluidized-bed bioreactor for run F-15 at 283 hr.

almost all metabolic activity ceased and ethanol productivity approached zero. However, when the pH was maintained above 4.0 by the addition of 0.1 *N* KOH at the first entry port up the column, there was significant ethanol productivity even at the upper part of the column. Multiple base-addition points may be necessary for optimum operation.

Biomass Loading

Ethanol productivity is heavily dependent on the concentration of the biocatalyst, in this case the *Z. mobilis* floc particles. A biomass concentration gradient was seen in all tests with biomass loadings as high as 108 g/l, on a dry weight basis (FIG. 6). The

TABLE 1. Bioreactor Carbon Balance for Run F-15 at 283 Hours

Carbon Source	Grams Carbon Per Hour
Input	
Glucose in feed	53.6 (±1.3)
Output	
Aqueous product:	
Biomass	0.3 (±0.1)
Carbon dioxide (calculated)	0.6 (?)
Ethanol	32.5 (±0.8)
Glucose	1.9 (±0.2)
Organic acids	1.4 (±0.1)
Off-Gas:	
Carbon dioxide	14.9 (±0.6)
Ethanol	0.3 (±0.1)
Total Output	51.9

level of biomass significantly decreased up the column as a result of the exceedingly high off-gas rate, which tended to disperse the solids and liquid. The gas content of the upper part of the reactor probably exceeded 50% of the column volume. Biomass concentrations were determined by neglecting the column void fraction due to the off-gas. Thus, the actual biomass loading on a column volume basis was probably somewhat lower than indicated in FIGURE 6, especially in the upper part of the reactor, and the corresponding specific ethanol production rate was probably somewhat higher.

Interestingly, the specific ethanol production rate based on biomass content tended to increase in the second section of the reactor, reaching a maximum value of approximately 4 g ethanol/dry g biomass · hr; then it decreased in the upper part of the reactor. This decrease may be due to a lower glucose concentration, a lower pH, or inhibitory effects of the ethanol product.

Multiple Feed Points

Most of the tests were made with a single feed entry point at the bottom of the column. However, it was found that higher biomass loading and overall column productivity could be achieved when the feed was introduced at two or more points. Feed was usually introduced at the column entrance and at the first entry port up the column. Both the biomass content in the upper part of the reactor and the overall ethanol productivity were increased with this mode of operation. These increases were probably due to an overall higher pH and carbon source availability, especially in the top portion of the bioreactor.

Column Material Balances

The carbon balance for a typical bioreactor is shown in TABLE 1. The balance of output as compared with input was 96.8% with the major effluent component being ethanol. When the usual assumed stoichiometry was considered,

$$C_6H_{12}O_6 \rightarrow 2C_2H_6O + 2CO_2 \qquad \textbf{(1)}$$

the ethanol production was 95.6% of theoretical, while CO_2 production was only 90% of theoretical for a glucose conversion of 96.3%. The unexpectedly low CO_2 production rate was experienced in all tests. This indicates that perhaps our monitoring and measuring techniques were not accurate, that CO_2 absorption in the aqueous product was greater than the calculated amount, or the assumed stoichiometry was not correct. Additional work will be required to evaluate these options.

Ethanol Productivity

Although the results presented here are preliminary in nature, the measured ethanol productivity gives an indication of the potential for such a system. As indicated in TABLE 2, ethanol productivity as high as 443 g/l · hr (with many values approaching 300 g/l · hr) was measured for the bottom section of the fluidized bed. Overall productivity, based on the entire active volume of the reactor, was as high as 120 g/l · hr.

It should be noted that overall ethanol productivity was strongly influenced by biomass loading in the upper part of the reactor. Higher glucose feed concentrations, pH control, and multiple feed introductory points also enhanced ethanol production. As might be expected, higher glucose conversions resulted in a somewhat lower ethanol productivity.

CONCLUSIONS

The flocculating strain of *Z. mobilis* (NRRL B-12526) has been shown to form stable floc particles that can effectively convert glucose to ethanol at rates in excess of 400 g/l · hr in a fluidized-bed bioreactor. However, the system is so reactive that the CO_2 off-gas makes it difficult to maintain a high biomass loading in the upper part of the reactor. This lower biomass content results in a decreased ethanol productivity, which approaches 100 g/l · hr for the entire active reactor volume when over 95% of the glucose is metabolized. The use of increased biomass loading and a glucose feed concentration in excess of 150 g/l would undoubtedly allow even higher levels of ethanol productivity to be achieved.

TABLE 2. Ethanol Productivity

		Feed		Ethanol Productivity (g/l · hr) in			
Run	Run Time (hr)	Flow Rate (l/hr)	Glucose Conc. (g/l)	Bottom Section	Overall	Ethanol Production (% of theoretical)	Overall Glucose Conversion (%)
F-2	283	0.72	121	396	81	99	73
F-3[a]	391	0.78	126	443	120	99	85
F-5[a]	331	0.48	104	173	68	97	94
F-15[a]	185	1.44[b]	97	298	93	98	91
F-15[a]	283	1.43[b]	93	294	91	96	96

[a]Base addition used for pH control.

[b]Dual feed entry points in tapered fluidized bed.

ACKNOWLEDGMENTS

The outstanding technical assistance of W. K. Alexander, S. L. Arnold, W. G. Chapman, G. D. Smith, and K. S. Whaley is gratefully acknowledged. This work was supported by the U.S. Department of Energy.

REFERENCES

1. Cysewski, G. R. & C. R. Wilke. 1978. Biotechnol. Bioeng. **20:** 1421.
2. Cysewski, G. R. & C. R. Wilke. 1977. Biotechnol. Bioeng. **19:** 1125.
3. Scott, C. D., C. W. Hancher & E. J. Arcuri. 1980. *In* Advances in Biotechnology. Moo-Young, Ed. Vol. **II:** 651. Proc. 6th Int. Ferment. Symp. London, Canada. July 20, 1980. Pergamon. New York.
4. Scott, C. D. 1982. Fluidized-bed bioreactors using *Zymomonas mobilis* for ethanol production. Presented at the American Chemical Society Meeting. Las Vegas, Nevada. March 31, 1982.
5. Rogers, P. L., J. J. Lee & D. E. Tribe. 1979. Biotechnol. Lett. **1:** 165.
6. Arcuri, E. J., R. M. Worden & S. E. Shumate. 1980. Biotechnol. Lett. **2:** 499.
7. Strandberg, G. W., T. L. Donaldson & E. J. Arcuri. 1982. Biotechnol. Lett. **4:** 347.
8. Gainer, J. L., D. J. Kirwan, J. A. Foster & E. Seyhan. 1980. Biotechnol. Bioeng. Symp. No. 10. p. 35.
9. Wada, M. J., J. Kato & I. Chibata. 1980. Eur. J. Appl. Microbiol. Biotechnol. **10:** 275.
10. Williams, D. & D. M. Munhecke. 1981. Biotechnol. Bioeng. **23:** 1813.
11. Supelco, Inc. 1982. Chromatography Supplies Catalog 20.
12. Narkis, N. & S. Henfeld-Furie. 1978. Water Res. **12:** 437.

Technology Developments in Biomass Alcohol Production in Japan: Continuous Alcohol Production with Immobilized Microbial Cells

MINORU NAGASHIMA, MASAKI AZUMA, AND SADAO NOGUCHI

Technical Research Laboratory
Hofu Plant, Kyowa Hakko Kogyo Co., Ltd.[a]
1-1 Kyowa-cho, Hofu city, Yamaguchi, 747 Japan

INTRODUCTION

The world's diminishing energy supply has created a growing anxiety about fuels, forcing industrial countries including Japan to strive for required quantities of energy, and in their all-out efforts for this purpose, to develop various kinds of alternative energy products to replace petroleum fuels.

Under these circumstances, the Research Association for Petroleum Alternatives Development (RAPAD) in Japan was established in May 1980, getting support from the Ministry of International Trade and Industry (MITI). RAPAD is organized by 23 member companies from such industries as the petroleum, fermentation, chemical, and engineering industries. It inaugurated research activities in June 1980, taking charge of a national project under a 7-year program (1980–1986). The amount of investment required for the project is estimated at around 38.8 billion yen (150 million U.S. dollars).

RAPAD's research activities cover the following three fields as shown in TABLE 1. The subjects on biomass utilization are as follows:

(1) Technology developments utilizing cellulose for ethanol or butanol with physical or chemical pretreatment;
(2) Technology for continuous ethanol fermentation using immobilized yeast cells; and
(3) Probing into other related items, such as recovering ethanol.

Studies are also being advanced on problems that may arise when biomass alcohol is used in a mixture with gasoline as motor fuel.

Continuous production of alcohol by immobilized yeast cells is one of the main themes of RAPAD's biomass utilization project. This work has been carried out by two groups in RAPAD since June of 1980 in a schedule to be completed in March, 1983. These two groups have been engaged in the exploitation of the immobilized yeast process, one group is studying the use of a specially designed artificial polymer (i.e., photo-cross-linking polymer), and the other group is studying the utilization of natural and/or synthetic materials as yeast carriers. The latter type of immobilization is the target of our group.

[a]As a participant of the Research Association for Petroleum Alternatives Development.

TABLE 1. Research Activities of RAPAD

Developing:
(1) Technology for manufacturing synfuels from syngas;
(2) Technology for upgrading tar sand bitumens and shale oils; and
(3) Technology for biomass conversion and utilization.

Several techniques have been proposed for continuous ethanol fermentation and some are applied to industrial production. A bioreactor composed of immobilized living yeast represents a new trend in these techniques for its high productivity and for the lowering of the initial investment and operational cost of the alcohol fermentation process.

The approaches for the project were made as follows:

(1) Selection of a carrier for immobilization of yeast cells and studies on the method of immobilization;
(2) Design of reactor and shapes of carrier;
(3) Selection of the most suitable strain for continuous ethanol production;
(4) Prevention of contamination;
(5) Studies on the maintenance of yeast viability; and
(6) Scale-up to pilot plant.

SELECTION OF CARRIERS

Living yeast cells were entrapped with various polymers. Various polymer beads containing living yeast cells were placed separately in glass columns, and cane molasses solution was fed continuously in each column for 5 days.

FIGURE 1 shows daily changes of alcohol productivity of the columns. Among the several polymers examined, Ca-alginate showed the best results. Carrageenan is comparable to calcium alginate, but it has undesirable characteristics during large-scale preparation and operation. Entrapping methods with hydrogel possesed high activities.

Calcium alginate was chosen as an entrapping material because of its high correlation with selection criteria shown in TABLE 2.

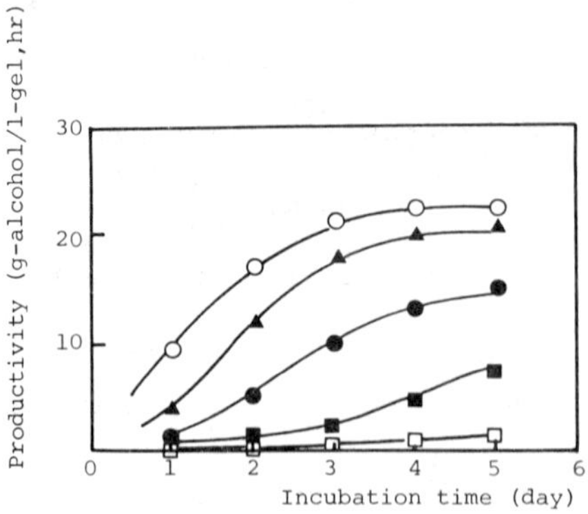

FIGURE 1. Ethanol productivity of immobilized yeast entrapped in various polymers: ○ Ca-alginate gel; ▲ polyacrylamide gel; ● porous epoxy resin; ■ porous polystyrene; and □ porous polyester microcapsule.

TABLE 2. Criteria for Selection of Supporting Materials for Microbial Immobilization

(1)	High carrier activity
(2)	Availability in quantity
(3)	Low cost of immobilization
(4)	Ease of scale-up of operation
(5)	Mechanical strength for long-life operation

FIGURE 2. A picture of prototype reactors.

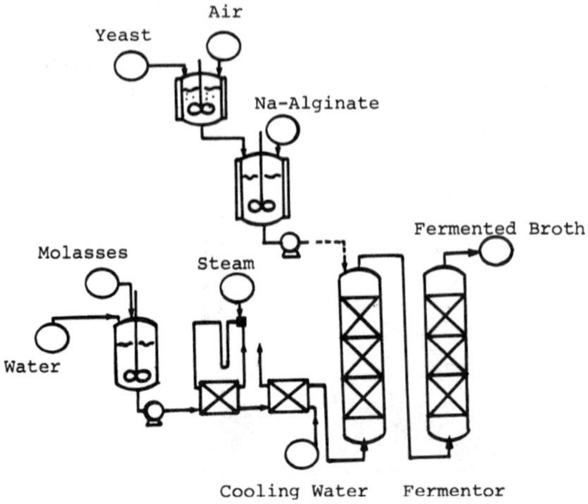

FIGURE 3. A tentative process flow sheet (PROCESS-1).

FIGURE 4. Immobilized cell beads just after the preparation in a 1-kl reactor.

DESIGN OF REACTOR

After bench-scale study, prototype reactors were constructed before pilot plant designation. The capacity of each reactor was one cubic meter. FIGURE 2 shows these three reactors. One is the tall tower type of reactor, L/D ratio can be adjusted from 7 to 10. The second is the short column type of reactor. L/D ratio can be adjusted from 1.5 to 2. The last is the rectangular type of reactor, shown behind two column reactors. The inside of this reactor was separated by many vertical plates and the reaction solution moved horizontally and snakewise from one end to the other.

According to the results of bench-scale experiments, tentative process flow was decided as shown in FIGURE 3. This process was designed so that the whole process can be operated aseptically. FIGURE 4 shows the immobilized-cell beads just after the

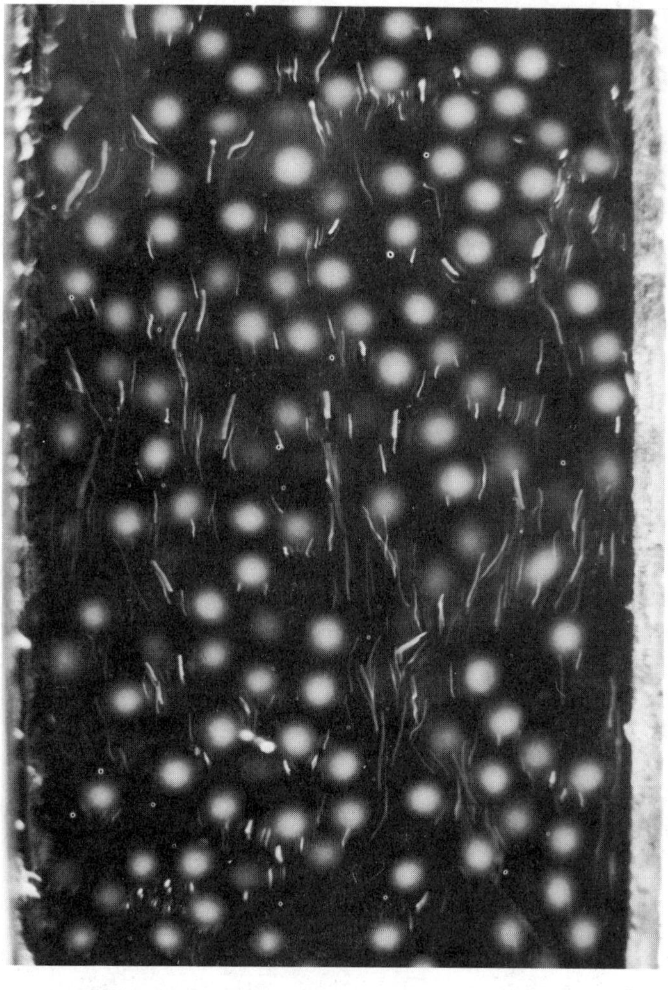

FIGURE 5. Fluidization of gel beads in a pilot reactor.

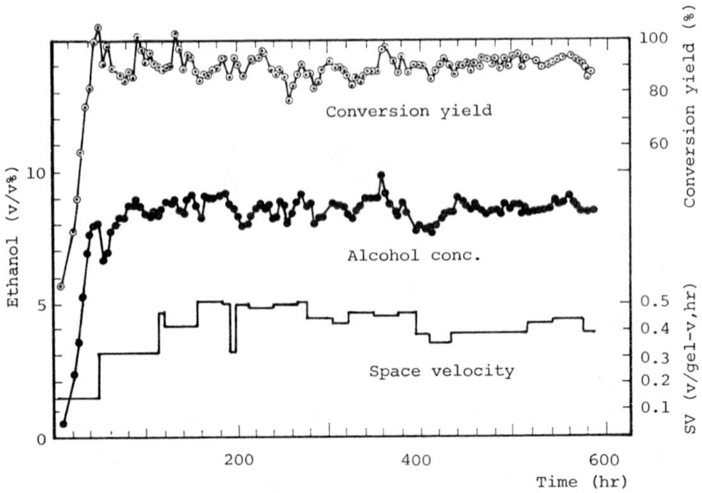

FIGURE 6. Continuous operation in semipilot plant (l-kl columns in series, strain: T-29).

preparation in a 1-kl-sized reactor. Continuous preparation of immobilized-cell beads with alginate was tried for the first time on a large scale, and conducted successfully without any trouble. FIGURE 5 shows the immobilized-cell beads during the pass of cane molasses solution. Good fluidization of beads was observed during fermentation.

Reaction characteristics were also studied for each column. The activities of immobilized yeast beads and productivity of each column were almost the same, but it

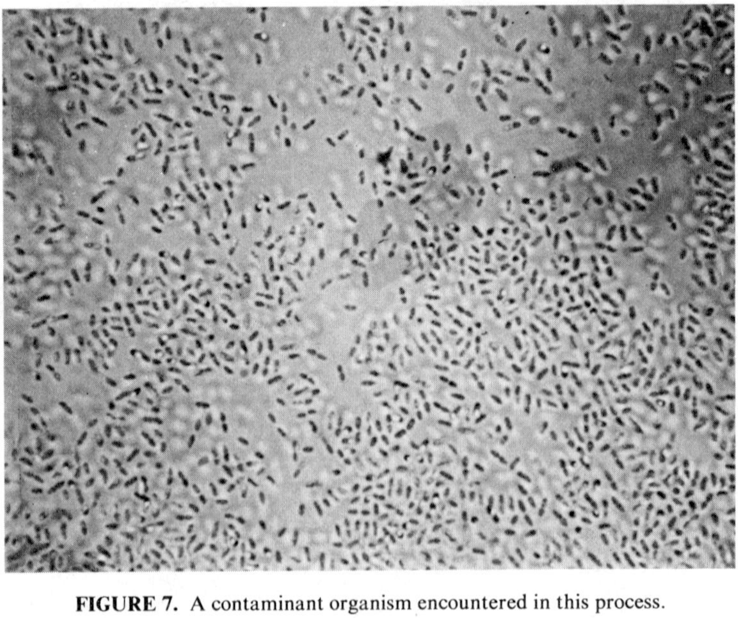

FIGURE 7. A contaminant organism encountered in this process.

was recognized that at least two columns must be connected in series to obtain a higher conversion yield owing to the strong mixing effect of evolving CO_2 gas. FIGURE 6 shows the results of continuous operation of the semi-pilot plant.

Sugar concentration of the inlet solution was about 15% and alcohol concentration of the outlet solution was about 8.5% (v/v). Conversion yield was more than 90%, and, productivity of alcohol was about 25 g/l-gel/hr.

PROCESS IMPROVEMENTS

Prevention of Contamination

In our experiments, operations sometimes suffered because of contamination by certain microbes. FIGURE 7 shows one of the contaminant organisms, *Acetobacter*.

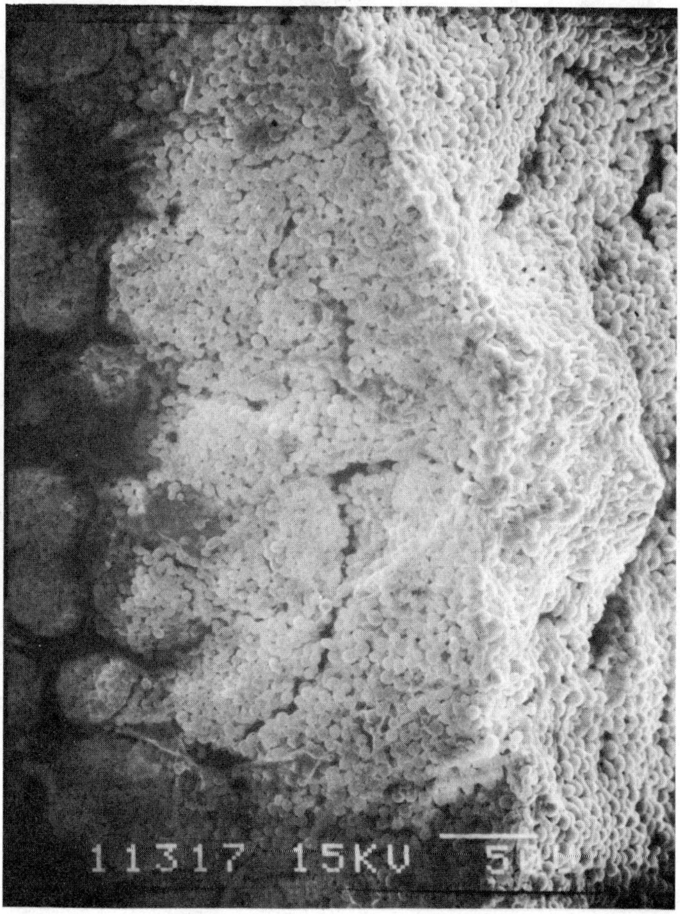

FIGURE 8. Scanning electron micrograph of the tentative immobilized gel beads.

In order to obtain a good fermentation yield and make a continuous process stable over the long term, it is necessary to prevent contamination. Studies on the isolated bacterial contaminants were made, and, as a result, operation at pH 4.0 with sulfuric acid or in the presence of some bactericidal substance to the inlet solution was also very effective. By employing such procedures, the contamination problem was almost eliminated and now the process can be operated without sterilization of the inlet medium.

Yeast Viability

It was recognized that the activity of immobilized yeast was gradually decreased during the bench-scale, long-run column operations. For the stabilization of long-life

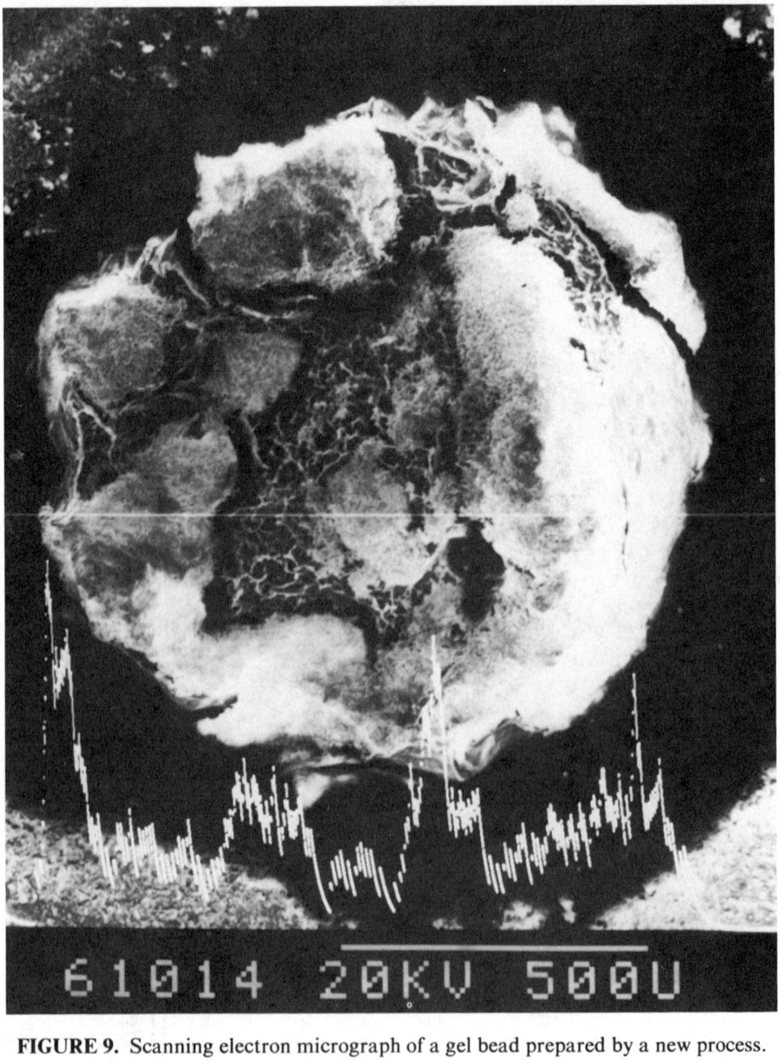

FIGURE 9. Scanning electron micrograph of a gel bead prepared by a new process.

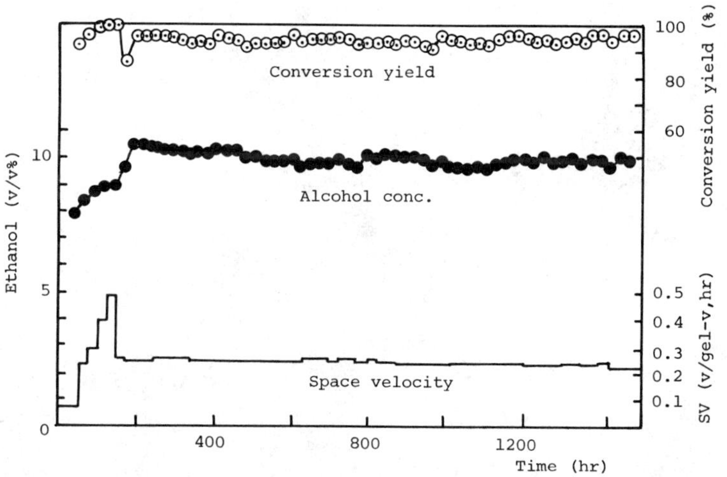

FIGURE 10. Continuous operation in laboratory for higher concentration of ethanol (200-ml columns in series, strain: T-620).

operation, this should be compensated for in these approaches. Maintenance of yeast viability was also investigated.

FIGURE 8 shows an electron micrograph of the cross section of an immobilized-cell bead. Yeast growth was observed to be richer in the outer layer than in the inside of the bead.

Yeast growth inside the carrier, which might be limited by substrate diffusion, was promoted by dissolved oxygen or certain sterols and unsaturated fatty acids that were required for yeast growth during fermentation. Also, the entrapment of such supplements together with yeast into gel beads enhanced yeast growth and attained higher productivity (30 to 50 g-enthanol/l-gel,hr). Moreover, aeration into the reactor enhanced the operational stability.

FIGURE 9 shows a cross section of a gel bead prepared by the new process. Yeast growth is observed to be abundant around the gel.

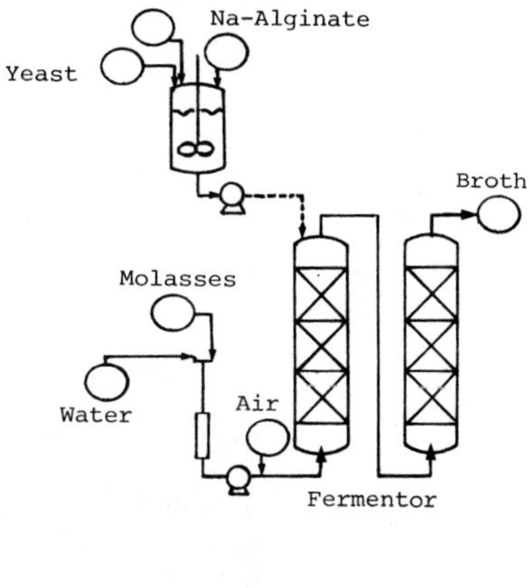

FIGURE 11. Process flow diagram of the pilot operation.

FIGURE 12. Pilot plant of continuous alcohol fermentation by immobilized yeast cells (Hofu lab).

Strain Improvement

An example of strain improvement is shown in FIGURE 10. Moreover, 10% of alcohol production was attained continuously by the new strain. In this experiment, all the technical skills hitherto obtained were applied and the productivity was increased twofold compared to the tentative process. Also, cane juice or starch hydrolysate were good raw materials for higher concentration of alcohol fermentation.

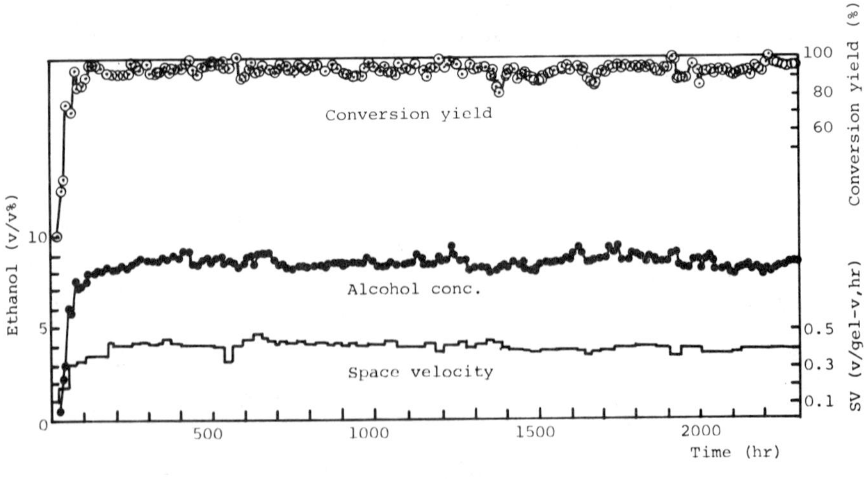

FIGURE 13. Continuous ethanol production by immobilized yeast in the pilot plant.

PILOT PLANT OPERATION

FIGURE 11 shows the new process. Now we can operate without pasteurization of medium. Seed fermenters are not required for full-scale operation over a 100-cubic-meter reactor. Our goal in these developments is simplification of the total system. According to the results of semi-pilot plant operations, a final pilot plant equipped with five column reactors was constructed in March, 1982. This pilot plant is composed of two channels of reactors, one with three columns and the other with two columns, both channels connected in series. The reactor volume of each channel is 2 kl in full capacity, so the total column volume is 4 kl. FIGURE 12 is a picture of our pilot plant. Pilot plant operation has been carried on since April of this year to confirm the results hitherto obtained. FIGURE 13 shows the results of pilot operation during these three months. As shown in this figure, the following results have been obtained:

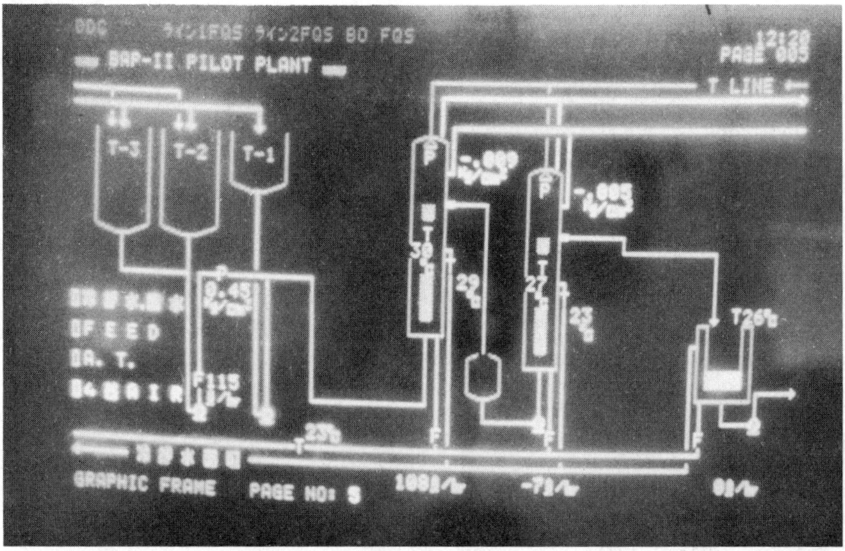

FIGURE 14. Minicomputor display of the pilot plant.

(1) 8.5 v/v% of ethanol was constantly produced from diluted cane molasses (sugar concentration 14 w/v%) at space velocity 0.4–0.5 for over three months at 30°C without any trouble.
(2) The productivity of ethanol based on the total column volume (full capacity of reactor columns) was calculated as about 20 g/l/hr. This means that 600 liters per day of pure ethanol can be produced using the 1 kl column reactor.
(3) The conversion ratio from sugar to ethanol was about 95% of the theoretical yield.

This pilot plant is connected with a computer system, and we are now aiming at completely automatic operation of the plant. FIGURE 14 shows a computer display of the process. A minicomputer with software packages specifically developed for display and process control was used. Various confirmation tests of this process are now in process.

(1) Full automatic control, on-line analysis of ethanol and sugar.
(2) Another process, easy handling with minimum control.
(3) A process available to various sugar sources.

CONCLUSIONS

TABLE 3 summarizes some advantages of our continuous fermentation process with immobilized yeast cells. This process may be industralized on a commercial scale in the near future.

SUMMARY

A project of developing a continuous production process for ethanol using immobilized yeast cells has been undertaken by this group as a member of RAPAD.

TABLE 3. Some Advantages of the Immobilized Yeast Process

(1)	Continuous fermentation with low operating and maintenance without skilled labor.
(2)	Low capital investment costs with higher productivity and higher yield.
(3)	Energy saving without sterile operation except gel preparation.
(4)	Applicable to a variety of liquid substrates.

The project was approached by first selecting carriers for immobilization, yeast strains, and environmental conditions for the operation of the process.

Calcium alginate was chosen as a carrier due to its compatibility with the immobilized-bed fermentation process. Various improvements were also made to guard against microbial contamination and for the maintenance of yeast viability for a stable, long-run operation.

A semi-pilot plant equipped with 1-kl column reactors was constructed and tests both for a large-scale immobilization and for a continuous production of ethanol were conducted successfully. A pilot plant of 4-kl total column volume was constructed, and it was confirmed that 8.5 v/v% alcohol could be constantly produced from diluted cane molasses for over three months.

ACKNOWLEDGMENTS

We express heart-felt gratitude to Professor Shuichi Suzuki for his help with the arrangements for this presentation.

Solid-state Fermentation of Cellulosic Residues[a]

ROBERT P. TENGERDY,[b,c] VINCENT G. MURPHY,[d] AND MARK D. WISSLER[b]

[b]Department of Microbiology
[d]Department of Agricultural and Chemical Engineering
Colorado State University
Fort Collins, Colorado 80523

Agricultural cellulosic residues, such as wheat straw, corn stover, and manure fibers may be enriched in protein for animal feed by solid-state fermentation (SSF) with cellulolytic fungi.[1-5] The advantage of SSF is that it yields a directly feedable product and that it can be performed under local farm conditions.

The limitations of SSF of cellulosic substrates with presently available microorganisms are low substrate utilization and low specific growth rates. Until genetic engineering solves these problems, the best attempt for increasing substrate utilization appears to be mixed culturing and for increasing growth rates, optimizing fermentation conditions. The ultimate limitation of fungal growth in SSF is the onset of sporulation; therefore, optimizing conditions should aim at prolonging vegetative growth as long as possible.

In the present paper, experiences gained in improving the SSF of cellulosic substrates are summarized. The substrates—wheat straw, corn stover and manure fiber—must be pretreated to make the cellulose available for cellulase attack. Pretreatment by steam at 180°C for 10 minutes proved to be the most efficient and most economic in our hands.[5]

The fungi yielding the most protein on the above substrates were *Chaetomium cellulolyticum* and *Trichoderma lignorum*. These fungi grow well even in the presence of N compounds associated with manure. In mixed culturing, *Candida lipolytica* was grown together with one of the molds.

Typical SSF fermentations of wheat straw and manure fibers are shown in FIGURE 1. The steam-treated substrates were adjusted to 25% w/w dry matter content with Mandel's salt solution,[6] and the C:N ratio was set to 10:1 by the addition of $(NH_4)_2SO_4$. The substrate was placed either into a pot fermenter where an approximately 3-cm layer of 50-g substrate was aerated through a stationary perforated plate,[7] or into a 3-l horizontal fermenter with rotating paddles inside containing 800 g of substrate.[4] The mold inocula were obtained by washing about 10^6 conidia from a slant into 100 ml of liquid medium containing glucose, salt, and filter paper and then shaking at 37°C for 48 hr. A 24-hour shake culture of *Candida lipolytica* in the same medium was used for yeast inoculum. The substrates were incubated with 1 wt% (dry basis) mycelium. In mixed culturing, 1% mold mycelium and 0.33% *Candida lipolytica* cells were used.

It is evident from FIGURE 1 that the best protein production was achieved in the

[a]This work was supported by grants ENG-75-18614 and INT-7812768 from the National Science Foundation; also by the National Academy of Sciences, USA, and the Academy of Sciences, Latvian SSR, USSR, under a USA-USSR Academy Exchange Program.

[c]Author to whom correspondence should be addressed.

mixed culture. Cellulose conversion proceeded with a cellular yield close to 0.5, the theoretical value; hence, most of the hydrolyzed sugar was utilized. In single fermentation, however, about 40–50 mg/g dry matter reducing sugar remained unused. This surplus sugar was utilized by the yeast in the mixed culture, resulting in more biomass and delaying the onset of sporulation of *Chaetomium cellulolyticum*.

In all three fermentations, cellulose degradation was limited to 33–40%, a range also reported in other SSF processes.[8–10] Since wheat straw and the more recalcitrant manure fiber gave similar degradation, the limiting factor is probably not substrate availability but cellulase production possibly regulated by glucose repression.[11,12] If so, this limitation can be removed in a two-stage fermentation, where cellulase production is separated from cellulose hydrolysis and fungal growth.

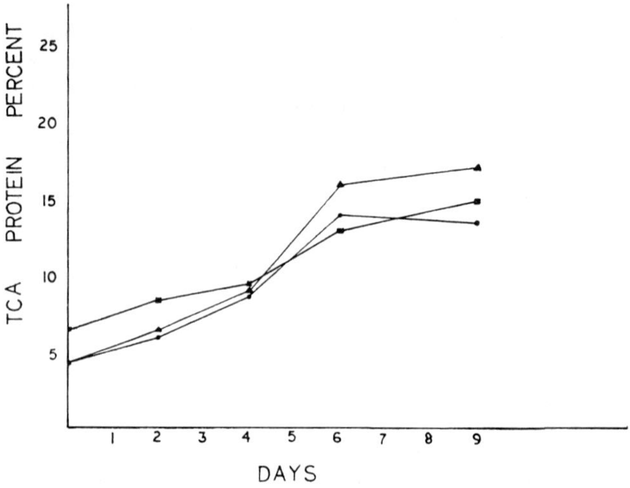

FIGURE 1. Protein production of SSF in wheat straw and manure fibers in single and mixed fermentation.

▲—*Chaetomium cellulolyticum* and *Candida* lipolytica on wheat straw

●—*Chaetomium cellulolyticum* on wheat straw

■—*Chaetomium cellulolyticum* on manure fibers

The assessment of true biomass produced in SSF is difficult, due to the insoluble plant N present. A new method based on ^{15}N abundance ratio measurement by mass spectrometry has been developed for SSF.[7] The method measures accurately the incorporation of $(^{15}NH_4)_2SO_4$ into fungal biomass, since the only insoluble ^{15}N in the fermented product is the newly synthesized fungal biomass.

A comparison of the ^{15}N method with the chitin assay[1,14,15] and the traditionally used Kjeldahl assay[13] of trichloroacetic (TCA) precipitable N is presented in TABLE 1. In this experiment designed to test the reproducibility of fermentation and sampling, the ^{15}N method gave comparable but slightly higher biomass values than the TCA method for both substrates. The relative uncertainty of the ^{15}N method was much lower than the TCA method, reflecting the inherent accuracy of the ratio measurement and the avoidance of making corrections for weight loss during fermentation. The ^{15}N method thus could serve for calibrating other biomass measurements and would be

TABLE 1. Comparison of Biomass Measurements in Solid State Fermentation of Wheat Straw and Corn Stover, with *Chaetomium cellulolyticum*[a,b]

		TCA Method			^{15}N Method		Chitin Method	
Substrate	Fermenter[c]	Total TCA Protein (mg/g)	Biomass Protein (mg/g)	Rel. Uncer.[d] (%)	Biomass Protein (mg/g)	Rel. Uncer.[d] (%)	Biomass Protein (mg/g)	Rel. Uncer.[d] (%)
Wheat straw	1A	49.93 ± 3.54	31.97 ± 3.90	12.21	35.24 ± 2.90	8.22	4.17 ± .55	13.24
	1B	53.93 ± 4.25	37.92 ± 4.87	12.84	39.63 ± 4.09	10.32	3.77 ± .60	15.96
	2A	48.01 ± 2.68	32.29 ± 2.68	8.28	36.05 ± 2.00	5.54	4.45 ± .35	7.89
	2B	47.02 ± 2.44	31.30 ± 2.44	7.78	34.38 ± 2.24	6.52	3.75 ± .99	26.29
	3A	57.78 ± 4.67	42.99 ± 4.88	11.35	45.69 ± 3.63	7.94	3.72 ± .55	14.82
	3B	56.27 ± 3.31	40.62 ± 3.30	8.14	43.88 ± 2.89	6.58	3.82 ± .91	23.81
Corn Stover	1A	70.24 ± 7.04	46.44 ± 7.04	15.16	54.26 ± 4.70	8.67		
	1B	71.07 ± 2.52	47.27 ± 2.52	5.34	50.11 ± 0.62	1.23		
	2A	64.08 ± 0.82	42.07 ± 0.82	1.95	43.59 ± 1.09	2.50		
	2B	63.82 ± 1.08	41.01 ± 1.08	2.63	43.34 ± 0.49	1.13		
	3A	58.00 ± 2.72	35.22 ± 2.72	7.74	41.25 ± 0.88	2.13		
	3B	66.56 ± 0.89	43.79 ± 0.89	2.04	50.09 ± 1.12	2.23		

[a] Ammonium sulfate enriched to 10% ^{15}N was added at 0.10 g per g substrate.
[b] Values in table are means ± standard deviation for five (wheat straw) or three (corn stover) replicate samples.
[c] Fermenters 1, 2, 3 were divided approximately in half (A, B) at the time of harvesting (7 days).
[d] $\text{Relative Uncertainty} = \frac{\text{Standard Deviation}}{\text{Mean}} \times 100$

indispensable for accurate growth rate studies in SSF. The chitin assay was totally disappointing in our hands, probably because the chitin content is variable in fungal cells of various stages of development in SSF, or because the extraction of chitin is inhibited in solid substrates.

SUMMARY

Wheat straw and washed manure fibers were converted into a protein-enriched animal feed supplement in solid-state fermentation with *Chaetomium cellulolyticum* or a mixed culture of *Chaetomium* and *Candida lipolytica*. Under optimal pretreatment and fermentation conditions, the maximum cellulose conversion was 33–40%, and the maximum final protein content was 15–17%. A new method for biomass determination in complex solid substrates, based on ^{15}N abundance measurement by mass spectroscopy, was also developed.

REFERENCES

1. Moo-Young, M., A. R. Moreira & R. P. Tengerdy. 1983. *In* The Filamentous Fungi. Vol. 4, Fungal Technology. J. E. Smith, D. R. Berry & B. Kristiansen, Eds.: 117–144. John Wiley & Sons, Inc. New York.
2. Cannel, E. & M. Moo-Young. 1980. Solid-state fermentation systems. Process Biochem. **15**(5): 2–7.
3. Chahal, D. S. & M. Moo-Young. 1980. Dev. Ind. Microbiol. **22:** 143–159.
4. Viesturs, U. E., A. F. Apsite, J. J. Laukevics, V. P. Ose, M. J. Bekers & R. P. Tengerdy. 1981. Biotechnol. Bioeng. Symp. **11:** 359–369.
5. Ulmer, D. C., R. P. Tengerdy & V. G. Murphy. 1981. Biotechnol. Bioeng. Symp. **11:** 449–461.
6. Mandels, M., L. Hontz & J. Nystrom. 1974. Biotechnol. Bioeng. **16:** 1471–1493.
7. Wissler, M. D., R. P. Tengerdy & V. G. Murphy. 1983. Dev. Ind. Microbiol. **24:** 527–538.
8. Latham, M. J. 1979. *In* Straw Decay and Its Effect on Disposal and Utilization. E. Grossbard, Ed: 131–137. John Wiley & Sons, Inc. New York.
9. Pamment, N., C. W. Robinson, J. Hilton & M. Moo-Young. 1978. Biotechnol. Bioeng. **20:** 1735–1744.
10. Zadrazil, F. & H. Brunnet. 1981. Eur. J. Appl. Microbiol. Biotechnol. **11:** 183–188.
11. Hulme, M. A. & D. W. Stranks. 1970. Nature **226:** 469–470.
12. Nisizawa, T., H. Suzuki & K. Nisizawa. 1972. J. Biochem. **71:** 999–1007.
13. Horwitz, W. 1975. Official Methods of Analysis of the Association of Official Analytical Chemists, 12th edition. Association of Official Analytical Chemists. Washington, D.C.
14. Swift, M. J. 1973. Soil Biol. Biochem. **5:** 321–322.
15. Aidoo, K. E., R. Hendry & B. J. B. Wood. 1981. Eur. J. Appl. Microbiol. Biotechnol. **12**(1): 6–9.

Transition from Acid Fermentation to Solvent Fermentation in a Continuous Dilution Culture of *Clostridium thermosaccharolyticum*

SANDRA L. LANDUYT, EDWARD J. HSU,[a,b] AND MEI LU

Department of Biology
University of Missouri-Kansas City
Kansas City, Missouri 64110

[a]*Institute of Botany*
Academia Sinica, Taiwan, R.O.C.

It has been previously reported that vegetative cells of *Clostridium thermosaccharolyticum* fermented glucose to CO_2, H_2, acetic, butyric, and lactic acids, but no significant amount of ethanol is produced.[1,2] However, as a result of repeated filtration and serial transfers into fresh glucose medium, cell division was interrupted resulting in synchronous growth and continuous elongation of individual cells well beyond normal cell length with a concomitant conversion of acetate to ethanol.[3] It is also well established that acetate is the first product of carbohydrate fermentation that upon accumulation can be converted to organic acids such as butyrate, propionate, and succinate or, due to the subsequent reduction in pH, be reduced to organic solvents such as acetone, butanol, and ethanol. This is true even in the so-called "solvent-producing strains" or sporulating cultures of *C. thermosaccharolyticum*.[4,5]

In this investigation, a continuous dilution culture was used to bring about a maximum dilution effect that resulted in a reduced growth rate and filamentous growth of cells. This morphological change and apparent interruption in cell division appeared to correlate with production of large amounts of solvents, particularly ethanol and butanol and may be the cause or the consequence of the solvent fermentation.

The reduction in growth rate was apparent by the increase in generation time from 2 hr in batch cultures to 4 hr in continuous dilution cultures for strain 3814, and from 2.5 hr in batch culture to 5.0 hr in continuous dilution cultures for strain SD 105 (FIG. 1, Curves A, C, B, and D, respectively). A stepwise increase in cell mass continued for nearly 2 generations (greater than 10 hr) with strain SD 105. Furthermore, there was a concomitant elongation at each step of 40–50× in 90% of the cell population (FIG. 2). The elongated cells showed increasing multiple septations (FIGS. 3 and 4) as the step-wise growth deteriorated.

Initially, much more ethanol (1800 mmol/OD or a 13,846-fold increase over the batch culture) was produced and dominated the entire fermentation period reaching

[b]Permanent address: Department of Biology, University of Missouri, Kansas City, MO. 64110.

the highest concentration of 0.048 moles at 6 hr (1,930 mmol/OD) when cells were shifted to a continuous dilution culture after precisely one generation of growth in a batch culture (FIG. 5). Small amounts of other organic solvents, as well as acetate and butyrate, were all formed simultaneously. Propionate and isobutyrate were not detected until 20 hr, followed by relatively steady concentrations.

It is important to note that the increase in ethanol concentration at 6 hr was proportional to the stepwise increase in cell mass and the filamentous growth. The fact that this elongation coincided with the peak in ethanol production strongly indicates that the morphological changes may be responsible for the solvent production without

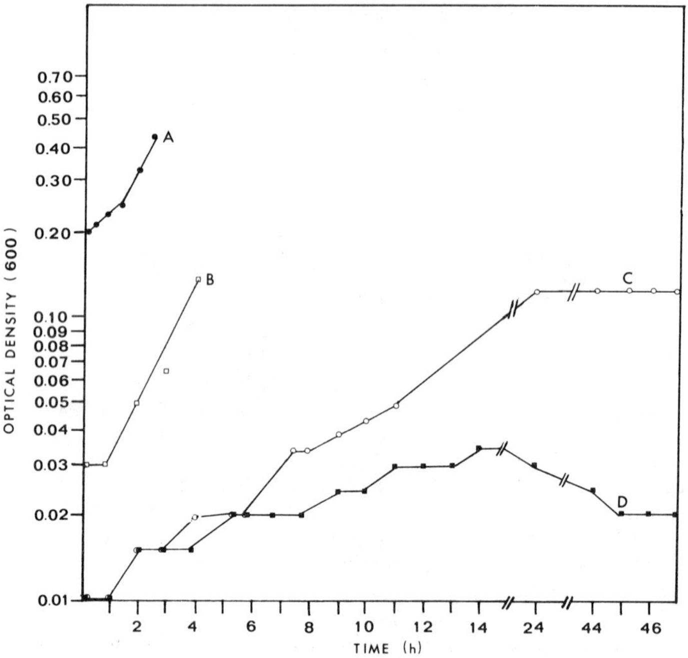

FIGURE 1. Growth curves of *C. thermosaccharolyticum*. Curve A (●) and Curve C (○) represent strain 3814 in batch culture and continuous dilution culture, respectively. Curve B (□) and Curve D (■) represent strain SD 105 in batch culture and continuous dilution culture, respectively.

a preceding acid production. The decline in ethanol production after 6 hr was accompanied by multi-septations along the entire filament (FIGS. 3 and 4) indicating a reversal of metabolism and a return to vegetative growth.

In conclusion, therefore, the transition from acid fermentation to solvent fermentation appears to correlate with a morphological change that results in a metabolic shift. The transition is not just a matter of reducing pre-existing organic acids but rather a consequence of gene expression that accompanies early stages of cell division or sporulation.

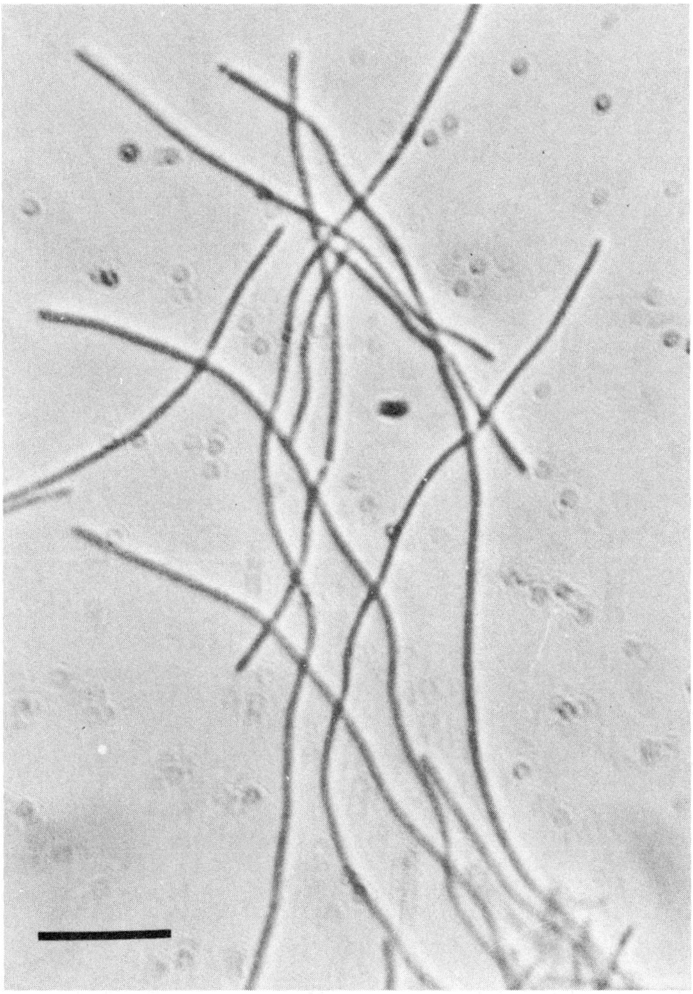

FIGURE 2. Highly elongated cells of *C. thermosaccharolyticum* strain SD 105 observed during the first two steps (FIG. 1, Curve D) in the continuous dilution culture. Bar represents 10 μm.

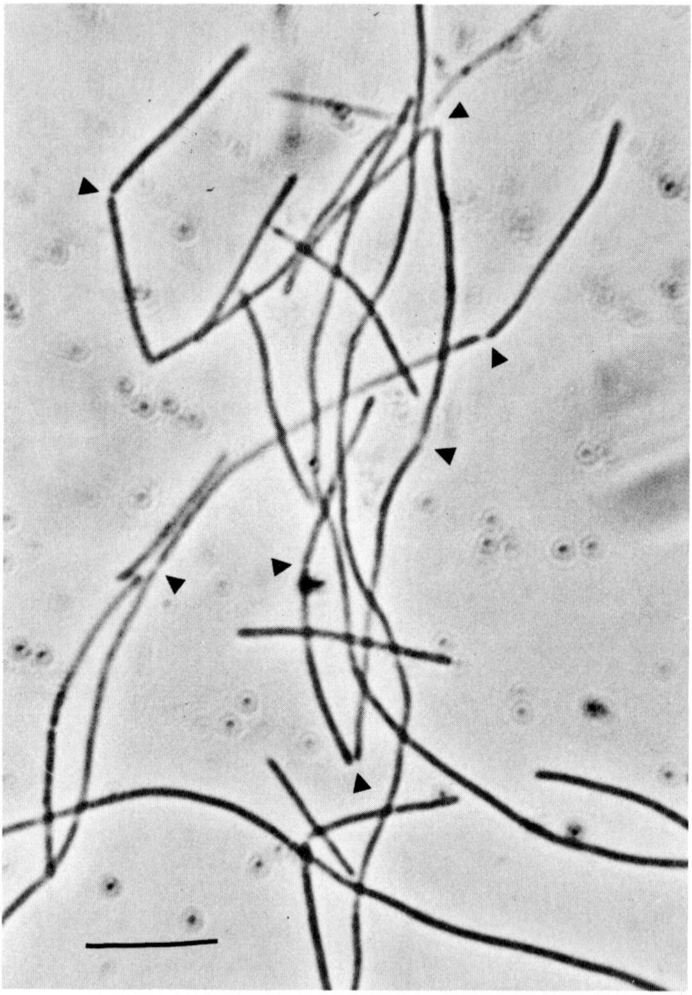

FIGURE 3. Elongated cells of *C. thermosaccharolyticum* strain SD 105 in continuous dilution culture. Note symmetric multiple septations (indicated by arrows) along the filaments that appeared as the stepwise growth deteriorated after 6 hr (FIG. 1, Curve D). Bar represents 10 μm.

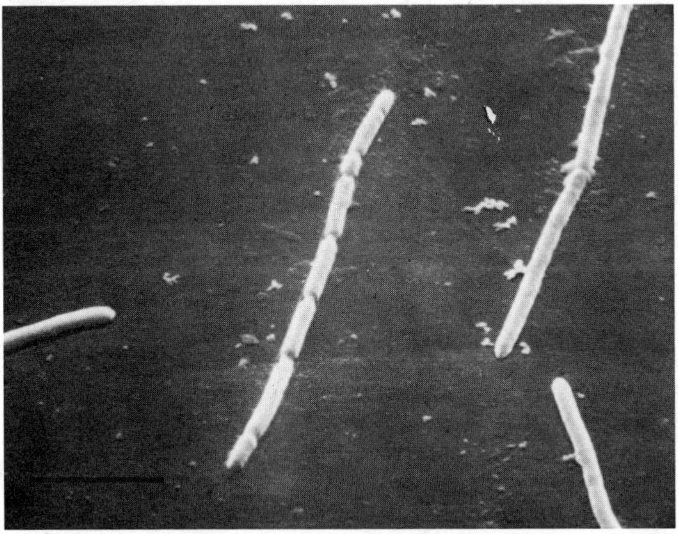

FIGURE 4. Scanning electron micrograph of elongated cells of *C. thermosaccharolyticum* strain SD 105 in continuous dilution culture. Note symmetric multiple septations along the filaments that appeared as the stepwise growth deteriorated after 6 hr (FIG. 1, Curve D). Bar represents 1 μm.

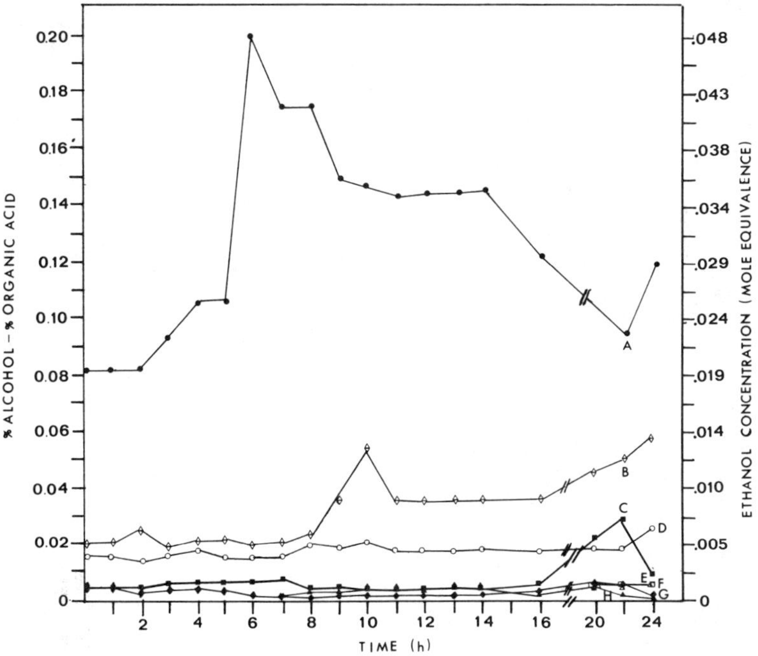

FIGURE 5. Concentrations of alcohols and organic acids produced during continuous dilution fermentation of 0.00522% xylan by *C. thermosaccharolyticum* strain SD 105, as determined by GLC analysis. Curve A (●) represents ethanol, Curve B (◇) represents butyric acid, Curve C (■) represents acetone, Curve D (○) represents acetic acid, Curve E (□) represents isobutyric acid, Curve F (△) represents propionic acid, Curve G (◆) represents butanol, and Curve H (▲) represents methanol.

REFERENCES

1. WIEGEL, J., L. G. LJUNGDAHL & J. R. RAWSON. 1979. Isolation from soil and properties of the extreme thermophile *Clostridium thermohydrosulfuricum*. J. Bacteriol. **139**(3): 800–810.
2. SJOLANDER, N. O. 1937. Studies on anaerobic bacteria XII. The fermentation products of *Clostridium thermosaccharolyticum*. J. Bacteriol. **34:** 419–428.
3. HOFFMANN, J. W., E. K. CHANG & E. J. HSU. 1978. Interruption in cell division by catabolite dilution producing synchronous growth of *Clostridium thermosaccharolyticum*. *In* Spores VII. G. Chambliss & J. C. Vary, Eds.: 312–317. American Society of Microbiology, Washington, D.C.
4. HSU, E. J. & Z. J. ORDAL. 1970. Comparative metabolism of vegetative and sporulating cultures of *Clostridium thermosaccharolyticum*. J. Bacteriol. **102**(2): 369–376.
5. GOTTSCHALK, G. 1979. Bacterial Metabolism. Springer-Verlag. New York.

Novel Immobilized-Cell Systems for the Production of Ethanol from Jerusalem Artichoke

ARGYRIOS MARGARITIS AND PRATIMA BAJPAI

Chemical and Biochemical Engineering
The University of Western Ontario
London, Ontario N6A 5B9, Canada

INTRODUCTION

Projected oil prices in the United States run as high as $80/bbl by 1985 and $100/bbl by the year 2,000,[1] despite the temporary price stability of oil due primarily to conservation measures. Since 1969, over a 13-year period, the prices of oil and natural gas have increased dramatically by a factor of about 16, while for the same time period, the average price of renewable carbohydrate materials has increased by a factor of about 4.[2,3] These escalating prices of nonrenewable oil resources have stimulated a worldwide interest in the use of fermentation ethanol as an alternative liquid fuel.[1–3] Ethanol can also be used as a starting raw material for the production of ethylene, a key material in the plastics industry. In carbohydrate-rich countries like Brazil, fermentation ethanol is emerging as an important strategic material.

A rapidly emerging new raw material of great potential is the Jerusalem artichoke (*Helianthus tuberosus*), which in the past received very little attention in the literature.[4,5] Most of the recent work on the production of fermentation ethanol from the Jerusalem artichoke has been reported by Margaritis *et al.*[6–15] The Jerusalem artichoke is a native plant of North America that is presently grown in very small quantities. The plant can grow as high as 10–12 ft tall and produces tubers in the ground. These tubers contain inulin, which is a polymer of about 30 fructose units with one glucose unit at the end of the molecule.[16] The advantages of the Jerusalem artichoke over other agricultural crops include the following: (a) it grows rapidly in poor-quality land with minimal fertilizer requirements and thus there is no competition for the good-quality land used for traditional food crops; (b) it possesses high tolerance to frost and various plant diseases; (c) it has the highest carbohydrate yields of any known crop reported so far, and these range between 3,400 to 6,000 kg sugars/acre/yr,[16–18] which correspond to an ethanol yield of about 1,800 to 3,000 kg ethanol/acre/yr; and (d) the solids left from extraction of sugars from the stalks and tubers have a good protein content and can be used as an excellent animal feed.

This paper describes the use of immobilized yeast cells of *Kluyveromyces marxianus* to produce ethanol continuously from tuber extract using a specially designed horizontal packed-bed bioreactor system.[19–21] The yeast cells were entrapped in small Ca-alginate beads of about 2 mm in diameter that contained extremely high cell concentrations. The immobilized packed bioreactor performance is also compared with that of an ordinary, continuous stirred-tank bioreactor containing free cells of *K. marxianus*.[11]

MATERIALS AND METHODS

The extraction of inulin sugars from the tubers of Jerusalem artichoke was done as described by Margaritis and Bajpai.[11,12] Free cells of *Kluyveromyces marxianus* UCD(FST) 55-82 were grown in 1-liter Bellco fermenters at 35°C having the following composition per liter of tap water: glucose, 100; yeast extract, 10; $MgSO_4$, 0.5; KH_2PO_4, 0.1; $(NH_4)_2SO_4$, 0.1 Free cells were harvested in the later exponential growth phase and concentrated by centrifugation at 6,000 rpm for 10 min. Concentrated cells were resuspended in 0.1% NaCl solution to which 1.6% Na alginate was added and mixed to form a thick suspension. The cell/Na-alginate mixture was then extruded as small drops using a specially designed apparatus into an ice-cold solution containing 4% $CaCl_2$. Small beads of about 1–2 mm diameter were formed and allowed to harden in the $CaCl_2$ solution for 30 min. During hardening, the Na^{2+} ions in Na-alginate matrix are exchanged with Ca^{2+} ions from the $CaCl_2$ solution. It is interesting to note that Na-alginate is water soluble while Ca-alginate is not. The beads were then transferred aseptically to a packed-bed column that was previously sterilized. The column had a total volume of 70 ml, of which 40 ml were the solid beads containing the immobilized cells and the remaining 30 ml was liquid volume. The immobilized cell concentration inside the Ca-alginate beads was 78.5 g dry wt. cell/liter bead volume, which corresponded to 45 g dry wt. immobilized cells/liter total bioreactor volume. At steady-state conditions, the following outlet concentrations were measured at a given dilution rate: total sugars using the anthrone reagent method,[12] ethanol by gas-chromatography,[12] and free cells in the liquid phase by the optical density method.[12] The liquid medium consisted of Jerusalem artichoke tuber extract containing 10.21% w/v total sugars, and supplemented with 0.05% v/v Tween 80, 0.01% oleic acid, and 0.01% cornsteep liquor. The pH was adjusted to 4.5 and the column was run at 35°C at different dilution rates ranging from 0.5 hr^{-1} to 3.7 hr^{-1}. In order to study the half-life of the immobilized-cell bioreactor, it was run continuously for 30 days at the same fixed dilution rate of 2.1 hr^{-1}.

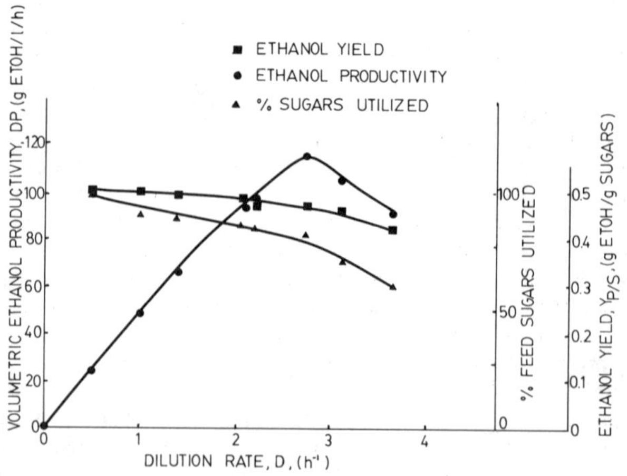

FIGURE 1. Ethanol yield, ethanol productivity, and sugars utilized at different dilution rates.

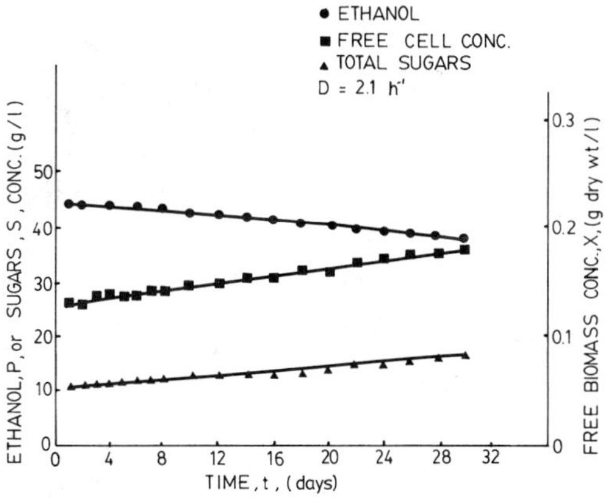

FIGURE 2. Ethanol, sugars, and free-cell concentration as a function of time at a fixed dilution rate of 2.1 hr^{-1}.

RESULTS AND DISCUSSION

FIGURE 1 shows the volumetric ethanol productivity, ethanol yield, and % sugars utilized as a function of dilution rate. These data show that the ethanol yield remained almost the same at about 0.48 g ethanol/g sugars used, while the volumetric ethanol productivity reached a maximum of 118 g ethanol/1/hr at 2.80 hr^{-1} dilution rate. Margaritis and Bajpai[11] used the same medium composition with free cells of *K. marxianus* in an ordinary continuous stirred-tank bioreactor and obtained a maximum volumetric ethanol productivity of 7 g ethanol/1/hr. Therefore, a comparison between a free and an immobilized-cell bioreactor system showed that the immobilized cell system gave ethanol productivities 17 times higher than the ordinary free-cell, stirred-tank bioreactor. Furthermore, the immobilized-cell bioreactor was operated at high dilution rates, that is, 3.7 hr^{-1} without cell washout. The washout dilution rate for the ordinary free-cell, stirred-tank bioreactor was found to be 0.41 hr^{-1}. The high dilution rates and high volumetric ethanol productivities are inherent advantages of the newly developed immobilized-cell systems when compared with the free-cell systems.

FIGURE 2 shows the outlet concentrations of ethanol, sugars, and free cells as a function of time at a fixed dilution rate of 2.1 hr^{-1}. These results show that in a 30-day period, there was only 15% loss of the original volumetric ethanol productivity, which corresponds to an estimated column half-life of about 100 days. The free-cell concentration in the liquid phase coming out of the bioreactor was less than 0.08 g dry wt. cell/l. These low free-cell concentrations indicate that the Ca-alginate beads were very stable and very few cells leaked out of the solid matrix. The high half-life values obtained in this study show that immobilized-cell bioreactor systems can be operated continuously for long periods of time without replacement. These advantages are strong indications of the important impact this new immobilized-cell technology will have on present fermentation technology, which is obsolete, being energy and capital intensive. More work is needed with large-scale, immobilized-cell systems in order to assess their economic feasibility and impact on free-cell fermentation technology.

SUMMARY

The inulin sugars extracted from the tubers of Jerusalem artichoke (*Helianthus tuberosus*) were fermented to ethanol in a continuous packed-bed bioreactor containing immobilized cells of *Kluyveromyces marxianus*. The packed-bed bioreactor was operated at dilution rates ranging from 0.5 hr^{-1} to 3.7 hr^{-1} and the effluent ethanol, total sugars, and free-cell concentration were measured. The maximum volumetric ethanol productivity was found to be 118 g ethanol/l/hr at a dilution rate of 2.8 hr^{-1}. The immobilized-cell bioreactor was operated continuously at a fixed dilution rate of 2.1 hr^{-1} for 30 days with 15% loss of its original ethanol productivity, which corresponds to an estimated half-life of approximately 100 days. The immobilized-cell bioreactor gave about 17 times higher volumetric ethanol productivity than an ordinary continuous stirred-tank bioreactor employing free cells of *K. marxianus*.

REFERENCES

1. Weaver, K. F. 1981. Natl. Geogr. **1**.
2. Margaritis, A. & J. Vogrinetz. 1980. Abstr. 6th Intl. Ferment. Symp. July 20–26. London, Ontario.
3. Margaritis, A. 1980. A.P.E.O. Dimensions. **1:** 21.
4. Boinot, F. 1942. Bull. Assoc. Chim. **59:** 792.
5. Underkofler, L. A., W. K. McPherson & E. I. Fulmer. 1937. Ind. Eng. Chem. **29:** 1160.
6. Margaritis, A., P. Bajpai & E. Cannell. 1981. Biotechnol. Lett. **3:** 595.
7. Margaritis, A. & P. Bajpai. 1981. Biotechnol. Lett. **3:** 679.
8. Margaritis, A. & G. E. Rowe. 1983. Dev. Ind. Microbiol. **24:** In press.
9. Margaritis, A., P. Bajpai & P. K. Bajpai. 1983. Dev. Ind. Microbiol. **24:.** In press.
10. Margaritis, A. & P. Bajpai. 1982. Biotechnol Bioeng. **24:** 941.
11. Margaritis, A. & P. Bajpai. 1982. Biotechnol. Bioeng. **24:** 1473.
12. Margaritis, A. & P. Bajpai. 1982. Biotechnol. Bioeng. **24:** 1483.
13. Margaritis, A. & P. Bajpai. 1983. Appl. Environ. Microbiol. **45**(2)**:.** In press.
14. Margaritis, A. & P. Bajpai. 1982. Appl. Environ. Microbiol. **44**(6)**:** 1325.
15. Margaritis, A., P. Bajpai. 1982. Submitted at CRC Crit. Rev. Biotechnol.
16. Fleming, S. E. & J. W. D. Groot Wassink. 1979. CRC Crit. Reve. Food Sci. Nutr., **12:** 1.
17. Chubey, B. B. & D. G. Dorrell. 1974. Can. Inst. Food Sci. Technol. J. **7:** 98.
18. Dorrell, D. G. & B. B. Chubey. 1977. Can. J. Plant Sci. **57:** 591.
19. Margaritis, A., P. K. Bajpai & B. Wallace. 1981. Biotechnol. Lett. **3:** 613.
20. Margaritis, A. & J. B. Wallace. Biotechnol. Bioeng. Symp. No. 12. In press.
21. Rowe, G. E. 1981. M.E.Sc. Thesis. University of Western Ontario. London, Ontario.

Novel Immobilized Bioreactor for Rapid Continuous Ethanol Fermentation of Cane Juice or Fruit Juices

SUSUMU FUKUSHIMA AND HIROYUKI HATAKEYAMA[a]

Department of Chemical Engineering
Faculty of Engineering
Kansai University
Osaka, Japan

[a]Technical Division
Takara Shuzo Co., Ltd.
Kyoto, Japan

This work is aimed at finding an excellent biocatalyst for alcohol fermentation of cane juice or fruit juices and at performing nonstop alcohol production in a novel bioreactor over a long period of time.

EXPERIMENT

Biocatalyst

The yeast strains used were *Saccharomyces cerevisiae montrachet* and *S. formosensis M-111* for cane juice, and *S. cerevisiae OC-2* and *S. cerevisiae W-3* for fruit juices. The biocatalyst, 1 mm in diameter and entrapping a single strain or a mixture of two strains with Al alginate, was prepared by a two-step method, in which the first step formed gel particles entrapping cells with Ca alginate, and the second step exchanged the Ca for Al ions. The cell content was 220 g(dry cells)/l(biocatalyst).

Cane Juice and Fruit Juices

Two series of cane juices were used. Juice A was obtained from fresh cane within one month after harvest. Juice B was from cane that was stored at 0°C for one year. Fruit juices were also used: mandarin orange and apple juices in Japan, and California Valencia orange juice. These concentrated solutions were stored at 5°C.

Bioreactor

A 3-stage, rhomboid bioreactor was designed by one of the authors (U.S. patent 4337315). The ratio of biocatalyst to working volume, v_S/v, was 0.30 and 0.45. The feed juice was adjusted at pH 2.8–3.2 by adding H_2SO_4. No agents for growing cells were added to the feed juice. The continuous operation was anaerobically performed at pH 2.8–3.2, 20 or 30°C during 20–200 days nonstop without risk of contamination.

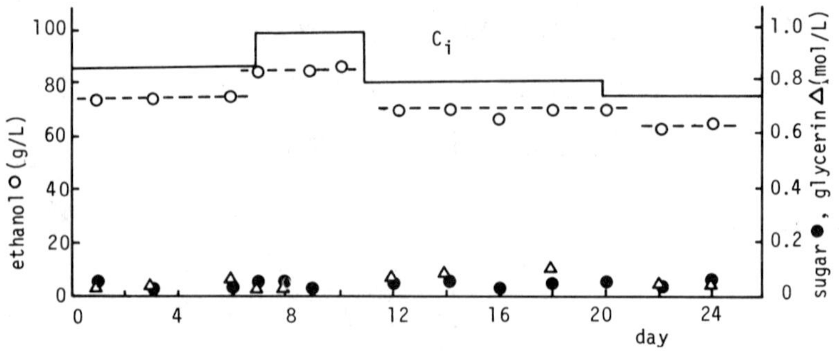

FIGURE 1. Continuous alcohol fermentation of cane juice A in a 3-stage rhomboid bioreactor packed with a biocatalyst entrapping a mixture of *Saccharomyces cerevisiae montrachet* and *S. formosensis M-111*. $v_S/v = 0.3$, $\tau = 6$ hr, $D_S = 0.56$ hr^{-1}

RESULTS AND DISCUSSION

Cane Juice

In the bioreactor that was packed with single-strain biocatalyst ($v_S/v = 0.3$), 78 g/l ethanol and 4 g/l glycerin were produced from juice A, with feed sugar concentration as monosaccharide C_i of 1.05 mol/l at the residence time τ of 6 hr. In the same performance, juice B gave one-half the ethanol and four times higher glycerin in comparison with juice A.

FIGURE 1 shows the observed data from juice A at $\tau = 6$ hr in the bioreactor packed with biocatalyst entrapping a mixture of two strains. The ethanol yield is 15% higher than that of a single strain. The dotted lines are the evaluated values from the simple simulation; the bioreactor is the ideal mixing tank, the reaction resistance producing ethanol from monosaccharide is predominant in comparison with diffusion and hydrolysis and the side reaction is negligible. The kinetic data were measured by the Warburg's apparatus. These lines agree well with the observed data.

Highly packaged bioreactor, $v_S/v = 0.45$, produced 9.5% (v/v) ethanol from a C_i of 1.1 mol/l as shown in TABLE 1. The alcohol productivity based on biocatalyst PD_S was 45 g/l · hr.

TABLE 1. Continuous Fermentation of Cane Juice A in a 3-Stage Rhomboid Bioreactor Packed with a Biocatalyst Entrapping a Mixture of *S. cerevisiae montrachet* and *S. formosensis M-111*[a]

Residence Time Based on Bioreactor τ (hr)	Dilution Rate Based on Biocatalyst D_S (1/hr)	Concentration P(g/l)		Productivity Based on Biocatalyst PD_S(g/l · hr)	
		Obs.	Simul.	Obs.	Simul.
2.4	0.93	53	51	49	47
4.5	0.49	95	95	47	47

[a] $v_S/v = 0.45$, $C_i = 1.1$ mol/l

Fruit Juice

The bioreactor—packed with the biocatalyst entrapping a mixture of different wine yeast strains—produced 10–13% (v/v) ethanol from Japanese mandarin and California Valencia orange juice at $\tau = 6$ to 20 hr during 7 months nonstop operation without any trouble. Furthermore, the similar bioreactor produced 17% (v/v) ethanol from Japanese apple juice when τ was 10 hr.

Therefore, for rapid continuous ethanol production from cane juice, it is necessary, in order to get superior results through this process, to use the juice obtained by pressing fresh cane. This biosystem is also expected to be useful for young fruit wine production.

Different Methods of Biomass Retention in Continuous Anaerobic Digestion

A. AIVASIDIS, AND C. WANDREY

Institute of Biotechnology
Nuclear Research Centre
D-5170 Jülich, Federal Republic of Germany

Anaerobic microorganisms usually grow very slowly and biomass concentrations of anaerobic processes are quite low. This results in large retention times necessary for digestion. In a continuously stirred tank reactor, liquid retention time is equal to the retention time for the biomass. Reduction of the retention time below the critical value involves bacteria washout. To avoid this instability, process configurations are desired for improved biomass retention at high influent flow rates. Different methods that, as is already known, promote high bacteria concentrations at short retention times as well as some new developments are summarized in TABLE 1.

These methods are described in the poster paper in detail and are illustrated with examples from the literature.

SEDIMENTATION

Biomass can be recovered from suspension by sedimentation using any natural flocculation capacity or by the addition of agents that enhance flocculation. In some cases, depending on the environmental conditions, bacterial granules have been observed. The anaerobic contact process consists of a reactor vessel and a sedimentation tank where the flocculated bacteria settle to the bottom and are returned by a recycle pump to the digester.

The basic principle of the Upflow Anaerobic Sludge Blanket (UASB) process is founded on the formation of anaerobic sludge with improved sedimentation properties. Reactor mixing is only effected by gas generation. A settler unit at the top of the reactor is designed to capture rising particles of biomass attached to the gas bubbles. It has been found that a significant part of the anaerobic sludge in the lower regions of the sludge bed appears in a "granular" form with better settleability in comparison to the "flocculated" sludge.

IMMOBILIZATION

Catalyst immobilization in chemical engineering is a general operation in order to combine the advantages of homogeneous and heterogeneous catalysts. Many techniques for living cell immobilization have been developed up to the present time; however the one most suitable for anaerobic waste treatment is adsorption by an inert support material. The biomass adheres to the carrier by electrostatic forces, Van der Waals and hydrophobic interactions, or by means of slime films. For illustration, an example is given for the case of immobilized *Methanosarcina barkeri* on carbon in a fixed-bed loop reactor on acetate degradation (FIG. 1).

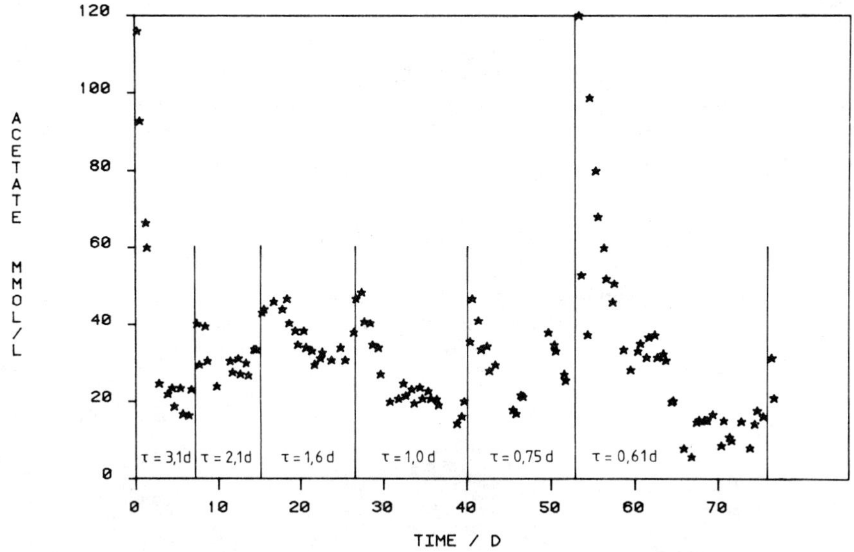

FIGURE 1. Performance of an anaerobic fixed-bed loop reactor on acetate degradation by means of a pure culture of *M. barkeri* attached on anthrazit N® by different retention times.

FILTRATION

This technique has not found any technical application for biomass recovery in anaerobic treatment processes. There are a lot of impediments, such as the special filters needed, large filter areas and high energy requirements because of the high volumetric flows, which prevent any practical use.

TABLE 1. Different Methods of Biomass Retention in Anaerobic Waste Treatment

1. Sedimentation
    - 1.1 Flocculation
        - 1.1.1 anaerobic contact process
    - 1.2 Pelletization
        - 1.2.1 upflow anaerobic sludge blanket (UASB) process
2. Immobilization
    - 2.1 Adsorption by an inert support
        - 2.1.1 fixed-bed reactor (FBR)
        - 2.1.2 fixed-bed loop reactor (FBLR)
        - 2.1.3 fixed-film reactor (FFR)
        - 2.1.4 fluidized-bed reactor (FBDR)
    - 2.2 Entrapment by an inert support
    - 2.3 Selective binding of cells by immobilized macromolecules
    - 2.4 Covalent or coordinate bonding of the cell to an inert support
3. Filtration

   Flotation

5. Centrifugation

FLOTATION

Concentration of solids by flotation has become increasingly important for waste water purification in recent years. Suspended solids with bad settlement characteristics are conveyed by rising highly dispersed gas bubbles at the surface and can be easily separated by foam removal.

CENTRIFUGATION

This separation method as an accelerated form of sedimentation has not yet found any application in biomass separation and recycling in continuous anaerobic digestion. One of the major impediments to its implementation is the high energy requirements for liquid recirculation.

Continuous Anaerobic Digestion with *Methanosarcina barkeri*

C. WANDREY AND A. AIVASIDIS

Institute of Biotechnology
Nuclear Research Centre Jülich
D-5170 Jülich, Federal Republic of Germany

INTRODUCTION

In spite of the increasing importance of anaerobic digestion, little is known about the kinetic data of relevant microorganisms. In order to obtain a dependable layout of waste water reactors and in order to identify limiting factors, it is essential to evaluate such kinetic data by means of appropriate measurements. Such experiments should be carried out continuously in a chemostat in order to establish reproducible conditions.[1] A pure culture should be used, so that besides the reaction conditions, the "catalyst conditions" are also well defined.[2]

Acetic acid is a suitable substrate since the methanation of acetic acid is thought to be the limiting step during the anaerobic digestion of many different residues. Acetic acid itself causes chemical oxygen demand (COD) in the waste water of important industrial processes, for example, in the paper and pulp industry.[3] Here in the condensate from the concentration of the spent sulfite liquor, acetic acid is found as almost the only carbon source. Such "distilled" waste water is very suitable for the transfer of results with model waste water to a case of practical importance. Furthermore, since only very few microorganisms can grow anaerobically with acetic acid as the only carbon source, such a culture is potentially autosterile. *Methanosarcina barkeri*[4] was chosen as a "catalyst" due to the frequent occurrence of this strain in many digestion plants.

$$CH_3COOH \xrightarrow{M.\ barkeri} CH_4 + CO_2 \qquad (1)$$

Methanosarcina barkeri (DSM 804) was grown on a mineral medium as given in TABLE 1. The element analysis of acetic acid grown *M. barkeri* is given in TABLE 2.

EXPERIMENTAL

Chemostat experiments were carried out with working volumes of 5, 10, 50, and 130 liters. Most of the data were obtained on the 5-liter scale, but the results could also be reproduced on a bigger scale. The space time yield on the 130-liter scale reached at least the corresponding values found on the 5-liter scale under all conditions. A flow scheme of the experimental setup is given in FIGURE 1.

The reaction conditions are given in TABLE 3. From the known initial or boundary conditions, all relevant data could be calculated by measuring the amount and composition of the biogas, as well as the amount of composition of the waste water (acetic acid). Furthermore, the dry weight of the catalyst in the outlet stream was determined. A constant residence time was guaranteed by synchronized substrate and product pumps.

TABLE 1. Mineral Medium for Growth of *Methanosarcina barkeri*

1000	mM CH_3COOH	Trace elements <10 μM
35	mM Na^+	(Co, Ni, Mn, Se, Zn, Mo)
10	mM NH_4^+	
5	mM K^+	
3	mM Ca^{2+}	Vitamins <1 μM
2	mM Mg^{2+}	(Wolfe's vitamins)
1	mM S^{2-}	
1	mM HPO_4^{2-}	pH_o = 3.5
0.85	mM L-Cystein	
0.55	mM Fe^{2+}	

TABLE 2. Element Analysis of Acetic Acid Grown *Methanosarcina barkeri*

Carbon	46.85%
Nitrogen	12.16%
Hydrogen	6.69%
Sulfur	1.05%
Phosphorus	0.91%

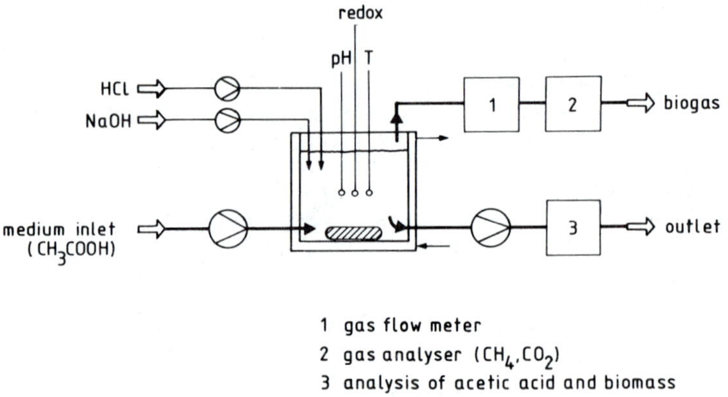

FIGURE 1. Flow scheme of the chemostat for continuous anaerobic digestion of acetic acid.

TABLE 3. Reaction Conditions

S_o =	1.0 mol/l
X_o =	0 g/l
pH_o =	3.5
S_e =	Reactor value at steady state
X_e =	Reactor value at steady state
pH_e =	6.3
T =	37°C

MATHEMATICAL MODEL AND EXPERIMENTAL DATA

At a steady state, the following balances are valid:

Material Balance for Substrate

$$0 = -\frac{1}{\tau} \cdot (S_o - S_e) + \frac{\mu_{max}}{Y_{max}} \cdot \frac{S_e}{k_m + S_e} \cdot X_e + m \cdot X_e \tag{2}$$

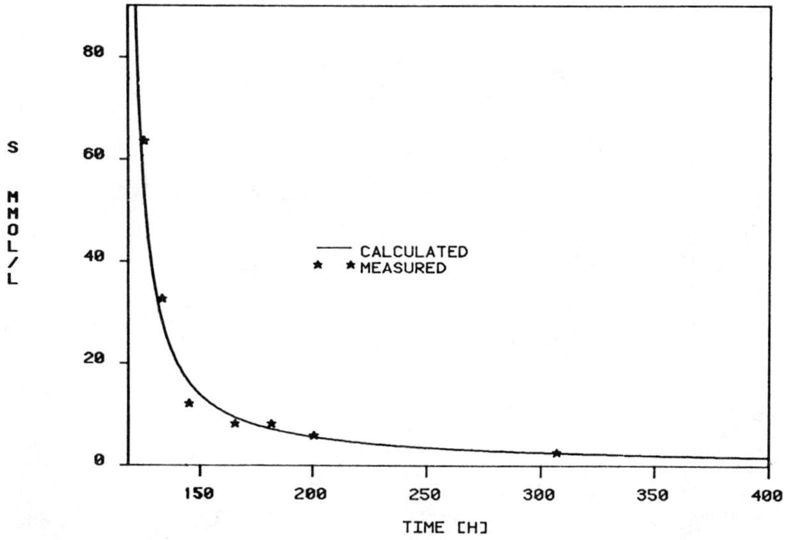

FIGURE 2. Substrate concentration at different residence times, $S_o = 1.0$ mol/l.

Material Balance for Biomass

$$0 = -\frac{1}{\tau} \cdot X_e + \mu_{max} \cdot \frac{S_e}{k_m + S_e} \cdot X_e \tag{3}$$

To determine the four model parameters (μ_{max}, k_m, Y_{max}, m), at least four independent measuring points are necessary for parameter determination. Here, steady states were measured at 7 different residence times in order to obtain the data basis for a nonlinear regression analysis.[5] FIGURES 2 and 3 show a comparison between experimental substrate and biomass data and the values predicted by the final model. While the model predicts the remaining substrate concentration fairly accurately, the feed for the biomass concentration is less satisfying. This is due to the very small maintenance requirements of *M. barkeri* (see below). From the different remaining substrate concentrations at different residence times, μ_{max} and k_m could be obtained according to FIGURE 4.

The individual growth rates are identical to the reciprocal residence times (dilution

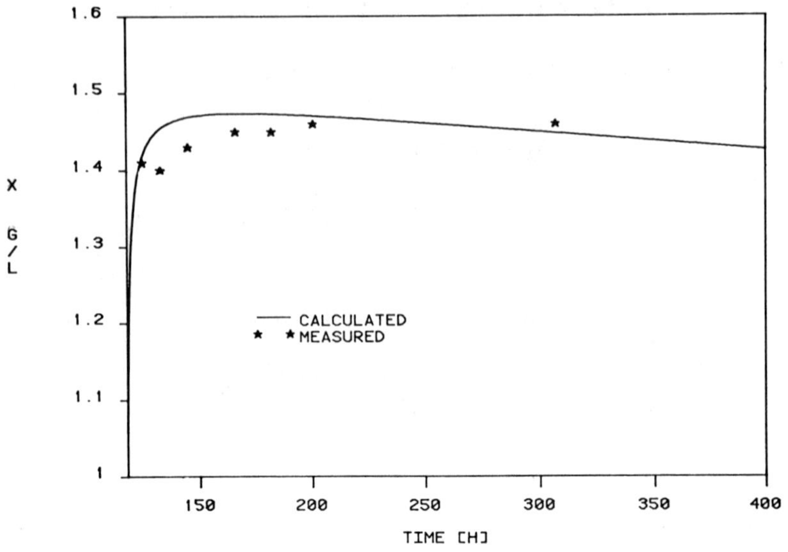

FIGURE 3. Biomass concentration at different residence times, $S_o = 1.0$ mol/l.

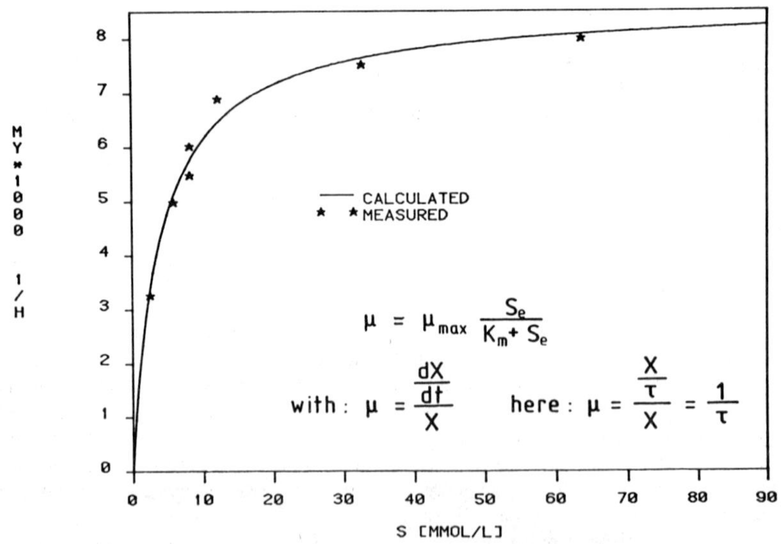

FIGURE 4. Specific growth rate, $\mu_{max} = 0.008593$ hr^{-1}, $k_m = 0.00401$ mol/l.

rates). The model parameters were obtained by nonlinear regression in order to avoid a distortion of the error distribution occurring during linear regression methods that use the reciprocal values of the original experimental data. Not only the specific growth rate can be regarded as a function of the remaining substrate concentration, but also the substrate consumption rate for biogas as shown in FIGURE 5. In spite of a different experimental basis (dry matter of biomass and biogas flow measurements, respectively), a very similar k_m value results. This fact indicates that the biomass formation is correlated with the biogas formation to a high degree. This can be understood via ATP generation during biogas formation as an energy source for carbon incorporation by the biomass.

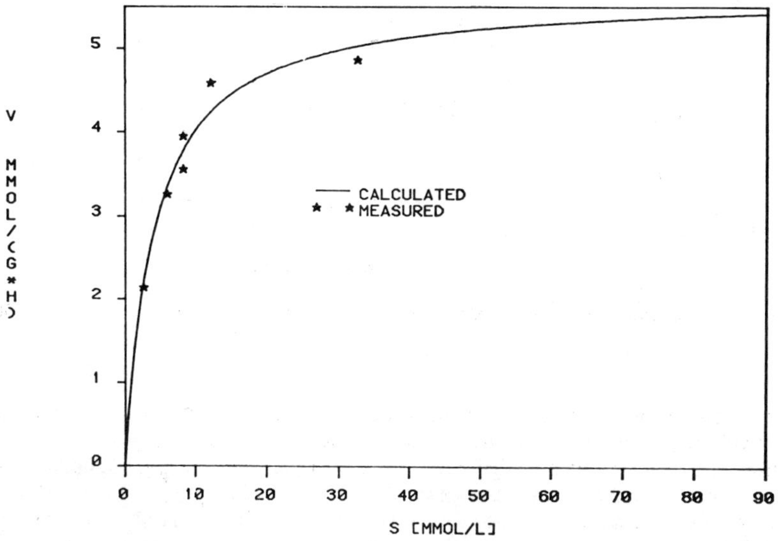

FIGURE 5. Substrate consumption rate for biogas, V_{max} = 0.00566 mol/(g.hr), k_m = 0.00407 mol/l.

From the measured biomass concentration at the different steady states (relating these values to the corresponding growth rates), Y_{max} and m can be obtained by means of the Pirt equation[6] (FIG. 6).

The scattering of the data is due to the low maintenance requirement of *M. barkeri*. Even without quantitative evaluation of the data, it becomes clear from FIGURE 6 that even at low growth rates (or long residence times), the biomass yield drops only slightly. It would have been desirable to have a further data point at an even lower growth rate (higher residence time). However, since it took several months to obtain a precise steady state for a residence time of 307 hr, it seemed not to be appropriate to go any further.

In TABLE 4, the model parameters obtained and some derivatives values are summarized. The generation time of 80.66 hr is obtained from the maximal growth rate. The critical residence time for washout is obtained by Equation 4:

$$\tau_{crit} = \frac{k_m + S_e}{\mu_{max} \cdot S_e} \tag{4}$$

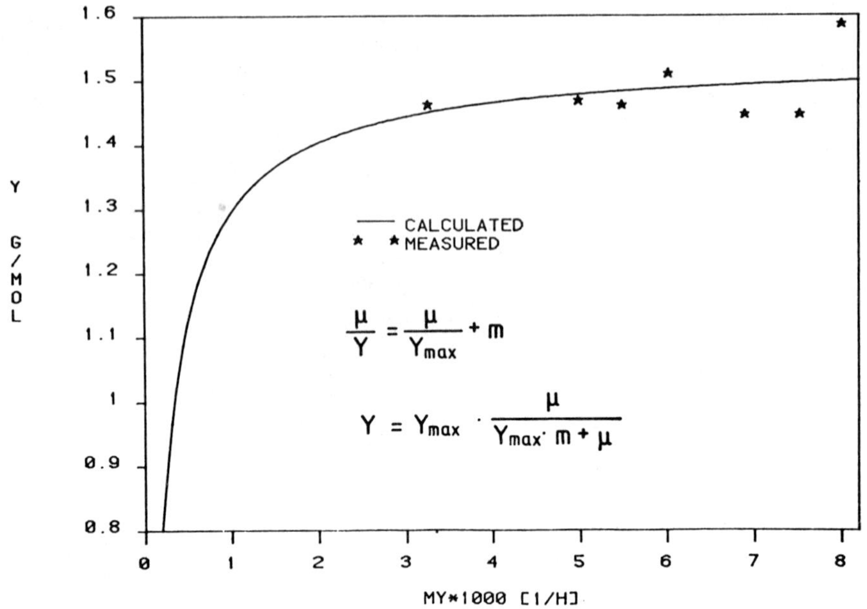

FIGURE 6. Biomass "yield" (per mole consumed acetate), $Y_{max} = 1.533$ g/mol, $m = 0.00012$ mol/(g.hr).

The residence time for maximal biogas production rate (124.24 hr) was obtained by numerical identification of the maximum in FIGURE 7. For this purpose, the mass balance for substrate and the mass balance for biomass has to be solved simultaneously as functions of residence time using the identified model parameters. As can be seen from FIGURE 7, the maximal space time yield can be obtained very close to the washout point, indicating the importance of a precise regulation of a waste-water reactor. Moreover—due to the favorably low k_m value—it can be seen that under such conditions, very high biogas yields can already be obtained. The optimum predicted by the model does not quite coincide with the experimental values, but one has to keep in mind that an extreme parameter sensitivity occurs at residence times below the residence time for the maximum. Here it was possible to reach a steady state at a residence time slightly below this value. The biogas production rate decreases at higher residence times due to substrate limitation. In the same course, the biogas yield

TABLE 4. Model Parameters

μ_{max} =	0.008593 (hr^{-1})
Y_{max} =	1.533 (g/mol)
k_m =	0.00401 (mol/l)
m =	0.00012 (mol/g · hr)
$t_{2/1}$ =	80.66 (hr)
τ_{crit} =	116.84 (hr)
τ_{max} =	124.24 (hr)

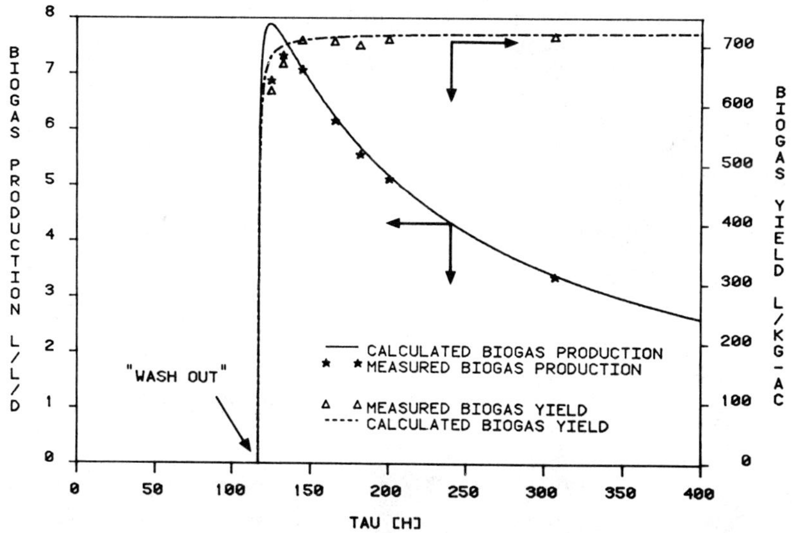

FIGURE 7. Biogas production rate and biogas yield, $S_o = 1.0$ mol/l.

increases only very slightly. The comparatively high biogas yield (more than 700 l/kg acetate) is due to the fact that acetic acid can be consumed practically quantitatively.

Since acetic acid was the only carbon source, it was possible to follow the fate of carbon during anaerobic digestion. If the amount of carbon incorporated into the biomass is related to the amount of carbon digested, one obtains the "incorporation

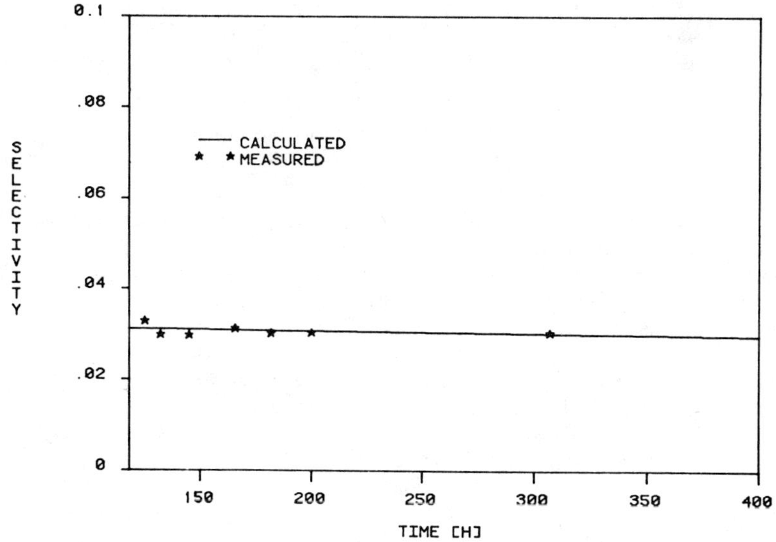

FIGURE 8. Incorporation selectivity.

selectivity" as shown in FIGURE 8. The model predicts these values very precisely. There is a decrease in incorporation selectivity with increasing residence time that is difficult to detect. Similarly the "maintenance selectivity" can be obtained from the substrate consumed for maintenance relative to the entire consumption of substrate at different residence times.

FIGURE 9 shows that with increasing residence time, more and more substrate is needed. The substrate consumption is related to the ATP generation needed for maintenance. The substrate consumption for the "repair" of cell mass can be neglected. Besides substrate consumption for incorporation and substrate consumption

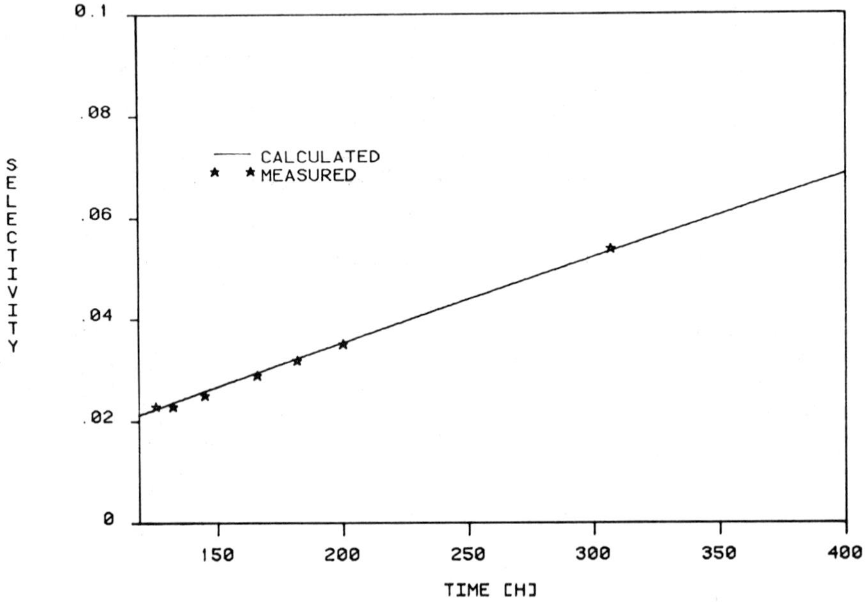

FIGURE 9. Maintenance selectivity.

to produce biogas and to deliver ATP for maintenance, there is substrate consumption to produce biogas and deliver ATP for the energy requirements during new cell formation. This portion—the selectivity for biogas for growth—adds up to 100% together with the incorporation selectivity and the maintenance selectivity. As shown in FIGURE 10, the selectivities are given as a function of conversion. A conversion of 94.06% represents the maximum in FIGURE 7, whereas another way to illustrate the carbon balance is given in FIGURE 11. This figure applies to the conditions at the maximum in FIGURE 7. The height of the arrows is proportional to the portion of the carbon coming from the substrate and going to biomass or biogas. Only about 3% of the total carbon digested is incorporated while about 97% is converted to biogas. The ATP generation in connection with the biogas formation is mostly used for biomas formation. A very minor portion is required for maintenance. This information can be extracted by means of the model parameters in TABLE 4.

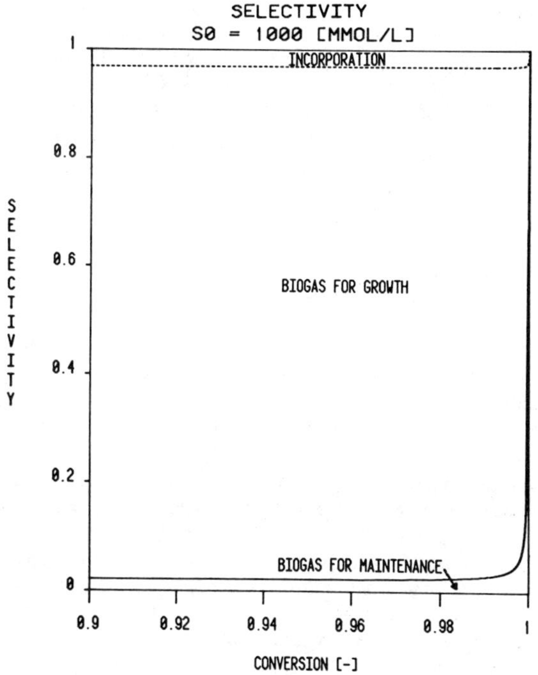

FIGURE 10. Selectivities.

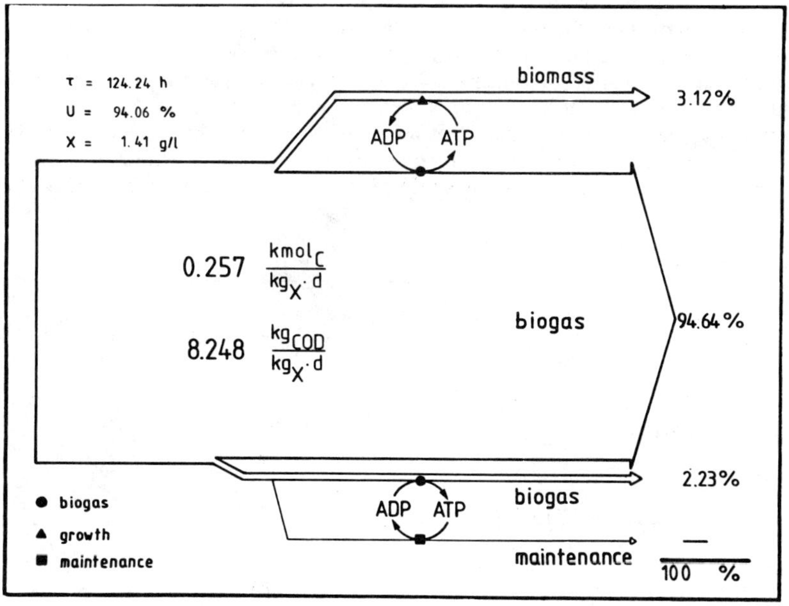

FIGURE 11. Carbon balance (anaerobic).

TABLE 5. Concentrations

$S_o = 1$ mol_S/l	$\triangleq 2$ mol_C/l	$\triangleq 2$ mol_{COD}/l
60 g_S/l	$\triangleq 24$ g_C/l	$\triangleq 64$ g_{COD}/l
$S_e = 0.059$ mol_S/l	$\triangleq 0.118$ mol_C/l	$\triangleq 0.118$ mol_{COD}/l
3.54 g_S/l	$\triangleq 1.416$ g_C/l	$\triangleq 3.776$ g_{COD}/l
$X_o = 0$ g/l		
$X_e = 1.410$ g/l		

STANDARDIZATION OF UNITS

Owing to the very different substrates used during anaerobic digestion, very different definitions are often used. According to Nyns,[7] all relevant values should be summarized as concentrations, rates, yields, and efficiencies (selectivities). For the conditions of the maximum in FIGURE 7, all values were calculated using the model parameters of TABLE 4.

In TABLE 5 the concentration values are summarized. The initial substrate concentration is given as moles substrate per liter as well as grams substrate per liter, moreover moles carbon per liter or grams carbon per liter are also specified. With respect to waste-water treatment, these values can be transferred to moles chemical oxygen demand per liter or grams chemical oxygen demand per liter.

The outlet substrate concentration can be specified in the same manner. In this case, this value corresponds to a conversion of 94.06%. The biomass concentration is zero in the inlet stream and the outlet concentration is given in grams dry biomass per liter.

Rates are specified in TABLE 6. First of all, the substrate consumption rate is given in moles substrate per gram dry biomass and day. This value can easily be transformed to kg chemical oxygen demand per kg dry biomass and day. With respect to enzymatic activities, a value in units per gram dry biomass can be specified. Using the biomass concentration, a substrate consumption rate can also be related to unit volume. Since these values depend on the biomass concentration and this value itself on the initial substrate concentration, the result does not directly indicate the biomass activity. This is why the first definition is preferred. The substrate consumption for biomass can be related to the entire substrate consumption; the substrate consumption for product

TABLE 6. Rates

R_S	$= 0.129$	$\frac{mol_S}{g_X \cdot d}$	$\triangleq 89.5$	$\frac{U}{g_X}$	$\triangleq 0.182$	$\frac{mol_S}{l \cdot d}$
	$\triangleq 8.256$	$\frac{kg_{COD}}{kg_X \cdot d}$			$\triangleq 11.64$	$\frac{kg_{COD}}{m^3 \cdot d}$
$R_{S/X}$	$= 0.031$	R_S				
$R_{S/P}$	$= 0.969$	R_S				
R_X	$= 0.193$	$\frac{g_X}{g_X \cdot d}$			$\triangleq 0.272$	$\frac{g_X}{l \cdot d}$
R_P	$= 0.250$	$\frac{mol_P}{g_X \cdot d}$	$\triangleq 5.60$	$\frac{l}{g_X \cdot d}$	$\triangleq 7.90$	$\frac{l}{l \cdot d}$

(biogas) can be similarly expressed. These values already indicate selectivities (see below).

The biomass production rate R_X must not be confused with μ_{max} since μ_{max} is a model parameter while the biomass production rate depends on the residence time. This value can also be related to unit volume, taking into account biomass concentration. The product production rate can be obtained from the biogas formed related to a unit weight of one gram dry biomass. Instead of using a value of moles product formed, the volume of biogas formed is often used (here the entire biogas formed and not only the methane portion is recommended). Such a value may be related to the unit weight of one gram dry biomass as well as to unit volume. For the same reasons (see above), the value related to one gram dry biomass is preferred.

Experimentally, 5.24 liters per gram dry biomass and day were found, indicating that *M. barkeri* exhibits a quite remarkable activity rate. In TABLE 7, yields and efficiencies (selectivities) are summarized. While yields have to be seen with respect to the entire substrate delivered, efficiencies (selectivities) are related to the substrate consumed. In this connection, it has to be pointed out that most of the "yields" often given in the literature are efficiences (selectivities) according to the definition given here. The values with respect to biomass and product formation may be calculated on a

TABLE 7. Yields and Efficiencies (Selectivities)

Y_{X/S_o}	=	0.0235	$\frac{g_X}{g_{S_o}}$	$\triangleq$	1.41	$\frac{g_X}{mol_{S_o}}$	$\triangleq$	0.0293	$\frac{g_{CX}}{g_{CS_o}}$
Y_{P/S_o}	=	0.911	$\frac{g_P}{g_{S_o}}$	$\triangleq$	0.681	$\frac{l_P}{g_{S_o}}$	$\triangleq$	0.911	$\frac{g_{CP}}{g_{CS_o}}$
$Y_{X/\Delta S}$	=	0.025	$\frac{g_X}{g_S}$	$\triangleq$	1.50	$\frac{g_X}{mol_S}$	$\triangleq$	0.0312	$\frac{g_{CX}}{g_{CS}}$
$Y_{P/\Delta S}$	=	0.969	$\frac{g_P}{g_S}$	$\triangleq$	0.724	$\frac{l_P}{g_S}$	$\triangleq$	0.969	$\frac{g_{CP}}{g_{CS}}$

gram or mole basis or on the basis of grams carbon per gram carbon. With the latter values, we obtain the relative yields for a given conversion or the dimensionless selectivities. The remarkably small value for biomass formation (experimentally up to 1.51 grams dry biomass per mole substrate consumed was found) is in good agreement with the theory developed by Roels *et al.*[8,9] Taking into account the thermodynamic values given by Roels per mole available electron, and furthermore the thermodynamic efficiency, a maximum value results of 2.16 grams dry biomass per mole substrate consumed. Since this theory does not refer to maintenance under the simplest assumptions, this is a maximum possible value, so that clearly, practical values below this magnitude have to be expected.

3% vss/g cod
0.03g vss/g cod
0.05g vss cod/g cod

REFERENCES

1. BASTIN, K. & C. WANDREY. 1981. Process dynamic aspects in continuous anaerobic digestion. *In* Biochemical Engineering II. A. Constantinides, W. R. Vieth & K. Venkatasubramanian, Eds. Ann. N.Y. Acad. Sci. **369:** 135–145
2. AIVASIDIS, A. & C. WANDREY. 1982. Continuous anaerobic digestion of acetate by means of a pure culture of *Methanosarcina barkeri*. Proc. 2nd Int. Symp. Anaerob. Digest. Travemünde. Elsevier Biomedical Press.

640

3. BRUNE, G., S. M. SCHOBERTH & H. SAHM. 1982. Anaerobic treatment of an industrial wastewater containing acetic acid, furfural and sulphite. Process Biochem. **17** (May–June): 20.
4. HIPPE, H., D. CASPARI, K. FIEBIG & G. GOTTSCHALK. 1979. Utilization of trimethylamine and other N-Methyl compounds for growth and methane formation by *Methanosarcina barkeri*. Proc. Natl. Acad. Sci. USA **76:** 494–498.
5. HOFFMANN, U. & H. HOFMANN. 1971. Einführung in die Optimierung. Verlag Chemie GmbH.
6. PIRT, S. J. 1965. The maintenance energy of bacteria in growing cultures. Proc. Royal Soc. London Ser B. **163:** 224.
7. NYNS, E.-J. & R. BUVET. 1982. Units of measurements. Proc. 2nd Int. Symp. Anaerob. Digest. Travemünde. Elsevier Biomedical Press.
8. ROELS, J. A. 1981. The application of macroscopic principles to microbial metabolism. *In* Biochemical Engineering II. A. Constantinides, W. R. Vieth & K. Venkatsubramanian, Eds. Ann. N.Y. Acad. Sci. **369:** 113.
9. ROELS, J. A. 1980. Application of macroscopic principles to microbial metabolism. Biotechnol. Bioeng. **22:** 2457–2513.

APPENDIX

SYMBOLS AND ABBREVIATIONS

k_m	(mol/l)	Michaelis-Menten constant
m	(mol/(g · hr))	maintenance coefficient
R_s	(mol/(g · d))	substrate consumption rate
$R_{s/x}$	(mol/(g · d))	substrate consumption rate for biomass
$R_{s/p}$	(mol/(g · d))	substrate consumption rate for biogas
R_x	(g/(g · d))	biomass production rate
R_p	(l/(g · d))	biogas production rate
S_o	(mol/l)	inlet substrate concentration
S_e	(mol/l)	steady-state substrate concentration
Δ_S	(mol/l)	$S_o - S_e$, substrate consumed
$t_{2/1}$	(hr)	generation time
T	(°C)	temperature
U	(−)	conversion
V_{max}	(mol/(g · hr))	substrate consumption rate for biogas
X_o	(g/l)	inlet biomass concentration
X_e	(g/l)	steady-state biomass concentration
Y_{max}	(g/mol)	maximal value of growth selectivity
Y_{x/s_o}	(g/mol)	biomass (growth) yield
Y_{x/Δ_s}	(g/mol)	biomass (growth) selectivity
Y_{p/s_o}	(l/g)	biogas yield
Y_{p/Δ_s}	(l/g)	biogas selectivity
μ_{max}	(hr^{-1})	maximal growth rate
τ	(hr)	residence time
τ_{crit}	(hr)	critical residence time
τ_{max}	(hr)	residence time at maximal productivity

Rapid Production of Methane with Immobilized Microbes

R. A. MESSING

Research & Development Division
Corning Glass Works
Corning, New York 14831

THOMAS L. STINEMAN

Research & Development Division
The Kroger Co.
Cincinnati, Ohio 45204

INTRODUCTION

The immobilization of high populations of microbes on a porous inorganic support was demonstrated to be a function of the pore diameters of the support, the dimension of the cells, and the mode of reproduction of the microbes.[1,2] Microbes that reproduced by fission could be loaded within the porous structure of the support to between 10^8 and 10^9 cells per gram if the ceramic contained pore diameters in the range of 1 through 5 times the major dimension of the cell. These studies predicted that the cells would immobilize on their minor dimension. Subsequent studies[5] indicated that this was the case, and in addition the cells that were immobilized on the support were closely packed within the pore structure. The hypothesis offered for the immobilization of the cell on a minor dimension was that the maximal surface remains available for the transfer of metabolites. Since no metabolites can be obtained from the support material, the cell must protrude into the solution.

Microbes that were immobilized within the pore structure of inorganic supports and subsequently placed in a continuous plug-flow reactor could not be washed out of that structure when very high flow rates were employed to deliver nutrients[3] (TABLE 1). In a study where 1 gram of immobilized *E. coli K12* composite was employed, the medium was delivered at 23 liters per hour and no washout occurred. In the study recorded in TABLE 1, it should be noted that at flow rates of 2.5 liters per hour, the cell count in the effluent was between 10^8 and 10^9 cells per ml. Therefore, not only could large populations of cells be immobilized within the pore structure, but also high populations of cells could be elaborated in the effluent. Another interesting aspect of these studies was that the cells delivered from the effluent exhibited a major dimension between 0.65 and 0.85 microns while the original culture immobilized contained cells between 1 and 6 microns in length. If these mini-cells are allowed to remain in the medium for more than 1 hour, they exhibit cell lengths of between 1 and 6 microns. It would appear that the rapid and continuous delivery of nutrients and removal of metabolic wastes promote early cell division with the production of uniform mini-cells.

Although cells may be loaded at levels of between 10^8 and 10^9 cells per gram in a porous ceramic material and these immobilized cells are capable of delivering 10^8 to 10^9 cells per ml, if nutrients are rapidly delivered and metabolic wastes are rapidly

TABLE 1. High Flow Rate *Escherichia coli K12* Microbial Generator[a]

No. Hours After:		Flow Rate	Bacterial Count, Cells	
Start	Rate Change	Ml/Hr.	Per Ml.	Per Hr.
0	0	100	—	—
0.25	0.25	100	4.9×10^7	4.9×10^9
15.5	15.5	100	1.3×10^9	1.3×10^{11}
16.5	1.0	250	7.2×10^8	1.8×10^{11}
17.0	1.5	250	1.5×10^8	3.8×10^{10}
18.0	1.0	500	1.9×10^8	9.5×10^{10}
19.0	1.0	1000	6.1×10^8	6.1×10^{11}
20.25	1.25	2500	4.1×10^8	1.0×10^{12}
21.0	2.0	2500	1.9×10^8	4.8×10^{11}
22.0	3.0	2500	1.1×10^9	2.8×10^{12}
23.0	4.0	2500	1.7×10^9	4.2×10^{12}
24.0	0.5	4200	4.0×10^6	1.7×10^{10}
25.0	1.5	4200	1.8×10^7	7.6×10^{10}
26.0	2.5	4200	4.2×10^7	1.8×10^{11}
27.0	3.5	4200	4.6×10^7	1.9×10^{11}
28.0	0.5	7680	5.2×10^7	4.0×10^{11}
29.0	1.5	7680	4.4×10^6	3.4×10^{10}
30.0	2.5	7680	3.7×10^7	2.8×10^{11}
31.0	3.5	7680	1.4×10^7	1.1×10^{11}
32.0	4.5	7826	9.7×10^6	7.6×10^{10}
33.0	5.5	7826	5.4×10^6	4.2×10^{10}
34.0	0.5	13333	7.3×10^6	9.7×10^{10}
35.0	1.5	13333	7.0×10^6	9.3×10^{10}
36.0	2.5	13333	2.8×10^7	3.7×10^{11}
37.0	3.5	13333	7.2×10^7	9.6×10^{11}
38.0	1.0	18947	1.3×10^7	2.5×10^{11}
39.0	2.0	18947	5.3×10^6	1.0×10^{11}
40.0	1.0	23000	2.2×10^7	5.1×10^{11}
41.0	2.0	23000	8.4×10^6	1.9×10^{11}

[a]From U.S. Patent 4,286,061, Reference 3.

removed, the requirement for utility has not been demonstrated. It appeared to us that an appropriate candidate for providing a demonstration of value was that of the anaerobic filter described by Young and McCarty.[4] We intentionally chose an anaerobic fermentation to avoid the problem of oxygen delivery. In addition to determining the functionality of the immobilized cells, it was desirable to design a reactor that would more efficiently convert the carbon in the feed to methane. In order to accomplish this goal, an additional horizontal stage was added to the anaerobic filter.

METHANE FROM SEWAGE[5]

The reactor for the study of sewage was similar to that depicted in FIGURE 1 except that the effluent outlet terminated horizontally from the second stage and only a single pressure gauge was inserted between the feed pump and the first stage. The total volume of the reactors utilized in this study was 120 ml and the liquid volume was 90 ml. Both stages were charged with controlled-pore ceramics having pore diameters between 0.8 and 30 microns. The carrier was either in a cylindrical form having dimensions of 2 mm by 6 mm or in the form of irregular slivers approximately 1 cm by 2 mm by 2–5 cm. Sewage adjusted to above pH 8.0 and varying between 800 and 2600 mg/1 COD was delivered to the reactor in an upward flow pattern with no recirculation. The temperature of the first stage was varied while that of the second stage remained at ambient (20–22°C) temperature. The first stage, the hydrolytic-redox, contained predominantly acid formers while the second stage contained predominantly *Methanobacter*. While the first stage was assembled vertically, the second was assembled horizontally. The fluid level of the second stage was maintained by a level controller that activates a gas pump to remove gas when the fluid level fell below the half-full mark. A check-valve was inserted in the effluent tubing to maintain pressure on the anaerobic stage. At residence times of between 2 and 5 hours, the COD was reduced by 77–88%. Approximately 45% of the total carbon delivered to the system was converted to methane. The gas delivered by the system contained greater than 90% methane with less than 5% carbon dioxide. Furthermore, under these conditions, no hydrogen sulfide was noted in the gas. In the case of sewage, it appeared as though the temperature optimum for the first stage was higher than 30° but certainly lower than 40°C. The optimum residence time appeared to be close to about 3.8 hours.

The efficient performance of the system appeared to be due to: (1) the optimum pore dimensions for both high cell accumulation and prevention of washout; (2) the more complete conversion of carbon dioxide that resulted from the elevation of pH to above 8.0 and the pressurization of the reactor by the check-valve insertion in the

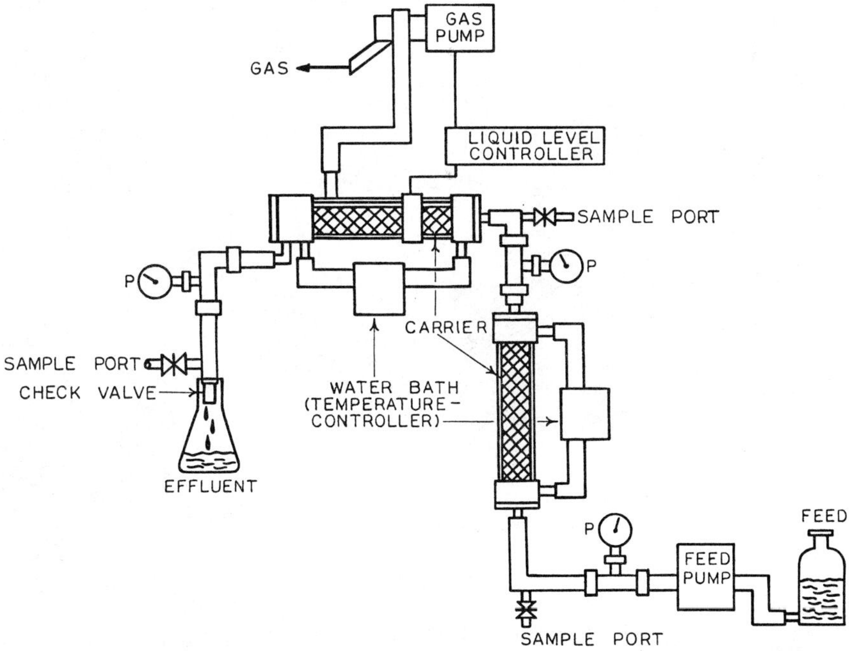

FIGURE 1. Poultry manure processor.

effluent stream; and (3) the gas–fluid interface maximization for removal of methane by the horizontal assembly of the anerobic stage and the maintenance of the level of about 50% fluid depth. The lower temperature of the second stage contributes to the retention of carbon dioxide in the fluid. Since the microbes are predominantly located in the fluid phase, in order to achieve a more complete conversion of carbon dioxide to methane, it is necessary to maintain that gas in the liquid. Thus, the three factors (high pH, pressurization, low temperature) contributed to the retention of carbon dioxide for further conversion to methane.

METHANE FROM CHICKEN MANURE

The processing of manure presents problems that were not encountered in our sewage studies. In addition to the substantially higher concentrations of carbon and COD, there are high quantities of insolubles and large particles. Thus for these studies, we were required to scale up the reactor size and to modify the support material such that it would allow passage of large particles without plugging. The modification for poultry manure processing involved the changing of position for the discharge of the effluent from the end to the bottom of the second stage to allow for passage of large particles and the prevention of plugging (FIGURE 1). In addition, provisions were made to monitor the pressure drop between the two stages and at the effluent tubing. Furthermore, a sampling port was constructed between the two stages. A further modification was the control of temperature in the second stage.

The two stages were custom-constructed. The internal diameter was 5.0 cm and the length was 112 cm. The temperature of each stage was controlled separately by means of individual circulating water baths. Brass Circle-Seal, 3 psi, ½-inch diameter check-valves were employed in the effluent tubing.

The laying-hen poultry manure was obtained from Egg City of California, Moorpark, California. The manure was stored at 4–6°C until used. The manure was diluted with water, ground with a Waring blender, and then filtered through burlap to remove the very coarse particulate material. The manure slurry was then adjusted to the desired feed concentration with water and gently agitated on an orbital shaker to ensure constant feed of suspended solids. The pH of the feed manure was not adjusted and was found to be between 6.8 and 7.3.

Four identical systems and three different carriers were employed for these studies. Two of the support materials were products of the Manville Products Corporation, Zelienople, Pennsylvania. They were insulating fire bricks, 9×4×2½ inches. One brick was designated as Johns-Manville 23SL fire brick and the other was Johns-Manville Microbiological Reactor Media (JM MRM) brick. The bricks were cut lengthwise, turned on a lathe to 4.9-cm diameter, and then four ¼″ holes were drilled through them lengthwise. The composition of the brick was alumina, silica, and calcium oxide. The third carrier was a product of the Aluminum Company of America, Alcoa Center, Pennsylvania. The latter carrier was an alumina bead, ½″ in diameter and identified as Alcoa controlled-pore alumina # P2311-75 (Alcopal) bead. The reactors were charged with 1640 grams of JM23SL in each of two reactors, 1650 grams of JM MRM in one system, and 2450 grams of the Alcopal bead in the fourth processor. The mercury porosimetry of these carriers are recorded in FIGURE 2. The JM23SL brick had an average pore diameter of 9 μ, a pore distribution 2–25 microns, pore volume of 0.84 cc/g, and a porosity of 67%. The JM MRM had an average pore diameter of 13 microns, pore distribution 2–35 microns, a pore volume of 1.22 cc/g, and a porosity of 75%. The Alcopal bead had an average pore diameter of 2 microns, pore distribution 0.1–9 microns, a pore volume of 0.25 cc/g, and a porosity of 36%.

The volume of the reactor system without the support was 4,350 ml including the tubing. The fluid volumes of the reactors were measured after the carriers were packed. Two were found to contain 2850 ml, while two were found to be 3000 ml.

Gas determinations were performed with a Perkin-Elmer Sigma One Gas Chromatograph equipped with a Sigma Ten Data System and a thermal conductivity detector. Total carbon determinations were performed on an Oceanography International Model 525 carbon analyzer. The COD was performed with an E.P.A.-certified procedure, the Ampule Method.

The seeding of two of the reactors, one containing JM23SL and the other containing Alcopal support, was performed with a sludge from a Cincinnati sewage treatment plant mixed with fresh poultry manure and circulated through the reactors for one day followed by dilute poultry manure. One reactor containing JM MRM was

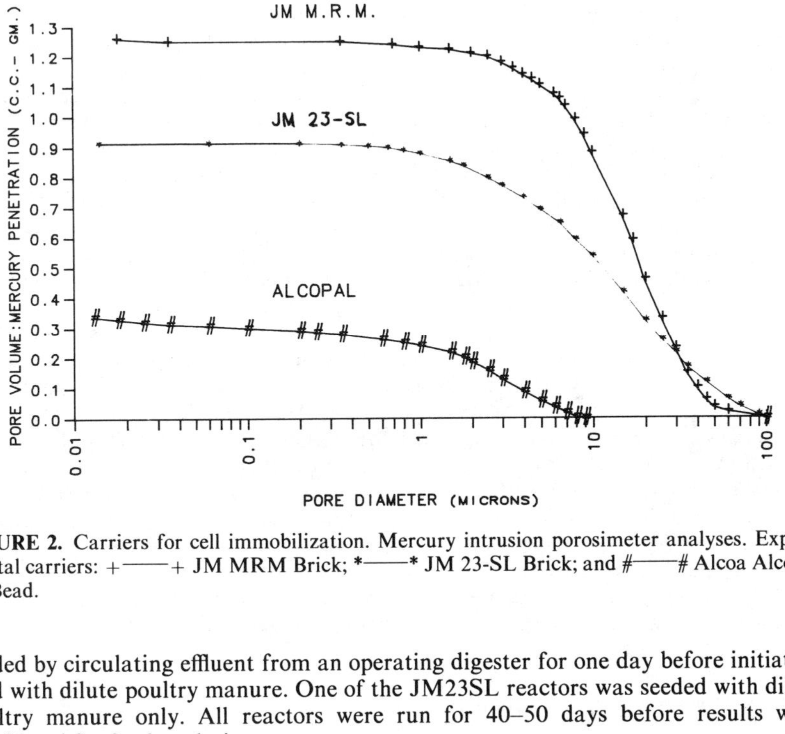

FIGURE 2. Carriers for cell immobilization. Mercury intrusion porosimeter analyses. Experimental carriers: +——+ JM MRM Brick; *——* JM 23-SL Brick; and #——# Alcoa Alcopal ½″ Bead.

seeded by circulating effluent from an operating digester for one day before initiating feed with dilute poultry manure. One of the JM23SL reactors was seeded with dilute poultry manure only. All reactors were run for 40–50 days before results were considered for final analysis.

Experiments were run utilizing three different carrier materials, residence times of 16–55 hours, temperatures for stage one varied between 20 and 40°C, temperatures for stage two between 20 and 35°C, and feed concentrations from 1000 to 15,000 ppm total carbon (TC).

The feed concentrations of the manure, initially, were monitored by chemical oxygen demand (COD). It was found that the reproducibility of the COD was poor. This was attributed to the high level of suspended solids in the feed slurry. Total carbon (TC) was found to be a better monitor of the feed levels for these reactors. COD was employed as a crude measurement for loading and dilutions. A sampling of the COD

TABLE 2. Chemical Oxygen Demand (PPM)

Feed	Outlet	% Change
5,800	1,870	−67.7
5,200	2,170	−58.3
15,080	8,340	−44.7
16,200	8,500	−47.5
43,680	33,800	−22.6
35,750	27,750	−22.4

results are recorded in TABLE 2. COD reductions range from 22% through 67% depending on the feed concentration. At approximately 5000 ppm COD, the reductions were the greatest. As we increased the COD through 35,000 ppm, decreases in reductions were noted with increased concentrations for a given residence time.

The two factors that were monitored closely over time were those of plugging and change in pore diameter distribution. The only plugging problem encountered was that at the entrance of the first stage. This was caused by the tubing rather than the system itself. The inlet and outlet pressures were monitored and typical data for pressure drops are recorded in TABLE 3. The largest pressure drop was noted across the first stage. After approximately six months of continuous operation, two systems were taken apart, one of the JM23SL carriers and the Alcopal carrier system. About 50 grams of the carrier was removed from the system for analysis. This used carrier material was dried at 100°C for three hours and submitted for porosimetry analysis and compared to the original starting material. No appreciable change in the pore diameter was noted (FIG. 3). There was an approximate 15% decrease in pore volume for both materials. This decrease may be attributed to cellular material within the pore structure.

The mode of seeding studies, FIGURE 4, indicated that the reactor seeded with a mixture of the effluent of a reactor operating on the manure and a dilute manure demonstrated an initial start-up more rapid than an unseeded system. Gas production started in the first five days in the seeded system while the unseeded system required an additional seven days to produce significant quantities of methane. The seeded reactor reached steady state in 25 days while the reactor seeded with manure only did not attain a steady state until the 40th day. These results indicate that steady state would be reached in a shorter period of time if the operating digester effluent were circulated for longer periods, 5–10 days when starting a new digester. Fresh manure, however, may be employed if no digester effluent is available.

After nine months of continuous operation with two systems and six months with the other two, the data were computer analyzed using multiple variable correlation

TABLE 3. Pressure Drop Information

Pressure 1 (psi)[a]	Pressure 2 (psi)[b]	Pressure 3 (psi)[c]
2.89	1.80	2.10
2.49	1.80	2.00
3.29	2.80	2.50
2.29	1.75	2.25
2.89	1.80	2.00
2.79	1.90	2.10

[a]Pressure 1: feed pressure into bottom of stage #1 adjusted for head pressure.
[b]Pressure 2: outlet pressure of stage #1 and inlet pressure of stage #2.
[c]Pressure 3: outlet pressure of stage #2.

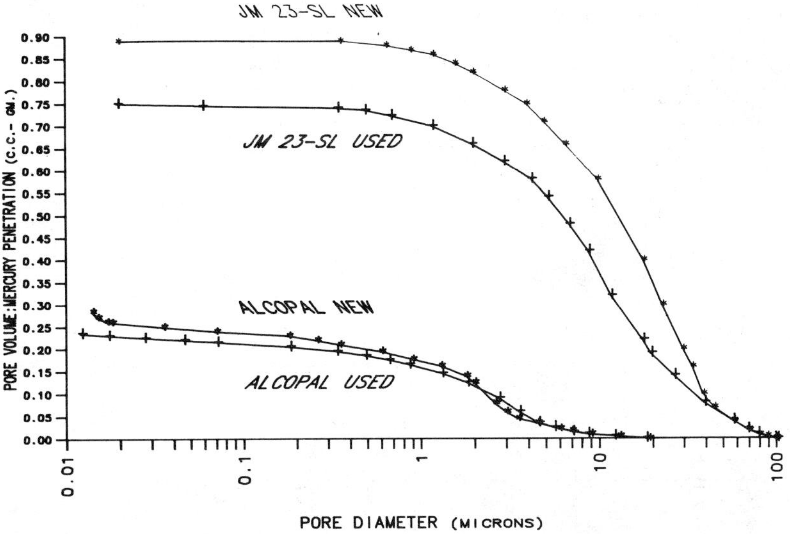

FIGURE 3. Carriers for cell immobilization. New carriers vs. used carriers (6 months continuous). Mercury intrusion porosimeter analyses. Experimental carriers: *——* JM 23-SL New; +——+ JM 23-SL Used; *——* Alcopal New; and +——+ Alcopal Used.

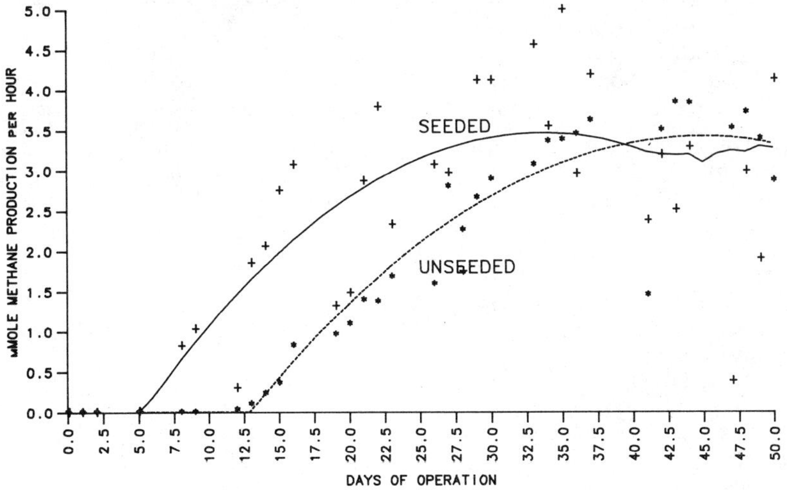

FIGURE 4. Immobilized cell digestor start-up. Seeded vs. unseeded start. Methane production mmole/hour. *-----------, unseeded start; +——, seeded start.

techniques. Separate analyses were performed for biogas production per hour and for methane content of the biogas. It was our intent to optimize all variables including any multivariable interactions for these two parameters.

The volume of biogas produced varied with the carrier material employed (FIG. 5). JM MRM brick exhibited the greatest biogas productivity. This carrier contained the highest pore volume of the three carriers evaluated (see FIG. 2). These results appear to indicate that as the pore volume increases, a concomitant increase in gas productivity was noted, and thus, there is an indication of a probable increase in cell loading and reproduction. The biogas production from JM23SL brick, which demonstrated pore volumes of 40% of that of JM MRM, yielded biogas at only 10 to 30% levels of that noted with MRM brick. The Alcopal carrier, which contained the lowest pore volume,

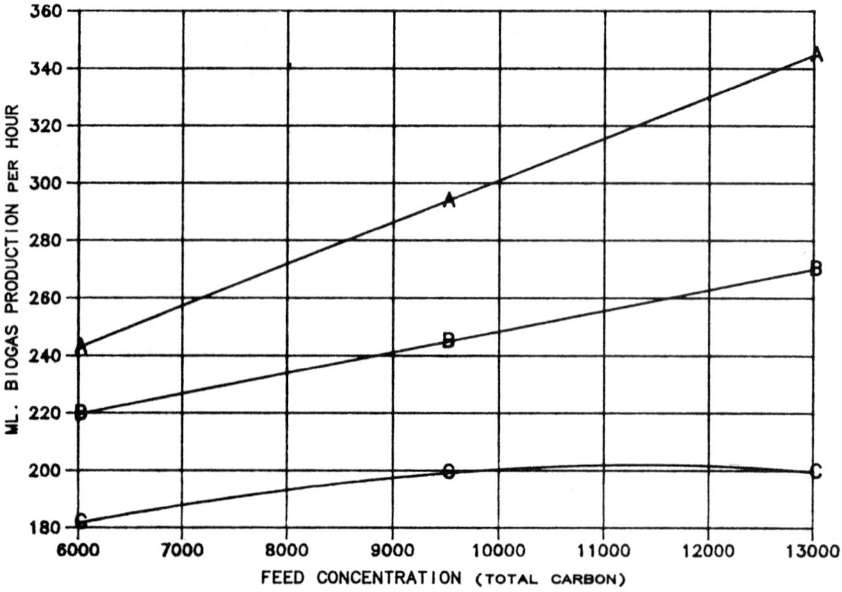

FIGURE 5. Milliliters biogas production per hour related to feed concentration and carrier type. Carrier types: A——A JM MRM Brick; B——B JM 23-SL Brick; and C——C Alcopal Bead. Condition settings: Residence time = 24.0 hr; Temperature #1 = 40.0°C; Temperature #2 = 25.0°C.

exhibited the lowest production rate of biogas. The results further indicate that as the feed concentration is increased, the effect is greater with the MRM than with the other two support materials.

The relationship between the residence time and the feed concentration (TC) may be noted in FIGURE 6. Gas productivity increases with residence time up to 23 hours. After this point, the productivity appears to diminish. However, another increase in productivity was noted after 50 hours of residence, which is probably due to the degradation of some of the recalcitrant carbon materials. The productivity of gas for these longer residence times was not pursued at this point. The production of gas may be directly related to the carbon concentration in the feed. The residence times

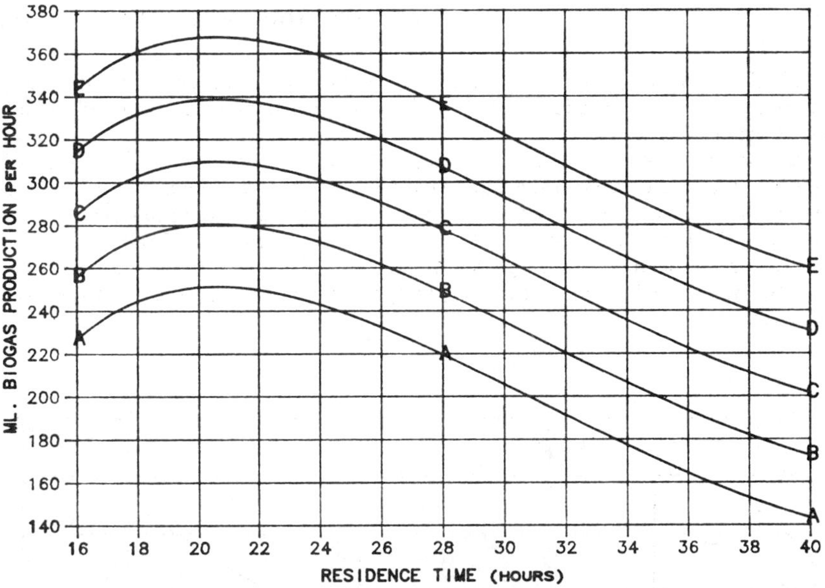

FIGURE 6. Milliliters biogas production per hour related to residence time and feed concentration. Feed concentration: A——A 6000; B——B 8000; C——C 10,000; D——D 12,000; and E——E 14,000. Condition settings: Carrier, JM MRM; Temperature #1 = 40.0°C; Temperature #2 = 25.0°C.

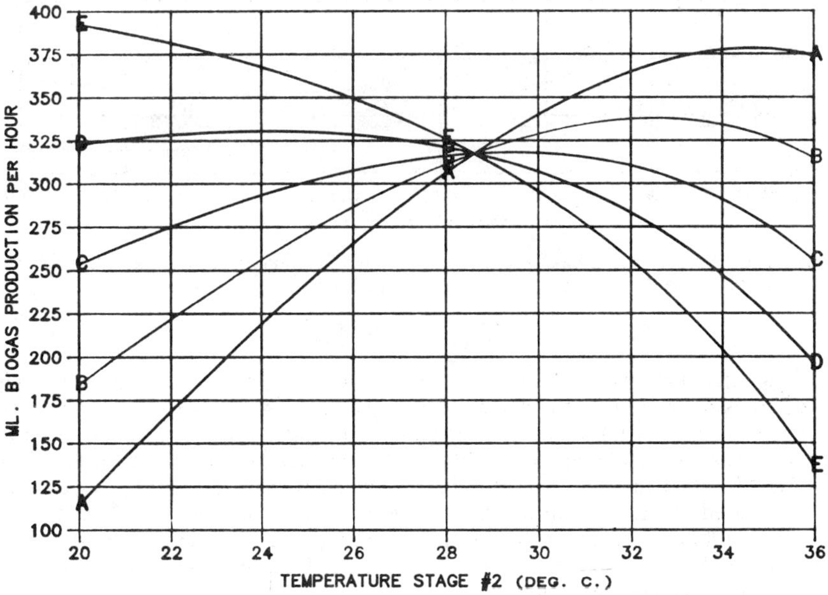

FIGURE 7. Milliliters biogas production per hour related to temperature stage #2 and feed concentration (total carbon). Feed concentration: A——A 6000 TC; B——B 8000 TC; C——C 10,000 TC; D——D 12,000 TC; E——E 14,000 TC Condition settings: Carrier, JM MRM; residence time = 24.0 hr; and temperature #1 = 40.0°C.

exhibited with poultry manure are substantially greater than those noted with sewage, but are consistent with the fact that the COD concentrations were considerably higher than those of the sewage.

An interesting but marked relationship between the temperature of the anaerobic stage and the feed concentration was noted (FIG. 7). As the temperature of stage two was increased at low levels of total carbon, the biogas increased; however, this was reversed at high concentrations of TC. At feed concentrations of 12,000 ppm TC, the temperature of the anaerobic stage begins to impact above 28.5°C. The temperature of stage two should be kept lower than this temperature if higher feed concentrations are employed.

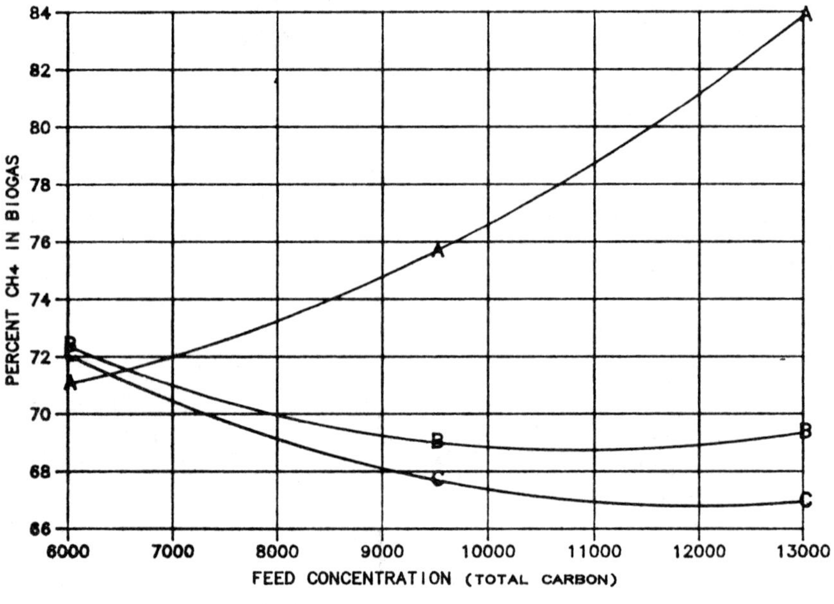

FIGURE 8. Percent CH_4 in biogas related to feed concentration and cell carriers. Carrier types: A——A JM MRM; B——B Alcopal; C——C JM 23-SL.

If quality methane is the objective of a program, then the quantity of biogas produced is not a sufficient parameter to be evaluated. For this reason, studies were initiated to compare the methane production with respect to the carrier utilized and concentration of manure employed (FIG. 8). The carrier with the highest pore volume, JM MRM, delivered methane in the highest concentrations. As the feed concentration was increased, the concentration of methane in the gas also increased. The other two carriers exhibited a reverse trend. In this case, the Alcopal carrier actually outperformed the JM 23SL carrier and did not appear to be a function of pore volume but rather that of surface contributions.

The level of methane increased as the residence time was increased above 14 hours (FIG. 9). In addition, when the feed concentration was increased above 5000 ppm TC, the methane content of the gas increased. Methane contents of greater than 88% were

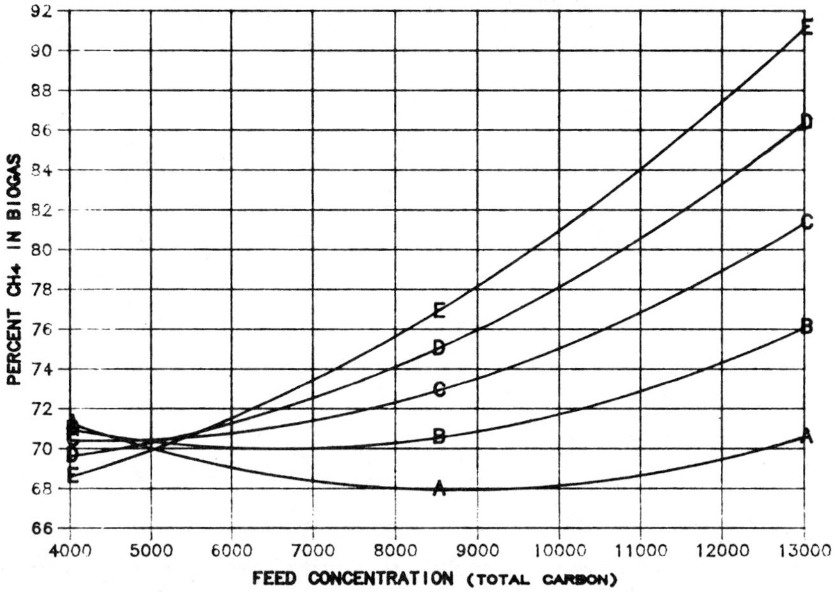

FIGURE 9. Percent CH_4 in biogas related to feed concentration and residence time. Residence time (hours): A——A 14.0; B——B 18.0; C——C 22.0; D——D 26.0; E——E 30.0. Condition settings: Carrier, JM MRM; Temperature #1 = 40.0°C; Temperature #2 = 25.0°C.

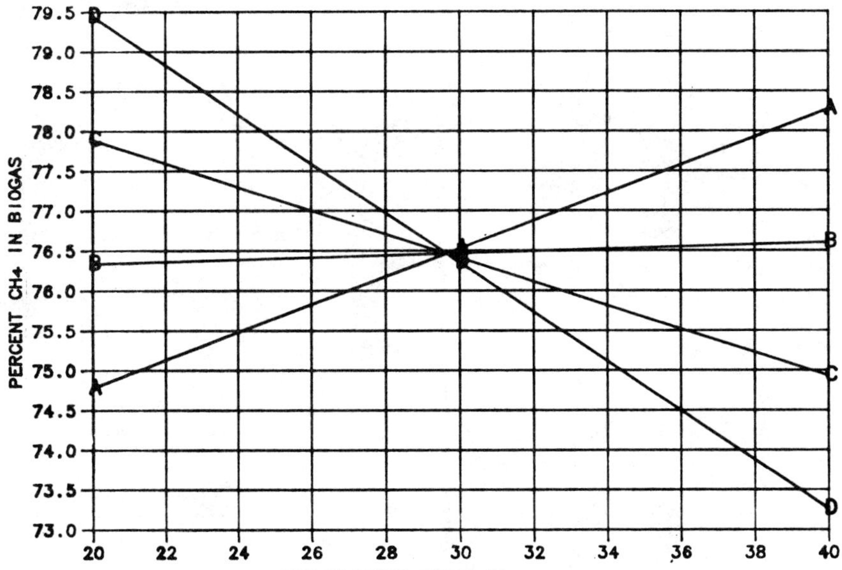

FIGURE 10. Percent CH_4 in biogas related to temperature stage #1 and temperature stage #2. Temperature stage #2: A——A 20.0°C; B——B 25.0°C; C——C 30.0°C; D——D 35.0°C. Condition settings: Carrier, JM MRM; residence time, 24.0 hr; feed concentration, 10,000 TC.

projected for residence times of 30 hours at total carbon concentrations of 13,000 ppm.

In the prior efforts with sewage, the indications were that less than 40°C was optimum for stage one, the hydrolytic-redox stage. With poultry manure, it was found that the rate of gas production increased as the temperature of stage one was increased; however, the optimum temperature for the hydrolytic-redox stage appears to interact with the temperature of the anaerobic stage (FIG. 10), when poultry manure was utilized. In this latter study as the temperature of stage one was raised to increase gas production, it was necessary to lower the temperature of stage two to below 25°C in order to increase the methane content of that gas. The lower temperature increases the solubility of carbon dioxide and supports the conversion by the microbes in solution or

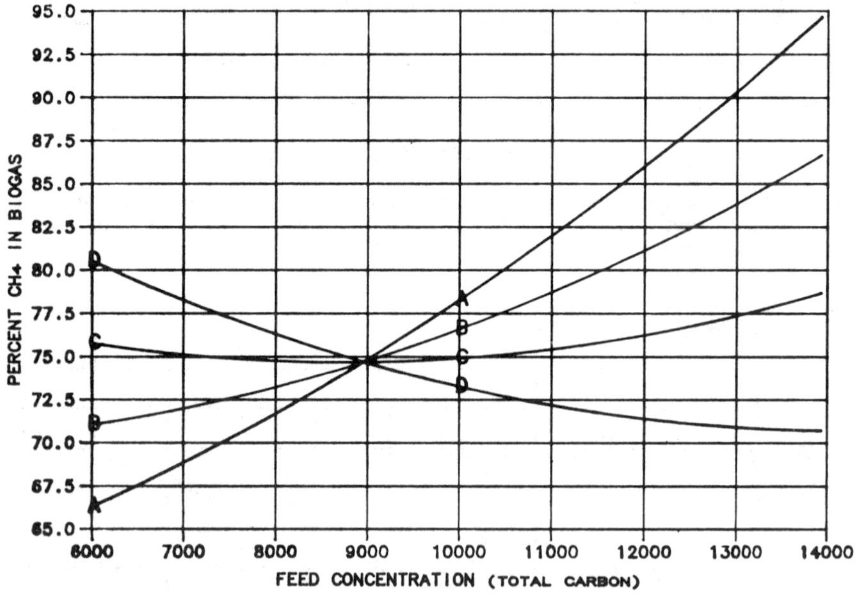

FIGURE 11. Percent CH_4 in biogas related to temperature stage #2 and feed concentration. Temperature stage #2: A——A 20.0°C: B——B 25.0°C; C——C 30.0°C; D——D 35.0°C. Condition settings: Carrier JM MRM; residence time, 24.0 hr; temperature #1, 40.0°C.

in the support to methane. As the temperature of the hydrolytic-redox stage was increased, the methane content decreased if the temperature of the anaerobic stage was not reduced. When temperatures of greater than 25°C were employed for stage two, the methane concentration was reduced, and the carbon dioxide concentration was increased as the temperature of stage one was increased.

The relationship between the feed concentration in terms of TC and temperature of stage two was determined (FIG. 11). Performance in terms of methane production increased as the loading increased above 9000 ppm COD if the temperature of stage two was reduced to below 25°C. At temperatures of greater than 30°C for the anaerobic stage, the percent methane in the gas decreased. When the carbon concentrations of the feed were above 10,000 ppm, the highest methane concentrations were achieved when the temperature of the anaerobic stage was set at 20°C. This further indicates that the increased carbon dioxide solubility in stage two at temperatures below 25°C increased the productivity of the system.

CONCLUSION

The two-stage sewage reactor modified to process poultry manure was scaled up and operated continuously for nine months. The carrier materials did not exhibit appreciable modifications over this period of time. When relatively high concentrations of poultry manure are fed to the reactor, the temperature of the hydrolytic-redox stage should be maintained at at least 40°C and the temperature of the anaerobic stage should be adjusted to 25°C or lower to achieve the highest productivity yield in terms of methane concentrations. When feed concentrations of greater than 10,000 ppm total carbon are utilized, the residence times should be between 22 to 24 hours for the nonrecalcitrant carbon. JM MRM was identified as the best carrier material of those tested for both the quantity of gas produced and the content of methane in the gas. The indications are that a carrier material with high porosity and high pore volume within the range of 2–35 micron pore diameters will outperform a material with similar pore diameters but exhibiting a lower pore volume and porosity.

The total volumes of each of the stages were equal in the reactor systems that we studied. Some of our efforts indicate that the volume of the hydrolytic-redox stage should be less than that of the anaerobic stage for superior performance. Additional studies must be performed to ascertain the optimum ratio for the two stages.

REFERENCES

1. MESSING, R. A. & R. A. OPPERMANN. 1979. Pore dimensions for accumulating biomass. I. Microbes that reproduce by fission or by budding. Biotechnol. Bioeng. **21:** 49–58.
2. MESSING, R. A., R. A. OPPERMANN & F. B. KOLOT. 1979. Pore dimensions for accumulating biomass. II. Microbes that form spores and exhibit mycelial growth. Biotechnol. Bioeng. **21:** 59–67.
3. MESSING, R. A., R. A. OPPERMANN, L. B. SIMPSON & M. TAKEGUCHI. 1981. Method For Continuous Culturing of Microbes. U. S. Patent 4,286,061.
4. YOUNG, J. C. & P. L. MCCARTY. 1969. The anaerobic filter for waste treatment. J. Water Pollut. Control Fed. **41:** R160–R173.
5. MESSING, R. A. 1982. Immobilized microbes and a high-rate, continuous waste processor for the production of high BTU gas and the reduction of pollutants. Biotechnol. Bioeng. **24:** 1115–1123.

Progress in Research toward Outdoor Biological Hydrogen Production Using Solar Energy, Sea Water, and Marine Photosynthetic Microorganisms[a]

A. MITSUI, E. J. PHLIPS, S. KUMAZAWA, K. J. REDDY, S. RAMACHANDRAN, T. MATSUNAGA, L. HAYNES, AND H. IKEMOTO

School of Marine and Atmospheric Science
University of Miami
Miami, Florida 33149

INTRODUCTION

In the early 1970s, the scientific community responded to the oil crisis by initiating new and vigorous efforts to develop alternative sources of energy. One of the approaches to dealing with this problem has been biological solar energy conversion. Within the limited amount of time since this initiative, some of these new biological solar energy conversion technologies are already making a contribution to the world's energy budget. However, most of the innovative technologies that first took root in the 1970s are still at the research rather than development stage.[1–3]

As part of this broad effort, our laboratory is exploring the potential contribution of biological solar energy resources from the marine environment.[4–6] With few exceptions, the role of marine photosynthetic organisms has taken a back seat to biosolar resources from the terrestrial environment. This is understandable, since there is a wide base of knowledge in conventional agriculture. However, limitations of fresh water and arable land may restrict the availability of biomass for conversion to energy, and hence it will be immensely advantageous to incorporate production systems based on salt water.[7,8]

Over the past ten years, an intensive effort has been made in our laboratory to study the production of hydrogen gas by marine photosynthetic microorganisms.[9–11] Experimentally most of this effort has concentrated on (1) isolation of hydrogen-producing marine photosynthetic microorganisms,[11–13] (2) revealing the biochemical mechanisms of hydrogen production,[9,10,13,14] and (3) controlling and enhancing hydrogen production capability through environmental and metabolic regulation.[14–20]

On the basis of our laboratory results, we are now making the transition towards the development of applied outdoor hydrogen production systems. This paper is an overview of our current concepts and experimental initiatives for establishing viable reactor systems in outdoor environments using solar energy and sea water.

[a]This work was supported by the grants to A. Mitsui from the U.S. National Science Foundation (No. CPE-8015142), Solar Energy Research Institute, U.S. Department of Energy (No. XR-9-8036-3), and U.S. National Aeronautics and Space Administration (No. NAS10-10531). Any opinions, findings, and conclusions or recommendations expressed in this publication are those of the authors and do not necessarily reflect the views of the granting agencies.

SELECTION OF MARINE PHOTOSYNTHETIC MICROORGANISMS FOR THE DEVELOPMENT OF HYDROGEN PRODUCTION SYSTEMS

Among the large numbers of marine blue-green algae and photosynthetic bacteria isolated in our laboratory, we have found strains capable of high rates of growth and large biomass yields. These strains are continuously being screened and tested for hydrogen photoproduction capability. These screening experiments are focused on the long-term accumulation of hydrogen, that is days, rather than short-term production. For applied systems development, stable hydrogen production for long periods of time and its accumulation in closed vessels is desirable. So far, over 150 strains of marine photosynthetic microorganisms have shown the capability for hydrogen photoproduction.

TABLE 1. Hydrogen Photoproduction, Acetylene Reduction and Dark H_2 Consumption in Different Groups of Marine Blue-Green Algae[a]

Blue-Green Algae[b]	No. of Strains Screened	H_2 Production (nmol/mg dw/hr)	C_2H_2 Reduction (nmol/mg dw/hr)	Dark H_2 Consumption (nmol/mg dw/hr)
Heterocystous Filamentous:				
Anabaena	15	6– 80	22–133	1– 75
Nostoc	12	38–148	76–230	10–102
Calothrix	6	4– 50	1– 96	0– 19
Scytonema	5	0– 41	2– 65	12–130
Nonheterocystous Filamentous:				
Microaerobic N_2 Fixing:				
LPP Group	8	3–260	5–252	0
Non-N_2 Fixing:				
LPP Group	2	0	0	0
Unicellular				
N_2 Fixing:				
Synechococcus	8	14–616	61–441	0– 23
Gloeothece	2	2– 10	71– 97	42– 80
Non-N_2 Fixing:				
Synechococcus	5	0	—	—

[a] Reddy and Mitsui, unpublished data.

[b] Five- to seven-day-old cells from early stationary phase were used for the experiment. Rates are averages of one-day incubation period. N_2-fixing strains were grown in medium without combined nitrogen as described previously.[14] Non-N_2-fixing strains were grown in media with a limited amount of nitrogen.[14] Initial gas phase for H_2 production assays was 100% argon. In dark hydrogen uptake assays, the gas phase contained 2% hydrogen, 5% oxygen, and 93% argon. Hydrogen, oxygen, and acetylene were measured as described previously.[14] Rates indicated in this table do not necessarily reflect peak values obtainable.

In general, nonheterocystous filamentous strains and unicellular, aerobic, nitrogen-fixing strains of blue-green algae are capable of producing hydrogen for longer periods of time and at higher rates than heterocystous filamentous strains (TABLE 1). This is largely attributable to the activities of the enzymes nitrogenase and uptake hydrogenase (TABLE 1). The most active hydrogen producers from the former two groups of blue-green algae are now being used for applied research.

Among the marine photosynthetic bacteria, strains belonging to the families

TABLE 2. Hydrogen Photoproduction by Marine Photosynthetic Bacteria[a]

Family	No. of Strains Capable of Hydrogen Production	Rate of Hydrogen Production (μmol/mg dry wt/hr)
Rhodosprillaceae	40	0.1–4.0
Chromatiaceae	40	0.05–3.0
Chlorobiaceae	3	0.02–0.26

[a] Ikemoto, Greenbaum, Ohta, Haynes, Matsunaga, Kumazawa and Mitsui, unpublished data.
[b] Cells from early stationary phase were used for screening experiments. Culture conditions and the method for hydrogen assays have been described elsewhere.[11,14–19] Rates expressed in this table are averages of 24–48-hour incubation period.

Chromatiaceae (purple sulfur bacteria) and Rhodospirillaceae (purple nonsulfur bacteria) are able to produce hydrogen at higher rates and for longer periods of time than members of the family Chlorobiaceae (green sulfur bacteria) (TABLE 2). One exceptional strain from Chromatiaceae and one from Rhodospirillaceae are now being used for applied research.

It is important to mention that our two best blue-green algal strains have lower hydrogen-producing activities than the photosynthetic bacteria. However, the blue-green algal strains can utilize sea water as the hydrogen donor for the overall hydrogen production reaction, whereas photosynthetic bacteria require either organic or inorganic (reduced sulfur) compounds for hydrogen production. Considering these advantages and disadvantages, our present hydrogen production research has been subdivided into the following four categories, according to the type of organism and the difference in using hydrogen donating substances.

(1) A two-step hydrogen production system using filamentous nonheterocystous blue-green algae and salt water as the hydrogen donor (FIG. 1).

(2) A one-step system using unicellular, aerobic nitrogen-fixing blue-green algae and salt water as the hydrogen donor (FIG. 2).

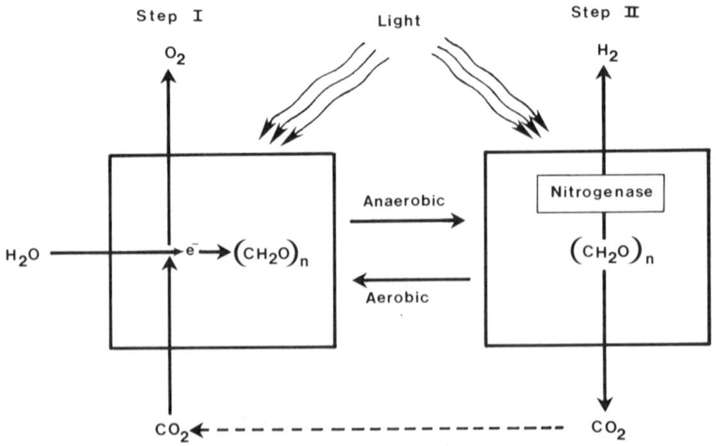

FIGURE 1. Scheme for two-step hydrogen production system using filamentous nonheterocystous blue-green algae and sea water as hydrogen donor.

(3) A one-step system using photosynthetic nonsulfur bacteria and organic waste materials as the hydrogen donor (FIG. 3).

(4) A one-step system using photosynthetic sulfur bacteria and inorganic wastes (like hydrogen sulfide) as the hydrogen donor (FIG. 4).

BIOLOGICAL HYDROGEN PHOTOPRODUCTION SYSTEMS

Two-Step Hydrogen Production System Using Filamentous, Nonheterocystous Marine Blue-Green Algae and Sea Water as Hydrogen Donor

Oscillatoria sp. Miami BG 7, isolated from the South Florida marine environment (Mitsui and Rosner, unpublished data) exhibited very high rates of hydrogen

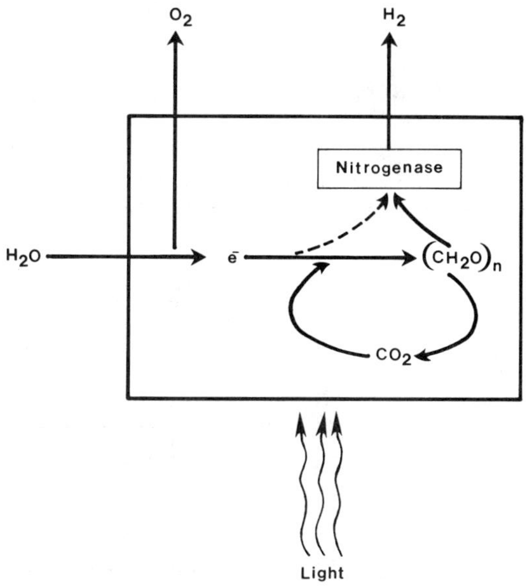

FIGURE 2. Scheme for one-step hydrogen production system using unicellular aerobic nitrogen-fixing blue-green algae and sea water as hydrogen donor.

photoproduction over a prolonged period of time.[9,10,14] The maximum rate observed so far is 0.54 μmol H_2/mg dry wt/hr. Recent results indicate hydrogen photoproduction values of 1.1 ml per 1 ml of blue-green algal suspension per day, in dense suspensions (FIG. 5).

The two-step hydrogen photoproduction system developed with this strain is depicted in FIGURE 1. In the first step, cells were cultured in a combined nitrogen-limited medium. During this culture period, the concentration of intracellular hydrogen donor substance, that is glycogen, increased dramatically, from an initial concentration of 20% to a final concentration of 65% of dry weight[14] (FIG. 6). Meanwhile the concentration of photosynthetic pigments like chlorophyll and phycobilins, which are responsible for oxygen photoproduction, decreased significantly, thus minimizing oxygen inhibition of nitrogenase (or hydrogenase).[14] Many green and blue-green algae are unable to produce hydrogen for sustained periods of time because

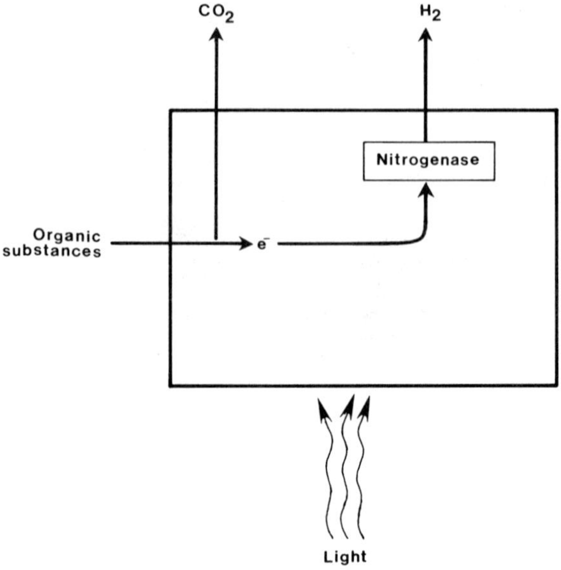

FIGURE 3. Scheme for one-step hydrogen production system using photosynthetic nonsulfur bacteria and organic substances as hydrogen donor.

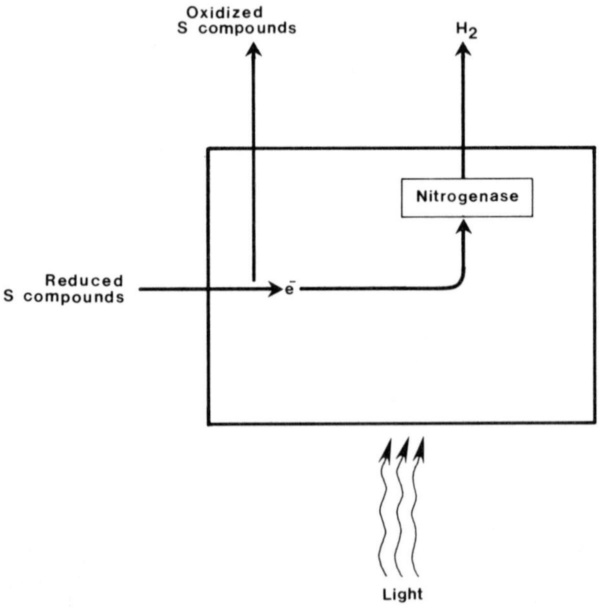

FIGURE 4. Scheme for one-step hydrogen production system using photosynthetic sulfur bacteria and inorganic waste as hydrogen donor.

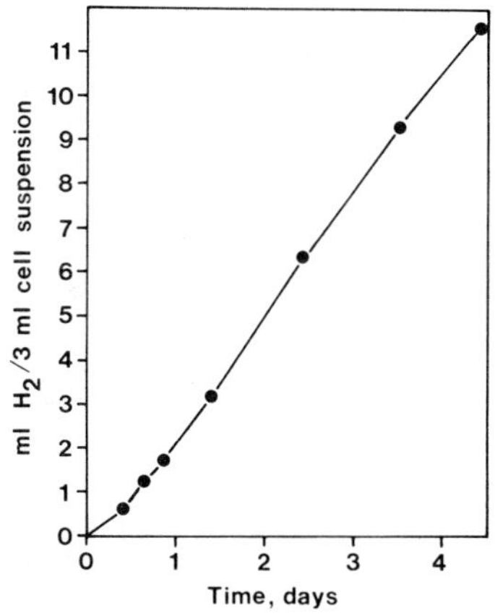

FIGURE 5. Hydrogen photoproduction by a 3-ml suspension of the blue-green alga *Oscillatoria* sp., Miami BG 7 (Phlips and Mitsui, unpublished data). Cell density of the suspension was 10 mg dry wt/ml. Culture conditions have been previously described.[14] The gas phase in the reaction vessel was collected and replaced with argon every 24 hours. The light intensity was 400 μEinsteins/m^2/sec and the temperature was 35°C. Hydrogen accumulation was measured by gas chromatography as previously described.[14]

the enzymes nitrogenase and hydrogenase, which are responsible for the terminal reaction of hydrogen photoproduction, are inhibited by photoproduced oxygen.

During the second step, the cultured cells were incubated under anaerobic conditions (argon atmosphere). Upon illumination, the accumulated intracellular glycogen was hydrolyzed to glucose and this was subsequently converted to hydrogen and carbon dioxide in a stoichiometric ratio of 2. Experiments yielded a value of 9.8 as the molar ratio of hydrogen production to glucose consumption (FIG. 7). In other words, 82% of the combustible energy held in the form of photosynthetically produced glycogen was converted to hydrogen.

Equations for step 1 and step 2 and the overall reaction are shown below. Overall inputs in these two steps are water and light and overall outputs are hydrogen and oxygen.

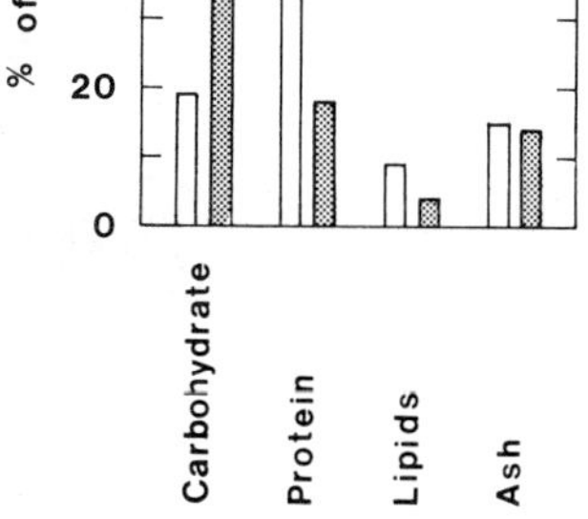

FIGURE 6. Carbohydrate, protein, lipids and ash content of 10-day-old, nitrogen-sufficient and nitrogen-limited cultures of *Oscillatoria* sp., Miami BG 7 (Kumazawa and Mitsui, unpublished data). Nitrogen-sufficient culture medium has been described elsewhere.[14] The nitrogen-limited culture medium was similar to the nitrogen-sufficient medium except for the amount of combined nitrogen source. Nitrogen-limited culture contained KNO_3, 5 mg and NH_4Cl, 5 mg/l. The culture method has been described elsewhere.[14] Carbohydrate, protein, lipids, and ash were measured, respectively, according to Dubois *et al.*,[28] Lowry *et al.*,[29] Williams,[30] and Larsen.[31]

Step 1: Aerobic Reaction

$$12H_2O + 6CO_2 \xrightarrow{\text{light}} C_6H_{12}O_6 + 6H_2O + 6O_2$$

$$n\, C_6H_{12}O_6 \longrightarrow \text{glycogen} + (n-1)H_2O$$

Step 2: Anaerobic Reaction

$$\text{Glycogen} + (n-1)H_2O \longrightarrow n\, C_6H_{12}O_6$$

$$C_6H_{12}O_6 + 6H_2O \xrightarrow{\text{light}} 6CO_2 + 12H_2$$

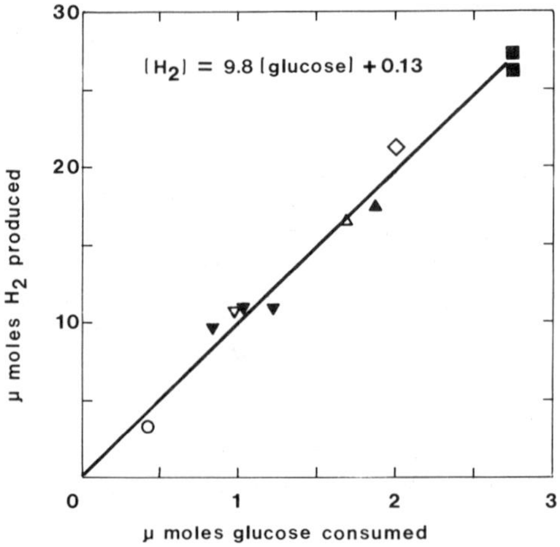

FIGURE 7. Linear relationship between glucose consumption and hydrogen photoproduction by nitrogen-limited cultures of *Oscillatoria* sp., Miami BG 7 (Kumazawa and Mitsui, unpublished data). 9-day-old (*closed symbols*) and 10-day-old (*open symbols*) cultures were used. Light intensity, temperature, and gas phase for the incubation were 30 μEinsteins/m^2/sec, 25°C and 100% argon respectively. The samples were analyzed after 1 day (0), 2 days (▽), 3 days (△), 4 days (◇), and 5 days (□) incubation periods for hydrogen production and glucose consumption. Each data point was measured to form individual subsamples (independent flasks). The relationship of the amounts of hydrogen produced and glucose consumed obtained by linear regression analysis was $H_2 = 9.8 \times \text{glucose} \pm 0.13$. The correlation coefficient (r) was 0.995. Methods for the analysis of glucose and hydrogen have been described elsewhere.[14]

Overall Reaction:

$$H_2O \xrightarrow{\text{light}} H_2 + 1/2\, O_2$$

The two-step hydrogen production principle established in the laboratory has also been successfully applied to outdoor reactor systems. Relatively high yields of biomass have been observed in outdoor growth chambers, using a natural-seawater-based culture medium (FIG. 8). But, perhaps even more significant is the observation that this strain of blue-green alga can grow equally well over a broad range of salt

FIGURE 8. Daily biomass yield of *Oscillatoria* sp., Miami BG 7 in outdoor culture systems (Ramachandran and Mitsui, unpublished data). A semicontinuous bench scale outdoor culture system (10-liter volume) with natural sea water enriched with nutrients[32] yielded an average daily biomass of 180 mg dry wt/1/day. The dilution rate was between 3.5 and 4 liters per day. The temperature ranged from 26°C to 32°C.

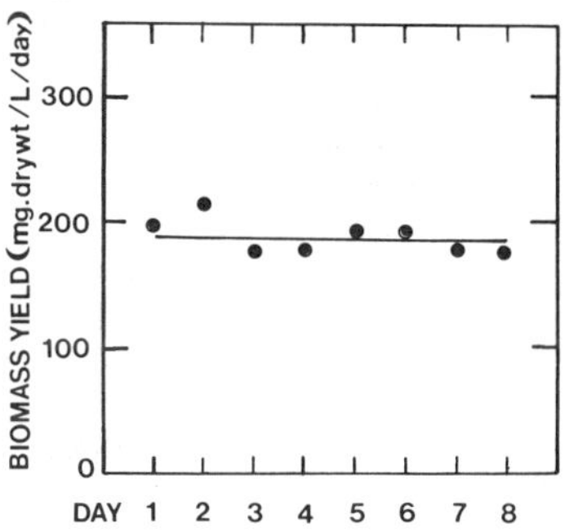

FIGURE 9. 5-liter outdoor bioreactor system for use with blue-green algae. The top photograph shows three 5-liter reactors containing suspensions of the marine blue-green alga *Oscillatoria* sp., Miami BG 7. The bottom photograph shows the environmental monitoring equipment used in the experiment.

concentrations (Ramachandran and Mitsui, unpublished data). This fact opens the door to the use of a wide variety of water types (e.g. brackish to hypersaline) in the cultivation of the strain BG 7.

In terms of hydrogen production, the second step in this system, BG 7 has proven to be quite adaptable to outdoor conditions. A series of five-liter bioreactors were designed in the general form of solar panels to permit efficient use of incident sunlight (FIG. 9). When cellular suspensions were transferred directly from culture chambers, sustained high rates of hydrogen production were observed (FIG. 10). As expected from the earlier results obtained in the laboratory,[9,14] hydrogen production occurred only during the day, at night the hydrogen concentration in the reactor remained constant and hydrogen uptake did not occur.

Another advantageous feature of this hydrogen production stage (step 2) is that it exhibits the same flexibility to different salt concentrations and salinities as growth. This is demonstrated by the fact that maximum hydrogen production rates are virtually the same for a wide range of salt concentrations and salinities (TABLES 3 and 4).

One-Step Simultaneous Hydrogen and Oxygen Photoproduction Using Aerobic Nitrogen-fixing Unicellular Marine Blue-green Algae and Sea Water as Hydrogen Donor

In contrast to the type of blue-green algae described in the previous section, there are groups that exhibit hydrogen production in the presence of oxygen. The most widely studied of these are the heterocystous forms, like the genera *Anabaena* and *Nostoc*. In these groups, some of the cells in the filament are modified into specialized cells called heterocysts. The morphological and physiological organization of these cells permit nitrogenase activity to continue under aerobic conditions. Unfortunately these same strains also are noted for high uptake hydrogenase activity (TABLE 1). Studies in our laboratory indicate that the freshwater strain *Anabaena cylindrica*, a well-known hydrogen producer, also exhibits high hydrogen uptake activity.[21,22] This uptake of hydrogen severely restricts the use of heterocystous algae in hydrogen production technologies. There is, however, another hitherto little-known group of aerobic nitrogen-fixing, unicellular blue-green algae that can produce hydrogen in the presence of O_2 but does not exhibit hydrogen uptake activity. Strains of the genus *Synechococcus* have been isolated from the subtropical Atlantic marine environment by our laboratory[8] (also: Rosner, Radway, Sprogis, Duerr, Leon, and Mitsui, unpublished data). Aerobic growth of these strains, in combined nitrogen-free media, exhibited a doubling time of less than 20 hours[23,24] (also: Leon, Kumazawa, and Mitsui, unpublished data). One of these strains, *Synechococcus* sp. Miami BG 043511, is being used in the study of hydrogen photoproduction.

The most remarkable feature of this strain is that without any special treatment, it not only can produce hydrogen at high rates (maximum rate 1.6 μmol H_2/mg dry wt/hr) but also simultaneously generates oxygen for up to 50 hours. This simultaneous hydrogen and oxygen production conforms to a 2 to 1 stoichiometry (FIG. 11). Although there are green algae, like *Chlorella* and *Chlamydomonas,* that exhibit such a phenomenon, hydrogen production is short lived in these species (i.e., in the order of minutes) unless special treatments are used.

In contrast to the nonheterocystous forms described in the previous section, this strain does not release carbon dioxide, indicating that electrons from water are being efficiently donated to H^+ to form molecular hydrogen directly or via the rapid refixation of carbon dioxide released during the breakdown of internal electron donor

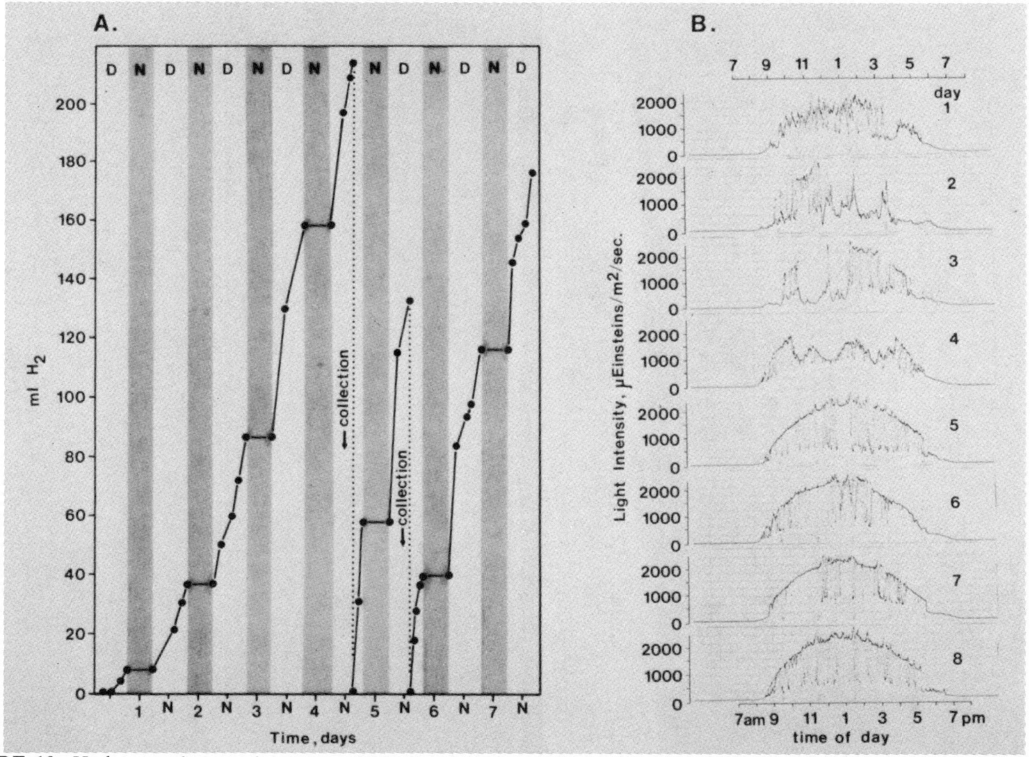

FIGURE 10. Hydrogen photoproduction in a 5-liter outdoor bioreactor by *Oscillatoria* sp., Miami BG 7 (Phlips and Mitsui, unpublished data. The reactor contained 4.4 liters of algal suspension. Cell density was 0.66 mg dry wt/ml. Variation in the intensity of incident sunlight is shown in part B. Temperature was maintained at 35°C. Accumulated hydrogen in the reactor was collected twice during the course of experiment and the gas phase was filled with argon. Hydrogen was measured by gas chromatography as previously described.[14] Total volume of hydrogen produced does not reflect the maxium volume obtainable in the vessel, because a low cell density was used (see FIG. 5).

TABLE 3. Effect of Salt Concentration on Hydrogen Photoproduction by the Marine Blue-Green Alga *Oscillatoria* sp. Miami BG 7)[a b]

Salt concentration (‰)	2.5	6	16	20	25	29	34	39
Maximum rates of H_2 production (μl H_2/mg dry wt/hr)	6.9	8.0	7.7	7.5	7.1	7.8	6.6	5.5

[a]Phlips and Mitsui, unpublished data.

[b]Cells for the experiment were taken from cultures containing the same salt concentrations as that used in the test. NaCl was added to A-N-NaCl medium, as described previously[14] to give appropriate salt concentrations. The hydrogen assays were run at a light intensity of 100 μEinsteins/m^2/sec and a temperature of 35°C. Values given are averages of replicate experiments. Hydrogen was measured by gas chromatography as previously described.[14]

TABLE 4. Effect of Different Sea Water Salinities on Hydrogen Photoproduction by the Marine Blue-Green Alga *Oscillatoria* sp. Miami BG 7[a,b]

Salinity (‰)	5	10	15	20	30	35	45
Rates of H_2 photoproduction (μl H_2/mg dry wt/hr)	5.9	6.4	6.0	6.0	5.8	5.2	3.9

[a]Ramachandran and Mitsui, unpublished data.

[b]Natural sea water was substituted for the artificial medium as the medium of growth and hydrogen photoproduction by *Oscillatoria* sp. Miami BG 7. Lower salinities were obtained by diluting sea water of 35‰ salinity with distilled water and for high salinities (i.e. 45‰) sodium chloride was added to sea water. Light intensity and temperature were 30 μEinsteins/m^2/sec and 30°C, respectively. Rates are averages of triplicates. Hydrogen was measured by gas chromatography as described previously.[14]

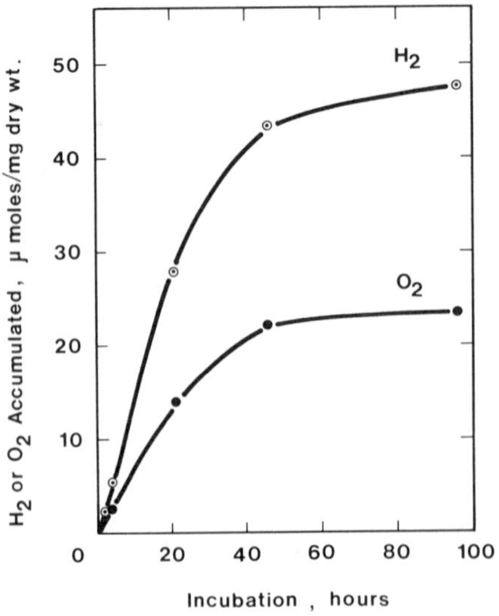

FIGURE 11. Simultaneous stoichiometric hydrogen and oxygen photoproduction by the aerobic nitrogen fixing unicellular blue-green alga *Synechococcus* sp., Miami BG 043511 (Reddy and Mitsui, unpublished data). 5-ml concentrated cell suspensions of 84-hour-old culture in A–N medium[14] were used for incubation. Cell density was 0.4 mg/ml suspension. Gas phase at zero hour incubation was 100% argon. Light intensity and temperature for incubation were 30°C and 120 μEinsteins/m^2/sec. Hydrogen and oxygen were measured by gas chromotography as previously described.[14]

compounds. The proposed scheme is illustrated in FIGURE 2. The overall input in this one-step hydrogen production system is water and light and the overall output is hydrogen and oxygen

$$H_2O \xrightarrow{\text{light}} H_2 + 1/2\, O_2$$

One-Step Hydrogen Photoproduction System Using Nonsulfur Photosynthetic Bacteria from Organic Wastes

Among the marine nonsulfur photosynthetic bacterial strains so far tested, the marine *Rhodopseudomonas* sp. Miami PBE 2271 has exhibited the highest rate of

FIGURE 12. 4-liter outdoor bioreactor system for use with photosynthetic bacteria. The reactor (inside the glass tank) shown in the left side of the photograph contains immobilized cells of the photosynthetic nonsulfur bacteria Miami PBE 2271. To the right of the reactor is a continuous gas monitoring meter.

hydrogen production (4 μmol H_2/mg dry wt/hr from malate). This strain was isolated from sea grass (*Thalassia*) leaves collected from the southern Florida marine environment in 1978 (Haynes and Mitsui, unpublished data). It utilizes a wide range of organic substances for hydrogen production.[18,19] This photosynthetic bacterium was immobilized in a series of thin agar plates and placed in a reactor (FIG. 12). Waste water from an orange-processing factory, diluted with sea water to give a final concentration of 430 ppm of total organic carbon (TOC), was used as the substrate for hydrogen photoproduction. Hydrogen was continuously produced for seven hours from

9:00 A.M. to 4:00 P.M. (FIG. 13). Total organic carbon in the reactor was reduced from 430 ppm to 270 ppm and biological oxygen demand (BOD) was reduced from 600 ppm to 40 ppm.

Although Miami PBE 2271 has a high hydrogen photoproduction capability, growth and biomass yield are only moderate. At present, many other photosynthetic nonsulfur bacteria are being screened for growth potential (Haynes and Mitsui, unpublished data) and hydrogen photoproduction (Kumazawa and Mitsui, unpublished data).

FIGURE 3 illustrates the hydrogen production system using nonsulfur photosynthetic bacteria. The overall input is solar energy and organic substances and the overall output is hydrogen and carbon dioxide. There are a wide variety of commercial and

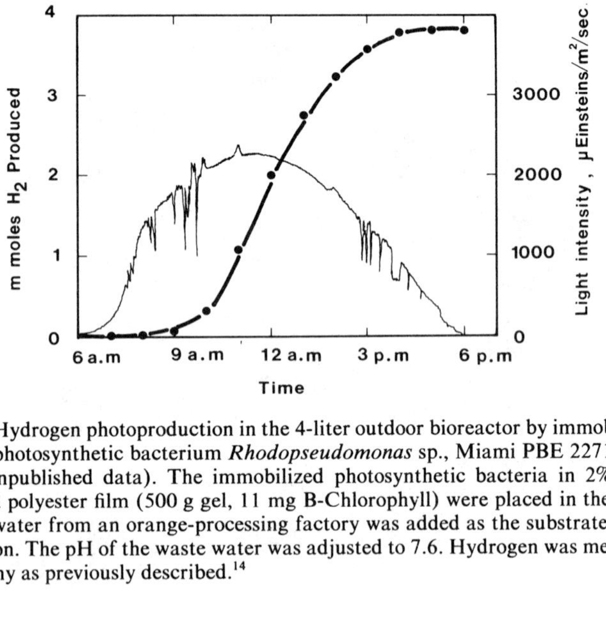

FIGURE 13. Hydrogen photoproduction in the 4-liter outdoor bioreactor by immobilized cells of the nonsulfur photosynthetic bacterium *Rhodopseudomonas* sp., Miami PBE 2271 (Matsunaga and Mitsui, unpublished data). The immobilized photosynthetic bacteria in 2% agar gel on agarose-coated polyester film (500 g gel, 11 mg B-Chlorophyll) were placed in the reactor. One liter of waste water from an orange-processing factory was added as the substrate for hydrogen photoproduction. The pH of the waste water was adjusted to 7.6. Hydrogen was measured by gas chromatography as previously described.[14]

public facilities that generate waste materials rich in organic substances suitable for growth and hydrogen photoproduction by nonsulfur photosynthetic bacteria. Since this waste must be treated before being released into the environment, the operation of product treatment facilities can be costly. Hence, developing a viable way of reducing the total organic substances (TOS) and/or the biological oxygen demand (BOD) of the waste and simultaneously generating energy would yield a very economically attractive package.

It is also important to note that hydrogen production by photosynthetic bacteria using organic substances is comparatively advantageous from an energetic point of view.[11] For example, in the case of glucose utilization, hydrogen photoproduction can yield 680 kcal/mol, while dark hydrogen production is limited to 227 kcal/mol and methane production is restricted to 586 kcal/mol.

One-Step Hydrogen Photoproduction System Using Sulfur Photosynthetic Bacteria from Reduced Inorganic Wastes

The marine photosynthetic sulfur bacteruim *Chromatium* sp. Miami PBS 1071 has the ability to grow rapidly (doubling time less than 2 hours) and produce hydrogen at high rates (maximum rate observed so far is 3 μmol/mg dry wt/hr),[11,15–18,25] (also: Frank and Mitsui, Ikemoto and Mitsui, unpublished data). This strain was isolated from the surface of sea grass (*Thalassia*) leaves collected from Biscayne Bay, Florida in 1974 (Mitsui, Hill, and Rosner, unpublished data).

This strain can photoproduce hydrogen using reduced sulfur compounds, such as sulfide and thiosulfate as electron donors. It can also grow well at a wide range of sulfide concentrations. This is an important ability since hydrogen sulfide is known to be toxic to many organisms. The processing of coal as well as many other industrial operations generates large quantities of sulfide-rich waste that must be eliminated from the emission systems. Sulfides are also found in sewage-polluted marine waters and organic-rich sediments and are responsible for bad odors.

Recent experiments indicate that this strain can grow well and produce hydrogen from hydrogen sulfide-containing sea water under outdoor conditions, without any inhibition by the high light intensity prevalent in the tropics. This strain also has a high respiration activity, so that small amounts of oxygen contamination do not affect either hydrogen production or growth. Hydrogen photoproduction systems using immobilized cells of this strain in sulfide-rich waste are now being studied in outdoor conditions.

Other than *Chromatium* Miami PBS 1071, strains with efficient sulfur-removing capability and high hydrogen production, as well as strains that produce hydrogen in ammonia-containing media, are also being explored (Ikemoto and Mitsui, unpublished data).

FIGURE 4 illustrates the one-step hydrogen production system using sulfur photosynthetic bacteria. The reduced sulfur compounds such as hydrogen sulfide are converted with light to oxidized sulfur compounds. If the oxidized sulfur compound is elemental sulfur, and if it is excluded by the cells in some strains, sulfur also could be removed from waste.

CONCLUSIONS AND FUTURE DIRECTIONS

The results of our laboratory's initial efforts to develop outdoor hydrogen photoproduction systems are encouraging. First, it is clear that hydrogen production systems can be operated at optimal efficiency in relatively uncontrolled outdoor systems. Second, there are certain metabolically unique strains, like nonheterocystous filamentous and unicellular blue-green algae, which hold special promise for future development of productive hydrogen-evolving systems. Finally, the existence of hydrogen-producing organisms like photosynthetic bacteria that can produce hydrogen from organic and reduced sulfur-containing waste, represent an energy source that could become economically feasible in the short run.

The question of whether biological hydrogen production is economically feasible remains, at this stage, a complex interaction of many factors such as future trends in energy prices, technological developments, and the ingenuity of the persons designing new hydrogen-producing systems.

It is clear, however, that in these types of systems, like many other alternative energy technologies, it is necessary to optimize economic output. The theme of our laboratory has long been that biological solar energy systems should take advantage of by-products and secondary applications in order to maximize the economic benefits acquired from the system.[4,5,7] In keeping with this principle, we are also engaged in the study of a variety of nonfuel applications of these organisms. For example, a number of the strains of blue-green algae in our collection of hydrogen producers have proven to be excellent feed for aquaculture.[5–7,26] There are also a number of strains that can produce useful chemicals like amino acids and peptides (Miyazawa and Mitsui, unpublished data; see also Ref. 26), and biologically active substances (Goodman and Mitsui, Reyes-Vasquez and Mitsui, Kusumi and Mitsui, unpublished data).

The success of the concept of biological hydrogen production is still a matter for future research to decide, but the prognosis is encouraging. There is no doubt, however, that the goal of developing a renewable energy based on solar energy and salt water is worth pursuing.

SUMMARY

During the past 10 years, an intensive effort has been made to collect and isolate numerous strains of blue-green algae and photosynthetic bacteria from tropical and subtropical Atlantic marine environments. These marine photosynthetic microorganisms have been cultured in sea water-based medium. Many biochemically and physiologically unique strains have been found. Among these are: (1) nonheterocystous hydrogen-producing strains in which uptake hydrogenase activity is missing; (2) unicellular blue-green algae capable of aerobic nitrogen fixation and hydrogen photoproduction, based on oxygen protection mechanisms for nitrogenase; (3) photosynthetic nonsulfur bacteria that grow and produce hydrogen using a wide range of organic substances; and (4) photosynthetic sulfur bacterial strains that exhibit high growth capability, biomass yield and hydrogen photoproduction capability from inorganic sulfur compounds such as sulfide. Based on the studies of the biochemical characteristics of these strains, outdoor biomass production and biosolar conversion studies are being successfully carried out in both free cell and immobilized cell systems. The results of our research indicate that there is a vast, largely untapped potential for the use of photosynthetic marine microorganisms in development of the hydrogen production technology.

ACKNOWLEDGMENT

We thank Mr. R. Cook, Plymouth Citrus Products, Inc., Florida for providing information on orange-processing waste water.

REFERENCES

1. Mitsui, A., S. Miyachi, A. San Pietro & S. Tamura, Eds. 1977. Biological Solar Energy Conversion. Academic Press. New York.
2. Mitsui, A. & C. C. Black, Eds. 1982. Handbook of Biosolar Resources. Vol. 1 Basic Principles. Parts 1 and 2. CRC Press. Boca Raton, FL.
3. Mitsui, A. 1979. Biological and biochemical hydrogen production. *In* Solar-Hydrogen Energy Systems. T. Ohta, Ed. Pergamon Press. Oxford and New York. pp. 171–191.

4. MITSUI, A. 1975. Multiple utilization of tropical and subtropical marine photosynthetic organisms. *In* Proceedings of the 3rd International Ocean Development Conference. I.O.D.C. Organizing Committee, Eds. Vol. **3**: 11–30. Seino Printing Co., Ltd. Tokyo, Japan.
5. MITSUI, A. 1980. Saltwater-based biological solar energy conversion for fuel, chemicals, fertilizer, food and feed. *In* Proceedings of Bio-Energy '80. World Congress and Exposition. Bio-energy Council, Ed. Washington, D.C. pp. 486–491.
6. MITSUI, A., R. MURRAY, B. ENTENMANN, K. MIYAZAWA & E. POLK. 1981. Utilization of marine blue-green algae and macroalgae in warm water mariculture. *In* Biosaline Research—A Look to the Future. A. San Pietro, Ed. Plenum Press. New York. pp. 216–225.
7. MITSUI, A. 1979. Biosaline research: The use of photosynthetic marine organisms in food and feed production. *In* the Biosaline Concept: An Approach to the Utilization of Underexploited Resources. A. Hollaender, J.C. Allen, E. Epstein, A. San Pietro & O. Zaborsky, Eds. Plenum Press. New York. pp. 177–215.
8. MITSUI, A. 1978. Marine photosynthetic microorganisms as potential energy resources: Research on nitrogen fixation and hydrogen production. *In* Proceedings of the 5th International Ocean Development Conference. I.O.D.C. Organizing Committee, Eds. Vol. **1**.(B1): 29–52. Seino Printing Co., Ltd. Tokyo, Japan.
9. MITSUI, A. & S. KUMAZAWA. 1977. Hydrogen production by tropical marine photosynthetic organisms as a potential energy resource. *In* Biological Solar Energy Conversion. A. Mitsui, S. Miyachi, A. San Pietro & S. Tamura, Eds. Academic Press. New York. pp. 23–51.
10. MITSUI, A., E. DUERR, S. KUMAZAWA, E. PHLIPS & H. SKJOLDAL. 1979. Biological solar energy conversion: Hydrogen production and nitrogen fixation by marine blue-green algae. *In* Sun II. K.W. Böer & B. H. Glenn, Eds. Vol. **1**: 31–35. Pergamon Press, New York and Oxford.
11. MITSUI, A., Y. OHTA, J. FRANK, S. KUMAZAWA, C. HILL, D. ROSNER, S. BARCIELA, J. GREENBAUM, L. HAYNES, L. OLIVA, P. DALTON, J. RADWAY & P. GRIFFARD. 1980. Photosynthetic bacteria as alternative energy sources. Overview on hydrogen production research. *In* Alternative Energy Sources II. T. N. Veziroglu, Ed. Vol. **8**: 3483–3510. Hemisphere Publishing Co., Washington, D.C.
12. MITSUI, A. 1976. A survey of hydrogen-producing photosynthetic organisms in tropical and subtropical marine environments. NSF Grant—Annual Report. Grant no. AER75-11171 pp. 1–68.
13. MITSUI, A. 1982. Nitrogen and hydrogen metabolism in marine tropical photosynthetic prokaryotes. *In* Abstracts of the 5th International Symposium on Photosynthetic Prokaryotes. Stanier-Cohen-Barire, Ed. B66. Bombannes-Bordeaux, France.
14. KUMAZAWA, S. & A. MITSUI. 1981. Characterization and optimization of hydrogen photoproduction by a saltwater blue-green alga, *Oscillatoria* sp. Miami BG7. I. Enhancement through limiting the supply of nitrogen nutrients. Int. J. Hydrogen Energy, **6**: 339–348.
15. MITSUI, A. 1981. Progress report: A study of hydrogen production by tropical marine photosynthetic bacteria for applied systems. (Duration: Nov. 1, 1979–Oct. 30, 1980). *In* Proceedings of the Review Meeting on Solar Hydrogen Production Program. SERI-DOE, Eds. SERI/SP-624-1095. pp. 1–19.
16. OHTA, Y., J. FRANK & A. MITSUI. 1981. Hydrogen production by marine photosynthetic bacteria: Effect of environmental factors and substrate specificity on the growth of a hydrogen-producing marine photosynthetic bacterium, *Chromatium* sp. Miami PBS 1071. Int. J. Hydrogen Energy 6: 451–460.
17. OHTA, Y. & A. MITSUI. 1981. Enhancement of hydrogen photoproduction by marine *Chromatium* sp. Miami PBS 1071 grown in molecular nitrogen. *In* Advances in Biotechnology. M. Moo-Young & C.W. Robinson, Eds. Vol. **II**: 303–307. Pergamon Press. Oxford and New York.
18. MITSUI,A., T. MATSUNAGA & S. KUMAZAWA. 1982. Progress Report: A study of hydrogen production by tropical marine photosynthetic bacteria for applied systems. (Duration:

Nov. 1, 1980–Oct. 30, 1981). *In* Proceedings of the Review Meeting on Solar Hydrogen Production Program. SERI-DOE, Eds. In press.
19. MATSUNAGA, T. & A. MITSUI. 1982. Seawater-based hydrogen production by immobilized marine photosynthetic bacteria. *In* Biotechnology and Bioengineering Symposium. **12**: 441–450.
20. PHLIPS, E. J. & A. MITSUI. 1982. The role of light intensity and temperature in the regulation of hydrogen photoproduction by the marine blue-green alga *Oscillatoria* sp. Miami BG7. Appl. Environ. Microbiol. **45**: 1212–1220.
21. REDDY, K. J., S. KUMAZAWA, S. IZAWA & A. MITSUI. 1981. Enhancement of hydrogen photoproduction in *Anabaena cylindrica* B629 by low concentrations of DCMU. Abstr. Annu. Meet. Am. Soc. Microbiol. K 102.
22. REDDY, K. J., S. KUMAZAWA & A. MITSUI. 1982. Comparative study of uptake hydrogenase activity and hydrogen photoaccumulation in blue-green algae. Plant Physiol. (Suppl.) **67**: 86.
23. DUERR, E. O. & A. MITSUI. 1980. Aerobic growth and nitrogenase activity of a marine unicellular blue-green alga, *Synechococcus* sp. Plant Physiol. (Suppl.) **65**: 160.
24. DUERR, E. O. & A. MITSUI. 1981. Aerobic growth characteristics of two marine unicellular nitrogen-fixing Cyanobacteria. Abstr. Annu. Meet. Am. Soc. Microbiol. K 85.
25. MITSUI, A. 1976. Bioconversion of solar energy in salt water photosynthetic hydrogen production systems. *In* Proceedings of the First World Hydrogen Energy Conference. T.N. Veziroglu, Ed. Vol. **II**(4B): 77–99. University of Miami Press. Miami, FL.
26. MURRAY, R. & A. MITSUI. Growth of hybrid tilapia-fry-fed, nitrogen-fixing blue-green algae in seawater. J. World Mariculture Soc. In press.
27. MITSUI, A. 1978. Nitrogen fixation with photosynthetic marine microorganisms. NSF Grant—Eighteen Months Progress Report. pp. 1–40.
28. DUBOIS, M., K. A. GILLES, J. K. HAMILTON, P. A. REBERS & R. SMITH. 1956. Colorimetric method for determination of sugars and related substances. Anal. Chem. **28**: 350–356.
29. LOWRY, O. H., N. J. ROSEBROUGH, A. L. FARR & R. J. RANDALL. 1951. Protein measurement with the Folin phenol reagent. J. Biol. Chem. **193**: 265–275.
30. WILLIAMS, J. P. 1978. Glycerolipids and fatty acids of algae. *In* Handbook of Phycological Methods: Physiological and Biochemical Methods. H. A. Hellebust & J. S. Craigie, Eds. Cambridge University Press. New York. pp. 100–107.
31. LARSEN, B. 1978. Brown seaweeds: Analysis of ash, fiber, iodine and mannitol. *In* Handbook of Phycological Methods: Physiological and Biochemical Methods. H. A. Hellebust & J. S. Craigie, Eds. Cambridge University Press. New York. pp. 182–188.
32. GUILLARD, R. R. L. & J. H. RYTHER. 1962. Studies of marine planktonic diatoms. I. *Cyclotella nana* Hustedt and *Detonula confervacea* (Cleve.) Gran. Can. J. Microbiol. **8**: 229–239.

Engineering Analysis of Potential Photosynthetic Bacterial Hydrogen Production Systems

ANN HERLEVICH, MICHAEL KARPUK,
AND HILDE LINDSEY

Solar Energy Research Institute (SERI)
1617 Cole Blvd.
Golden Colorado 80401

Numerous photosynthetic bacteria (PSB) have been observed to evolve hydrogen when supplied with appropriate organic substrates and maintained in anaerobic conditions with controlled temperatures.[1] In these conditions, the bacteria follow an alternative metabolic path in which hydrogen is a waste product. Hydrogen production has been demonstrated in the laboratory with a sunlight-to-hydrogen conversion efficiency of 5%.

The bacteria are capable of metabolizing a wide variety of substrates, but the highest hydrogen evolution rates are obtained with organic acids. The effluents from several industries, such as the food processing, pulp and paper, chemical, and plastics industries, are potential sources of such organic compounds. The bacteria are capable of removing nearly all soluble organics and heavy metals from the waste stream. The resulting effluent is of tertiary quality and can be used directly as irrigation water or returned to rivers or streams after bacteria are filtered out. PSB plants may benefit from significant economic credits by providing waste stream cleanup.

An engineering analysis was performed at SERI to examine system requirements for the application of this biological process to large-scale hydrogen production. A system capable of producing 28,000 m^3 of hydrogen per day was conceptually designed. Two cases were examined: the first with 5% conversion efficiency, the second with 10% efficiency. The higher efficiency represents the potential for hydrogen production as estimated by SERI microbiologists.

The major plant design parameters are shown in TABLE 1. FIGURE 1 shows the steps in the hydrogen production process.

The design of the solar bacterial reactor is the most critical element of the system design because of its effect on process efficiency and high fraction of system cost. The

TABLE 1. Design Assumptions for PSB Hydrogen Production Plant

Parameter	Near-Term Application	Long-Term Application
Plant capacity	28,000 m^3/day	28,000 m^3/day
Conversion efficiency (sunlight to lower heating value of hydrogen)	5%	10%
Insolation	0.24 kW/m^2	0.24 kW/m^2
Collector area	2.8×10^5 m^2	1.4×10^5 m^2
Substrate concentration	5 g/l H_2O	5 g/l H_2O
Water processed	3.8×10^6 l/day	3.8×10^6 l/day

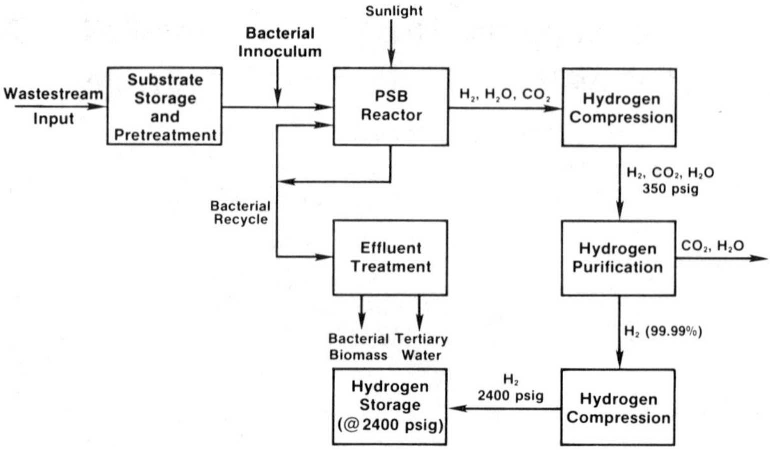

FIGURE 1. Process diagram: photosynthetic bacterial hydrogen production system.

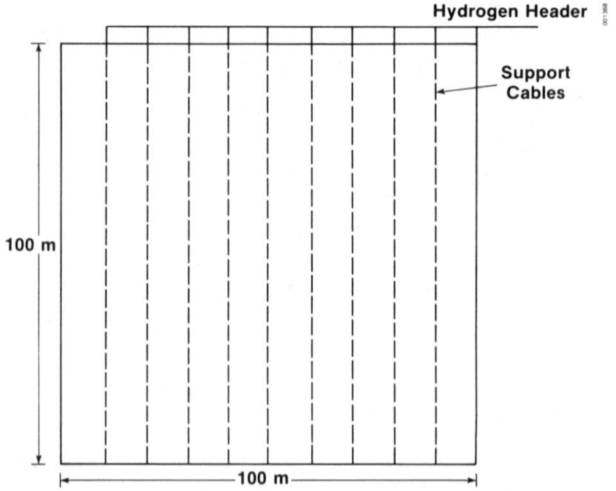

FIGURE 2. Deep pond reactor, top view.

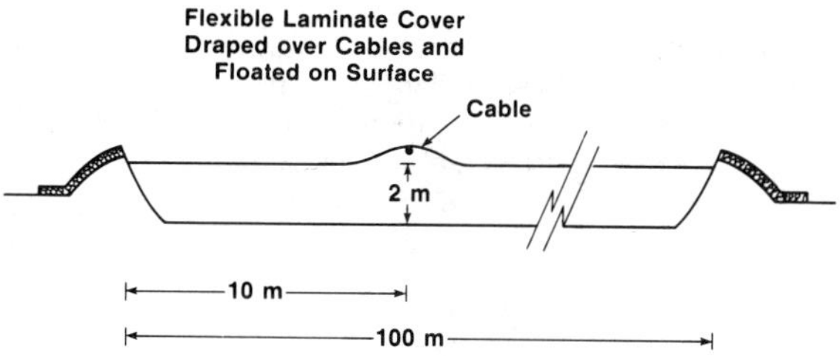

FIGURE 3. Deep pond reactor, side view.

design requirements considered for the reactor were low cost, temperature control, transparency of reactor cover, ease of hydrogen collection, and impermeability to hydrogen. Residence time and hydrogen production rates will vary depending on reactor design, insolation, and ambient temperature.

Several reactor designs that could potentially meet these criteria were analyzed and were judged on their potential cost. A critical assumption was the availability of a transparent hydrogen-impermeable material at $\$4/m^2$, a cost believed reasonable by potential materials manufacturers. One of the designs, the deep pond reactor shown in FIGURES 2 and 3, was selected for further study because of its low cost of $\$21/m^2$.

In addition to studying the reactors, the entire hydrogen production system was sized and costed. Equipment for reactor temperature contol and associated costs were not included. Capital and operating costs for a plant with 5% conversion efficiency are shown in TABLE 2.

The projected first-year cost of hydrogen from this plant at $83.30 per MWhr is not competitive with the cost of hydrogen produced from natural gas or coal at today's prices (about $51.20 per MWhr). The projected cost of the hydrogen from the 10% efficient plant at $53.60 per MWhr is, however, competitive with the current price of hydrogen from conventional sources.

The production of hydrogen from photosynthetic bacteria appears technically and economically feasible from an engineering viewpoint. In particular, two conceptual designs show promise and warrant further engineering research. Innovative methods of

TABLE 2. Near-Term Plant Capital and Operating Costs

	Capital, $	O & M, $/year
Solar bacterial reactor[a]	$5,961,000	$ 119,000
Substrate storage	302,000	6,000
Effluent treatment	30,000	3,000
H_2 Compression	400,000	166,000
H_2 Purification	987,000	57,000
H_2 Storage	—	50,000
Land	500,000	—
	$8,180,000	401,000
Cost of capital (0.25 fixed charge rate)		$2,045,000
Annual operation cost		$2,446,000
First-year cost of hydrogen		$83.30 per MWhr ($24.40 per 10^6 Btu)

[a]Based on deep pond reactor design.

temperature control should be investigated. In addition, development of materials appropriate for use in the reactor should continue. Research into the efficiency of the biological process and substrate pretreatment requirements must also continue. If these problems are attacked in parallel, a useful technology for hydrogen production from renewable sources will become available in the near term.

REFERENCES

1. WEAVER, P. F. 1981. Photoconversion of organic substrates into hydrogen using photosynthetic bacteria. Proceedings of Energy from Biomass and Wastes V Conference. January 26–30. Lake Buena Vista, FL; Institute of Gas Technology. Chicago, IL. pp. 489–97.

Modeling of Immobilized Glucoamylase Reactors

J. M. S. CABRAL, J. M. NOVAIS, AND J. P. CARDOSO

Laboratório de Engenharia Bioquímica
Instituto Superior Técnico
Universidade Técnica de Lisboa
1000 Lisboa, Portugal

INTRODUCTION

The enzymic hydrolysis of starch has been modeled, assuming that the irreversible hydrolysis of each maltooligosacharide follows a simplified Michaelis Menten equation, based on a multichain mechanism and the reversible hydrolysis of maltose.[1,2] Other models have been applied considering the inhibition of this reaction by glucose.[3,4]

In the starch-immobilized glucoamylase system, the kinetic model formulations include internal and external mass transfer effects. However, due to the substrate heterogenity, the kinetics of this system are not simple to describe and several deviations from the models have been observed.[5]

In this communication, a model based on experimental results of the hydrolysis of soluble starch and maltodextrins is developed.

The performance of immobilized glucoamylase in packed- (PBR) and fluidized-bed reactors (FBR) is also compared in this study.

KINETIC MODEL

For an isothermal batch reactor (BR) or plug flow reactor (PFR), the design equation for the hydrolysis of starch, assuming Michaelis Menten kinetics, is:

$$S_0X - K_{mi} \ln (1 - X) = 0.162\, k_{2i}\, \zeta$$

where ζ, is the normalized residence time ($E_T \cdot t/V$ for a BR and E_T/Q for a PFR) and K_{mi} and k_{2i} are the intrinsic kinetic constants for the immobilized enzyme.

However, this equation has little application, even when diffusional and dispersion effects are included, as it does not model immobilized glucoamylase reactors.

A model that takes account of mass transfer limitations and kinetics was developed by the authors.[5,6] For BR and PFR, an integrated equation can be written in the form:

$$\zeta = \alpha X - \beta \ln (1 - X)$$

where α and β are parameters, involving the intrinsic kinetic constants and mass transfer effects, and which depend on linear velocity, particle size, temperature, and initial substrate concentration. They can be calculated from the straight line obtained by plotting ζ/X against $-[\ln (1 - X)]/X$.

In the absence of mass transfer limitations, it can be seen that: $\alpha = S_0/0.162k_{2i}$ and $\beta = K_{mi}/0.162\, k_{2i}$.

EXPERIMENTAL

Materials

Glucoamylase was obtained as AG 150, a gift from NOVO Industri A/S, Denmark. Controlled-pore silicas were used as enzyme carriers. $CPSiO_2$ 1170 Å, a gift from Corning Glass Works, U.S.A., was used on soluble starch hydrolysis and

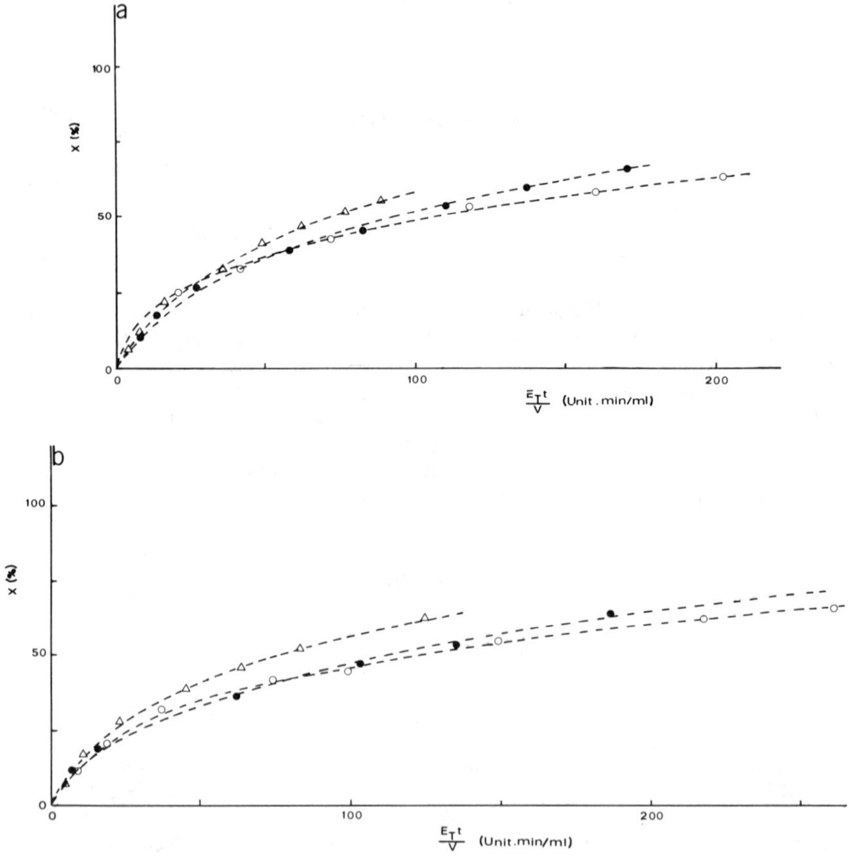

FIGURE 1. Influence of flow rate on soluble starch hydrolysis in immobilized glucoamylase reactors. (a) FBR—Flow rate (ml/min): (○) 8.8; (●) 12.9; (△) 21.5. (b) PBR—Flow rate (ml/min): (○) 9.4; (●) 14.0; (△) 24.0. (---) integrated equation model.

Spherosil XOB 015 (+100–200 μm) from Prolabo, France, was used on production of high-dextrose syrups.

Soluble starch, from B.D.H., was used as substrate at 1% w/v in 0.02 *M* acetate buffer pH 4.5, at 45°C.

Corn syrup DE40, a gift from COPAM, Portugal, was used in the production of high-dextrose syrups at 30% w/w in solution, pH 4.5 at 50°C.

TABLE 1. Influence of Flow Rate on α and β

Reactor	Flow Rate (ml/min)	Linear Velocity (cm/min)	α	β	r
FBR	8.8	11.2	−544	547	0.9968
	12.9	16.4	−276	334	0.9911
	21.5	27.4	−180	235	0.9884
PBR	9.4	12.0	−567	588	0.9917
	14.0	17.8	−357	418	0.9826
	24.0	30.6	−252	291	0.9928

Immobilization Technique

Glucoamylase was immobilized on alkylamine derivatives of titanium-(IV)-activated inorganic supports according to a method developed by the authors.[7,8]

Results

Influence of Flow Rate and Particle Size on Soluble Starch Hydrolysis

The influence of flow rate on the conversion of soluble starch by immobilized glucoamylase particles (+354–500 μm) in total recirculated column reactors is shown in FIGURE 1. FBR presents a slightly better performance than PBR. The integrated equation model describes the hydrolysis and the values of parameters α and β are shown in TABLE 1 as well as the correlations obtained for the equation model.

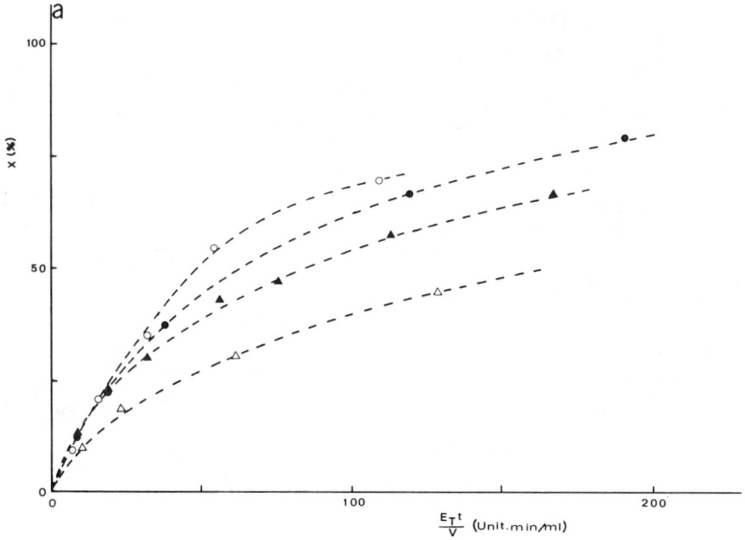

FIGURE 2. Influence of particle size on soluble starch hydrolysis in immobilized glucoamylase fluidized-bed reactors. (○) +177–250 μm; (●) +250–354 μm; (▲) +354–500 μm; (△) +500–707 μm. Flow rate (ml/min): (a) 9.7; (b) 12.9; (c) 18.8 (---) integrated equation model.

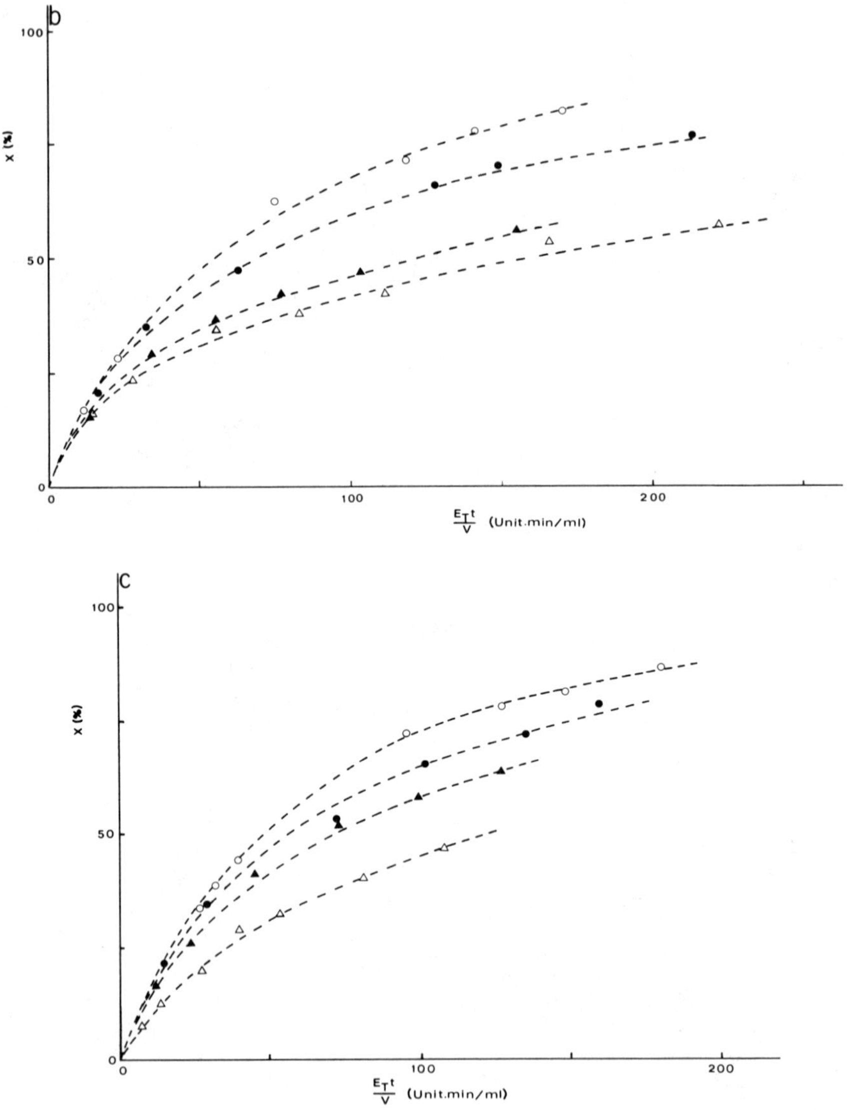

FIGURE 2. (continued)

TABLE 2. Influence of Particle Size on α and β

Particle Size (μm)	Linear Velocity (cm/min) 12.4			16.4			23.9		
	α	β	r	α	β	r	α	β	r
+177–250	−84	139	0.9987	−67	130	0.9963	−53	112	0.9903
+250–354	−220	259	0.9960	−139	192	0.9990	−120	170	0.9867
+354–500	−526	552	0.9932	−295	334	0.9958	−191	243	0.9909
+500–707	−774	781	0.9880	−499	579	0.9935	−345	426	0.9769

The influence of linear velocity (U) on α and β can be written by the following equations:

$$\text{FBR: } \alpha = -157 - 6.93 \times 10^5 \times U^{-3.1} \qquad \beta = 193 + 1.56 \times 10^5 \times U^{-2.5}$$

$$\text{PBR: } \alpha = -206 - 8.54 \times 10^4 \times U^{-2.2} \qquad \beta = 166 + 1.07 \times 10^4 \times U^{-1.3}$$

The influence of particle size on the hydrolysis of soluble starch by immobilized glucoamylase preparations in FBR is presented on FIGURE 2, for three linear velocities. From these results, the parameters α and β were calculated (TABLE 2). These parameters can be correlated with the linear velocity by the equations shown in TABLE 3.

For FBR, the following general equations can be derived:

$$\alpha = 46 - (1.08 \times 10^{-4} \times U^{2.35} - 3.94 \times 10^{-2}) \times \text{dp}^{(6.39 \times 10^4 \times U^{-4.35} + 1.17)}$$

$$\beta = 55 - (3.32 \times 10^{-6} \times U^{-2.44} - 1.45 \times 10^{-3}) \times \text{dp}^{(1.38 \times 10^3 \times U^{-2.85} + 1.51)}$$

In the absence of mass transfer limitations,

$$\alpha_{\text{dp}\to 0} = S_0/0.162\ k_{2i} = 46 \text{ and } \beta_{\text{dp}\to 0} = K_{mi}/0.162\ k_{2i} = 55$$

from which the intrinsic kinetic constants can be obtained as follows:

$$k_{2i} = 1.34\ \mu\text{mole min}^{-1}\text{Unit}^{-1} \text{ and } K_{mi} = 12.0\ g\ 1^{-1}$$

Influence of Linear Velocity on Production of High-Dextrose Syrup in Immobilized Glucoamylase Reactors

The production of high-dextrose syrups from a corn syrup with DE 40 was accomplished in FBR and PBR, with total recirculation of substrate solution.

TABLE 3. Influence of Particle Size and Linear Velocity on α and β

Linear Velocity (cm/min)	α	β
12.4	$30 - 5.36 \times 10^{-4}\,\text{dp}^{2.29}$ (r = 0.9999)	$55 + 8.87 \times 10^{-5}\,\text{dp}^{2.57}$ (r = 0.9999)
16.4	$52 - 3.78 \times 10^{-2}\,\text{dp}^{1.50}$ (r = 0.9988)	$63 + 1.60 \times 10^{-3}\,\text{dp}^{1.99}$ (r = 0.9996)
23.9	$55 - 1.47 \times 10^{-1}\,\text{dp}^{1.23}$ (r = 0.9977)	$48 + 6.20 \times 10^{-3}\,\text{dp}^{1.67}$ (r = 0.9979)

The influence of linear velocity is shown in FIGURE 3. FBR shows a superior performance when compared with PBR. The kinetic model describes, generally, this system and allows the calculation of parameters α and β (TABLE 4). From these values, the following equations can be derived:

$$\text{FBR: } \alpha = -2326 - 1884 \times U^{-2.2} \qquad \beta = 1311 + 2067 \times U^{-0.7}$$

$$\text{PBR: } \alpha = -3761 - 1061 \times U^{-2.8} \qquad \beta = 1680 + 2902 \times U^{-0.5}$$

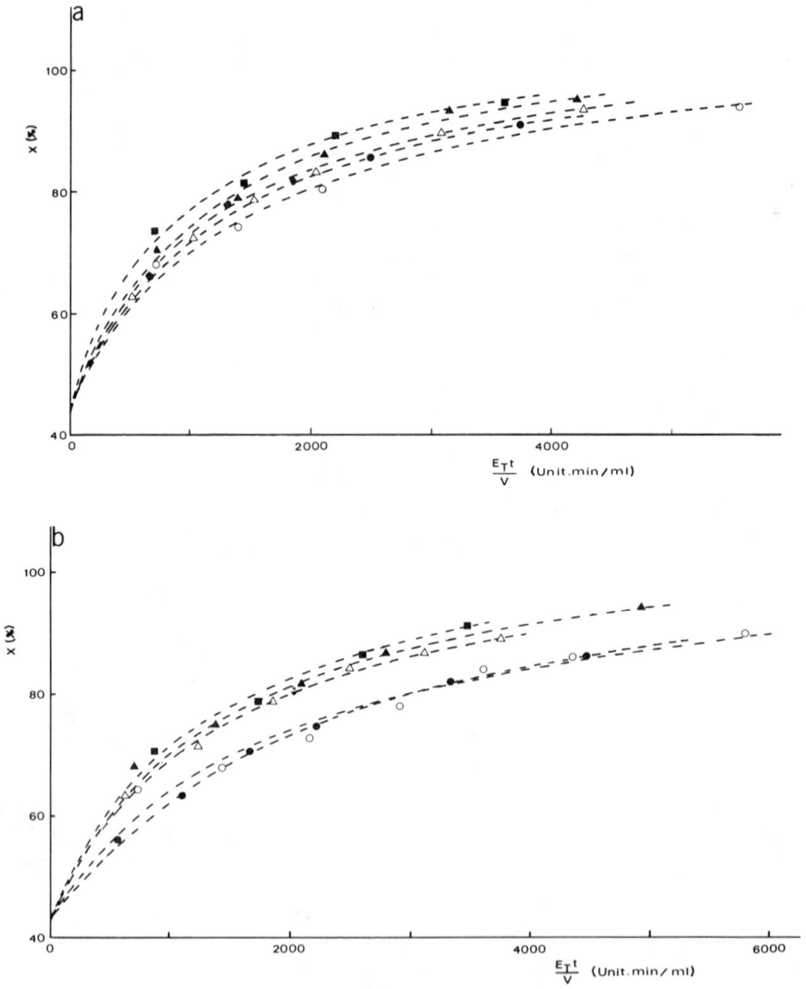

FIGURE 3. Influence of flow rate on production of high-dextrose syrups in immobilized glucoamylase reactors. (a) FBR—Flow rate (ml/min): (○) 0.675; (●) 1.11; (△) 1.34; (▲) 1.81; (■) 2.71. (b) PBR—Flow rate (ml/min): (○) 0.703; (●) 1.06; (△) 2.86; (▲) 4.49; (■) 5.93. (---) integrated equation model.

TABLE 4. Influence of Flow Rate on α and β

Reactor	Flow Rate (ml/min)	Linear Velocity (cm/min)	α	β	r
FBR	0.675	0.859	−4959	3597	0.9928
	1.11	1.41	−4287	3042	0.9829
	1.34	1.71	−2910	2617	0.9989
	1.81	2.30	−2606	2070	0.9894
	2.71	3.45	−2467	2199	0.9772
PBR	0.552	0.703	−6623	5146	0.9590
	0.831	1.06	−4598	4513	0.9962
	2.24	2.86	−4205	3393	0.9902
	3.53	4.49	−3535	2890	0.9903
	4.66	5.93	−3675	3019	0.9900

CONCLUSIONS

(1) In both fixed- and fluidized-bed reactors, internal and external mass transfer limitations were observed.
(2) The fluidized-bed reactor always shows a superior performance compared with the fixed-bed mode, owing to a higher availability of the enzyme and possibly different degrees of channeling in the two beds.
(3) The kinetic model presented described the results of the hydrolysis of both soluble starch and corn syrup 40 DE, in both forms of reactors.
(4) This model allowed the calculation of the intrinsic kinetic constants and modeling of the enzymic starch hydrolysis.
(5) For the production of high-dextrose syrups (93 DE) the linear velocity was unimportant from 2.3 cm/min in the fluidized-bed reactor, while for the fixed-bed reactor, this velocity was 4.5 cm/min.

REFERENCES

1. Lee, D. D., G. K. Lee, P. J. Reilly & Y. Y. Lee. 1980. Biotechnol. Bioeng. **22**(1): 1–17.
2. Swanson, S. J., A. Emery & H. C. Lim. 1977. Biotechnol. Bioeng. **19**(11): 1715–1718.
3. Weetall, H. H. & N. B. Havewala. 1972. Biotechnol. Bioeng. Symp. **3:** 241–266.
4. Engasser, J. M., J. Caumon & A. Marc. 1980. Chem. Eng. Sci. **35:** 99–103.
5. Cabral, J. M. S., J. P. Cardoso & J. M. Novais. 1981. *In* Proceedings of the 3rd International Chemical Engineering Conference, CHEMPOR'81. T. R. Bott, Ed. Vol **1:** 134–143. CHEMPOR'81, Póvoa de Varzim, Portugal.
6. Cardoso, J. P. & M. B. Costa. 1983. Biotechnol. Bioeng. **25**(3): 745–759.
7. Cabral, J. M. S., J. M. Novais & J. P. Cardoso. 1981. Biotechnol. Bioeng. **23**(9): 2083–2092.
8. Cabral, J. M. S., J. F. Kennedy & J. M. Novais. 1982. Enzyme Microbiol. Technol. **4**(5): 337–342.

Extractive Bioconversions in Aqueous Two-Phase Systems[a]

BÄRBEL HAHN-HÄGERDAL,[b] ELIS ANDERSSON,[b]
MATS LARSSON,[c] AND BO MATTIASSON[c]

[b]*Applied Microbiology*
[c]*Pure and Applied Biochemistry*
Chemical Center
University of Lund
S-220 07 Lund, Sweden

Bioconversions are often regulated by product inhibition or an equilibrium between product and substrate. It is therefore desirable to extract the product from the reaction mixture. Few solvents are known to be biocompatible, especially not with cells and microorganisms. Enzymes tend to accumulate and unfold at the interphace when organic solvent/water systems are used. In this context, aqueous two-phase systems, which are formed either when two solutions of water-soluble polymers are mixed or when a solution of a water-soluble polymer is mixed with a salt solution, have the advantage that both phases hold more than 90% water and, thus, are biocompatible.[1]

The number of aqueous two-phase systems to be designed is almost unlimited, which means that it should be possible to find a phase system suitable for any bioconversion.[2] The physical-chemical characteristics of an aqueous two-phase system are determined by the polymers, the molecular weight of the polymers, salts, ionic strength, temperature, and pH. By varying these parameters it is possible to influence the partition of biocatalysts, substrates, and products in a desired direction.

Cells and microorganisms are preferentially partitioned to the bottom phase in aqueous two-phase systems, whereas enzymes can be partitioned almost completely to any phase depending on the physical-chemical characteristics of the phase system. Small-molecular-weight compounds are more evenly distributed between the phase.

When biocatalysts are partitioned to one phase of an aqueous phase system they can easily be recovered in analogy with biocatalysts immobilized to solid supports. However, in aqueous two-phase systems, biocatalysts remain soluble, which prevents inhibitory concentrations of products to build up in close proximity to the biocatalysts.

In order to make biotechnical processes competitive with already existing processes, it is important that they do not involve expensive chemicals. The utilization of polymer/salt systems is one way to reduce polymer costs. Another solution is the utilization of a soluble polymer that can act both as a substrate and a phase component.

CONVERSION OF BENZYLPENICILLIN TO 6-AMINOPENICILLANIC ACID

In a phase system composed of 8.9% (w/w) PEG 20,000 and 7.6% (w/w) potassium phosphate, penicillin acylase (PA) is enriched in the bottom phase with a

[a]This project was supported by the Swedish National Board for Technical Development.

partition coefficient, $K < 0.01$. In this system, 100 g/l bezylpenicillin (BP) was converted to 6-aminopenicillanic acid (6-APA). Within 4 hours, 40% of the substrate was converted using an enzyme concentration of 0.3 g/l (6300 IU/mg enzyme), at which point the top phase was removed and replaced with new top phase again holding 100 g/l BP. Using the initially added enzyme, these batch conversions were repeated four times over a 100-hour period, each time starting with fresh top phase (FIG. 1). The productivity decreased from 3.16 to 0.89 μmol 6-APA/mg enzyme × min, which is in the same range as has been reported for penicillin acylase immobilized to solid supports.

CONVERSION OF STARCH

When α-amylase was added to a mixture of 20% PEG 6,000 and 10% starch, the enzymatic action initially resulted in the formation of a phase system. However, the

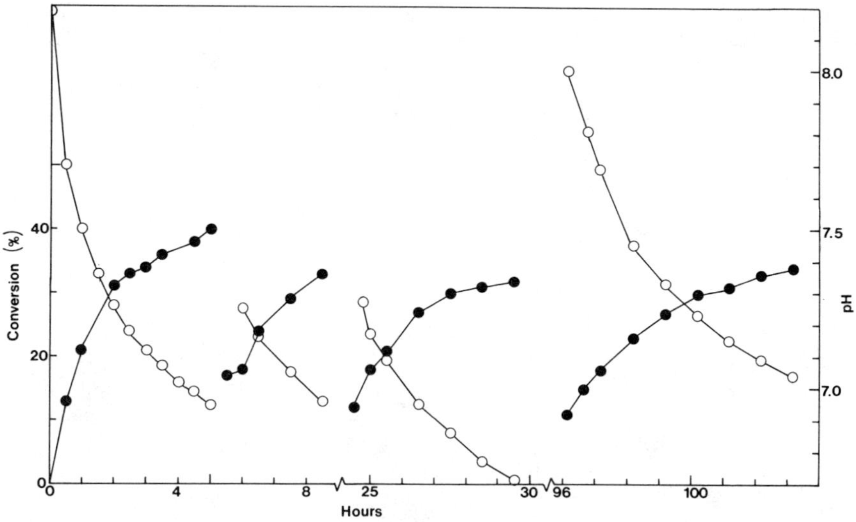

FIGURE 1. Conversion of benzylpenicillin to 6-aminopenicillanic acid in an aqueous two-phase system composed of PEG 20,000 and potassium phosphate using 0.3 g/l (6,300 IU/mg) penicillin acylase.

production of low-molecular-weight sugars was accompanied by a dramatic change in volume ratio between the phases, which proceeded until no phase separation could be observed. The phase system can be stabilized by continuously removing top phase and simultaneously adding fresh top phase containing starch. This was demonstrated in a system where α-amylase was supplemented with amyloglucosidase and baker's yeast, *Saccharomyces cerevisiae*. Ethanol was continuously produced in a phase system composed of 10% (w/w) PEG 6,000 and 15% (w/w) starch in a mixer-settler-type reactor. The PEG top phase holding ethanol and oligosaccharides was continuously withdrawn and fresh PEG holding 10% (w/w) starch was added at a rate of 30 ml/hr. The system was stable for 24 hours before it collapsed. When more starch was added, the phase system could be restored and could again be continuously operated for 24 hours.

In conclusion, aqueous two-phase systems composed of polymer/salt mixtures as well as mixtures of a polymer and a water soluble polymer substrate can be utilized in bioconversions. In repeated batch conversions, productivities similar to those for biocatalysts immobilized to solid supports were achieved. Extractive bioconversions in aqueous two-phase systems should be regarded as soluble, immobilized systems[3] and offer the advantage that additional biocatalyst can be added without disturbing the system and that the immobilization is cheap, simple, and reversible.

REFERENCES

1. MATTIASSON, B. & B. HAHN-HÄGERDAL. 1983. CRC Press. In press.
2. ALBERTSSON, P-Å. 1971. Partition of Cell Particles and Macromolecules, 2nd Ed. Wiley-Interscience. New York.
3. HAHN-HÄGERDAL, B. MATTIASSON, E. ANDERSSON & P.-Å. ALBERTSSON. 1982. J. Chem. Tech. Biotechnol. **32:** 157–161.

Use of Perfluorochemicals for Oxygen Supply to Immobilized Cells[a]

B. MATTIASSON AND P. ADLERCREUTZ

Pure and Applied Biochemistry
Chemical Center, University of Lund
S-220 07 Lund, Sweden

Immobilized aerobic cells require a good oxygen supply in order to function better than the corresponding free cells; however, the low solubility of oxygen in water limits the amount that can be introduced and dissolved in the medium.

In order to improve aeration, different approaches have been tested. By placing the immobilized cells in an airlift fermenter, improved oxygen transfer was achieved.[1] A tempting approach for overcoming the low O_2 solubility has been to generate the oxygen *in situ*. This can either be achieved by adding peroxide to the medium and letting coimmobilized catalase generate oxygen within the solid support, in close proximity to the aerobic cells[2] or by using photosynthetic oxygen production catalyzed by coimmobilized algae.[3]

The present report deals with an alternative method for improving the oxygen supply to immobilized cells. In this procedure, substances capable of binding oxygen are added to the medium. The properties of these compounds resemble those of hemoglobin and they have actually been used as artificial blood, both in animals and humans. They are called perfluorochemicals and are characterized in that all hydrogen atoms are replaced by fluorine. They are reported to be biocompatible and may thus be a good choice when organic solvents are used together with biological material. The perfluorochemical FC-72 (from 3-M, St. Paul, MN, USA) was emulgated in buffer with Pluronic F-68 (from AB Montoil Stockholm, Sweden) as emulgator and the emulsion was mixed into the substrate. The emulsion obtained is very stable. The medium was saturated with air or pure oxygen. The immobilized-cell preparation used to evaluate this method of supplying oxygen was *Gluconobacter oxydans* immobilized in alginate[2] and packed into a column. The reaction studied was the conversion of glycerol to dihydroxyacetone.

The emulsion-substrate mixture was passed through the column and the amount of dihydroxyacetone generated was determined. In order to evaluate the yield of oxygen, the amount added in the emulsion had to be quantified. It turned out not to be possible to use polarographic oxygen electrodes since they measure the *partial pressure* and not the total amount of oxygen in the system. Instead, a specific modification of an enzyme thermistor had to be developed.[4] For our purpose of oxygen evaluation, a thermistor filled with immobilized glucose oxidase (E.C. 1.1.3.4) and catalase (E.C. 1.11.1.6) was used. When using perfluorocompounds in oxygen-saturated medium, up to 17 times higher oxygen content was recorded, as compared to the case using only buffer equilibrated against air.

To evaluate the optimal density of the emulsion in the substrate feed, a series of experiments were performed. FIGURE 1 illustrates that there is a linear relationship between the dihydroxyacetone produced and the volume fraction of FC-72 in the

[a]This project was supported by the Swedish National Board for Technical Development.

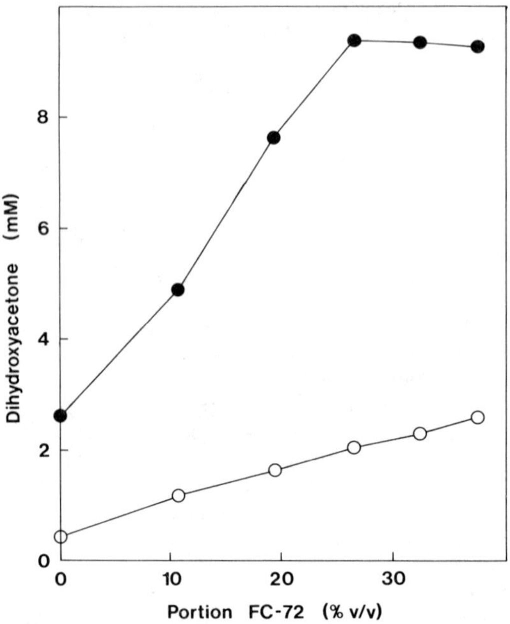

FIGURE 1. Effect of the amount of perfluorochemical (FC-72) in the medium on the production of dihydroxyacetone by immobilized *Gluconobacter oxydans*. The emulsions were saturated with air (○) or oxygen (●).

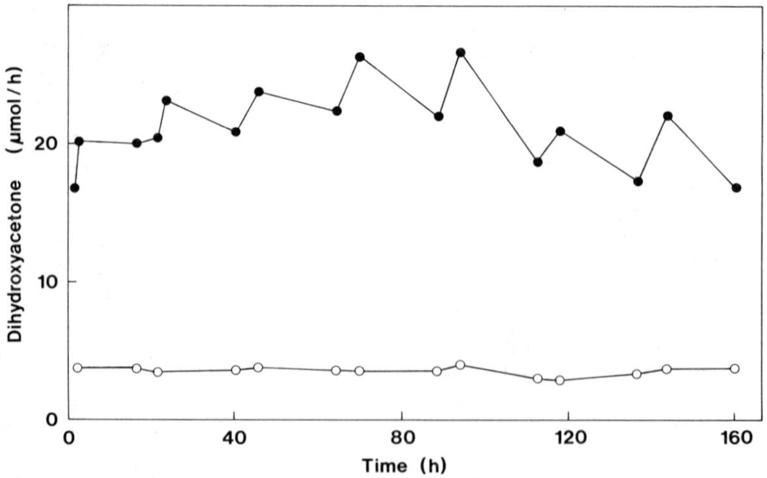

FIGURE 2. Operational stability of immobilized *Gluconobacter oxydans* in a perfluorochemical emulsion. A 32.4% emulsion of FC-72 in 0.1 *M* succinate buffer (pH 5.0) containing 0.25 *M* glycerol and 10 m*M* $CaCl_2$ was continuously pumped through a column containing 105 mg immobilized bacteria. Pluronic F-68 was used as emulgator. The flow rate was 0.10 ml/min. Samples were taken regularly from the effluent of the column (●). An identical column was treated in the same way except that FC-72 and the emulgator Pluronic F-68 were omitted (○).

emulsion. This observation was valid when the system was equilibrated with air up to 35% and also when pure oxygen was used up to 30%.

On the basis of reports in the medical literature, a high degree of biocompatibility between the perfluorochemicals and the cells was expected. In a long-term experiment, a column filled with *Gluconobacter* entrapped in alginate produced a constant amount of dihydroxyacetone over a week's time period (FIG. 2). At present, some practical points have still to be resolved. The perfluorochemical FC-72 is rather volatile and must thus be kept in a closed system. Furthermore, FC-72 is too expensive to be a realistic alternative when setting up large-scale aerobic bioprocesses. However, there are cheaper fluorinated hydrocarbons on the market.

In conclusion, this report clearly shows that perfluorochemicals are biocompatible, and thus are useful reagents when an oxygen carrier is needed. Furthermore, the immobilized cellular bioconversions can be performed at increased efficiencies due to the improved oxygen supply.

REFERENCES

1. TRAMPER, J. & W. J. J. VAN DEN TWEEL. 1981. Sec. Eur. Congr. Biotechnol. Abstr. p. 161.
2. HOLST, O., S.-O. ENFORS & B. MATTIASSON. 1982. Eur. J. Appl. Microbiol. Biotechnol. 14: 64–68.
3. ADLERCREUTZ, P. & B. MATTIASSON. 1982. Enzyme Microbiol. Technol. B **36:** 651–653.
4. ADLERCREUTZ, P. & B. MATTIASSON. 1982. Acta Chem. Scand. **16:** 165–170.

Biocatalysis in Water-immiscible Organic Solvents: The Use of Immobilized Living Microorganisms

JOSÉ M. C. DUARTE

QUATRUM—Empresa Nacional de Quimica Orgânica, SARL
Av. João XXI, 10—6º. D
1000 Lisboa, Portugal

INTRODUCTION

The use of enzymes and microorganisms to catalyze the transformation of water-insoluble substrates has been receiving some attention since the early works of the groups of Cremonesi[1] in Italy and Lilly[2] in the United Kingdom and is now established as a field of biotechnology.

Recently Lilly[3] wrote a review of the main parameters that influence the design of such systems in which reference is made to some aspects of the work done by Duarte[4] in his laboratory.

One of the main problems of these systems is the difficulty in keeping the reaction going for the length of time that would make it attractive for industrial operation. Recently, Omata *et al.*[5] reported the optically selective hydrolysis of DL-menthyl ester by immobilized yeast cells in organic solvent systems; as already noticed by Duarte and Lilly,[6] the operational stability of immobilized cells was much superior to that of the free cells.

However, the initial rate of reaction is much smaller for the entrapped-cell system than for free cells; this effect was observed in the *Nocardia*/cholesterol system and was explained as a limitation due to mass transfer of the reactants.[7]

Until now the use of water-immiscible organic solvents to transform valuable water-insoluble chemicals has been practically applied to single-step reactions that were not dependent on the metabolic activity of the cells as is the case with energy-dependent co-factor regeneration. This fact limits the scope of the technique because it leaves out most of the potentially interesting types of reactions that microorganisms are able to perform (e.g., hydroxylations and dehydrogenations).

This is the case with steroid transformation, one of the most important industrial applications of microorganisms. Cholesterol has for some years been regarded as an important raw material for the production of cheap steroids; commercial production of AD and ADD from cholesterol is now well established.[8] A process that would involve the use of an organic solvent could drastically reduce the size of the fermentation needed by increasing the cholesterol concentration. Here, it is shown that a *Nocardia* strain may be used to perform complex reactions, as is the case with cholesterol oxidation in the presence of organic solvents.

MATERIALS AND METHODS

Nocardia rhodochrous (NCIB 10554) cells grown on cholesterol and a mineral medium were entrapped in two different types of supports as shown on TABLE 1.

TABLE 1. Shake Flask Experiments on the Conversion of Cholesterol by *Nocardia* Grown in a 5-Liter Fermenter

	Flask Number		
	1	2	3
Support	4% Alginate	4% Alginate	8% PAA
Energy Source	2% Glycerol	1% Acetate	YE/Gly[a]
Substrate (in TCE)	Cholesterol (50mg/ml)	Cholesterol (50mg/ml)	Cholesterol (50mg/ml)
Nocardia[b]	200 ml	200 ml	200 ml

[a]YE/Gly: 1.2% yeast extract + 0.6% glycerol in Tris buffer.
[b]200 ml of the fermentation medium as centrifuged (29°C) and the cell paste immediately entrapped.

The substrate cholesterol was dissolved in 1,1,1-trichloroethane (TCE), which was previously shown to be the best for use with this system.[6]

The transformation reaction was conducted on shake flasks in an orbital shaker at a temperature of 29°C. The total volume of the reaction was 100 ml: 50 ml of TCE (containing the substrate) + 20 ml of an aqueous solution (containing a carbon source) + 30 ml of gel with the entrapped cells. The flasks were agitated at 200 rpm.

The reaction was followed by measuring the decrease in the cholesterol concentration and analyzing for possible steroid intermediates (by the HPLC technique).

RESULTS AND DISCUSSION

It was previously observed (unpublished) that there was an initial high activity of cholesterol degradation by cells grown on cholesterol; this activity was increased when a carbon energy source was added to the reaction medium. Glycerol and acetate were tried as potential energy sources with the alginate gel, and for the polyacrylamide gel, glycerol supplemented with yeast extract was used; this was not tried with alginate to avoid possible adverse effects on the ionic gel structure.

The system of flask 3 showed the best capacity for cholesterol conversion. After 100 hours, more than 50% of the cholesterol had disappeared from the medium while cholestenone-4 production only accounts for less than 10% of the cholesterol initially present; under other conditions,[6] the cholestenone production matches the cholesterol

TABLE 2. Steroid Formation (μg/ml) as Measured by HPLC

		Reaction Time (hours)					
		2.5	19	66	90	114	138
AD	Flask 1	22	21	28	49	67	97
	Flask 2	18	22	29	86	89	128
	Flask 3	16	31	—	290	380	360
ADD	Flask 1	—	230	56	403	—	—
	Flask 2	—	56	160	100	—	—
	Flask 3	—	—	—	60	—	—
C146	Flask 1	—	—	35	45	45	43
	Flask 2	—	29	20	69	45	49
	Flask 3	28	54	94	231	240	710

that disappeared. The rates of cholesterol transformation by flasks 1 and 2 were about 6 and 17 times slower, respectively, than that on flask 3.

Other steroids detected by the HPLC analysis are shown in Table 2. Androst-4-en-3,17-dione (AD) seemed to begin accumulating after 90 hours of reaction; androst-1,4-diene-3,17-dione (ADD) was observed only sporadically. An unidentified compound, probably an intermediate on the cholestenone-AD pathway, with a retention time on the HPLC column of 146 seconds, was designated by C-146; it was accumulating mainly in flask 3 after 114 hours, when the cholesterol transformation rate seemed to be practically stopped. C-146 may be the product of a limiting reaction on the cholesterol degradation to AD not yet recognized.

From the results, it was concluded that *Nocardia rhodochrous* cells (grown in cholesterol), when entrapped on a polyacrylamide or alginate gels (with agar similar results were obtained—unpublished), are able to degrade cholesterol further than cholestenone even in the presence of an immiscible solvent (TCE).

The intermediates detected did account in the most, for 20% of the cholesterol disappeared. This strongly indicates that in the conditions of the experiments cholesterol was, at least during the first 100 hours of the reaction, being oxidized to carbon dioxide and water; the accumulation after this period of some intermediates of the cholesterol oxidation suggests that immobilized living cells (other evidence supports that there is growth inside the gel) may be used in the presence of water-immiscible organic solvents to perform complex, energy-dependent transformations. This is being applied to study the production of AD and other steroid compounds, by the use of the appropriate mutants.

ACKNOWLEDGMENTS

I thank Professor M. D. Lilly, who was my "Ph.D." supervisor when this work was done, for his important contribution to it. I also want to thank the Calouste Gulbenkian Foundation and my company, QUATRUM, in Lisbon, who supported first my stay in London and then my trip to Santa Barbara, California. Lastly, I thank Dr. K. Venkatasubramanian and the Engineering Foundation for inviting me to this conference.

REFERENCES

1. Cremosi, P., G. Carrea, L. Ferrara & E. Antonini. 1974. Eur. J. Biochem. **44:** 401.
2. Buckland, B. C., P. Dunnil & M. D. Lilly. 1975. Biotechnol. Bioeng. **XVIII:** 815.
3. Lilly, M. D. 1982. J. Chem. Technol. Biotechnol. **32:** 162.
4. Duarte, J. M. C. 1982. Ph.D. Thesis. University of London.
5. Omata, T., N. Iwamoto, T. Kimura, A. Tanaka & S. Fukui. 1981. Eur. J. Appl. Microbiol. Biotechnol. **11:** 199.
6. Duarte, J. M. C. & M. D. Lilly. 1980. Enzyme Engineering, Vol. 5. Plenum Press, New York. p. 363.
7. Duarte, J. M. C. Enzyme Engineering, Vol. 6. Plenum Press, New York. p. 157.
8. Marscheck Jr., W., S. Kraychy & R. Muir. 1972. Appl. Microbiol. **23:** 72.

Syntheses of Optically Active Amino Acids by the Combination of Chemical Methods and Microbial Techniques

KENZO YOKOZEKI, CHIKAHIKO EGUCHI, AND YOSHIO HIROSE

Central Research Laboratories of Ajinomoto Co., Inc.
Kawasaki-shi, Kanagawa, Japan

Novel processes for the production of optically active amino acids were successfully developed combining chemical methods and microbial techniques.

These processes brought about optically active (L- or D-) amino acids from the corresponding racemic substrates, which were the intermediates in the chemical syntheses of amino acids.

PRODUCTION OF AROMATIC L-AMINO ACIDS[1]

Chemical synthetic and chemico-enzymatic flows for the production of L-tryptophan are shown in FIGURE 1. *Flavobacterium aminogenes* nov. sp. AJ-3912, isolated from soil, showed a strong ability to hydrolyze L-5-indolylmethylhydantoin and to produce L-tryptophan.

Substrate of D-isomer was inferred to be easily racemized under the conditions of enzymatic reaction from the fact that L-tryptophan was equally produced from both substrates in L- and DL-form, at a molar yield of more than 50% (actually 82%). The racemization was inferred to be catalyzed chemically from the fact that the racemization occurred equally in reaction conditions without cells. Though this strain had the tryptophan degradation pathway (Trp oxygenase), L-tryptophan production reached 100% molar yield using a mutant AJ-3940 whose tryptophan degradation pathway was blocked.

In addition, this strain showed the ability to produce L-phenylalanine and L-DOPA from the corresponding hydantoin compounds.

PRODUCTION OF D-AMINO ACIDS[2]

The microorganisms that can hydrolyze hydantoin compounds and produce D-amino acids were isolated from soil. *Pseudomonas hydantoinophilum* nov. sp. AJ 11220 was selected as a representative strain that can produce D-amino acids. This strain showed a strong ability to produce D-*p*-hydroxyphenylglycine from DL-5-(*p*-hydroxyphenyl)hydantoin. Substrate of the L-isomer was inferred to be easily racemized under the conditions of an enzymatic reaction from the fact that D-

p-hydroxyphenylglycine was produced at a molar yield of more than 90% from substrate of L-isomer as well as D-isomer. The racemization was inferred to be catalyzed chemically from the fact that the racemization occurred equally under reaction conditions without cells. This strain showed wide substrate specificity to produce various D-amino acids from the corresponding hydantoin compounds.

PRODUCTION OF L-CYSTEINE[3]

The microorganisms that can hydrolyze DL-2-aminothiazoline-4-carboxylic acid, intermediates in the chemical synthesis of L-cysteine, and produce L-cysteine were isolated from soil. The representative strain, *Pseudomonas thiazolinophilum* nov. sp., showed a strong and stable ability to produce L-cysteine from 2-aminothiazoline-4-carboxylic acid.

This strain had the activity of cysteine degradation (cysteine desulfhydrase), and the L-cysteine formed was partially degraded. L-Cysteine production reached 100% molar yield with the addition of hydroxylamine, cysteine desulfhydrase inhibitor. This fact indicates the racemization of substrate of the D-isomer. The racemization was inferred to be catalyzed by enzyme from the fact that the racemization was not catalyzed chemically.

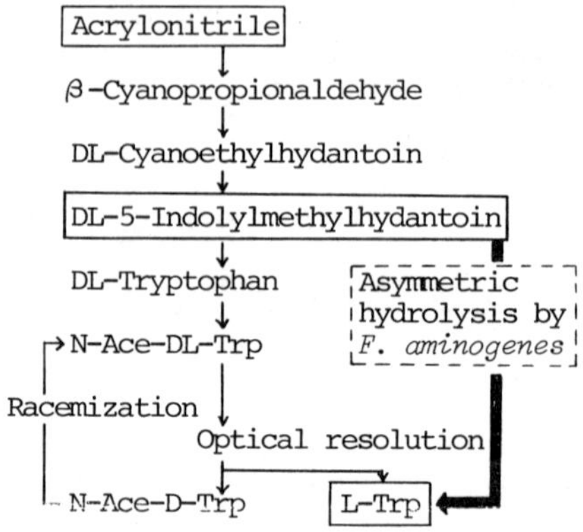

FIGURE 1. Flows for the production of L-tryptophan.

SUMMARY

The newly developed processes combining the advantages of chemical methods and microbial techniques brought about optically active amino acids (D- or L-) from racemic compounds. In all cases, the other isomer of substrates (L- or D-) that cannot be catalyzed by hydrolyzing enzymes were easily racemized under reaction conditions. Then, optically active amino acids were produced quantitatively from the racemic compounds.

REFERENCES

1. SANO, K., K., YOKOZEKI, C. EGUCHI, T. TAMURA, I. NODA & K. MITSUGI. 1977. Agric. Biol. Chem. **41:** 819.
2. YOKOZEKI, K., S. NAKAMORI, C. EGUCHI S. YAMANAKA & F. YOSHINAGA. 1981. Proc. Annu. Meet. Agric. Chem. Soc. Jpn. p. 18.
3. SANO K., K. YOKOZEKI, T. TAMURA, N. YOSHIDA, I. NODA & K. MITSUGI. 1978. Appl. Environ. Microbiol. **34:** 806.

Continuous 6-APA and 7-ADCA Production Using Semacylase® (Immobilized PEN-V Acylase)

STINA GESTRELIUS, BJARNE HELWIIG NIELSEN, AND HENRIK MØLLGAARD

NOVO Industri A/S
DK 2880 Bagsværd, Denmark

Semisynthetic penicillins and cephalosporins are produced from natural penicillins. The removal of the acid side chain from the natural β-lactam nucleus is accomplished either chemically or enzymatically by penicillin acylases. Semacylase® (SP 217)[1] is an immobilized amidohydrolase specific for (*p*-hydroxy-) phenoxyacetylamides, such as V-penicillin (V-pen) and phenoxyacetyl-desacetoxycephalosporanic acid (V-DCA).

Semacylase particles have the shape of short cylinders which have good pressure stability. The pH activity profile is broad around pH 7 and the stability of the enzyme in 0.2 *M* acetate or phosphate buffer at 50°C is maximal between pH 5 and 7. During catalysis, however, the pH within the particles decreases relative to the bulk-pH due to the acid produced and diffusion limitation. Therefore the optimum bulk-pH for application is above 6.5.

The apparent K_m-value is about 10 m*M* (0.4% w/v) for V-pen and V-DCA in 0.2 *M* phosphate pH 7.0 (sieve fraction 180–425 μm, dry particles). The maximal activity with V-DCA is one-fifth of the activity with V-pen.

Substrate inhibition is quite low while inhibition from the products is pronounced. Assuming an ordered sequential reaction mechanism with the β-lactam nucleus as the first product released, the inhibitor constant for 6-APA and phenoxyacetic acid is about 0.20 *M* (noncompetitive inhibition) and about 0.05 *M* (competitive inhibition), respectively. These values were determined in 0.2 *M* phosphate buffer pH 7.0 for the sieve fraction from 180 μm to 425 μm.

The combined effect of product inhibition and reverse reaction is illustrated in FIGURE 1 for high degrees of conversion. The reduction in the effective utilization of the enzyme activity is expressed as the dimensionless factor, τ, which is the ratio between the observed time needed for a desired conversion and the calculated time for total conversion with maximal initial activity. The efficiency decreases for higher degrees of conversion and for higher substrate concentrations since the product inhibition and the reverse reaction grow more significant. The optimal substrate concentration for Semacylase in a stirred tank is therefore 4–6% w/v V-pen and 2-4% w/v V-DCA, when maximal enzyme efficiency is required.

The influence of the reverse reaction depends on the equilibrium of the reaction, which again depends on the β-lactam concentration and the pH. As shown in FIGURE 2, the equilibrium degree of conversion, which is the upper limit for the enzymatic cleavage of the β-lactam, is decreasing with increasing substrate concentrations and with decreasing pH. Thus a nearly complete degree of conversion with high substrate concentrations requires a high pH value. However, the pH must be close to the neutral region to minimize the nonenzymatic decay. The figure shows that V-pen is more favorable as a substrate for 6-APA production than G-pen.

Application of Semacylase in packed-bed reactors requires that outlet pH is held

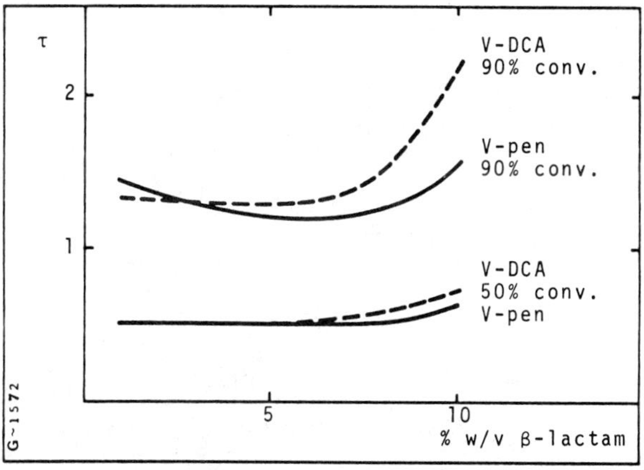

FIGURE 1. τ as a function of substrate concentration. τ was measured from progress curves at the specified degrees of conversion in 0.20 *M* phosphate buffer pH 7.5 at 35°C. Sieve fraction: 250–700 μm.

above 6.5. With a substrate feed in 50 m*M* phosphate, pH 7.8, this allows a maximal conversion of 1% w/v β-lactam per reactor passage due to the phenoxyacetic acid produced. Under these conditions, the initial activity with 4% w/v V-pen is 0.4 g 6-APA/g Semacylase/hr and the accumulated productivity (until 25% residual activity) about 750 g 6-APA/g enzyme (Novo enzyme information IB 236b-GB). The enzyme half-life is 1500–2000 hours. With increasing substrate concentration, the half-life and accumulated productivity decrease, for example, more than 50% for 12% w/v V-pen.

Three percent w/v V-DCA as substrate gives an initial activity and accumulated productivity of 0.05–0.10 g 7-ADCA/g/hr and 50 g 7-ADCA/g, respectively, and a

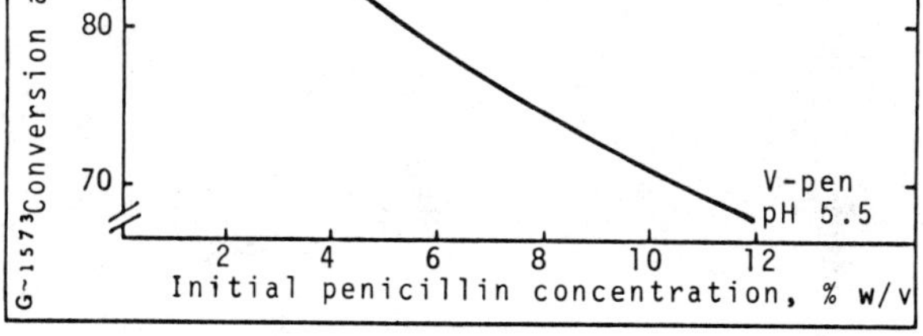

FIGURE 2. Degree of penicillin conversion at equilibrium at pH 7.0 and 5.5 as a function of penicillin concentration. The pH-independent equilibrium constant is 2.83 m*M* for G-Pen[2] and 1.28 m*M* for V-Pen.[3]

half-life of about 450 hours. The overall yield of *crystalline* 6-APA and 7-ADCA is above 85%.

Crude V-pen extracts can be used instead of purified V-pen but this results in lower activity and a shorter enzyme half-life.

A process for the conversion of 1 ton of V-pen per day has been proposed (Novo report F-815131b). It consists of 6 reactors in a merry-go-round arrangement with pH-adjustment in between. Alternatively, the enzyme can be used in, for example, stirred tanks or recirculated packed beds.

REFERENCES

1. GESTRELIUS, S. 1982. Appl. Biochem. Biotech. **7:** 19–21.
2. McDOUGALL *et al.* 1982. Enzyme Microb Technol. **4:** 114–115.
3. PETERSEN, J. 1982. M.Sc. Thesis. Technical University of Denmark. Copenhagen, Denmark.

Gel Entrapment of Enzymes in Cross-linked Prepolymerized Polyacrylamide-Hydrazide

A. FREEMAN,[a] T. BLANK, AND B. HAIMOVICH

Center for Biotechnology
The George S. Wise Faculty of Life Sciences
Tel-Aviv University
Tel-Aviv, 69978, Israel

Gel entrapment of enzymes in crosslinked synthetic gels has been carried out mostly through copolymerization of acrylamide and bisacrylamide, in the presence of the entrapped enzyme. In many cases, the immobilized enzyme suffers a considerable loss of activity, mainly due to denaturation caused by the monomers and the heat evolved during the polymerization reaction. In many cases, continuous leakage of the entrapped enzyme was observed.[1]

A new method for the immobilization and stabilization of enzymes, under mild conditions, by means of gel entrapment in a crosslinked synthetic gel was developed. The method is based on dissolving the enzyme in an aqueous solution of prepolymerized, linear polyacrylamide, partially substituted with acylhydrazide groups, and crosslinking with glyoxal (as shown schematically in FIG. 1). This method is a further development of the procedure orginally worked out for whole cell immobilization,[2] and may serve as an alternative to gel entrapment through copolymerization of water-soluble monomers in the presence of the enzyme.

The backbone synthetic polymer is readily prepared by polymerization of acrylamide (or copolymerization of acrylamide and other vinyl monomers), followed by controlled hydrazinolysis (~5% substitution). The polymer is then separated and stored. Following dissolution of the polymer (>10% w/v), the enzyme is added and crosslinking is effected by contacting the viscous solution with 1–5% glyoxal solution. Gels could be obtained either as a thin layer, (mounted on a solid glass support) or as a gel "noodle," which could be readily fragmented, to particles of about 0.25 mm in diameter.

Gel entrapment of enzymes using the new method was demonstrated and characterized for two enzymes of analytical importance: acetylcholine esterase and glucose oxidase.

Yields of active entrapped enzyme were higher, as compared to gel entrapment via copolymerization of acrylamide/bisacrylamide. Sixty-four percent of the added acetyl choline esterase enzymatic activity were immobilized using the new method, while only 35% survived the entrapment using the direct polymerization method. Similar results were obtained for glucose oxidase (54% vs. 13%).

Perturbation of kinetic parameters was relatively mild: the pH optimum for entrapped acetylcholine esterase was only about 0.5 pH units higher than the optimum for the free enzyme. Apparent K_m values of 2.8 $10^{-4}M^{-1}$ and 1.9 $10^{-2}M^{-1}$ were

[a]Correspondence should be sent to: Dr. A. Freeman, Center for Biotechnology, The George S. Wise Faculty of Life Sciences, Tel-Aviv University, Tel-Aviv, 69978, Israel

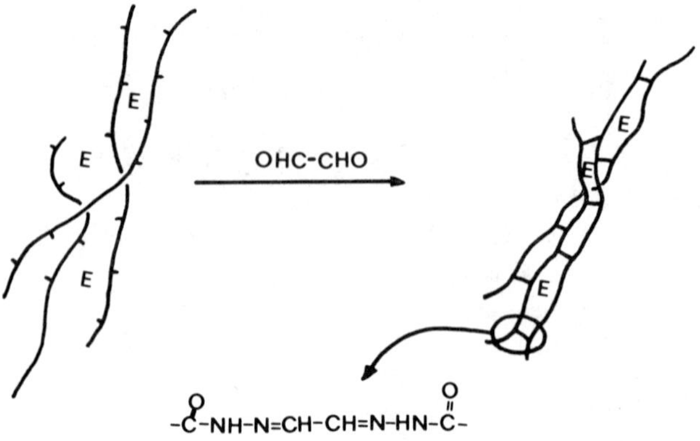

FIGURE 1. Gel entrapment of enzymes through crosslinking of prepolymerized polyacrylamide-hydrazide by glyoxal.

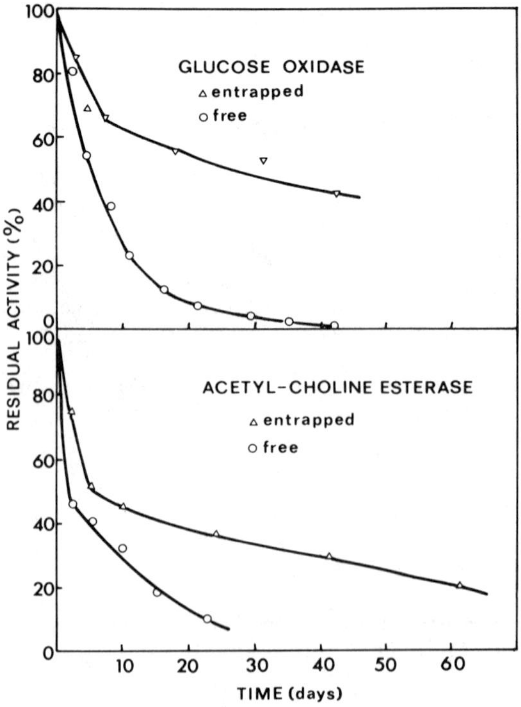

FIGURE 2. Temperature stability of free and entrapped enzymes, incubated at 37°C.

recorded for the immobilized acetylcholine esterase and glucose oxidase, respectively (K_m values for the soluble enzymes: 2.4 $10^{-4}M^{-1}$ and 2.2 $10^{-2}M^{-1}$, respectively).

A temperature stability profile of the two immobilized enzymes was significantly improved. Survival of the entrapped enzymes during continuous incubation at 37°C is shown in FIGURE 2. The data shows that the two entrapped enzymes were stabilized against heat denaturation.

Leakage of entrapped acetylcholine esterase from gels prepared from acrylamide/bisacrylamide was significant: 12% of the entrapped activity leaked to the buffer within 24 hr. No leakage was observed for the enzyme immobilized according to the new procedure. A similar ratio was found for leakage tested under continuous washing of gel particles packed in a column.

In conclusion, the new method offers improved answers to some of the problems often encountered with gel entrapment of enzymes: high retention of entrapped enzymic activity, minimal leakage, good mechanical and chemical properties, and minor kinetic perturbations. In addition, the method allows for significant improvement of thermal and storage stability. By the introduction of various comonomers into the prepolymerized backbone, it may be anticipated that the mechanical and chemical properties of the gels obtained could be monitored.

It is expected that the developed system will be useful especially for the preparation of enzyme films and membranes (e.g., for enzyme electrodes) as well as enzyme-containing beads.

REFERENCES

1. KOCH-SCHMIDT, A. C. 1977. *In* Biomedical Applications of Immobilized Enzymes and Proteins. T. M. S. Chang, Ed. Vol. **1:** 47–67. Plenum Press. New York.
2. FREEMAN, A. & Y. AHARONOWITZ. 1981. Biotechnol. Bioeng. **23:** 2747–2759.

Index of Contributors